Scientific Method for Ecological Research

Scientists tend to take the thought processes that drive their research for granted, often learning them indirectly by observing first their supervisors and then their colleagues. This book emphasizes the advantages of being explicit about these thought processes and aims to help those undertaking ecological research to develop a critical attitude to approaching a scientific problem and constructing a procedure for assessment. The outcome is a text that provides a framework for understanding methodological issues and assists with the effective definition and planning of research. As such, it is a unique resource for anyone embarking on their research career. It also provides a valuable source of information for those more experienced researchers who are seeking to strengthen the methodology underlying their studies or who have an interest in the analysis of research methods in ecology.

DAVID FORD is a Professor in the College of Forest Resources at the University of Washington, USA. His teaching spans courses such as Scientific Method, Ecological Modeling, Spatial Processes in Ecology, and Ecology of Managed Forest Ecosystems.

Scientific Method for Ecological Research

E. David Ford

College of Forest Resources, University of Washington

CAMBRIDGE
UNIVERSITY PRESS

PUBLISHED BY THE PRESS SYNDICATE OF THE UNIVERSITY OF CAMBRIDGE
The Pitt Building, Trumpington Street, Cambridge, United Kingdom

CAMBRIDGE UNIVERSITY PRESS
The Edinburgh Building, Cambridge CB2 2RU, UK
40 West 20th Street, New York, NY 10011-4211, USA
477 Williamstown Road, Port Melbourne, VIC 3207, Australia
Ruiz do Alarcón 13, 28014 Madrid, Spain
Dock House, The Waterfront, Cape Town 8001, South Africa

http://www.cambridge.org

First published 2000
Reprinted 2001, 2002

Typeset in Garamond 11/13pt [VN]

A catalogue record for this book is available from the British Library

Library of Congress Cataloguing-in-Publication Data
Ford, E. D. (Edward David)
Scientific method for ecological research/E. David Ford.
     p.   cm.
Includes bibliographical references and indexes.
ISBN 0 521 66005 X (hardcover)
1. Ecology – Research – Methodology.   2. Science – Methodology.
I. Title.
QH541.2.F66   2000
577′.07′2 – dc21   99-30065 CIP

ISBN 0 521 66005 X hardback
ISBN 0 521 66973 1 paperback

Transferred to digital printing 2004

For Rosemary

Contents

11 Strategies of scientific research in ecology   309

12 Use of mathematical models for constructing explanations in ecology   351

Preface

My motivation for writing this book was to provide a text for new researchers in ecology, giving a framework for understanding methodological issues, and helping them to define and plan research. In the late 1980s I started teaching a graduate course in research methods at the University of Washington. My faculty colleagues were concerned that students were having difficulty writing research plans. At the same time statistical consultants were spending much time asking students to clarify their study objectives and logic of investigation before statistical advice could be given. Discussions with colleagues at other universities suggested that these were not unusual circumstances.

Problems with ecology and its methods are also encountered by established researchers and there have been substantial criticisms of the subject and its research methods. Concern has been expressed that there has been a lack of progress in ecology, that no general theory has emerged, that ecological concepts are inadequate, and that ecologists fail to test their theories (Chapter 16). While some of this criticism may be justified, the students' needs required me to look beyond it and seek ways of being constructive. This has required making two distinctions that are fundamental to the way this book is written.

The first distinction is between methods of *reasoning* such as how we use logic, construct a hypothesis and develop a theory, and techniques of *investigation*, such as an experiment, or a survey, or constructing a simulation model. Much of this book is about methods of reasoning for ecological research and how they can best be used with particular techniques of investigation. Analysis of methods of reasoning leads to answers to the question "How can we make scientific inference in ecology?" Ecologists have tended to define and answer this question in terms of what they actually do in their work rather than the processes of reasoning they use.

The second distinction is between different types of concept we use in ecology. For some concepts we make measurements or observations but others can be defined only through abstract reasoning in the process of theory construction. Difficulties with this latter type of concept arise in even the most practical problems. For example, I have taught a number of students involved in *restoration* ecology, sometimes applied to *wetlands*. Neither of these concepts have definitions that are simple, general and easy to apply consistently to all particular cases. Government agencies who want *restoration* may fund research to show how it could be done, but may not define precisely what is required, or may define only certain aspects. Researchers writing grant or contract proposals for this work must communicate with these agencies on the agencies' own terms and may set practical objectives and suggest methods that are considered to be successful in other circumstances. Difficulties in defining precisely what would constitute *restoration* of a different *wetland* in new particular instances may become apparent only during detailed research planning.

It is because of our difficulties with definitions, and so with the relationship between general ecological theory and particular instances, that we have to be precise about the methods of reasoning we use and how we make scientific inference. Some ecologists have taken refuge from difficulties about method by emphasizing the values of particular techniques of investigation, e.g., experiments are the key to successful research because they can be used to test hypotheses, or in criteria about data, e.g., questions must be asked that can be answered with a significant P value. Such emphases and repeated application of a particular research technique or approach can be a solution for the individual scientist working in an established research program (see Chapter 11). But they imply that only certain types of question can be tackled. Faced with a class of graduate students, all with interesting but different types of research, I found a more comprehensive approach was necessary.

This book deals first with problems as they are most usually encountered by beginning researchers and then progresses to wider issues in ecological research and the consequences of social systems on the research process. Section I presents techniques for conceptual and propositional analysis that can be used to define a research question, develop statements defining the investigations to be made, and specify how they will answer the question. Section I defines different properties that measurements can have and describes the range of experiments used in ecology and the types of question for which they are appropriate. Section I also defines the types of logic that scientists use, particularly when falsification is possible, the importance of causal arguments, and why contrasts are so important. The requirement for exploratory work to define an effective statistical hypothesis test is described and the distinction is made between statistical and scientific inference.

Research planning starts with analysis but it must also incorporate how a comprehensive assessment will be made of work to be done. This leads to Section II and defining scientific inference, particularly for the types of concept we use in ecology. The approach taken uses work from the philosophy of science that has focused on defining scientific explanation and its assessment through developing explanatory coherence. Section II describes how scientific inference is developed and refined by groups of scientists working in research programs.

This book defines a methodology for ecological research in terms of methods of reasoning and principles for applying them. The essential principle is continuous application of *just and effective criticism.* A consequence of depending upon criticism is that explicit account must be taken of social processes in research. These are discussed in Section III. Social studies of the research process have advanced remarkably but scientists are sometimes impatient with social analysis of their subject, or even angered if studies suggest that what may actually be going on does not meet their ideal. But the consequences of these findings, and the details they describe, illustrate how difficult it can be to achieve just and effective criticism. Because we depend upon abstract reasoning in ecology and can not assume that scientific questions can be resolved in a straightforward way by measurement, then we must be explicit about use of criticism. The result of Section III is a four-stage definition of the process of criticism.

Some experienced researchers have little patience with making a formal description of the research process, believing strongly that research is best taught by the student taking the role of apprentice in following their procedures and methods. After long practice, experienced researchers may find the logic of what to do next obvious and attempts to take a cautious and refining view of all the arguments can seem a waste of time, pedantic, or even threatening. However, when you are at the outset of research, some of the assumptions made by established researchers are not obvious and the process described here can be helpful. This book does not suggest that the close relationship between supervisor and student can be replaced. Far from it. Such a relationship is essential. However, it is important to be explicit about the difficulties that we all can have with ecological research.

In addition to being a text for new researchers this book, I hope, provides a reference for definitions of terms and components of the debate over method in ecology. As a subject develops and encounters new types of problem so too must its methods evolve. Some critics of ecology have been led by their criticism to suggest restrictions either in what the aims of the subject should be or in the techniques of investigation that should be used. The methodology of *Progressive Synthesis,* defined in Chapter 15, is not restrictive in either of these ways. However, it does require that researchers be explicit in how they define scientific inference and that they analyze carefully their ideals for what research should achieve and how those ideals influence research.

Acknowledgements

I am grateful to successive classes of students taking the course QERM521, whose diverse research problems have repeatedly challenged my ideas about ecological research. My special thanks are due to Ashley Steel and Denise Hawkins, who, when they were graduate students, allowed me access to their first thoughts about a research problem and have permitted me to publish these ideas and their development. This is courageous of them. As discussed in the book, the scientific world has conventions about how research must be presented that are not designed to illustrate how ideas develop. I am also grateful to Linda Brubaker, Bob Teskey, and David Janos for painstaking reviews of the use I made of their work, to Bruce Menge, Stephen Carpenter, and Bob Paine for helpful insights into the progress of their research, and Joshua Klayman and Paul Thagard for valuable comments on the use I made of their approaches to the analysis of science and its methods.

I particularly wish to thank David Chart of the Department of History and Philosophy of Science, Cambridge University, for his review of my use of ideas from philosophy of science. His detailed comments were always sympathetic to the purpose of the book and I hope my responses are adequate.

I am indebted to Carol Perry, both for her book on technical writing (Perry 1991), and for her extensive and repeated editing of Section I. Writing about how we can analyze problems is challenging and she consistently persuaded me towards accessible text.

I am most appreciative to colleagues at the University of Washington: to Bob Francis, John Skalski, and Bob Stickney for initial ideas and encouragement; to Jim Agee, Bob Edmonds, Tom Hinckley, Dave Petersen, Joel Reynolds, and Doug Sprugel for ideas and motivating comments; and to Ray Hilborn and Walt Dickoff for their positive and constructive comments during peer review. I wish to thank Peter Guttorp and the National Center for Research in Statistics and the Environment for support in writing Chapter 12.

Foundation work was done during a sabbatical at Harvard University and the University of Lancaster. I am grateful to David Foster and Peter Diggle, respectively, for their hospitality and to Harvard University for a Bullard Fellowship whilst I was there. At Harvard a reading group of Emery Boose,

Jana Compton, David Janos, and Jason McLachlan kindly commented on the text. I am particularly grateful to Sally Hollis for her comments and suggestions on Chapter 8.

My thanks are due to Alan Crowden of Cambridge University Press for his perseverance, support at crucial times, and arranging for critical reviews, particular by Craig Loehle, who made many helpful suggestions, and two anonymous referees.

The support, encouragement, and tolerance of my wife Rosemary have been selfless and given so willingly and freely.

1 Component processes of ecological research

Summary

Ecologists starting research for the first time have many questions about the scientific method. This chapter gives examples of these questions and classifies them into three groups: those concerned with the analytical process of research; those concerned with the special problems of ecology, particularly how we synthesize knowledge and develop ecological theory; and those concerned with the social aspects of research. The chapter describes how these groups of questions will be answered in successive sections.

SECTION I

Developing an analytical framework. A logical framework is required for research. Assumptions must be clearly stated as axioms, and questions must be formed as postulates, i.e., statements that can be investigated and then classified as true or false, or assigned a probability. This framework is the foundation for the research you do and the basis for making scientific inference.

SECTION II

Making a synthesis for scientific inference. Ecological and environmental research present special problems of scientific methodology. Ecological systems are open to multiple influences and vary in ways that limit the types of investigation used and the generality of scientific inference that can be made. Some important concepts have to be defined by a theory.

SECTION III

Working in the research community. Social interactions influence the scientific method. Science is, in part, a social activity that can influence how, and what, problems are researched.

SECTION IV

Progressive Synthesis is defined as a methodology for developing ecological theories and examining how coherent they are as explanations.

A distinction is drawn between what scientists do in research (the progress they make) and how it should be done (the processes that should be used). Progress comes from continuous dialogue between our theories about the world and the measurements we make that result in the synthesis of new or improved theories. To energize this dialogue, we use four processes: creativity, definition, assessment, and, most important, critical analysis. Critical analysis is a foundation for creativity because it reveals what may be wrong or deficient in definitions and assessments and what must be reconstructed.

1.1 Questions about the process of scientific research

Scientific discovery has a particular satisfaction. The joy of discovery – "Here is something new!" – is heightened by the realization that what has been created through research increases or changes the knowledge developed by other scientists. Furthermore, a new discovery may make it possible to solve a practical problem. These components, a discovery that is relevant to other knowledge and that may have practical value, make science an exciting and rewarding creative activity.

Many ecologists starting research have questions about the scientific method and how to apply it, whether it really works for ecology, and how the scientific community functions collectively in making discoveries (Table 1.1). Those who gave the questions for Table 1.1 were embarking on research in basic ecology, the conservation and management of natural resources, and environmental sciences. Despite the diverse nature of their research topics, the questions show repeated concerns about three components of the research process:

(1) *How to develop a conceptual and logical framework for discovery and assessment* (Questions 1–21, Table 1.1). Many students have a practical question they wish to answer or a subject they wish to research. However, it can require considerable analysis to define how a piece of research should proceed, what should be measured, how investigations should be carried out, and what process should be used for making scientific inference.

(2) *How to approach the particular difficulties associated with ecological and environmental research* (Questions 22–25, Table 1.1). Ecological systems present particular difficulties to the research scientist. Their variability can make them difficult to sample, but this variability is not simply a nuisance. It is a fundamental characteristic of ecological systems! For example, at the organism and population levels, genetic variability and plasticity in development are inherent characteristics. At the community or ecosystem level, differences in local environment or history can make an ecological system unique in an important way that

Table 1.1. *Questions asked by ecology students at the start of a graduate course in the scientific method. These questions are complete and as they were written by students. Questions 16 and 17 contain misuses of the terms "hypothesis" and "prove". See Chapters 3, 7, and 8 for correct definitions and examples of use*

I. Questions about research planning and the scientific method

1. How can I come up with a question?
2. I have ideas and topics but I don't have a project.
3. How can I refine a question so that an answer can be obtained?
4. I have a project. Can I find an innovative way to look at the problem? Can I find a way where others have not?
5. When starting research how do we know what analogies to use? How to get new ideas?
6. Am I making a significant contribution to the body of knowledge?
7. I have my project – but have I missed anything out?
8. How will I know whether I am duplicating other research?
9. How will I know whether my project is relevant or significant to a theory or a practical question, or just a stupid idea?
10. How can I conduct a literature review? How will I know what research is going on, what has been done? What about research in a foreign language?
11. How will I know whether my project is too big or too small?
12. How can I confine the question so that it is a Master's topic and not a lifetime of study?
13. How will I know when to stop (as an individual) given the cyclical nature of science?
14. How do I limit the project so that it can be finished in two years?
15. How will I analyze my data and write my thesis? How will I know what to conclude? Have I got there?
16. You make a hypothesis, and then you try to prove it. What if it is incorrect?
17. We start by trying to prove a hypothesis, but we end up trying to fit a new hypothesis to the data?
18. Will I collect bad data, or data insufficient to answer the question?
19. Some questions have answers. Some problems are exploratory. How do you present the exploratory work? This is difficult because the scientific culture expects answers.
20. What types of control or experimental design will I need?
21. How can I interpret my results in an unbiased way so that I do not massage my results to fit any preconceived viewpoint?

II. Questions about research in ecological science

22. How can I ask the right question and measure the right things? How can I sample a whole ecosystem? How can I avoid samples being influenced by unusual events?
23. How do we extrapolate from research on a limited system (laboratory or field plot) to give an understanding of a whole ecosystem?
24. How can I find a place (site) where a question can be answered? There may be difficulty matching a proposed theory with practical reality.
25. Which techniques are the best for measuring the response of a tree to environmental factors? For example, the development of a bioassay?

Table 1.1. (*cont.*)

III.	**Questions about working in the research community**
26.	Can I get funded?
27.	Funding!
28.	I am interested in funding and individual recognition.
29.	How do I choose a committee that will give me a broadly based opinion?
30.	How can I get my committee to stop suggesting things to do?
31.	What if my committee steers me incorrectly?
32.	How can I analyze, integrate, and present to the outside world?
33.	How can I choose a research topic of value? Or if you have chosen a research topic how can you persuade someone to fund it?
34.	How can I make research relevant to current topics of theoretical and practical interest?

must be taken into account in a research study. Because of this variability, research in ecology can involve two phases of activity: (a) discovering a phenomenon or process and developing a theory that explains it, and (b) understanding how important that phenomenon or process is in different situations.

(3) *How to work within the scientific and natural resource management communities* (Questions 26–34, Table 1.1). Supervisors, major professors, and associated faculty and colleagues influence a student's research through their distinct research perspectives, which the student must identify and weigh. Local, state, or national agencies frequently fund ecological and environmental research, and social factors influence who receives grants or contracts from them. In addition, much ecological research investigates practical problems of environmental management with the objective of providing solutions. Almost inevitably, any proposed solution will have an economic or social impact on someone or may run counter to his or her ideals. Even the possibility that this may happen can influence what research is proposed and how it is conducted.

1.2 Scientific methodology

This book presents a methodology for scientific research.

METHODOLOGY

A system of techniques of investigation, methods for applying the techniques, and general principles for how the methods should be used in scientific inquiry.

Techniques of investigation used in research, such as experimentation, survey, or constructing computer models, must be applied using methods that ensure an effective process of scientific reasoning. Principles, such as the continuous use of criticism and being explicit in definitions and methods of assessment, must be used consistently for all techniques and methods.

The first objective of this book is to provide a structure for the process of scientific reasoning and research planning. The second is to illustrate particular problems in ecological research that determine how different techniques of research can be used and show how careful we must be in our methods of using these techniques. The third is to discuss how social influences can affect scientific research. In this respect ecological research faces some particularly difficult challenges when a problem is important for resource or environmental management.

1.3 Distinction between progress and process in scientific research

The foundation for answering the questions in Table 1.1 is to recognize that scientific research has two parts. (1) What scientists do – the techniques they apply to solve problems and make progress. (2) How scientists think – the thought processes scientists use in analyzing a problem.

(1) *What scientists do.* Scientists analyze problems and then make investigations and/or conduct experiments to discover new information. They order and explain this information in a synthesis by producing theories of how the world works. Theories are the result of a dialogue between what we think we know (our previous theories), and what we see or measure in the world (our data). It takes effort to become proficient in particular techniques of investigation, e.g., ecological field experiments, vegetation survey and analysis, and to understand how theories are constructed with information obtained by using them. Not surprisingly the research questions we choose, and the way we analyze them, may be strongly influenced by the techniques of investigation we have mastered. The theories we construct, or rate as most valuable, may be biased towards information obtained from the techniques of investigation with which we are most familiar. The challenge we face is that theory and data are interlinked (Fig. 1.1), and this circular nature of investigation can strongly influence what we find out – even when we think we are approaching new problems. This is why problem analysis is so important at the start of research.

(2) *How scientists think.* The repeated dialogue scientists conduct between THEORY and DATA requires critical analysis and creativity. Definition and assessment must be in constant interaction: we define our ideas

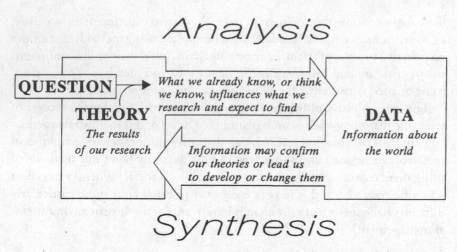

Analysis

QUESTION → *What we already know, or think we know, influences what we research and expect to find*

THEORY

The results of our research   DATA — *Information about the world*

Information may confirm our theories or lead us to develop or change them

Synthesis

Fig. 1.1. The cycle of scientific investigation. Although scientists use data to assess theories, the type of data they collect is determined by those theories. Analyzing a scientific theory so that it can be questioned by data is a major challenge for the scientist.

as a theory, develop new ideas from the theory, assess those ideas using measurements, and redefine the theories in a new synthesis. However, the interaction between definition and assessment must be pushed, prodded, and cajoled by critical analysis. Critical analysis is the fundamental process we must continually apply to make definitions exact and unambiguous, and assessments unbiased (Fig. 1.2).

Criticism is an essential foundation for creativity. This may at first seem counterintuitive; we often associate criticism with destruction, the opposite of creativity. But discovering something new may first require realizing that a current theory or set of observations is defective or inadequate in some way.

Scientists themselves are most concerned with what they do in research. They tend to take the thought processes that drive their research for granted. Except for statistical testing, which is one component of assessment, these processes are less discussed than the results of scientific research. Most scientists learn the use of these thought processes indirectly by observing research progress. This book emphasizes the advantages of being explicit about those processes. Ecology is an interesting, but difficult, subject. Many of its valuable concepts either require many measurements to define them or can not be measured directly but are defined by theory. This book defines a method of reasoning for analyzing ecological concepts, how a new synthesis is made following research, and how that synthesis can be assessed.

1.4 Section I: Developing an analytical framework

Doing research is a different activity from learning by course work. This becomes apparent when research planning starts. Section I of this book provides answers to Questions 1–21 (Table 1.1) in the form of a description

Fig. 1.2. Critical analysis and creativity must motivate the cycle of scientific advance. Critical analysis must ensure rigor in definition of problem, theory, and assessment of data. Creativity can occur during both theory definition and data analysis and is frequently stimulated by critical analysis.

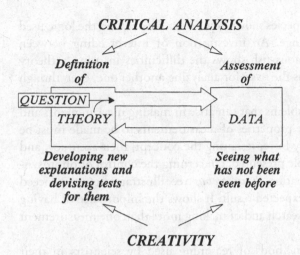

of the process of research that can be followed, particularly at the planning stages when the task is analysis of the problem and how to solve it.

Chapter 2 discusses in detail Questions 1–21 (Table 1.1) as they relate to the five processes of research planning: (1) defining a research question; (2) applying creativity to develop new research ideas; (3) ensuring the proposed research has relevance to prior scientific knowledge; (4) ensuring the proposed research is technically feasible and can be completed with available resources; and (5) determining how conclusions can be drawn.

Chapter 3 details the key to research planning: making a conceptual and propositional analysis by defining three types of knowledge (Fig. 3.1): *axioms*, what we assume we know; *postulates*, what questions we have but stated in a propositional form so they can be classified as true or false using the research; and *data statements*, which define the data needed to classify the postulates as true, false, or acceptable with a certain probability. The hardest work can be in determining exactly what can be taken as an axiom and what must be researched as a postulate. The key to this is defining the individual concepts that make up axioms and postulates.

Chapter 4 illustrates conceptual and propositional analysis with an example, an investigation of the use made by small mammals of piles of woody debris cast up alongside a river after floods. Conceptual and propositional analysis is a progressive process that may take weeks or months to complete. This very active phase of research brings together published information in the scientific literature, information known about a field site or species, and knowledge of what can be done practically.

Chapter 5 presents a first discussion of the development and use of ecological theories to illustrate the importance of analysis. Though we continually seek generalities in our theories, we must modify them to investigate specific situations. Development of a theory to describe the post-glacial invasion of Alaska by tree species shows how both a theory, and the logic used to investigate it, can change. An investigation of interbreeding between related species of anadromous fish shows the difficulties in using a theory developed for one system as the basis for analyzing another one, even though the species are related.

Chapter 6 illustrates problems that can arise in making measurements and planning experiments. The properties of measurements to be made must be analyzed for their efficiency in representing the concept, their accuracy, and their precision. The example presented, concerning the control of photosynthesis rate of foliage on a mature *Pinus strobus* tree, illustrates how to proceed when encountering an unexpected result. It shows the importance of having competing postulates in research and of making more than one measurement of a new concept.

Chapter 7 introduces methods of reasoning used by scientists in their research and difficulties associated with them. Deduction and the rules of logic are descibed and the general inductive approach followed in research is illustrated. Conditions when it is possible to falsify a postulate are described, and why that can be an important procedure, and the logic for having competing postulates, rather than just one, is illustrated. In ecology we depend upon causal reasoning; different types of causes are illustrated.

Chapter 8 distinguishes between scientific and statistical inference. The tests and procedures that statistical science has produced are based on statistical theories whose assumptions must be understood before they can be applied. This is particularly important for ecological and environmental research, where some assumptions used in statistical theory may not hold in important ways. This chapter emphasizes the need for exploratory analysis to establish whether, and how, a statistical test can be constructed.

Chapter 9 concludes Section I by describing different individual scientific philosophies and how these influence research by motivating different types of analysis of scientific problems. Philosophies are neither right nor wrong, but are sets of ideas about how research is, or should be, conducted. A philosophy can bring both strengths and weaknesses to a research investigation. Some philosophies do make assumptions about the nature of what will be found out in ecology.

1.5 Section II: Making a synthesis for scientific inference

Ecology is characterized by use of concepts such as *ecosystem*, *community*, and *population* and properties such systems have such as *diversity*, *stability*, or *persistence*. These concepts can not be measured directly; we have to construct theory to define them. In Chapter 10 these are classified as *integrative* concepts and we build theory about them using both *functional* concepts, that define how organisms and the environment interact directly, and *natural* concepts, that define organisms and environmental factors or conditions. This classification acts as a heuristic for the development of ecological knowledge. We also have to recognize that our theories about *integrative* and *functional* concepts can take time to develop and, explicitly, that any theory may not be universally true. For example, a theory that explains high and low species diversity between examples of one type of ecosystem may not explain it between examples of another. We should not simply assume generality in the theories we propose. The idea that theories have domains of application is introduced in Chapter 10 and is used to illustrate how theories may develop and be reconciled with other, apparently conflicting, theories.

Ecological theories, particularly those defining integrative concepts, are complex and they can not be rejected by single, critical, investigations. We seek to know whether such theories are good explanations. In Section II a definition is made of what constitutes an explanation, and the idea of explanatory coherence for making an assessment is introduced; examples are given in each chapter of the section.

The use scientists make of ecological theory, and their concern about how to develop it, determine research strategy. Chapter 11 discusses some different research strategies described by philosophers who have studied science. They are used to illustrate what is critical in the different tasks of establishing a new theory and stimulating the development of a growing theory. The chapter discusses further applications of the conceptual classification given in this section and the development of explanatory coherence in different situations.

Difficulties in both the analysis of ecological systems, and the use of integrative concepts for making a synthesis, have led to the use of quantitative models. Chapter 12 illustrates assumptions of different types of model and discusses the difficulties those assumptions cause for using models to represent, or even replace, theory.

1.6 Section III: Working in the research community

In this book, attention is paid to the relationship between what we do as scientists and the social environment in which we work. A critical approach is the essential requirement for a research scientist; but this can be strongly influenced by colleagues, research sponsors, and sometimes environment and natural resources managers. Scientists must be aware of those influences and their consequences. Socially imposed restrictions on criticism and innovation can be the most difficult problem a young scientist encounters. This book stresses the need for scientists to develop the logical basis for their research and in this way to become an integral part of a community of scholars. Social and educational factors, however, can influence attempts to become part of a community.

Chapter 13 discusses the types of social influence in science and their effect on planning and conducting research and in preparing scientific papers for publication. The social difficulties that may be encountered in research are not simply something extra to be coped with. They are due to the very nature of research as an imprecise activity – where recent results are still insecure and we can not be certain of the best way ahead.

Chapter 14 illustrates particular difficulties faced by doing ecological research in an applied context where a natural resource is managed or impacts on people's economic or cultural well-being. Decisions about environmental policy and the management of natural resources are made from particular standpoints. Scientific research can influence these decisions but the ways this might occur most effectively still seem uncertain and can be influenced by the standpoints of scientists, managers, and policy-makers. Examples are given where scientific analysis was central to a particular resource management policy, and where it was excluded. In both cases major changes became inevitable: in the first as society's demands changed; in the second when environmental catastrophes occurred.

1.7 Section IV: Defining a methodology for ecological research

At appropriate points in the first three sections there are discussions of the strengths and weakness of well known techniques of investigation. A number of innovations are introduced for analyzing ecological research problems and making synthesis of results to develop new theory. Chapter 15 is an integrated description, presented as a methodology for ecological research, of *Progressive Synthesis*. This methodology has three principles:

> There should be continuous application of just and effective criticism.
> Precision is required in defining axioms and concepts, postulates and data statements, and theories.
> Explicit standards must be used to examine the relation between theory and data.

These principles are applied to a method for reasoning designed for making inference about ecological concepts and theories. This method focuses on how change can take place through concept definitions and the construction of theory, use of comparisons, and development and assessment of scientific explanations.

Chapter 16 reviews criticisms of ecological research: that there has been lack of progress, no general theory has emerged, ecological concepts are inadequate, and that ecologists fail to test their theories. These criticisms are based upon ideals about what types of knowledge we should be finding out in ecology and what are the most appropriate techniques of investigation. They have frequently been formed through observing sciences with different types of problem from those encountered in ecology. In contrast, Progressive Synthesis focuses on the method of reasoning that should be used in ecological research and how appropriate techniques of investigation can be applied.

1.8 Synopsis of methodological problems facing a new researcher in ecology

(1) Frequently the processes of scientific research are not made explicit to graduate students (Stock 1985, 1989). This book tackles this deficit directly by illustrating how to refine a question into a program of measurement and methods for making scientific inference.

(2) You have to master multiple processes to make progress in research. Simultaneously, you have to define what has been done, devise questions and research to answer them that will advance knowledge, construct investigations that will actually work, and most likely learn what procedures of statistical inference to use. This complete process of scientific research is difficult the first time you attempt it: as when riding a bicycle, you must learn to balance, watch the road ahead, and pedal, all at the same time. This book shows how to define a scientific objective, even for complex problems in ecology, and this provides direction that enables you to balance your different activities.

(3) As an undergraduate, you may have read original scientific papers and learned to debate opposing views. In research, you come closer to the center of discussion and debate, where opposing views have not been reconciled, and sometimes are not even fully defined, so no reconciliation may yet be possible. Then arguments are fierce, and social factors can play an important role in deciding who follows what argument. This book defines these issues to help you to identify the ideals that individuals and organizations may have and how those ideals may affect research objectives, methods, and interpretations.

(4) Ecology has some unique problems in the method of scientific reasoning for its important concepts. When ecologists express differences about the value of particular techniques of investigation they use, it is not always clear whether the resultant debates are about the scientific details or the philosophies behind their use of particular techniques. This book shows why different types of research problem require different techniques of investigation and presents a methodology for reasoning about ecological concepts and theories.

1.9 How to use this book to develop your research skills

The challenge of ecological research is not just in what you must learn but also in how you must think. The focus in this book is on developing a critical attitude about a scientific problem and constructing a procedure for assessment. The book will have the most practical value if you can follow four suggestions:

(1) *Adopt a scientific question you wish to investigate.* Establishing the logical basis for an investigation and determining what type of scientific inference should be made are best learned by applying the techniques of Section I to a real problem. Do not be concerned if your ideas are not yet crystal-clear – this book is designed for you! The techniques presented help to sharpen general ideas into specific research plans.

(2) *Practice the development of a "logic for discovery" with other students.* Set up small discussion groups with three or four other students working on similar research problems. It is most important to learn the art of criticism, particularly self-criticism, without being personally destructive either to your ideas or to those of others. It is essential to learn, in the planning stage, that an idea that you had was incomplete and to know what it feels like to improve that idea through further work, or to reject it and adopt another. You must learn how to treat an idea not as part of your own identity but as something to be turned over, poked at, and examined to see how it fits with other ideas.

(3) *Do not be apprehensive about, or dismissive of, philology or philosophy.* Philology is the science of language. Much science is about defining terms and much progress in ecology is made by developing definitions of concepts. Concepts start imprecise, become more precise as a research plan takes shape, and then sharpen even more by the end of the research. This process of definition and redefinition requires work – and extensive critical analysis. This book deliberately sets out to give you a framework for defining ideas, and deciding how certain you are about their definition. Words are all we have with which to express ideas in science. So being exact in science means being exact about how we define and use words. A philosophy is a system of thinking about things. We all have philosophies, though we may not think of them as such. The philosophy we hold can have a major influence on the way that we first cast a scientific problem – and it is frequently our philosophy that we have to question rigorously. This book examines different philosophies held by scientists and how they influence the approach to research.

(4) *Understand the strengths and weakness of science as a social activity.* Part of developing as a scientist is to associate with groups of people with interests in similar problems and frequently with similar approaches to solving them. For students starting research, "socialization" into a particular scientific group is very important if you are to absorb a

detailed understanding of problems and techniques. However, because the basis of good science requires developing a critical approach, you may sometimes find yourself challenging the very ideas, and ideals (philosophies), that the members of the group you have joined hold most dear.

1.10   Further reading

Mahoney (1976) gives a strong, and sometimes vehement, description of the logical difficulties of the scientific method and, particularly, scientists' failure to confront them. Stearns (1986, 1987) advises graduate students how to proceed in their research; Huey (1987) amplifies and debates that advice. The relationship between research student and supervisor is discussed in a publication of the Council of Graduate Schools (Anon. 1990), and lists of practical suggestions are given for both students and supervisor. Smith (1990) gives a guide for graduate students in the sciences to most practical things graduate students must do and be involved in and Locke *et al.* (1993) describe and analyze effective dissertation proposals.

The two questions "How can we find things out?" and "How do we know whether our answers are correct?" are the problems of developing a heuristic and an assessment procedure. Through his description of the types of theory constructed and whether, and on what grounds, they have been accepted, McIntosh (1980) shows how methodological problems associated with heuristics and assessment have been central to the development of ecology. His account is valuable as a background for understanding why some questions in ecology have been studied repeatedly and apparently not resolved.

Introduction to Section I:

Developing an analytical framework

Ecological research involves defining, then continuously redefining, ideas and objectives as new results are obtained. Experienced researchers debate the meanings of the terms they use, the significance or interpretation of results obtained, what should be done next, and how it should be carried out. They engage in each of these activities over an extended period as their research proceeds. However, when you start research your process of questioning and debate is intense as you make a detailed analysis of a research problem (Fig. I.1).

Section I focuses on these analytical procedures. The key planning process is the first one – defining the research question. How to do this is described in Chapters 2 and 3. A research plan is not simply a list of what you will observe or measure, or the experiments or surveys you will make. It must define the questions you are asking and the foundation for those questions. It must specify the objectives of the survey, experiment, field observation program, or other procedure that you will follow. Most importantly, it must specify how you will make scientific inference from what you do.

When you start research and have to make a comprehensive analysis of a research question and how you will tackle it, it is only reasonable to ask "What should the definition look like?" and "How will I know when my

Fig. I.1. Research analysis requires defining the following: the question being asked; the relevant theory that might be used to answer it; the techniques of investigation for the research, e.g., survey, experiment, model; the effectiveness, accuracy, and precision of measurements; and the method of inference to be used.

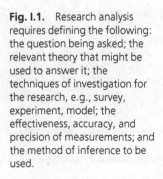
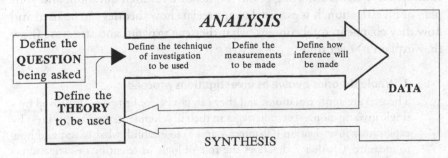

analysis is sufficient?" Section I presents techniques for making an analysis, describes how to present it, and discusses the choices you will have to make between approaches.

SCIENTIFIC ANALYSIS

Scientific analysis results in:

(1) Definition of the theory to be used to answer a specific question.
(2) Definition of the postulates to be examined.
(3) Definition of the techniques of investigation to be used.
(4) Definition of measurements to be made.
(5) Definition of how inference will be made.

There are seven points to recognize about scientific analysis:

1. Knowledge exists in three states

Axioms, i.e., statements of what scientific information you will base your study on, defined from the literature and local, or particular, knowledge of the site, system, or organisms that you will study.

Postulates, i.e., questions you have, but stated in a propositional form so that you can subsequently classify them as true, false, or acceptable with a certain probability, by your investigations.

Data statements, which define the data needed to classify the postulates as true, false, or acceptable with a certain probability and the techniques of investigation that will be used to collect them.

A written plan must contain this classification of your initial knowledge. To do this you must make a conceptual and propositional analysis. The details of this process and its logical basis are given in Chapter 3. A practical example, which illustrates the process of development in one person's conceptual and propositional analysis as she planned her research, is given in Chapter 4. If you understand practical examples more readily than theoretical descriptions, then you may prefer to read Chapter 4 before Chapter 3.

When starting an investigation there is always some theory, however rudimentary, that is a starting point. To define a research question, and your plan of investigation, it is essential to appreciate how theories can be used and how they change in total content (what they can explain) and structure (how they explain it) over time.

2. Whole theories evolve as investigations proceed

Theories are approximations, and they can not be confirmed or disproved by a single investigation. At certain stages in their development, theories progress by exploratory investigation (Chapters 4 and 5) to establish what to test and how to measure. Chapter 7 describes the role of logic in scientific investigation –

particularly, the role of disproof. Chapter 8 distinguishes scientific inference and statistical inference.

A first requirement is to classify your knowledge but the question arises "How can this be done effectively?" An important way in which scientific knowledge advances is through the introduction of new concepts and the gradual refinement of their meaning as they are applied in different situations.

3. Explicit and careful definition must be made of concepts

Often concepts are used in scientific papers without an explicit definition and implied definitions can vary between scientists. It is essential to be critical about the meaning of any concept you use, to make a careful and explicit definition, and to compare that definition and use with others. Two classifications of concepts are given. One deals with how much knowledge we know about a concept and its status in a scientific analysis (Chapter 3). The other deals with the type of use we make of the concept in ecology (Chapter 10). Asking how a concept fits into these classifications is useful when one is making a critical analysis.

As you proceed in the planning process you construct postulates. For field- or laboratory-oriented projects the usual objective is to make measurements to test postulates. Section I describes problems that can be encountered in both measurement and assessing postulates.

4. Assessing postulates is the business of science, but there may be no definitive test

There are three reasons for this. First, a postulate may be constructed either to confirm (induction) or falsify part of a theory, but the logic of neither induction nor falsification is invincible (Chapter 7). Logically, a postulate can be assessed only as part of a whole theory. Second, any test depends upon measurement, and measurements of ecological quantities or conditions are usually limited (Chapter 6). Third, we spend much research effort in ecology making postulates sufficiently precise that we can develop tests for them. This phase of research is exploratory. How a postulate can develop is discussed in Chapter 8.

5. Measurement is essential in science, but there may be no absolute measurement of an ecological quantity or condition

Measurements are rarely exact representations of ecological concepts, and different measurements can inform about a concept in different ways. The idea of measurements as concepts is introduced in Chapter 3 to emphasize that measurements do not always represent concepts accurately but instead are usually limited in some aspects of their effectiveness, accuracy, or precision and must be weighed and considered as partial representations. Principles of measurement are defined and illustrated in Chapter 6.

As you proceed with research planning you will be faced with choices about approaches to take and methods to use. Two things should be realized (see 6 and 7 below).

6. No single technique of investigation – whether experiment, survey, field descriptions, or analysis of patterns in existing data – is always superior

Certain techniques of investigation are more appropriate at certain times (Chapter 9). The task of the scientist is to recognize when each is most appropriate according to the current needs of the investigation, not to adhere to one procedure through habit. The practical examples in Chapters 4, 5, and 6 ask different types of question and illustrate how different methods may be appropriate.

7. Different ecologists have different philosophies about what should be studied and how to study it

Although there may be debate about the "best" technique of investigation (experiment versus field investigation, mathematical modeling versus statistical description) such debates often reflect the different types of question that people want to ask, or the types of scientific explanation with which they will be satisfied. There may be no single correct answer to the type of either question or scientific explanation that is preferable. These choices can be determined by individual philosophies. Principal philosophies and the techniques of investigation and methods for applying them are identified in Chapter 9.

In Section I most attention is focused on questions that can be studied using inductive and deductive logic and the causal-statistical method, and where the objective is to classify a defined proposition as true or false, or ascribe a probability to it. More complex questions that use integrative concepts, which can not be measured directly, must be studied by constructing scientific explanations that explicitly recognize conceptual change as the outcome of scientific inference, and these are discussed in Section II.

2 Five processes of research planning

Summary

This chapter outlines the process of research planning and details the chapters in Section I that give further information about each of the five component processes.

(1) Defining a research question.

It is essential to define scientific knowledge and questions as propositions, statements that can be assessed as true or false or assigned a probability. The propositions that define the knowledge you are prepared to accept as a basis for the research are axioms, and those that define questions are postulates.

(2) Applying creativity to develop new research ideas.

Creativity can be "seeing what others before have not seen," whether stimulated by new observations, new measurements or new analysis. To do this, it is important to recognize discrepancies, to be critical.

(3) Ensuring the proposed research has relevance to prior scientific knowledge.

It is essential to decide what you can consider as established knowledge. This requires a literature review specifically aimed at defining the axioms of your project and the concepts they use. In this review you must guard against self-confirmatory bias, i.e., too readily accepting research that aligns with your own thinking.

(4) Ensuring the proposed research is technically feasible and can be completed with available resources.

The help of your graduate committee or research supervisor and other researchers is essential in this process, but a critical approach must also be applied to their advice. Even experienced scientists can overlook the possibility of unexpected problems.

(5) Determining how conclusions can be drawn.

Scientific inference and statistical inference must be clearly distinguished. The prerequisite for gauging scientific inference is to establish the logic of investigation through a conceptual and propositional analysis. Statistical inference is the application of statistical theory and practice to the analysis of one or more statements defined by a conceptual analysis.

Questions 1–21 (Table 1.1) are analyzed in terms of the five component processes. Whereas each of the five is distinct, with its own technique, it may be used repeatedly or "out of sequence" to produce a research plan.

2.1 Introduction

Questions 1–21 of Table 1.1 reflect concerns about how to proceed with each of the five processes of research planning:

(1) Defining a research question (Questions 1–3).
(2) Applying creativity to develop new research ideas (Questions 4–5).
(3) Ensuring the proposed research has relevance to prior scientific knowledge (Questions 6–10).
(4) Ensuring the proposed research is technically feasible and can be completed with available resources (Questions 11–14).
(5) Determining how conclusions can be drawn (Questions 15–21).

These five processes are rarely applied in a strictly ordered sequence. Nor is each used only once. Research planning is an iterative procedure in which questions of definition, relevance, feasibility, and scientific inference are continually assessed, each in the light of decisions made in other stages. For ease of presentation the principal techniques associated with these processes are considered in sequence. In Chapter 4, an actual research problem is analyzed, to illustrate the five processes in action.

But how long does the planning process take? It continues until you have completed a conceptual and propositional analysis of your research problem. This results in the following:

- A defined set of axioms. These should be justified from the scientific literature or previous investigations, or may define certain assumptions about a site, organism, or ecological system. Each concept should be defined.
- A defined set of postulates that you can investigate. Each new concept should be defined.
- A data statement for each postulate, written in three parts. The first part defines what techniques of investigation will be used, e.g., experiment, survey, etc., and justifies the approach. The second defines each observation or measurement and gives its limitations. The third defines how statistical inference will be made, if that is possible.

2.2 Process 1: Defining a research question

Students' questions from Table 1.1
1. How can I come up with a question?
2. I have ideas and topics but I don't have a project.
3. How can I refine a question so that an answer can be obtained?

Some people are drawn to research because they have a burning practical question they wish to resolve; some by what they have already learned about a particular subject. Ecology, resource management, and environmental science fascinate people. They want to make a contribution through research but are not sure how to start. Questions 1–3 are typical of their initial concerns.

Some people have scientific questions that are fundamental to the subject but may be too general to be researched directly. For example, the questions

> *What will be the effect of climate change on forests?*
> *How can we increase the runs of native salmon on the Columbia River?*

are important for resource and environmental management, and scientifically challenging, but very broad in scope. Equally interesting but broad questions occur in ecological theory:

> *How can the provision of shelter control animal communities?*
> *How is the balance of tree species distribution controlled in tropical forests?*

As they stand, these four questions can not be researched without further analysis that breaks them into components. Each question implies other questions. For example, the question about the influence of climate change implies further questions about the type and magnitude of such change and what we mean by "effects" on forests, e.g., does that mean on timber yield, or species composition? It is possible to define these component questions by rigorously defining each term, i.e., concept, in the first question asked. Usually, the research community has already broken down such general questions as these into research areas.

2.2.1 Origins and types of research questions

Generally, the research of graduate students starts with more precise questions. Sometimes a question arises out of the analysis of previous data (see Chapter 8) that you may be shown. But typically, practical ecological research questions arise in four ways. You may:

(1) observe a natural phenomenon that interests you and decide to explore its relevance to an existing theory,

(2) attempt to apply a theory to a new situation, which is particularly exciting if you put the theory under a stringent examination,

(3) resolve an apparent discrepancy in a theory or between two theories, or

investigate an event out of line with what theory predicts, or

(4) apply a new technique of measurement or data analysis to define and answer new types of question.

This is not an exclusive list. Research may also start with more specific requirements, e.g., a requirement to characterize a species life history, or to construct a simulation model of an ecological system. In both of these examples, it is likely that some theory will be applied – e.g., based on the life histories of related species or how a community functions – and that theory is the starting point of the research.

Problems in resource management, environmental management, and conservation start with practical questions, but answering them usually requires scientific investigations of one of the three types listed. For example, a question such as:

How can we restore the vegetation of a degraded wetland?

depends upon the theory of what constitutes a wetland and how it functions as an ecological system. You would be concerned with applying the theory to a particular situation(s) and would have to determine whether the theory used was actually appropriate. So this practical problem is of a type (2) research question. Resource management and conservation issues are important components of the coarse woody debris study described in Chapter 4, a type (1) question. The control of species integrity in the face of introductions (Chapter 5) is a type (3) question.

1. Interest in observed natural phenomena

A research project may arise through a combination of a person's interests. At the start of her Master's research Ashley Steel observed the piles of logs thrown up after a major flood along a bank of a river running from the heavily forested Cascade Mountains in Washington State. She asked the question:

I wonder how animals use the piles of logs deposited by floods along a river plain?

Three aspects of Steel's knowledge and experience stimulated this question:

(a) Her observation and amazement at the tremendous volumes of wood deposited in piles, particularly the large size of some pieces and their potential to remain for a long time on the riverbank.

(b) Her knowledge that deposition of wood within rivers and streams is important in determining habitat and breeding sites for fish and other aquatic animals[1]. It seemed possible that piles of logs may provide

---

[1] A description of this theory, and its defining references at the time it was written, are given in Chapter 4. References are deliberately not given in this chapter. What is important here is to describe the origin of different types of investigation, rather than to chart the scientific basis of each one individually.

important habitat for terrestrial animals on shingle banks and bars along the river's edge.

(c) Her experience as a qualified white-water rafting instructor. Piles of logs are usually removed from along edges of rivers used for recreation. Steel appreciated the tension between the requirements of this popular recreational activity and wildlife habitat.

This problem is analyzed in Chapter 4.

2. Application or test of a theory

A research project may try to work out how a theory may apply to a new situation, and that can lead to a study of its deficiencies. Patricia M. Anderson and Linda B. Brubaker have described the development of a theory for vegetation changes in central Alaska during the late Quaternary. The first stage was application of a theory, based on observations in central Canada and other regions of the world, that there had been a uniform northward advance of the tree line 6000 years ago. The theory for vegetation change in central Alaska developed through five stages:

Stage 1: Rejecting a simple postulate.
Stage 2: Exploring for patterns of spatial and temporal changes.
Stage 3: Introducing axioms from tree ecology.
Stage 4: Increasing the precision of the theory.
Stage 5: Working towards explanations that are coherent with meteorological theories.

The research is still active in stages 4 and 5.

At one stage in the development of the theory, there were discrepancies and inadequacies in measurements; these were points for further scientific investigation. Critical analysis enables you to define what stage a theory is in, and so what type of investigation is most likely to be successful. Of course not all theories go through the same stages illustrated by the example of vegetation change in Alaska – but they do all go through stages where the most pressing problems may change from assessment, to generating new theory, to developing new measurements or methods of investigation.

This problem is analyzed in Chapter 5.

3. Resolution of a discrepancy between theory and observation

A research project may attempt to resolve an apparent discrepancy between a theory and an observation. Denise Hawkins found this necessary in a study of genetic isolation in trout. Generally, the geographical ranges of western North American trout species do not overlap, so hybridization between them would not be expected to occur. The exception is that rainbow trout coexist with cutthroat trout throughout much of their native distributions. This does not, however, lead to hybridization. These two trout species can live in sympatry without losing their species integrity; so there is a theory that reproductive isolation is maintained by behavioral characteristics such as mate preference and/or nest site selection. But when non-native rainbow trout are introduced to stock lakes or rivers for recreational fishing, the cutthroat are then susceptible to hybridization – even to the point that some people consider the viability of the cutthroat population to be at risk. This observation appears counter to the theory.

This also raises practical questions such as whether, or under what conditions, the introduction of other non-native species is appropriate. The process by which the two native populations of rainbow and cutthroat can coexist is an important scientific problem in its own right. Understanding the basis of this coexistence may be essential to understanding the management problem, which prompts the research question:

> How is species integrity maintained in western North American species of trout with overlapping geographical ranges?

This problem is analyzed in Chapter 5.

4. Applying a new technique of measurement or method of data analysis

Advances in measurement technique, e.g., radiotelemetry, satellite imaging, photosynthesis measurements, DNA analysis, and advances in the application of data analysis procedures, e.g., more robust methods of statistical analysis, or new statistics for spatial analysis, can open up new fields of investigation. Questions can be answered that previously had no solution. However, it is most important to focus on defining the new question, and not simply to make measurements with a new technique, or find some data that can be analyzed differently. Following fashions can lead to difficulties because once the measurements are made, or the analyses are complete, you still have to draw conclusions and that requires that you framed an effective question to start with.

A scientific problem opened up by development of a measurement technique is discussed in Chapter 6.

2.2.2 Analysis of questions

Each of the three research questions exemplifying types (1), (2) and (3) provides a starting point for research, but none can be answered in its stated form. Each question must be broken down into a sequence of statements that can be researched. Three techniques make research questions precise enough to be investigated:

A. Transforming assumptions and questions into propositions

The first task of analysis is to translate your overall question into a series of propositions, i.e., statements of logically linked concepts about what is known and what is to be investigated. Using concepts that can be named and defined is important, although it frequently requires a number of attempts. Steel's question:

I wonder how animals use the piles of logs deposited by floods along a river plain?

is developed into the following propositions.

Propositions of what is assumed

    Axiom 1:   Debris piles are an important component of stream ecosys-
                tems
    Axiom 2:   Debris piles in forested areas provide habitat and/or forage
                for small mammals.
    Axiom 3:   Small mammals use riparian areas.

Note the change of "piles of logs" into the more inclusive debris piles and the restriction of "animals" to small mammals. Both of these concepts required further definition as part of the planning process (Chapter 4).

Propositions of what is to be researched
First phase of the research.

    Postulate 1:   Debris piles in the study area can be classified as old piles
                    or new piles and as piles in the forest or piles on a cobble
                    bar. Differences exist in debris piles that are distributed
                    over the study area.
    Postulate 2:   Small mammals use debris piles in riparian areas for
                    denning, feeding, and/or shelter.

Second phase of the research.

    Postulate 1:   Small mammal use of old debris piles is different from
                    small mammal use of new debris piles.
    Postulate 2:   Small mammal use of piles that are at the forest edge is
                    different from small mammal use of piles that are on a
                    cobble bar.

Whether these are adequate postulates depends upon two things: whether the concepts they use, such as debris piles and small mammal use, are consistent with use in the scientific literature; and/or whether data can be collected to measure them and then test the postulate.

B. Reviewing propositions in relation to the scientific literature

The scientific literature should be used to assess the degree of certainty of the axioms, the statements considered as known. This assessment involves critical questioning of both the statements and the literature. You question three things at once: whether your statements are correct; whether the literature is consistent; and whether the literature contributes a basis for your axioms.

In this example, it was important to try to define debris piles and, particularly, how small mammals might use them. What is it about debris piles that might be particularly advantageous to small mammals? At the time of planning, no literature was found about that point, but a considerable amount was found about the way that woody debris may be an important component of streams as habitat for fish, and that many animals do use large wood on forest floors as habitat. In this investigation, the concept of "use" was being extended from a general definition, based on other investigations, into a specific, measurement-based definition for this particular ecological situation.

C. Ordering propositions into those supported by the literature or direct observation, the axioms, and those that must be researched, the postulates

It was known that coarse woody debris existed in piles, rather than random pieces, and that in streams these created important habitat for fish species. Of course, if the information that the literature can provide is exhausted but you have not justified the first view of the axioms, then you will have to rethink. You may have to reconsider the complete axiom as a postulate, or possibly split it into an axiom with reduced scope, which can be justified from the literature, and a postulate. It is at this stage of analysis that you may incorporate local knowledge or information from preliminary investigation. In this example, direct observations made by walking over cobble bars indicated a substantial difference between new and old woody debris piles.

Postulates must be logically connected to the axioms. Usually a postulate will contain at least one idea (concept) that is part of an axiom, but other ideas that are new. In the example, the concept of *debris piles* occurs in both Postulates 1 and 2. At the start of the study, it was not certain what measurements would determine a *debris pile*, though there was the general idea that the measurement should indicate something about animal habitat.

This approach of making a painstaking logical analysis may seem dry, lacking the romance and bravura of asking speculative questions. It does have the great benefit of focusing on critical analysis of the literature. By critical, I mean asking both: (a) whether the literature is consistent within itself, i.e., do all the published papers contribute to a coherent and logical theory, or are there gaps and contradictions; and (b) whether the theory is relevant to the new research. In this particular example, some scientists may consider theory about the consequences of woody debris piles in streams for fish irrelevant for its effect on land for small mammals. For them, the extension of the theory and its concepts would be too great. The objective of conceptual and propositional analysis is to refine speculative questions and connect them to present knowledge in a detailed way, but you will still need to exercise judgement about what is, and is not, important.

Research questions such as those those posed by Steel and Hawkins are broad. This is typical at the start of research. The background to a broad question is a synthesis of many research results, and it is essential for new researchers to analyze the literature to define the basis of that existing synthesis. Writing out the components of a theory as propositions, and then asking yourself whether you can accept them as axioms, gives focus to a literature review. As literature reviews proceed, then alternatives or qualifiers present themselves much more readily if you use the proposition form than the question form. Most importantly, propositions can be analyzed to see whether each of their terms can be measured or sampled, or some reasonable assumption made. Conceptual and propositional analysis helps you to transform your fascination with a subject into statements that are logically connected, that are part of a theory, and that can be researched. As such, it provides the method to give specific answers to Questions 1–3 of Table 1.1:

HOW CAN I COME UP WITH A QUESTION?

By observing natural phenomena, applying a theory to a new situation, or resolving a theoretical discrepancy – though these are not exclusive. What is important is that you analyze any general question to make it specific.

I HAVE IDEAS AND TOPICS BUT I DON'T HAVE A PROJECT

Ask a question about those to which you do not know the answer – then turn that question into a proposition and follow the process of conceptual and propositional analysis.

HOW CAN I REFINE A QUESTION SO THAT AN ANSWER CAN BE OBTAINED?

Use the same technique as in answering the previous question. Turn your research question into a proposition. For example:

> *I wonder how animals use the piles of logs deposited by floods along a river plain?*

becomes,

> *Animals use the piles of logs deposited along a river plain.* (True or false?)

And that sparks the following questions: What do you mean by *use*? Which animals should be considered? Why is this an important scientific question? For example, does use of woody debris piles imply greater presence of animals on the site than if piles were not there? So then you define *use* and rewrite the proposition.

2.3 Process 2: Applying creativity to develop new research ideas

Students' questions from Table 1.1

4. I have a project. Can I find an innovative way to look at the problem? Can I find a way where others have not?
5. When starting research how do we know what analogies to use? How to get new ideas?

To resolve how to proceed requires creativity – not only in analyzing previous knowledge and developing precise and insightful questions, but in other activities as well. A creative solution to a measurement or sampling problem, or a logistic difficulty, can each make new research possible.

Analyzing what needs to be known in a research problem is different from assimilating existing knowledge from textbooks and the scientific literature. It is during the process of planning research that a scientist makes the transition from receiving knowledge to creating it. This is a very important development, yet one that is little discussed among researchers. To many people who start research, the requirements for this transition seem hidden – part of the mystique of science.

Scientific creativity involves "seeing what others before have not seen". This is not solely the province of famous scientists but is accessible to anyone. Moreover, scientists can be creative in many phases of research. Contradictions may be detected when one analyzes the scientific literature. A set of linked statements that make up a theory may not hold when applied to a field situation not previously investigated. New measurement techniques, or more effective sampling procedures or experiments may reveal new and illuminating details about a process. During data analysis, an abnormal result may indicate a previously unrecognized process (see the discussion of exploratory analysis, Chapter 8). These are all creative acts. Frequently, "finding a way where others have not" (Question 4) depends first upon seeing a discrepancy in a way that others have not.

Creativity in science can be stimulated by the continuous application of criticism. A critical approach is essential if an investigator is to be open to observations or results that do not fit preconceptions or be prepared to measure or investigate things that other scientists have not. In particular, if the use of a theory is to be extended, perhaps from one site or population to another, then the assumptions of the theory must be critically appraised as they apply to the new conditions. From a subjective standpoint, creativity may seem the complete antithesis of criticism. However, in the development of scientific research, creativity and criticism are complementary.

Two processes involved in creativity in science may seem contradictory. First, current theory must be understood in detail. The investigator must accept the theory, to some degree, but simultaneously a critique suggesting that the theory is not absolutely correct must be developed. Sometimes the most innovative research is the result of exposing a theory to a trial that it fails, or passes only partly. The innovation to develop such trials depends upon understanding the theory and recognizing its weakness. As Question 5 implies, analogies are valuable, but they contain some part that is appropriate and some part that is not. Using analogy requires broad knowledge, as well as imagination, to apply to the problem. Many ecologists use mathematical models in their research and the function of a model as an analogy is discussed in Chapter 12.

To give specific answers to Questions 4 and 5 of Table 1.1: do not wait for ideas or innovation to come to you! Make a conceptual and propositional analysis and ask what is inadequate in each concept, and why. Look at measurements and see whether the theory you use would be improved by greater accuracy, or application in a different situation. Searching for effective analogies can be helped if you know the history of the theory you use, and from what it was developed. Remember that an analogy is a caricature of a system, not a perfect representation, so in seeking an analogy you may have to strip the process to the essentials.

2.4   Process 3: Ensuring the proposed research has relevance to prior scientific knowledge

Students' questions from Table 1.1
6.   Am I making a significant contribution to the body of knowledge?
7.   I have my project – but have I missed anything out?
8.   How will I know whether I am duplicating other research?
9.   How will I know whether my project is relevant or significant to a theory or a practical question, or just a stupid idea?
10. How can I conduct a literature review? How will I know what research is going on, what has been done? What about research in a foreign language?

You can use the scientific literature to construct and analyze a set of propositions. However, conducting a literature review is not always easy. It requires that you:

(1)   understand the strengths and weakness of the scientific publication system;

(2) develop a procedure for reading scientific papers to criticize them for
     how the research was carried out and conclusions were drawn – this is
     different from reading them for content; and
(3) appreciate the constant tension between the general nature of the theory
     that ecologists attempt to develop and the specific and local nature of
     the systems they actually research.

1. Understanding the publication system

The strength of the scientific literature is peer review, i.e., the process by
which other scientists read and criticize scientific papers before publication. It
is important to know which journals in your field are peer reviewed. Through
questioning authors' techniques, results, or interpretations, peer reviewers
may improve the quality of the work, and the reliability of a paper. The very
existence of peer review should stimulate scientists to question their plans,
and the conduct and analysis of their research.

However, the peer review process is by no means perfect, and it is
important that you appreciate its imperfections so that, as a reader, you can be
critical in an effective way. Problems with the scientific literature are due to
two things.

(a) We all tend towards a self-confirmatory bias, a tendency to seek confir-
     mation of what we think we already know rather than to explore the
     possibility of being wrong. Peer review may not detect all the conse-
     quences of this because the options of what might have been studied
     may not be apparent to the reviewers, and journal editors may not
     request comments on that.
(b) Scientists organize themselves into groups of like-minded people who
     tend to be selected by journal editors to review each other's papers.
     Their criticisms of details may be rigorous and effective, but their
     criticism of fundamentals may be muted. These issues are discussed in
     Section III.

2. Reading a scientific paper for method rather than content

Analysis of techniques of investigation, measurements, and how inference was
made, should influence your interpretation of a research paper. Chapter 13
lists a set of questions that are not about the subject of the paper but about the
conduct of the research and its reporting style. When you read a scientific
paper, you may read quickly to find the results and conclusions, and this leads
to being less critical of methods. A paper that is important for your research,
for example it may provide support for an axiom, should be reread specifically
to analyze the technical procedures – and this does require a more measured
approach than reading for content. For example, no study will have used
exactly the same subject as you intend to use and you need to determine

whether the differences may have influenced the application of methods. You also need to be sure that a method you think has been used was actually followed in the way you define it.

3. Appreciating the tension between general theory and on-the-ground research

Though published ecological and environmental research most frequently focuses on specific places or particular organisms, authors may attempt to draw more general conclusions from the specific instances investigated. However, the circumstances that led an author to draw particular conclusions may not apply in precisely the same way to your new investigation. Identifying what, of a body of theory, may be assumed to hold for a new situation and what has to be reexamined can require considerable work by the investigator. Hawkins' study of breeding mechanisms (Chapter 5) typifies this problem.

To give specific answers to Questions 6–10 of Table 1.1, the essential thing is to connect your reading with your research; to move from general reading, which is valuable but not sufficient, to directed reading, aimed at constructing a supportable set of axioms. You must list the axioms of your research and examine, for each one, whether the papers you read support it. Do not expect your set of axioms to be correct or complete on the first attempt.

You will be making a contribution to the body of knowledge if you are examining a postulate that no one has previously examined. Sometimes, reexamining a postulate in a new community or situation, or for a different species, is also valuable. The importance of your contribution can be assessed only by considering how the theory may develop. If your axioms form a complete and justified set, and if your postulates follow from them, then you are not missing anything.

Ensuring that you have covered the literature is important, and ways of judging that are discussed in Chapter 13. But the literature is not the only indicator of what is going on in research. Society meetings, workshops, and personal contacts are also important; for these, you need guidance from committee members and other scientists with whom you work. With regard to literature in a different language, some journals are translated into English. For an index of translations see Anon. (1994). However, translations are not always easy to obtain and may introduce errors. If you think an important body of work is being reported in another language, you may have to make special efforts to learn the language.

2.5   Process 4: Ensuring the proposed research is technically feasible and can be completed with available resources

Students' questions from Table 1.1.

11.   How will I know whether my project is too big or too small?
12.   How can I confine the question so that it is a Master's topic and not a lifetime of study?
13.   How will I know when to stop (as an individual) given the cyclical nature of science?
14.   How do I limit the project so that it can be finished in two years?

There are both logical and technical aspects to this stage in research planning. The most important result of research planning is to identify and assess what is known. After that, it becomes much more apparent exactly what can be judged as a contribution to knowledge by a new investigation, or what must be determined before a practical management question can be answered. The important process then is to write out the data statements required for each postulate so that you can assess the resources required to make measurements, both in time and facilities.

But analyzing the logic of the investigation and specifying the required measurements take you only part of the way. Social aspects of science then become important. A student's committee should have on it scientists who can estimate how long a study should take, or just what can or cannot be achieved with specific resources in a limited time. However, for any project, different levels of satisfaction can be sought. Mature scientists may see further ahead into what might be achieved and so expect more than a research student can attain in a reasonable time. This is why it is essential to present your supervisor or major professor with a detailed plan – perhaps more detailed than he or she expects or asks for. Don't let them guess what's involved, describe it!

Writing a research plan is important to establish the logical basis of your work. But your idea of what can be attained almost certainly will change when you first go into the field to start measurement. Steel and Hawkins both discovered that, as have many other students, myself included. However, the existence of a written plan that defines the logic of the investigation is invaluable when deciding how research should be refocused when practical problems are encountered.

The best course of action is to lay out your intentions and how they will be realized in detail. In the excitement of pursuing a new investigation, you may minimize the effort required. Answering specific questions with a clearly defined scientific inference is more valuable as a research experience than overreaching and achieving only partial and ill-defined success in answering complex questions.

To give specific answers to Questions 11–4 of Table 1.1: you will not know for certain whether your project was too large or small (Question 11) until you have finished! You can do two things (and this also answers Questions 12 and 13). First, lay out the data required to answer each postulate. Data statements are described later in this chapter, in Chapter 3, and again with examples and further discussion in Chapters 4 and 8. Second, assess whether, if these data statements were complete and the postulates confirmed or rejected, adequate progress would have been made. For this, you need to determine expected thesis requirements. For example, Steel's Master's research comprised four component aspects of research, the field work for which was conducted in parallel over a nine-month period. She investigated the woody debris piles as potential habitat, measured their microclimate, and made a survey of debris pile use by both birds and small mammals. She then synthesized these components. She was not required to conduct follow-up studies on either of the questions she raised, or on other sites, in order to examine the generality of her results; such follow-up investigations would be more appropriate for a Ph.D. study. Each university has its own standards, and balance between research and course work, and it is the student's responsibility to explore them and understand faculty interpretations. Find out which recent theses are respected as good pieces of work and read them.

2.6 Process 5: Determining how conclusions can be drawn

Students' questions from Table 1.1

15. How will I analyze my data and write my thesis? How will I know what to conclude? Have I got there?
16. You make an hypothesis, and then you try and prove it. What if it is incorrect?
17. We start by trying to prove a hypothesis, but we end up trying to fit a new hypothesis to the data?
18. Will I collect bad data, or data insufficient to answer the question?
19. Some questions have answers. Some problems are exploratory. How do you present the exploratory work? This is difficult because the scientific culture expects answers.
20. What types of control or experimental design will I need?
21. How can I interpret my results in an unbiased way so that I do not massage my results to fit any preconceived viewpoint?

It is the problem of inference that most haunts graduate students. To conduct research and then still not be able to answer the motivating question is most unsatisfactory. It may also threaten the award of a higher degree – if the failure was due to inadequate planning!

The essential prerequisite for avoiding such difficulties is to develop a data statement for each postulate. It is most important to realize that many different data statements can be constructed for any given postulate. The data statement has three parts:

(1) *Specifying the type of investigation to be used.* Different types of technique of investigation may be chosen, e.g., survey, experiment, and field observation of a process.

(2) *Specifying the conditions of the investigation and the measurement details.* In this part, a series of practical decisions must be made, e.g., for an experiment, what number of treatments, how many replicates, and what response variables. Each decision has its consequences, and a critique can be developed for each decision.

(3) *Specifying any statistical hypotheses to be used and calculations to be made.* Statistical tests – e.g., *t*-test, *F*-ratio test, analysis of variance (ANOVA) – all require that the data used meet certain requirements (Chapter 8).

The data statement is an essential part of any research plan. It lays out your choice in the type of investigation, your practical decision in how to carry out your choice of investigation, and the requirements for testing statistical hypotheses if that is what your investigation calls for. While the logic of your research depends on how you construct axioms and postulates, and delineate between them, your results hang on how you develop each data statement.

2.6.1 Developing a data statement: An example

As an example, consider the problem of how plants may die in an abandoned pasture, whether through herbivory or competition from other plants (Reader 1992). The first axiom is:

> Axiom 1:　In closed sward communities, plants encounter competition from other plants that reduces their growth.

According to Reader (1992), the evidence supporting this axiom is provided by many experiments where plants that have been freed from competition have grown more than those experiencing it have. But Reader suggests that removing neighboring plants may also reduce herbivory on the isolated plant by reducing food and shelter for herbivores. Reader suggests that herbivory, apparently well established in closed sward communities, is a confounding factor in plant-removal experiments.

> Axiom 2:　In closed sward communities, plants experience herbivory.

In this study an assumption was:

> Axiom 3:　The effect of herbivory is sufficiently large to be detectable relative to the effect of competition.

The question that concerns Reader is whether herbivory is a confounding factor with competition. So he postulates:

Postulate: Isolating plants from neighbors reduces herbivory as well as competition.

This postulate was made possible by an experiment that Reader envisaged of protecting plants from herbivory by enclosing them in cages of small mesh size (6 mm). Cages could enclose both plants growing in swards and those with neighbors removed.

Data statement: Part One

The first part of a data statement for the postulate specifies the type of investigation to be used.

An experiment: There were two types of treatment:

(1)   protection and no protection from herbivory, and
(2)   release and no release from competition.

Data statement: Part Two

The second function of the data statement is to define the measurements to be made and, in this particular case, the details of the experimental conditions and to specify the limitations of the actual investigation relative to the general statement given by the postulate. For each section (Table 2.1 (a) through (e)) of this part of the data statement, I offer criticisms that may raise questions or suggest alternative procedures. These criticisms are not intended to be carping: I find Reader's (1992) experiment effective in what it set out to do. My point is that actually doing something requires making a series of decisions about treatments, sampling, or measurement. But these very decisions influence the results and so must limit the conclusions drawn.

Data statement: Part Three

In this example, statistical hypotheses are specified.

Statistical Hypothesis 1:   The overall hypothesis tested by an analysis of variance (ANOVA) is that mean plant survival does not differ among the alternative treatments.

Requirements of
statistical theory: Samples, in this case the mean plant survival for each replicate, are expected to come from a normal population and sample variances should be equal. To examine Statistical Hypothesis 1, between a series of treatments, $H_0: \mu_1 = \mu_2 = \mu_3 \ldots$, ANOVA is robust and operates well even with considerable heterogeneity of variance, i.e., where the assumption of equal variance is not met exactly.

Table 2.1. *Conditions of the experiment and measurements made, and a critique of them, for a herbivory and competition effects experiment conducted by Reader (1992)*

	Conditions of the experiment and measurements	Critique
(a)	The experiment used two levels of neighbor removal: neighbors were left intact (control), or neighbors were removed from a 1 m² plot	The two levels may seem reasonable given that the prime object of the experiment is to examine the interaction with herbivory, but it is possible that removing all plants created special conditions that are quite different from within the sward in a way not attributable simply to reduced competition. For example, neighbor removal from the 1 m² plots may markedly change microclimate or nutrient status
(b)	The experiment used three levels of plant caging: (i) A full cage of a wire mesh (6 mm) cylinder 25 cm tall and 10 cm diameter, closed at the top and mounted on a plastic tube driven 4 cm into the ground and with 4 cm remaining above the ground (ii) A "half-cage" designed to simulate the reduced light that the full cages produce, i.e., 15%. A 1.2 m × 1.2 m piece of mesh horizontally spread 25 cm above the ground attached to the tops of wooden stakes (iii) No caging	Manipulative experiments in ecology may have both a desired effect, in this case stopping herbivory, and an undesired effect, in this case the shading of plants by the cages. It is not possible to examine the desired effect without the undesired effect occurring. In this particular case, the effect of the cages in producing shade simulates at least some of the effect that would be expected from competition. The half-cage, designed to look at only the shading effect, may in turn also cause other, unforeseen effects, e.g., possibly changing rainfall amount and chemical composition. So the experiment makes assumptions about the effectiveness of the treatment that can not be completely tested. For complex experiments, it is possible to write out a list of assumptions as the axioms of the experiment
(c)	Each combination of neighbor removal and caging was used at the bottom of each of five hills (replicates)	The complexity of the experiment, the available resources and human effort, and the needs of the statistical testing to be used in hypothesis testing usually determine the number of replicates. The greater the number of replicates, the greater the opportunity for resolving possible differences between treatments. In this case, the number of replicates seems to have been effective

Table 2.1. (*cont.*)

Conditions of the experiment and measurements	Critique
(d) Seeds of the three species were grown in a greenhouse for three weeks and transplanted to the field in the first week of June, when naturally occurring seedlings were of the same size. Four seedlings of each species were planted, 25 cm apart, into each 1 m² plot in a 4 × 3 grid with randomly assigned locations	The use of transplanted seedlings is one of the most important decisions made. If the effect of competition in the natural sward occurs prior to the time when naturally occurring seedlings achieved the same size as transplants, then the experiment is not effective. That possibility was not tested. The decision to use seedlings was necessary for the type of experiment planned. The chances of finding naturally growing seedlings may have been small, particularly at the bottoms of hills where these species were comparatively rare. If found, they were likely to have been variable in size and perhaps in situations that were difficult to manipulate with neighbor removal or caging. Indeed, it would probably take a very different type of investigation to determine the occurrence and fate of germinating seeds
(e) Measurement was of number of plants surviving, between 0 and 4, for each species for each replicate	Measuring complete plant survival rather than perhaps trying to measure herbivory on individual seedlings assumes that the treatment effects will be substantial, which in this case they were. Measuring complete plant survival is the more stringent measure that properly weights the results of the investigation against finding an apparent effect based upon estimates of partial damage

As an example, after 16 weeks for *Medicago lupulina*, ANOVA results indicated that both experimental treatments, removing neighbors and caging plants, had an effect on plants that was significant at $P < 0.0001$. That is, the probabilities of these being chance occurrences are less than or equal to 1 in 10 000. There was also an interaction between the two treatments, with the same probability.

Statistical Hypothesis 2: A further analysis of the data looks at pairwise comparisons, with the hypothesis that there is no difference between their means.

So there are a series of null hypotheses, $\mu_1 = \mu_2$, $\mu_1 = \mu_3$, $\mu_2 = \mu_3$, etc. for each combination. The probability of observing differences between means by chance can be calculated.

Requirements of statistical theory:

Multiway comparisons, e.g., $\mu_1 = \mu_2$, $\mu_1 = \mu_2$, etc., are more sensitive to the requirement of homogeneity of variance (Zar 1996).

Survival for *Medicago lupulina* after 16 weeks is shown in Fig. 2.1. Mean survival was 4.0 (all plants survived) for caged plants with neighbors present or neighbors removed, and were greater than for all other treatments. This indicates that protection from herbivory by caging was more effective in producing survival than was removing neighbors. Treatments with neighbors removed, with no cage or half cage, had means that were not themselves different (mean survival 2.4, $P < 0.05$) but that were significantly greater than those for both non-caged and half-caged treatments with neighbors present. This suggests that removing neighbors increased survival but not as much as did caging plants.

Can we use this statistical inference to accept that herbivory is a confounding factor with competition and that isolating plants from neighbors reduces herbivory as well as competition? We certainly could not ignore this result. But the use we make of it, the scientific inference, depends upon the view we take of the conditions of the experiment and their limitations. Was the choice of the experiment appropriate, e.g., should a more extensive survey of herbivory on plants have been done first? Were the practical decisions reasonable, e.g., is it reasonable to investigate the effects on transplants? The fact that a statistically significant hypothesis test was obtained has to be interpreted in the full light of choices and decisions made before confirming the postulate. Chapter 8 discusses the role of exploratory analysis in investigating these types of question and how exploratory analysis can help to sharpen the alternatives that may be compared in statistical testing. In the Introduction to Section II the process of making scientific inference in ecology will be described and examples given through the chapters in that section.

2.6.2 Using statistics to illuminate the problem, not support a position

It is essential to recognize the distinction between scientific inference and statistical inference. The difference is similar to that between strategy and tactics. Making scientific inference involves the logic and limitations of the

Fig. 2.1. The mean number of plants surviving in each of the six experimental treatments for *Medicago lupulina*. The standard error of the mean is shown for week 16. Treatment means for week 16 with the same lower-case letter do not differ significantly, $P < 0.05$, Tukey's test. (From Reader 1992, with permission.)

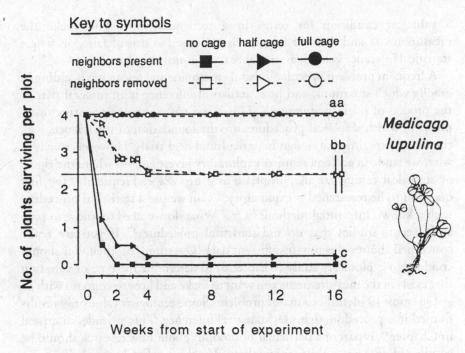

Key to symbols

complete investigation (see Introduction to Section II). Statistical inference is the application of statistical theory to the analysis of one or more of the postulates. The statistical hypothesis has limitations as it applies only in the particular circumstances where, when, and how the measurements were made. The extent to which you accept the hypothesis as an indication of the truth or probability of the postulate depends upon the compromises that you may have made in forming the hypothesis in the first place, and in making measurements. You cannot discount a hypothesis once it has been investigated, but you may need to repeat an investigation with different measurements, or at another time or place; i.e., you may need to investigate another data statement for the same postulate before coming to a scientific inference.

Question 15 (from Table 1.1) is answered when one or more postulates have been confirmed or shown to be false, or the research has shown that they need to be redefined. Questions 16 and 17 illustrate typical, but incorrect, uses of the word "hypothesis". A hypothesis is investigated. Hypotheses are not fitted to data. Data must be obtained that can classify the hypothesis as true or false with a specified probability. This is not semantics. When you construct a set of measurements to investigate a hypothesis, your intent must be to make measurements as accurate as possible, not to select the measurement that will support what you think should happen. In this book, the term "hypothesis" refers to a more precise and limited statement than a postulate. A hypothesis is constructed so that you can measure each term or assume a value or condition for terms in a recognized way. You make the measurements, and then the hypothesis is assessed as true or false, or, where statistical inference is used, as probably true or probably false.

A frequent problem in ecological and environmental research is deciding in exactly which situations, and how, statistical inference is an integral part of the process of scientific inference. This is less true for some scientific disciplines in which statistical procedures are the foundation of the science, e.g., the use of experimental design in agricultural field trials. However, much of what we study in ecology requires exploratory investigations where the rigors of statistical testing are inappropriate as a first stage. Frequently, the first question to be researched is exploratory: "Can we use a statistical procedure with a known inferential method?" Or, "What do we need to know to plan and execute studies that do use statistical procedures?" If you are really concerned about collecting insufficient data (Question 18), then you should conduct an exploratory study (Chapter 8) to determine the type of variation that exists in the measurements you want to make and how to contend with it.

Question 19 identifies a major problem. Some ecologists value only results framed in a procedure that uses statistical inference. This attitude, discussed in Chapter 9, is part of a particular philosophy about how research should be conducted. It is not a philosophy followed in this book. Chapter 4 illustrates that developing a research plan can require considerable exploratory work just to construct postulates correctly. All too frequently such studies are not reported in the scientific literature. Sometimes, fully formed hypotheses seem to spring from the page, complete and mature! This may add to the mystique of science, but it does not help us to understand the scientific process or even, sometimes, the question being investigated.

Questions 20 and 21 require specific answers that can be given only knowing the individual problems. If unsure, you should consult a statistician while planning your study. The way to avoid bias is to investigate and lay out in advance what your critical measures and procedures are.

2.7 Further reading

Through this book I stress the importance of criticism and in this chapter I suggest that it is important in stimulating creativity. This is not all that is involved. A large amount has been written on creativity in science. Root-Bernstein (1989a) examines many different strategies and contrasting ideas in a practical way and is an excellent source of further references; see also Root-Bernstein (1984, 1989b). Two stimulating lines of thought about creativity in science are Holton's (1973, 1981) concept of thematic analysis and the use of analogy. Holyoak and Thagard (1995) discuss use of analogy in making discoveries.

3 Conceptual and propositional analysis for defining research problems

Summary

Scientific research requires that we place the subjective process of developing new ideas into a logical framework of challenge and questioning through debate and the collection of new data. This is a continuous process (Fig. 3.1), and the iteration between axioms and data is driven forward by four processes – *imagination, investigation, comparison,* and *deduction.* These develop knowledge through three states: (1) *axioms* that already have been researched and refined, (2) *postulates* that you wish to research, and (3) *data statements* that describe how data will be collected and used to assess a postulate.

To begin in research you must make a *conceptual and propositional analysis.* Axioms and postulates must be defined and a description given of the data statements required to examine the postulates. The key to these definitions is a critical analysis of the concepts they use. The component concepts provide a network of connections among axioms and between axioms and postulates. Concepts can be classified according to their status in the function of research, i.e., whether they are well researched, are recently constructed from imagination, or are measurements. Classifying concepts allows you to define and describe your knowledge about the subject you intend to research and what you intend to measure. The way to logical and statistical analysis is then open.

3.1 Introduction

The growth of scientific knowledge is sometimes compared to the building of a structure brick by brick. This analogy stresses the progressive nature of science as new investigations are made. But it does not accommodate the reevaluation and change that can take place as we develop scientific understanding.

An alternative analogy, more appropriate for the approach taken in this book, is to consider science as developing a network of knowledge, defined by our theories, that is constantly tested and examined, broken and reconstructed, as well as extended to answer new questions. Breakage may arise because of some important trial or practical use, or deficiencies in a theory may

Fig. 3.1. The progress of science develops through continuous iteration between axioms and data driven forward by four processes.

become apparent through carefully analysis of scientific papers. The way in which the knowledge network is extended depends upon the research we do and the purposes we use it for. This "network" analogy implies we may not discover scientific knowledge in an organized, encyclopedic way; for instance, practical problems may determine what we investigate, or the development of a new technique or instrument may lead to a new emphasis. Nor can we know everything there is to know about a subject. We may only say we know sufficient to answer a set of questions to a particular level of accuracy.

In Chapter 1, a distinction is made between the *progress* scientists make, i.e., what they find out, and the *processes* they use to make progress. Scientific progress is the discovery of new information and production and development of theories and their application. The scientific process is motivated either by lack of information, discrepancies within a theory, or by a problem that requires the application or development of a theory in a new way. Chapter 2 describes five processes of research planning and illustrates how questions about research could be investigated through each of these processes. In practice, research planning has to synthesize all of these processes.

This chapter shows how this synthesis can be achieved and specifies an objective for it at the planning stage of research. It first describes important constituents and properties of theories that must be defined by a research plan and that will show what progress has been made and what is intended for investigation. It then shows why, and how, a *conceptual and propositional analysis* of a theory is used to define these constituents and properties. It takes the questions that arise in research planning and synthesizes answers into a practical foundation for the process of scientific research.

In this chapter I define a series of terms and show how they specify theories and the questions to be answered and so enable research progress. Flew (1984) and Dancy and Sosa (1992) give fuller definitions of some terms and their origins and synonyms. My purpose is to provide a consistent set of definitions that can be used in planning and conducting ecological research. In Chapter 4 these terms are put to work in an example. I find that about half the scientists who use this process prefer to read Chapter 4 before this chapter in order to understand in some detail what the method produces.

3.2 Constituents and properties of theories

Scientific research is commonly believed to increase our certainty about the universe and how it works. What science produces is information, organized through use of theories, which contain ideas of varying degrees of certainty. The challenge of science is that we use subjective thoughts and ideas, our imagination, to initiate the process of discovery that we hope will result in more objective knowledge. The task of research is to ensure that new ideas find their proper place, and that previous ideas are changed where necessary, so developing the whole theory.

SCIENTIFIC METHOD

The purpose of *scientific method* is to place the subjective process of developing new ideas into a logical framework of challenge and questioning to develop objective knowledge.

In practice scientists themselves rarely discuss how this is done. They frequently identify what I call techniques of investigation, e.g., experimentation or survey, as scientific methods. In practice the complete process, which includes how questions are chosen and scientific inference is made, requires a *methodology*. The arguments that sometimes break out over which is the better "method" (i.e., technique of investigation) are usually about how they can be used in making scientific inference or values of the types of question they can be used to investigate.

OBJECTIVE KNOWLEDGE

Objective knowledge is knowledge that has been researched and then scrutinized by, and debated among, scientists. Such knowledge is independent of a single person, and therefore not subjective. In contrast to some common uses, the term "objective knowledge" does not imply absolute or permanent knowledge but simply the most reliable current knowledge. The content of objective knowledge changes as scientific investigation continues, new information is discovered, and the network of knowledge both grows and changes.

Notice that these definitions of scientific method and objective knowledge both focus on the process of scientific investigation.

THEORY

A *theory* has two parts:

The working part of a theory is represented as a logical construction comprising propositions, some of which contain established information (axioms) while others define questions (postulates), as diagramed in Fig. 3.2, discussed in the following text, and summarized in Table 3.1. The working part of a theory provides the information and logical basis for making generalizations.

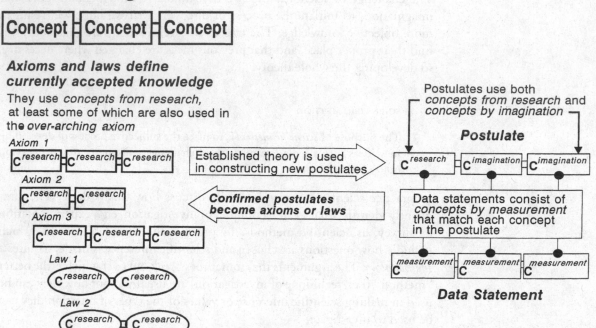

Theories also contain a motivational and/or speculative component that defines a general direction for investigation or type of question that can be answered. An over-arching axiom or postulate frequently specifies this component.

Theories are neither completely speculative, as implied in some older definitions (see Dancy and Sosa 1992), nor completely certain. For example, we refer to Charles Darwin's theory of the origin of species (the motivating question) through natural selection (for which Darwin provided evidence). Over time, as research has progressed, both the type of question being asked about evolution and the information in support of the theory have changed, and the theory has been both qualified and expanded to the point where Darwin's theory is distinct as the origin of present theories. Darwin's theory and its successors provide explanations, though to different questions. Some theories make very specific predictions (in the form of postulates) as well as explain and examine how these can be used to test the theory.

Scientific research extends or reconstructs scientific theory, or may develop a new theory. In order to do this, the component parts of a theory must be recognized and defined. The most basic component is the concept.

Fig. 3.2. Diagrammatic representation of the relationships between the components of the working part of a theory. Questions may come from practical problems, new observations, or discrepancies in the theory. An over-arching postulate rather than an over-arching axiom may motivate the theory. Three types of concept are represented: concepts from research, $C^{research}$; concepts by imagination, $C^{imagination}$; concepts by measurement, $C^{measurement}$. For definitions, see the text and Table 3.1. Axioms and laws are connected by common use of individual concepts.

Table 3.1. *Summary of the components of theories, their roles, and the types of uncertainties associated with them*

Component	Role	Uncertainty
Concept	Any term that can be defined and is used in an axiom or postulate	Definitions may not be agreed upon and constant among ecologists. Definitions may have to be made relevant to particular investigations or sites – and so have an element specific to a particular problem or circumstance
Axiom	A proposition assumed to be true and used in defining the body of knowledge to be used	Sometimes axioms are found not to be true and have to be qualified, reclassified as postulates, or even dismissed. This may be a useful result of research, but it can be difficult to detect that an axiom is not correct or is incorrectly specified
Over-arching axiom	Defines the overall body of knowledge to be used	The concepts used rarely have a definition that is both comprehensive and precise. The over-arching axiom must be further defined by the axioms of the particular study
Postulate	A proposition to be investigated by measuring its concepts	It can take considerable research to develop a postulate to the point where it can be classified as true, false, or probably true or false. Research into concept definition and measurement may be necessary first; during which time the value of the proposition may remain uncertain
Over-arching postulate	Specifies a broad question that can not be answered by a single investigation	At the start of research it is not certain that a set of postulates can be constructed to provide adequate scientific inference for an over-arching postulate. Much research may be required in defining an over-arching postulate, rather than just answering it

Table 3.1. (*cont.*)

Component	Role	Uncertainty
Law	An empirically derived relationship that is universally consistent	There may be no theoretical understanding of why the relationship exists, which makes use of the law uncertain
Data statement	The specifications for the data collected to test a postulate	Constructing a data statement requires choices about the method of investigation and decisions about techniques. These choices and decisions are part of the imaginative process of science
Hypothesis	A data statement constructed to give a logical test of a postulate	It is usually possible to construct more than one hypothesis to test a postulate, depending on measurements, sampling, and other technical aspects of the science. The result of just one hypothesis test may be insufficient to classify a postulate definitely as true or false or ascribe a probability that will be agreed upon by other scientists

CONCEPT

A *concept* is any object or idea to which we can give a name and define, and so enable things to be understood in a particular way.

Not all ecologists may define a concept in the same way, and some concepts come to have different meanings as their use changes. For example, *migration* and *photosynthesis* are both concepts for which there are general definitions. *Migration* is the long-range movement of animals but may be defined differently when one is considering the once-in-a-lifetime return of some salmon species to their hatching place, as opposed to the repeated annual migrations of some bird species. *Photosynthesis* can be defined as the fixation by plants of carbon dioxide into organic carbon compounds using light energy. But photosynthesis by C3 and C4 plants has different patterns of light saturation, and different biochemical pathways fix CO_2. These are different *kinds* of photosynthesis, and if these differences are relevant to the proposed research, they must be included in definitions.

Concepts have to be defined and understood in the context of the theory with which they are used. For use in research, concepts require explicit definition that delineates between different situations in which they may be used. Particular care must be taken in the many cases in ecology where a concept also has a common language meaning so that the definition delineates between the meanings. Each concept is defined in terms of other concepts or is equivalent to some measurement that itself must be defined. The chains of definitions make up the theory. This approach of insisting on definition forces you to be precise, which is crucial as you move from talking about something in the general sense to researching and measuring it. In actually doing the research you will have to make choices about the measurements to be taken, and those choices must be reflected in your definitions.

In the general definition of *migration*, we assume that both *long-range movement* and *animals* are measurable concepts. In the general definition of *photosynthesis*, we assume the terms *light energy*, *organic carbon compounds*, and *carbon dioxide* can be measured, and *species* or *parts* that are themselves concepts can further define *plants*. For the new researcher, the problem of definition may not be trivial. In his review of the background of ecology and its recurring scientific problems, McIntosh (1985) notes "A traditional problem of ecology has been that ecologists, like Lewis Carroll's Humpty Dumpty, often use a word to mean just what they choose it to mean with little regard for what others said it meant. This tendency has not disappeared." At the very least, you must make explicit definitions in your own work – and be able to relate your meanings to those of other scientists so that what is common and what is different between uses can be distinguished.

The relationships between the constituent concepts define the way a theory is represented. To determine these relationships requires classifying concepts according to their properties and usually defines a hierarchy of knowledge in which some attributes are shared and some are not:

(1) Concepts can be related as the same *kinds* of thing. For example, *once-in-a-lifetime return* and *annual migrations* are both *kinds* of migration; C3 and C4 processes are both *kinds* of photosynthesis.
(2) Concepts can be related as *parts* of things. For example, a *leaf* is part of a *tree*; the *light* and *dark reactions* are both *parts* of *photosynthesis*, and *migration* may be *part* of the *life cycle* of an *animal*.
(3) Concepts can be related such that one *has the property* of another. For example, a *leaf* has the property of *photosynthesis*.

Further classifications of concepts that are important in constructing ecological theories are discussed in Section II. Concepts classified as kinds of things, parts of things, and having particular properties are discussed later in 3.4.

PROPOSITION

A *proposition* asserts a relationship between a set of concepts. A proposition can be classified as true or false using propositional logic, or frequently, in ecological research, as probably true or probably false using statistical inference. There are two types of proposition: those that we consider sufficiently true to use as the basis of future research (axioms), and conjectures (postulates), which we investigate.

For example:

> *Mature sockeye salmon migrate from the ocean to fresh water and spawn.*
> *Green plants fix carbon dioxide into organic compounds using visible radiation.*

are both propositions. On the other hand:

> *Plants are green so they can photosynthesize.*

is not a proposition. It is an assertion expressing a supposed purpose. The "so they" means that the statement can not be tested; as a statement, there is no way to show it as true or false. (This type of statement is teleological; see Chapter 9.) The types of logical relationship that propositions can define are described in Chapter 7.

AXIOM

An *axiom* is a proposition assumed to be true on the basis of previous research, observations, or information, and is used in defining the working part of the theory that is the foundation for the research. Typically, an axiom specifies that something does or does not occur, or that one thing does or does not influence another; or it defines a mathematical relationship.

Most research is based upon a series of logically connected axioms. The purpose of explicitly stating the axioms of the working part of a theory is to see exactly what the required assumptions of the research are and so what must be postulated to test the theory or answer a question. The two examples of propositions are both axioms – but for each, the component concepts should themselves be defined by a series of axioms and so make them refer to a precise situation where a postulate is being considered.

One of the difficulties about scientific research is knowing where to start making detailed definitions, i.e., just what is needed as the working part of a theory. We use terms such as "ecosystem theory," "population dynamics theory," or "disturbance theory" in a general way to refer to bodies of knowledge that we expect will answer certain types of question. The basic axiom(s) of such a theory is an over-arching axiom(s).

OVER-ARCHING AXIOM

An *over-arching axiom* is a fundamental proposition, used as an axiom, which states broad assumptions of the theory and cannot be challenged directly by single investigations.

Over-arching axioms identify that part of the network of knowledge that the investigator thinks is important for the problem being studied. The acceptance of an over-arching axiom is due to the collected and collated knowledge from many studies, and revising an over-arching axiom involves a major reassessment of theory. The previous examples of axioms, about sockeye salmon migration and photosynthetic carbon fixation, are general in nature, and both are of the type used as over-arching axioms. We may continue to research how migration takes place, how genetic variation in migration manifests over time, or the role of olefactory sensors in fish migration and how migration may be affected by chemical or other environmental factors. However, the answers obtained will not challenge the over-arching axiom that *mature sockeye salmon migrate from the ocean to fresh water and spawn.* Even if there is a case where migration of sockeye salmon does not occur, it is likely that this will be treated as a distinct contrast from the normal situation and an explanation will be sought for it in relation to the accepted theory, e.g., explainable disruptions to the process that the theory suggests may cause migration. The exception is most likely to be used to further the explanatory capacity of the theory – not as a simple disproof.

POSTULATE

A *postulate* is a conjecture written in the form of a proposition. It is untested, or considered sufficiently uncertain to be the subject of further direct investigation. Even though there may be inconsistency in the literature about a theory, you can identify that and construct a postulate(s) to investigate it.

Previous postulates can become axioms on which new postulates are based. For example, at one stage:

Mature sockeye salmon spawn in their natal tributary.

was a postulate that built on the axiom,

Mature sockeye salmon migrate from the ocean to fresh water and spawn.

i.e., not only do sockeye salmon return to fresh water, they return to the specific segment of fresh water where they hatched. The precise meaning of the postulate depends on the definition of the concept *natal tributary*, i.e., to what detail a river system might be divided. Testing this postulate required tagging young salmon as they left their natal tributary and identifying where they spawned on returning.

OVER-ARCHING POSTULATE

An *over-arching postulate* is a general question, stated in propositional form, which can not be answered by a single investigation. It is a speculation motivating a program of research and the development of a theory.

In Chapter 11 a program of research is discussed that has been motivated by a contrasting pair of over-arching postulates:

> *The primary control of trophic interactions in ecological systems is resources.*
> *The primary control of trophic interactions in ecological systems is predation.*

These postulates have been studied in many different habitats, most often singly though sometimes as a contrasting pair, and one of the most important tasks for investigators has been to translate them into specific postulates for the particular ecological system they intend to investigate.

Within most ecological theories there are usually some concepts that cannot be measured directly, or can be measured only for particular instances, or may have to be measured in more than one way to obtain a full appreciation (Chapters 5 and 6). Consequently there is uncertainty about how generally a theory may apply – as well as possible doubts about its content. These contrast with laws that are restricted in scope and most frequently are based upon quantitative relationships between measurements.

LAW

> A *law* is based on an empirical relationship between two or more concepts, established by measurement, and asserted to be universally true.
> A *law* can be used as a rule of inference.

An example is the expansion of water as it cools prior to freezing. This is found where, and whenever, measurements are made and so is considered to be universally true. It is used as a rule of inference in explaining why water pipes and car radiators crack in freezing weather. In this case the underlying processes (molecular and energetic) are understood.

In some cases, a consistent empirical relationship, i.e., based on measurements, is established, but there is no a satisfactory explanation for it. For example, Taylor (1961) analyzed data for 24 animal populations and proposed a simple power law:

$$s^2 = am^b$$

where s^2 is the variance at each population density, m the mean density based on plots of $\log s^2$ vs. $\log m$, and a is the calculated empirical relationship. A number of attempts have been made to explain the reasons for the biological and ecological processes that may cause the relationship (see Yamamura

1990). Biogeographical relationships also provide broad generalizations of this type but, generally, laws have not been enduring in ecology (McIntosh 1980), i.e., repeated investigations have revealed differences in the relationship.

The physical sciences and chemistry have many laws, but many, if not all, have exceptions to them. The precise relationships framed as laws are discovered under particular conditions and we can use them elsewhere only with the caveat of "other things being equal" (Cartwright 1983, particularly pages 44–55). The interesting thing about ecology is that the conditions between investigations are often sufficiently different to have an influence on the results. There are two related reasons for this:

(1) Difficulties with measuring the same concept consistently across different situations. Many of the things that ecologists are interested in have to be measured in different ways in different circumstances and the different measurements have some part in common and some part that is different.

(2) The requirement that the relationship is exactly the same in different conditions. An example of this difficulty is seen in the $-3/2$ law of self-thinning. In single-species plant communities with closely spaced individuals, plants compete and, as a result, some die while the survivors continue to gain in weight. For a number of cases, a relationship was observed (Yoda *et al.* 1963) between the density of survivors, p, and their mean weight, $\bar{w}$,

$$\ln{(p)} = c - \frac{3}{2}\ln(\bar{w})$$

where c is calculated empirically and may vary between species and growth conditions. Yoda *et al.* write "The quantitative relation . . . is so universally found in various plants that it may be termed 3/2 power law of self-thinning." Begon *et al.* (1986) propose an extension of this thinning law to animal populations with a finite rate of food supply (for a discussion, see Latto 1994).

Some researchers found instances in plants where the relationship departed from $-3/2$ (Weller 1987, Zeide 1987). These discrepancies were acknowledged by some scientists as genuine exceptions, so they did not accept the empirical relationship as constant and rejected the *law*. However, some ecologists continued to assert the generality of the $-3/2$ law (e.g., Lonsdale 1990, and a reply by Weller 1991).

There were many attempts to explain the $-3/2$ relationship (e.g., White 1981) and to assert its more general significance (e.g., Westoby 1981) for ecological theory. Explanations have focused on the relative expansion of plant crowns, thought to determine the competition process, and increase in plant weight (see Yoda *et al.* 1963, Norberg 1988). None of the explanations has been completely accepted. To the extent that there are differences in the self-thinning relationship, i.e., the relationship varies around the value of $-3/2$, then an acceptable explanation should be able to account for both the main effect and these differences. But this would be a theory rather than a law.

We are interested in the possibility that laws exist. (1) Because they demand an explanation and, since they are supposed to be universally true, then the explanation should be universal. (2) Because if they are true then they can be used as a rule for inference in providing explanations. Laws must be *general*, they must not contain special provisions or exceptions for particular individuals or groups. They must tell what would happen *if . . . then . . .* They tell what is physically necessary, possible or impossible (Salmon 1990).

In Fig. 3.2 I represent laws as components of theories because they provide information about a phenomenon that has been quantified and that needs to be explained. The difference between a law and an axiom is in degree of certainty and universality. A law needs no additional confirmation and applies everywhere. An axiom is assumed to be true for the particular research being conducted and if the research is confirmatory then the axiom will gain additional likeliness.

What information should be used to define a theory?

CODIFIED KNOWLEDGE

The scientific literature contains *codified knowledge* – that is, knowledge that has a recorded structure and is used to define axioms and concepts that determine the logic of the theory.

UNCODIFIED KNOWLEDGE

For practical investigations there is also *uncodified knowledge* – that is, knowledge about the particular field site, population, species, or environmental conditions that informs how a theory may be applied in a particular circumstance and that you must discover for yourself through preliminary research.

If a question requires research at a particular site or with a particular species, then the theory you need may have to have particular knowledge incorporated into it about that site or species, which may not exist in the scientific literature. For example, if you wish to study some aspect of salmon migration through experimentation, you may need to determine the general numbers of a species migrating to the river you are considering, and perhaps the conditions of the spawning habitat. This may require preliminary observations. This uncodified knowledge will define some axioms that you need in your study. It may determine how you can use a theory and the types of postulate you can construct. An important task of a conceptual and propositional analysis is to bring together these two types of knowledge, codified and uncodified, and examine where the uncertainties are in both types. There may also be uncodified knowledge in how certain measurements are made, e.g., aspects of setting an animal trap or things that must be done to obtain a measurement of leaf photosynthesis, that are not written in any scientific paper but may have an important influence on what you can do.

DATA STATEMENT

A *data statement*

(1)   defines the scientific procedure to be used in investigating a postulate,
(2)   specifies the measurements to be made for each concept of a postulate, and
(3)   specifies the requirements of the data for any statistical test to be applied.

The purpose of the data statement is to define the assessment procedure for deciding the logical outcome of a postulate: to reject or confirm it outright if the assessment procedure uses propositional logic, or to reject or accept it with a specified probability if the assessment procedure uses statistical inference. Where the intended tests are found by research to be inadequate, a postulate may be reconstituted as one or more different postulates using further conceptual analysis, or revised and reexamined with further exploratory research (Chapter 8). In many cases, exploratory research is required before a test can be designed for a postulate.

Once a proposition is accepted as an axiom, then the data statement(s) used in deciding it is really part of that axiom, although it is rarely written out in full in this form. These data then are actually part of the theory. A comprehensive analysis of any theory (which is what you should do at least when you first start working with it) should involve examination of the foundation data statements as well as examination of the logical construction of the theory. Reassessment of the data statement(s) can result in reclassification or reconstruction of propositions within the theory.

The role of the data statement is to define the context and conditions under which data will be collected. Acknowledgement and understanding of the context and conditions are essential during the interpretation of your results, by both you and others. It is a contradiction to say, "Let the facts speak for themselves!" Data do not exist in a logical vacuum. Data are not neutral. Sometimes people refer to data as facts, but there is no such thing as "pure fact," particularly in an active field of research. Facts exist only when there are criteria for establishing them, rules governing how they will be investigated and how they are to be used – and this is what the data statement must do.

The iterative nature of scientific investigation means that we always collect data with some proposition in mind, however basic that proposition may be. For example, if you start an investigation by making a survey, perhaps trying to establish what is present at a site or looking for some basic relationships, you will still frame that survey with previous information, i.e., axioms, in mind. A survey may seem to assume nothing more complicated than that you will, or will not, find what you are looking for – but your ideas (postulates) about such things as the frequency of occurrence or distribution will influence how you proceed (construction of your data statements.) These ideas must be made explicit as postulates, and their concepts defined.

HYPOTHESIS

The term *hypothesis* is reserved for use where a specific test is designed for a postulate. The test may be examination of a logical outcome, or it may be statistical. In a statistical argument, the construction of hypotheses takes particular forms (Chapter 8). A *hypothesis test* must be specified in a data statement, and it may be possible to construct more than one hypothesis test for a postulate.

A *hypothesis* must be constructed with great care. It requires positive action. In this book, I use the word *hypothesis* only where a data statement can be formally assessed, using either propositional logic (Chapter 7) or, as is more usual in ecological, environmental, and natural resources research, statistical inference (Chapter 8). Statistical tests are carried out through data statements, and it can take much careful investigation to construct a data statement for a postulate so that a hypothesis can be tested. For one postulate, there may be a number of different possible data statements depending on the measurement and sampling choices (Chapter 6) and so there may be a number of possible hypotheses.

What you gain by constructing a hypothesis is the use of propositional logic or statistical theory in the arbitration of whether a postulate should, or should not, be accepted. That is extremely valuable. But to make that gain you have to:

(1)   choose a particular method of investigation,
(2)   decide on the details of your investigation (its treatments, sampling design, measurements, etc.), and
(3)   ensure that the data obtained meet the requirements of the statistical procedure you want to apply.

Further, you must realize that your choices and decisions will make your hypothesis test more limited in scope than the postulate – the hypothesis test is only one piece of evidence about a postulate. In some cases, postulates can be constructed that can not be tested directly, particularly those using integrative concepts (discussed in Section II).

Fig. 3.3. Three meanings of the word hypothesis, as sometimes construed by others, and how each meaning is referred to in this text as a specific concept.

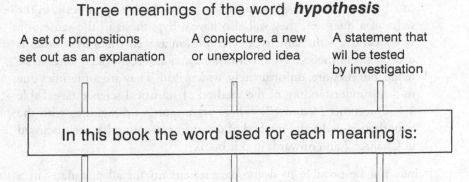

Three meanings of the word *hypothesis*

A set of propositions set out as an explanation

A conjecture, a new or unexplored idea

A statement that wil be tested by investigation

In this book the word used for each meaning is:

theory   *postulate*   *hypothesis*

In this book essential distinctions are made between *theory*, *postulate*, and *hypothesis* (Fig. 3.3). The use of *hypothesis* for all three meanings leads to major confusion in ecological research.

(1) The word *hypothesis* is **not** a synonym for *theory.* Unfortunately, this widespread use, sanctioned by most dictionaries, leads to confusion in research. It obstructs a proper conceptual analysis of the research problem and leaves many scientists baffled about the correct role of statistical testing in research (see Table 1.1, questions 16 and 17). The use of hypothesis to describe a theory can lead to attempts to construct formal statements H_0, H_1, etc., about a theory. This, in turn, leads to attempts to use statistical inference on postulates that have not been developed to the level where data statements can be constructed. For example, there can be no straightforward inference for the statements:

H_0: Climate change will have no effect on forests.
H_1: Climate change will have an effect on forests.

These are over-arching postulates, not hypotheses, and use of H_0 and H_1 in this way repeatedly leads to confusion. There can be scientific inference about these postulates. It would be based on hypotheses that tested component postulates of a theory about climate change, occurrence, and influence. This is an obvious example – but less obvious examples are frequently encountered in research planning.

(2) The word *hypothesis* is **not** a synonym for *postulate.* Frequent use is made of hypothesis for the questions of research. This meaning and use are sanctioned by the hypothetico-deductive method of scientific reasoning but there are drawbacks to it. Again it leads to confusion. In this

case because some people assume that if they have a hypothesis, in the sense of a question, they will also have a hypothesis in the sense of a statistical test – the same type of confusion as seen when hypothesis is used to mean theory. These confusions may not seem likely ones to be made, but they are, unfortunately, widespread. They are sometimes due to a misunderstanding of the method of statistical science (see Table 1.1, Questions 16 and 17). There are important difficulties with the hypothetico-deductive method of scientific reasoning that are discussed in Chapter 7 and onwards in this book.

It may not be possible to devise data statements for all postulates of a theory, at least not simultaneously. Some postulates may define important questions that cannot be tested by hypothesis immediately. That does not devalue their importance in the whole theory, but it may mean that you must undertake the hard work of exploratory analysis (Chapter 8) to construct data statements that make a hypothesis for a postulate. A particular data statement may be questioned, e.g., because of the measurement technique used, or sampling or experimental design employed, so that the effectiveness of a hypothesis may be questioned rather than the postulate for which it was constructed.

3.3 Conceptual and propositional analysis

The meaning of both axioms and postulates depends upon how you define the concepts they use. Some concepts used in axioms are themselves defined by other axioms. Some concepts used in postulates must be defined by measurements. Of course, how the concepts are arranged in constructing a postulate is important but, ultimately, the meaning of your research depends upon concept definitions. For this reason, the first stage of research planning, where you define what you will do and the knowledge on which it is based, is termed a conceptual and propositional analysis. The two processes of defining concepts and using them to construct axioms and postulates go together.

The purpose of such an analysis is threefold:

(1) To define what you know about the problem by bringing together codified and uncodified knowledge into a set of axioms.
(2) To specify your questions as postulates that can be answered true or false or assigned a probability.
(3) To construct data statements that develop specific hypotheses used to assess postulates.

In summary the purpose is to define the working part of the knowledge network that you will use, and how it must be extended to answer your question.

Conceptual and propositional analysis as a method has received less attention than techniques of investigation. For example, whole books have been written on experimental design, but these usually assume that you have already decided why you wish to conduct the experiment, what your treatments should be, and what you will measure as a response variable. Similarly, there are many types of data analysis for different types of data, e.g., different types of surveys, multivariate statistics, categorical data analysis. However, descriptions of data analysis do not delineate the initial problem that required study. To start out with the idea that you are going to use a particular data analysis technique – even if there is a handy computer program for it – can warp your approach to the real problem.

Northrop (1983) states:

"Nothing is more important, therefore, than to realize that the initiation of scientific inquiry constitutes a stage of scientific procedure by itself, with a specific scientific method appropriate to itself and different from the scientific methods appropriate to the later stages of inquiry. The task of the first stage of inquiry is the analysis of the problem."

Every research project starts out with a mixture of things known, things unknown, and things partly defined (Fig. 3.4). As a result there are three phases in conceptual and propositional analysis: identifying the principal issues, classifying concepts according to their states in the progress of research, and examining the complete research procedure. The following discussion outlines conceptual and propositional analysis as a critical process requiring creativity and imagination to resolve the difficulties exposed. The discussion assumes that conceptual and propositional analysis is being applied continually to structuring and restructuring axioms, postulates, and data statements. The three phases are described sequentially but may be applied in various ways depending on the nature of the problem encountered.

3.3.1 Phase One: Identifying the principal issues

The list of questions under Phase One (Fig. 3.4) illustrates how iteration within the component planning phases takes place. For example, in attempting to make a specific inquiry about a general question, you have to choose a site or organism, which leads to further questions. To keep the complete planning process moving in a directed way, you must establish an initial set of axioms (and possibly laws), postulates and preliminary data statements, and then target the successive questions that arise during the five planning processes onto them. This allows you to focus your thoughts towards action while hammering out the critical position and resolving research details. You must, of course, be prepared to modify or perhaps even discard your initial ideas as you proceed with the five planning processes.

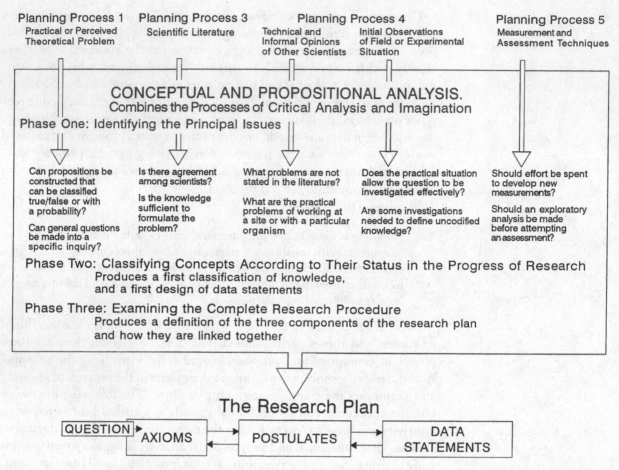

Fig. 3.4. In research planning, a mixture of ideas, problems and techniques is encountered during Planning Processes 1, 3, 4, and 5. Conceptual and propositional analysis is used to identify and define these by defining and classifying concepts, and producing and refining a set of axioms, postulates, and data statements in a research plan. Planning Process 2, applying creativity, is used throughout a conceptual and propositional analysis. For specific details of the purpose and methods of each phase, see the text.

3.3.2 Phase Two: Classifying concepts according to their status in the progress of research

The work of conceptual analysis is driven forward by your critical examination of each concept. Is the concept restricted to general use in ecological theory – or is there a specific definition for your study? In ecology, we use many terms that have a well-understood, non-scientific meaning, which makes our subject easily accessible – but also can lead to poor research, if we fail to distinguish between the meanings explicitly. Concepts initially related as the same kinds of thing, as parts of things, or as having a particular property may undergo changes in their status within a theory as research progresses, and understanding this process can help you to see where the most effective research can be applied. Has sufficient work produced an agreed definition of a concept – or do definitions differ among researchers and, if so, why? Is there an accepted measurement for a concept – or must you examine alternative definitions to see how to make a measurement?

An important decision to make in building a theory is when to change a postulate to an axiom. Concepts used in a postulate – the subject of intense investigation – and thought of as conjectural must have become less conjectural to be used in an axiom.

Concepts can have four types of status: *concepts from research, concepts by imagination, concepts by measurement,* and *concepts by intuition.* In later chapters, the superscripts *res, imag,* and *meas* respectively indicate the status of concepts as deriving from research, by imagination, or by measurement.

CONCEPT FROM RESEARCH

Concepts from research are the ideas we use in describing the more established parts of a theory, i.e., the axioms.

Concepts from research reach this status in two ways: through direct tests with data, or repeated successful use in scientific argument that tests related or component concepts. All research projects use some concepts from research; they build either on what other scientists have done or on what previous research a scientist has made on the same subject.

When you start research as a graduate student, your most likely source of concepts from research is the scientific literature and other work conducted by the research group with which you are associated. For concepts in the literature, you must examine consistency among definitions and uses, and whether your use is identical or different.

Many theories have axioms intended to apply in more than one specific instance. However, the more general a theory, the more difficult it may become to specify precise linkages between the axioms of the theory and site- or species-specific questions of your research. For a theory to be subject to research, and so to logical development, it must be possible to resolve the linkages between postulates and data by specific investigations. If you are going to use a general theory in a practical investigation, you have to make site- or species-specific definitions for general concepts.

For example, consider the development of a possible general theory represented by the over-arching axiom that coarse woody debris (CWD) in riparian zones gives a physical structure that provides habitat for animals (Chapter 4) (Fig. 3.5). The origin of the theory is research indicating that CWD provides habitat structure within stream courses, which provides an axiom. In the new research, piles of CWD on stream banks were postulated to provide shelter or nesting and denning sites and to act as food sources for small mammals and birds. We could link these two theories about

coarse woody debris ⇒ *structure* ⇒ *habitat.*

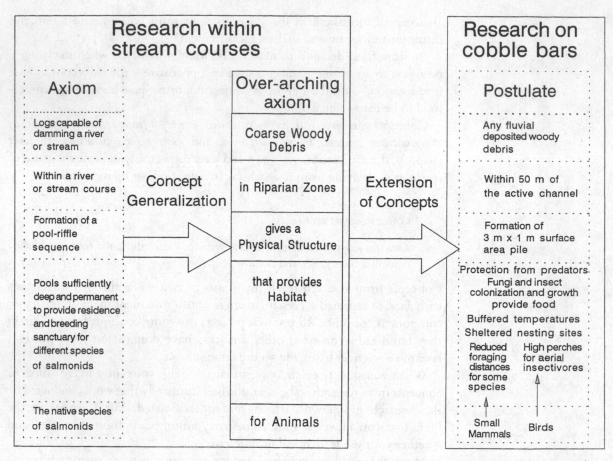

Fig. 3.5. A comparison of concept definitions for two applications of a general postulate about the importance of coarse woody debris for animals in riparian zones. The over-arching axiom depends upon the work done in forming the axiom shown, and other related axioms that build up the whole theory. The postulate for the cobble bar requires that the concepts for use are extended to a new situation, and this requires redefinitions.

together only in a general way (where ⇒ means *implies*). In the two instances, stream courses and stream banks, structure has a different meaning, and habitat is different in terms of what the animals do within the structures formed by the CWD. So a general theory that

> *Coarse woody debris in riparian zones gives a physical structure that provides habitat for animals.*

has meaning only if definitions of the concepts *structure* and *habitat* are themselves general. The concepts must have different meanings in specific cases (Fig. 3.5). For example, in a stream or river, for CWD to create a habitat for fish, there should be pieces large enough to lodge and alter the water flow pattern, whereas on stream banks, CWD could be any sizes of wood that form piles.

It is not precisely clear what is meant by *structure* in either example of Fig. 3.5. What exactly is it about riffle pool sequences and 3 m × 1 m piles of wood that provides the habitat requirements? While *structure* has been researched, and so is a concept from research, it must be possible to define what is and is not known about it. If you were to continue research that used the same definitions of *structure*, then you would be using them as concepts from research. If you decide to research the concepts – e.g., to do an experiment to determine the minimum size of log pile that can be used as a habitat for animals – then you would have to create new concepts by imagination, perhaps about what size or property of log piles was important for what particular animal functions.

Riparian area is a concept from research used in Chapter 4. It is a defined zone known to have particular ecological features through being along streams or riversides where water and terrestrial systems interact. Its definition in Chapter 4 is based on forest practice regulations: "A riparian management zone is 50 meters wide on either side of a stream that has fish in it (or of a river with or without fish)". Regulations specify 50 m because forest management practices, such as felling, have the greatest impact on streams when close to them. However, for some research processes, this definition of *riparian zone* might not be appropriate. For example, tree felling in a catchment greater than 50 m from the watercourse may affect the quantity and quality of water draining to that watercourse. In this sense the concept *riparian zone* has the same utility, and drawbacks, as a concept such as *ecosystem* (see Chapter 10).

CONCEPT BY IMAGINATION

Concepts by imagination are ideas used in developing postulates. They may arise through logical reasoning from within the theory under examination, through comparative reasoning with other theories, or by considering something that has not been measured before.

For example, you wish to develop a theory about the control of growth of Species A. Growth of this species may be imagined sufficiently close to that of the taxonomically related and better researched Species B, which has a well-defined theory for its growth. To extrapolate the theory from Species B to Species A, an assumption must be made: that taxonomic criteria defining the two species as closely related can be used to predict growth-control processes. If you are prepared to accept that the theory of growth control for Species B can be used in your research of Species A, then its concepts are concepts by research for your study. If you wish to check some aspects of the theory by repeating investigations on Species A already made on Species B, then the concepts used in that research will be concepts by imagination. You would use other information, perhaps uncodified knowledge, in deciding whether to accept that the theory for Species B is applicable, or repeat some investigation. For example, if Species A inhabited an environment very similar to that of Species B, you might be more inclined to accept the concepts as concepts from research. Of course, it is valuable to be able to use axioms of an established theory when investigating a new problem. But it may be that when the individual concepts from that theory are questioned, one by one, both strengths and weakness are appreciated and the need for exploratory work to test the applicability of the theory is recognized.

CONCEPT BY MEASUREMENT

Concepts by measurement are data used to examine the logical outcome of a postulate. It is most important not to assume that measurements represent exactly the concepts for which they are designed. Measurements may be limited in effectiveness, accuracy and precision and more than one concept by measurement may give information about the same concept from research or concept by imagination.

Why ever should we consider measurements as concepts? Remember that a concept is a defined abstraction. And that is what a measurement is. It is an abstraction because it represents, but **is not**, the thing being measured, and at the frontier of research this distinction is particularly important. The technical processes of measurement frequently are complicated, and what is measured may only partly represent what the concept from research or by imagination specifies. Measurements may fail to represent a concept by:

(1)   being ineffective, where measurement is only a partial representation or an approximation;
(2)   being inaccurate because of difficulties, usually technical, with the measurement or sampling system, that may cause bias; and
(3)   being imprecise because the measurements have a large variance.

Consider measuring the photosynthetic uptake of CO_2 by a forest. Obtaining measurements is not easy. Yet this could be important for estimating world CO_2 balance and its effect on possible global climate change. Suppose we wish to construct a computer model that could simulate the effects of changes in temperature, humidity, and light on forest CO_2 uptake – and to base this model on measured relationships. Different measurements have different advantages and drawbacks, and the type of measurement and its required accuracy depend upon the intended purpose.

(1)   We could use cuvettes closed over pieces of foliage, typically a leaf or, for conifers, a shoot segment with foliage. An air stream is passed through the cuvette, and the change in CO_2 concentration recorded. Leverenz and Hällgren (1991) reviewed this technique. Despite much research in developing cuvette systems the measurement is not an entirely effective representation of CO_2 uptake. Enclosure changes the environment around the foliage, particularly temperature and humid-

ity, so conditions inside are different from those outside. Because measurement alters what is being measured, an absolute measure of CO_2 uptake by a piece of foliage under natural conditions cannot be obtained. Beyond difficulties with the cuvette per se, using a cuvette system requires measuring samples of foliage, and variation in the canopy microclimate and tree physiology may make sampling a complex task (Linder and Lohammer 1981). From measurements of environmental conditions inside the cuvette as well as photosynthesis, cuvette data can be used for building a theory of how radiation, temperature, and humidity influence photosynthesis. Total canopy photosynthesis can then be estimated if these environmental factors and the total amount of foliage are measured or estimated for the whole canopy.

(2) Because of difficulties in measuring foliage photosynthesis and integrating measurements into a whole-canopy estimate, we might try to estimate CO_2 uptake directly by the eddy correlation technique (Jarvis and Sandford 1985). Changes in CO_2 concentration and vertical wind speed are measured over short time periods so that the actual down draught of CO_2 just above the canopy during daylight hours can be calculated. We would hope that this technique may be an effective measure of CO_2 flux at the point of measurement, i.e., the rate at which CO_2 is moving from the atmosphere into the canopy, because the measurement is a complete representation, and it does not alter the process being measured. However, instruments may not respond quickly enough to the very small and rapid changes in CO_2 concentration necessary for this technique (Jarvis and Sandford 1985) and so may be inaccurate. Despite its precision at a point, compared with the indirect measurements using cuvettes, a single eddy correlation instrument is still a sample of the uptake of CO_2 from the atmosphere and so, unless replicated, may be not be effective. Even then there must be a considerable area of similar forest upwind of the measurements. Furthermore, eddy correlation measurements made above the forest canopy do not estimate CO_2 released from the soil and taken up in photosynthesis and additional measures would be required to estimate that.

So, despite the potential importance of measuring CO_2 uptake, and despite continued research to improve these these techniques, we can only make approximations. This is typical in ecology. Frequently, when wishing to measure the structure or function of a whole community, we lack efficiency or accuracy, or meet with sampling difficulties. Considerable research may be required in designing measurement systems and defining what relationship measurements have to the concept they represent. If you simply assume that a measurement represents a particular concept effectively, you may be avoiding a major part of the scientist's craft. Achieving technical mastery of a complex measuring device, such as a photosynthesis measuring system or telemetry for animal movement, and making it accurate and precise, require great skill. But making it accurate does not absolve the investigator from considering its efficiency in representing the concept.

The important thing is for a measurement to be:

(a) effective: Is the concept represented completely? Is the right thing measured?
(b) accurate: Is it measured well, and without bias?
(c) precise: Are measurements repeatable with low variance?

All three need to be assessed in answering the whole question, and you may have to use combinations of measurements to achieve a satisfactory assessment.

There may be several data statements for a postulate that use different measurement devices (Chapter 6). Sometimes it may be difficult to obtain a measurement that effectively represents a concept by imagination. For example, trapping animals next to a pile of CWD as an indicator of animal *use* of debris piles is not a particularly effective measurement because it gives no indication of what the animals were using the piles for, though it may be a reasonable start. The problem here is with the concept "use" rather than the measurement. It might be necessary to change the concept as measurements are changed. For example, intensive trapping combined with marking captured animals may give information on home range; repeated measurement of recaptured animals may give information on growth or breeding success. So instead of *use*, the more specific concepts such as population density or even breeding success might become appropriate.

CONCEPT BY INTUITION

Concepts by intuition are non-specialist ideas used in a theory. These concepts are taken from a non-scientific context, usually have no rigorous definition, and may be important at the start of an investigation. They are often made more precise as research programs develop.

Color is a concept by intuition. For example, attempts to estimate the nutrient status of foliage by assessing foliage color led to a call for standardization of colors. Hamilton (1960) proposed standardization based on Munsell Color Charts (Munsell 1976). This is not unscientific simply because it depends upon the human eye. Of course, colors can be defined by a wavelength band of radiation, but in this case providing a visual standard of color is sufficient, particularly where many assessments must be made. Surveys involving measures such as color standards cannot simply be dismissed as non-scientific because they do not use more precision, e.g., combinations of wavelength standards. They are likely to be more useful than precise measurements that cannot be made extensively.

As research proceeds, there is frequently an attempt to define concepts by intuition more precisely, or to refine them for automated measurement. For example, Hadley and Mower (1990) evaluated tree canopy color of video images using a pixel by pixel assessment and compared the color classifications against Munsell standards.

3.3.3 Phase Three: Examining the complete research procedure

When you have constructed a set of axioms, postulates, and data statements, it is valuable to reexamine them using broader questions than applied at the start of the conceptual analysis (recall Fig. 3.4). At this stage, the overview given by a diagrammatic representation, see 3.4, can be particularly valuable.

AXIOMS

Is your theory based on a similar theory but developed for a different situation? If so, have you devised adequate test procedures to see whether it really applies as you wish it to (Chapter 5)? Has propositional logic or statistical inference been used in the research that provides the information for your axioms? This may influence the certainty that you should apply to the axioms.

POSTULATES

Make a list of competing postulates and compare them with those you have adopted (Chapter 7).

DATA STATEMENTS

Can you really assess your postulates – or should you first conduct exploratory analyses (Chapter 8)? If you plan an experiment, have you determined the correct balance among treatments, controls, and replicates (Chapter 6)? Are your proposed measurements likely to provide the correct balance between efficiency, accuracy, and precision so you can classify your postulates correctly?

3.4 Representing theories as networks

In order to classify relationships between concepts and propositions the structure of a theory can be represented diagrammatically as a network by using different symbols for concepts with different status (Fig. 3.6).

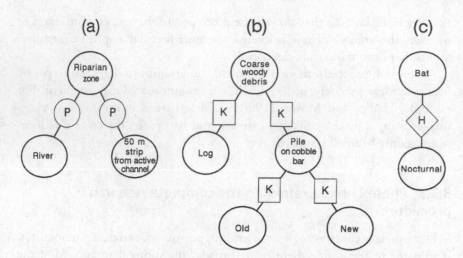

Fig. 3.6. Diagrammatic symbols used to represent different relationships between concepts. (a) Concepts related as parts of things are connected through (P). (b) Concepts related as kinds are connected through a K. (c) The relation between the concept and a property it has is connected through ◇H◇.

Relationships between concepts can be described by considering the concepts as *kinds* or *parts* of things (Thagard 1992). For example, the *zone within 50 m of active channel* and the *river* itself are parts of a *riparian zone* (Fig. 3.6a). Both a *log* and *pile on cobble bar* are *kinds* of *coarse woody debris* (Fig. 3.6b). Research may establish relationships or that some concepts have particular properties, for example that a *bat, has the property* of being *nocturnal* (Fig. 3.6c).

These kind and part connections between concepts make up the vertical, single-line connections of the diagrammatic representation of the theory. Frequently, as in these examples, the *part, kind* and *having the property* relationships are deduced by observation or through some very basic property that needs no detailed investigation, e.g., by definition, coarse woody debris refers to all fallen wood of large diameter. These connections between concepts define the hierarchical ordering of concepts and axioms and postulates to which they belong. This ordering will connect the over-arching axiom to axioms and postulates. Propositions that use lower levels of parts or kinds will be further removed from the over-arching axiom. For example, *old piles of woody debris* and *young piles of woody debris* (Fig. 3.6b) are both kinds of piles of woody debris. Investigation of postulates that use them may tell something about piles on cobble bars, but a further synthesis bringing in information from other studies, e.g., logs on forest floors, is required to define woody debris more completely.

While concepts may be connected vertically in a way that emphasizes hierarchical organization, horizontal connections can be made to represent use of concepts in axioms and postulates (Fig. 3.7). These vertical and horizontal connections illustrate that the growth of theory requires two types of work: research into postulates that shows the relationship between very different types of concepts; and research into the nature of similarity and

Fig. 3.7. Representation of the over-arching axiom (solid horizontal bar) of the theory that coarse woody debris (CWD) provides structure that defines a habitat for animals, and a postulate (open horizontal bar) for application to cobble bars. Bar-like connections join the concepts. Thick solid lines represent concepts from research. Thin dashed lines represent concepts by imagination. (Small mammals were considered to be a concept by imagination in the postulate because at the outset of the investigation the species actually occurring at the site had not been determined.)

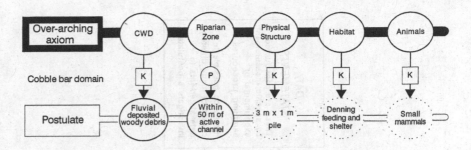

dissimilarity of comparatively close concepts, e.g., those that are both either kinds or parts of a parent concept.

The value of constructing the network diagram for a theory is that it gives you a visual representation of the knowledge structure with which you are working, and insight into what might be done next to see whether the network can be extended or simplified. Of course, in the process of constructing a network you must define the axioms of your study and determine how they are linked through the different relationships between concepts.

3.5   What can be gained from a conceptual and propositional analysis?

Conceptual and propositional analysis helps you to distinguish the type of progress already made in an area of research, and define what might be done next so that you can establish your work in the cyclical research process (Fig. 3.8) of increasing and refining the body of axioms. Four processes are used, imagination, deduction, investigation, and comparison, to develop ideas between the three states of knowledge, axioms, postulates, and data statements (Fig. 3.8).

Science is a continuous process in which what is discovered may not only extend a theory but also challenge at least some part of its construction. Concepts by measurement, those that describe the results of investigations, must be matched against concepts by imagination, those that extend theory. Almost inevitably, there is a need for refining concepts by imagination before raising their status to concepts from research. Because concepts used in postulates are frequently connected through part and kind relationships with concepts used in axioms, this in turn may demand adjusting the meaning and relative importance of concepts from research already used in constructing the original concepts by imagination. In the excitement of finding something new, the need for reexamining the base of the theory can easily be overlooked. Conceptual and propositional analysis is a tool that, through its reliance on criticism, can help to avoid this trap.

A complete conceptual and propositional analysis of a scientific problem is a time-consuming procedure if rigorously pursued, i.e., all concepts fully

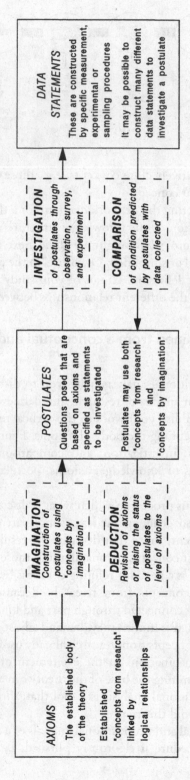

Fig. 3.8. The relationship between the three states of knowledge and four processes of the scientific method that drive research.

defined, axioms and postulates written out completely in the terms of propositional logic (Chapter 7). However, conducting a thorough conceptual and propositional analysis resolves the following four practical problems.

3.5.1 Deciding whether you can assume something or must investigate it

The very act of defining a concept, which involves subjecting it to criticism, can help to identify it as the component of either an axiom or a postulate. Classifying ideas as different types of concept can help to remove personal attachment to an idea and give you enough distance to be objective. It is all too easy to think of concepts as concrete, or not even to define them. Three points about this must be stressed.

(1) In ecology, the definitions of so many of the concepts we use have attributes relating to a specific application. The classification of concepts in this chapter emphasizes the need to define these different attributes.

(2) A concept can change its status as a result of research, i.e., being the new component of a postulate, the subject of a measurement, and maturing to be part of an axiom. As a concept is used in these different ways, part of its definition is retained, but the precise definition at each status will be different.

(3) Ecologists use many integrating concepts, e.g., *niche*, *community*, necessary for the development of the subject – but their level of abstraction puts these concepts beyond our ability to measure them precisely. In this sense, they are not "real," although they may be essential to the development of ecological theory. For example, an important aspect of the struggle with both the concepts *niche* (Schoener 1989) and *community* (Keddy 1987, Shipley and Keddy 1987) has been to find sets of functional concepts that define them. This process, which is particularly important in ecology, is discussed further in Chapter 10, where additional categorizations of concepts are introduced.

3.5.2 Understanding logical relationships between different pieces of knowledge

Structuring the body of theory as a series of axioms and postulates allows you to see the linkages that may exist between concepts. If you review the literature, you may well find that different scientists use concepts with different shades of meaning. As you read different scientific papers, each in isolation, a natural empathy with the author, or with the interpretation the author emphasizes, may blind you to these differences. Writing down the axioms, postulates, and concept definitions can reveal differences in usage. Practical classifications of concepts also may be useful in conducting such an analysis.

3.5.3 Assessing how complete a theory is

Conceptual and propositional analysis can be particularly valuable to gain understanding of how fully specified a theory is. Interestingly, the conceptual analysis of the CWD problem (Chapter 4) revealed only one concept from research and ten concepts by imagination. On this basis, the theory is essentially conjectural and much of its rationale depends upon it having a parallel structure to research in a different habitat. This is different from, but not better or worse than, the examples in Chapters 5 and 6 for which there are large bodies of theory and many concepts from research.

It is important to establish: which propositions are treated by authors as axioms but are not well researched and really state a "point of view"; which propositions are well-studied axioms; which concepts are basic to the whole theory; and which concepts appear frequently in postulates but are assumed rather than measured. Sattler (1986) notes that philosophers of science make a distinction between the intention of a concept, its defining set of properties, and the extension, its domain of applicability. He suggests that, in biology, concepts are frequently applied beyond their defined limits. This certainly happens in ecology. We base theories on research from a few specific cases and then extend their application without specific research to define the concepts used in the theory (Section II).

3.5.4 Knowing when to start practical investigation

An important question when developing a research plan is knowing when to stop analyzing and start investigating. Chapter 2 gives criteria that can be expanded here. Generally, you reach this stage when:

(1) you have constructed one or more postulates that are critical to the theory you are investigating, or wish to apply in a practical problem;
(2) you have a data statement with defined measurements of the concepts that make up the postulate(s); and,
(3) a procedure for assessing the postulate(s).

Of course, not all research investigates new postulates through making measurements. For example, some scientists may spend their efforts in developing a new measurement. In that circumstance, we may say that (2) is the research, but that it will subsequently contribute to a wider investigation. Conceptual and propositional analysis may continue during the practical parts of an investigation. At that stage, a data statement will have been formulated and a decision made as to whether propositional logic (true/false classification) or statistical inference (probabilities) will be used. But an exploratory phase, in which what is actually measured is compared to the intended concepts, may be warranted. At that point, it may be necessary to refine measures and revise the initial concepts and how they are used.

3.6 Conclusion

Scientific research is a complex mental activity. We have to conduct a personal dialog between moving forward using imagination, investigation, comparison, and deduction and yet maintaining a critical posture, pressing for exact definition and effective assessment. The important thing to recognize is that critical analysis does not retard the investigation but, through questioning, increases its quality and may even make a project possible where it previously was not.

Conceptual and propositional analysis is the foundation method for Progressive Synthesis (Chapter 15). Postulates are not neutral but are influenced by previous theory in ways that can bias their investigation, e.g., in the definition of concepts and the measurements used for them. Neither may a postulate be confirmed or rejected by a single investigation. Any single "test," i.e., where a hypothesis is constructed for the postulate, is made under very specific conditions and need not give an absolute and final result for the postulate (Chapters 6, 7, and 8).

3.7 Further reading

Popper develops the description of theories as nets "designed by us to catch the world" (Popper 1982). However, he notes that theories are not just instruments, that we do aim at truth, and that theory testing is the essential process we use for that (Popper 1982). Thagard (1992) gives a detailed analysis of how concepts change as research proceeds and these changes in definition change theory content and structure.

Popper (1968) discusses axiomatic systems and concepts. In his book, axioms are used to define empirical or scientific theories (see Popper 1968, pages 71–75). Popper's (1968) discussion of universal and individual concepts is recommended. Universal concepts are such as "dictator", "planet", "H_2O", whereas "Napoleon", "the earth" and "the Atlantic" are individual names. The extent to which some ecological concepts can be considered to be universals, and how we should define such concepts, are central to much ecological research and are discussed in Section II. A series of articles in Mellor and Oliver (1997) present detailed discussions of issues in forming, defining, and using concepts and what they represent.

Other philosophers have differing views about what concepts are and particularly whether they must have constant definitions in all applications, as Weitz (1988) argues, or whether they are open and definitions may change over time, which Hull (1988) describes as a principal result of scientific investigation. Thagard (1992) gives a diagrammatic taxonomy of the main views on concepts.

Cherrett (1989) includes a series of articles that examine the history and definition of some frequently used ecological concepts. Kiester (1980) describes how natural history observations are used and developed into concepts with theoretical content.

4 Development of a research plan

Summary

This chapter is a case study of the application of the five processes of research planning and the development of a conceptual and propositional analysis. A research plan does not spring from the mind complete and perfectly formed at the first attempt. Research planning involves refinement and reorganization of ideas.

Ashley Steel started her planning with ideas that piles of coarse woody debris (CWD) washed up on a riverbank by a flood would provide shelter and habitat for birds and small mammals. Not all wildlife biologists agreed with her. The process of developing a research investigation meant that she had to specify and organize her initial ideas into a theoretical framework and make exploratory investigations of uncodified knowledge.

Steel constructed hypotheses and tested whether animal use of different types of CWD piles was different. As typical of field investigation, the results led to further research and, particularly, the need to refine concepts that can be measured more effectively.

After completing her Masters research, Steel reflected that two things in particular would have aided this planning process: (1) a pilot study, which would have improved field measurements; and (2) more extensive criticism from scientists with areas of specialist knowledge, which would have helped the conceptual analysis.

4.1 Introduction

This chapter describes the evolution of a research plan. It starts with the imprecise, incomplete first ideas of an investigator and shows how these evolve into a plan with defined objectives. The aim of planning – a set of axioms, postulates, and data statements – was kept continuously in mind although it was not achieved at the first attempt.

Each of the five processes involved in research planning (Chapter 2) is identified as it is used:

(1) Defining a research question.
(2) Applying creativity to develop new research ideas.
(3) Ensuring the proposed research has relevance to prior scientific knowledge.
(4) Ensuring the proposed research is technically feasible and can be completed with available resources.
(5) Determining how conclusions can be drawn.

You must start with Process 1, and planning is not finished until Process 5 is complete. But you should not think of applying each process just once, nor necessarily in the sequence listed here. In this example, the five processes are discussed in the sequence 1, 4, 3, 2, and 5, although aspects of Processes 4 and 2 interrupt the investigator's consideration of Process 1 before 4 and 2 are fully developed. These five processes together provide a deliberate, conscious questioning of the assumptions made, of the questions asked, and of the ideas and terms used in describing both assumptions and questions. A satisfactory plan is usually the result of many revisions.

4.2 Process 1: Defining a research question

4.2.1 The first description

The first task is to write a Background and Introduction statement of your questions and why they interest you. This first written description, for private use or discussion with a sympathetic critic, is a way for you to explore why you view a research problem in a particular way.

Steel made just such a description at the start of planning a Master's research topic (Box 4.1). The text has deliberately not been edited since the planning process. I am extremely grateful to Ashley Steel for permission to illustrate the development of her ideas and for her reflective criticisms of them after she had completed her thesis. Our scientific culture emphasizes the value of the scientific paper written for publication in a peer-reviewed journal, and that rarely shows the development process.

No literature was cited in the text of Box 4.1, and this is deliberate. At this stage, ideas come from ecology foundation classes, field observations, suggestions of supervisors or others experienced in the research area, and the scientific literature (Fig. 3.4). A conceptual and propositional analysis reviews all these ideas. If, at the very start of writing a plan, you attempt to support all ideas with literature citations then either only those ideas supported by the literature, the codified knowledge, will be written down, or the interpretation of the literature may be biased to support what it really does not. If either of these happens, the planning document is incomplete because the assumptions from field observations and other less formal sources, the uncodified knowl-

---

Box 4.1. Steel's first description of background and introduction

Small mammal use of the coarse woody debris piles on the north fork of the Skykomish River

Although there has been much research on the function of coarse woody debris (CWD) in small streams, there has been little research on the function of large piles of woody debris in or near large rivers. In a natural system, there are many debris piles. These piles may be an important link between upland, riparian, and aquatic communities. Forest management may impact on the distribution of these piles; yet, the effects of forest management on these piles are unknown. This study proposes to investigate large piles of woody debris on the north fork of the Skykomish River in Washington.

There are many important questions, answers to which will be necessary for both a more complete understanding of river ecology and an improved management plan. My project will be a first step in answering one of these questions. I will focus on the following problem: do large piles of woody debris supply critical resources for upland communities by providing habitat for small mammals and/or food resources for insectivorous birds? I will document the postwinter colonization of debris jams by small mammals and postwinter use of debris jams by birds that feed on resident insects. The pattern of small mammal colonization will demonstrate whether debris jams are used when small mammal populations are at their lowest (i.e., optimal habitat) or whether debris jams are used only when small mammal populations are high (i.e., suboptimal habitat). The pattern of bird use will be an indicator of the importance of woody debris habitat for insects and also an indicator of the dependence of higher levels of the food chain on the insect resource. I will gain information that describes the attributes of debris piles that make them more or less suitable as habitat for insects and mammals by comparing the colonization of piles that are at the edge of the forest versus piles on cobble bars and the colonization of new piles (1990 flood) versus preexisting piles.

edge, which are frequently at the very heart of the research, remain inadequately defined, and their significance may be overlooked.

Many observations and ideas came together to act as the basis for this research. Steel rafted the Skykomish River in Washington State. A recent flood had deposited large amounts of woody debris on the river banks. Although the river is protected with State of Washington Wild and Scenic Status, there was talk of dynamiting piles of woody debris to increase access and safety for rafters. There already exists a considerable body of published research about the effects on fish habitats of removing coarse woody debris (CWD) within rivers. What Steel wished to examine was whether CWD is also important for wildlife along riverbanks, an extrapolation of the published work. Such extrapolation, though not a detailed analogy, is a form of inductive reasoning (Chapter 7), i.e., if something is important in one place, it will be important in another. Within streams and rivers, CWD lies in piles of different sizes that interact with sediment movement, and suitable environments form for different animal uses. Steel's proposed role for debris piles along riverbanks was that they provide shelter and food for small mammals and birds; further, they may be food sources for insects eaten by birds.

4.2.2   Initial development of a theory for the problem

The second paragraph of this first description (Box 4.1) shows concern with making this very general question into one that can be tackled practically. Two devices have been used: first, the idea that CWD has a *critical* role at some stage in the development of bird and small mammal populations and, second, the idea that bird use of CWD piles will be an *indicator* of insects in the debris. With these two constructions, Steel started to develop a theory about use of CWD piles. But note that the two ideas of *critical* and *indicator* are not defined.

Broad, undefined concepts are often used at this stage in developing a research proposal to investigate an observed field phenomenon. *Critical* and *indicator* are holding concepts.

HOLDING CONCEPT

A *holding concept* expresses the existence of a phenomenon or relationship without defining it sufficiently to be the subject of practical research. An important objective of conceptual and propositional analysis is to refine the description of holding concepts to the point where they can be studied. Typically, holding concepts assert:

(1)   an undefined collective description of a set of natural concepts (*indicator* in the example),

(2)   the existence of a functional concept (*critical* in the example), or,

(3)   a functional concept that defines or qualifies an integrative concept (important in Axiom 1: see later, Box 4.3).

Much of our work in ecology is to refine and develop the meaning of such concepts – to gain in precision. As you start to grapple with a research problem, you are likely to use general descriptions such as *indicator, critical,* and *important*. This type of concept can help you to maintain a complete view of your problem and then work at defining the parts, i.e., each successive axiom and postulate. What you must not do is start the practical aspect of research without defining such words.

The second paragraph of the first description (Box 4.1) contains a further question about the occurrence of CWD piles. Do piles increase insect biomass and other characteristics that could be of importance for bird and small mammal use? This establishes a possible basis for a comparative investigation – examine pile use in different environments and see whether they provide different habitat use. Something of general interest may be found out from observing differences between piles in different environments.

Clearly, Steel was developing a mental picture of how this ecosystem works based partly on previous studies and partly on her own field observations. At the very start of planning she had not considered CWD to be grouped in piles. This was the first concept she developed during an early visit to the river. She was generating a theory based upon analogy and extension of the meaning of the concepts already used in the study of habitat structure within rivers and trying to adapt and change them. (See Fig. 3.4 and accompanying discussion in Chapter 3.)

4.2.3 First definitions

After the first written description, definitions must be made more precise. So Steel specified the assumptions that she was making and the statements that she wished to investigate (Box 4.2). The three assumptions (Box 4.2) imply that there is a set of relationships, and that it may be important for riparian zones of forest ecosystems as a whole. But these assumptions are not specific. They imply the existence of a scientific problem, and must be made more precise. The specification of assumptions must lay out the logic of a study. Then if you investigate and find them wrong or irrelevant, any incorrect logic in the assumptions can be identified. If the logic is clear, then all results give some information. It can be more valuable to produce a result that questions a commonly held assumption than to provide a further confirmatory result.

Steel revised the assumptions, and Statement 1, as given in Box 4.2 into the axioms listed in Box 4.3. Recall that an axiom is a proposition assumed to be true and used as a foundation in your study. However, it should be stated in a way that can be challenged or supported by the research that you conduct. The challenge to an axiom is neither direct nor complete. For example, no matter what the result of this research, the results of previous research into the effects of CWD in streams and rivers means that Axiom 1 is unlikely to be completely wrong. However, if animal use of debris piles on land was neither greater than, nor different from, other habitats surrounding or close to those

---

Box 4.2. Steel's first written statement of assumptions and statements

Assumptions

(1) Woody debris is an important component of stream ecosystems.
(2) Natural river systems have many large piles of woody debris.
(3) Small mammal and bird communities are important links in the food web supporting forest ecosystems.

Statements to be investigated

(1) Large piles of woody debris are an important component of river ecosystems.
(2) Large piles of woody debris integrate aquatic, riparian, and upland communities.
(3) Large piles of woody debris provide critical resources for small mammals.
(4) Large piles of woody debris provide critical resources for bird communities (insectivorous birds).
(5) The location of a pile will influence its suitability in providing critical resources.
(6) The age of a pile will influence its suitability in providing critical resources.

---

piles, then you would rewrite Axiom 1 and replace stream ecosystems with within stream courses.

The remaining problem with Axiom 1 is the use of *important* as a holding concept. Axiom 1 is the over-arching axiom of the study – it defines the field of interest and connects the study to others, e.g., investigations within streams.

4.2.4 First consideration of Process 4: Ensuring the proposed research is technically feasible and can be completed with available resources

By this stage in her research planning, Steel realized the practical difficulty of investigating both bird and small mammal use of CWD piles in the available time, particularly if different types of debris pile were to be compared. So she dropped birds from the study; the second and third axioms concentrate just on small mammals (Box 4.3). At this stage, Steel was prepared to sacrifice an attempt at a more comprehensive study to obtain detailed, precise answers to fewer questions. This was an important decision. Steel judged that an investi-

---

> **Box 4.3. Steel's first written translation of assumptions into axioms**
>
> Axiom 1: Debris piles are an important component of stream ecosystems.
> Axiom 2: Debris piles in forested areas provide habitat and/or forage for small mammals.
> Axiom 3: Small mammals use riparian areas.

gation of different types of CWD pile for species from only one phylum was logically stronger than, say, a study of species from two phyla for only one type of CWD pile. The value of using contrasts will be repeatedly illustrated in this book. The features of contrastive (sometimes called comparative) investigations are summarized in Chapter 15.

4.2.5 First consideration of Process 2: Applying creativity to develop new research ideas

The development of the research plan, even to this stage, has been a creative process. The use of holding concepts to structure natural history ideas into propositions was the first stage in developing a scientific approach. Establishing such a structure requires the realization that a natural history description may be necessary to motivate the study but not sufficient to establish the details of research. Most importantly, the holding concepts pave the way for critical analysis. Restricting this study to small mammals was also a creative decision motivated by critical analysis. It focused the work and enabled Steel to develop detailed plans. The power of the comparative method, which depends upon sharpening a contrast, influenced Steel in these decisions.

Later in the planning process, Steel added birds back into the study, although she commented, after completing the research, that she wished she had been more exhaustive in analyzing the small mammals – even at the cost of investigating birds! Some decisions like this one, about whether or not to include birds in the study, cannot be regulated by information available before the investigation. They can be arrived at subjectively – which is where people's different philosophies have an important impact (Chapter 9). For example, one person might say that this study would be more "ecological" if it contained birds and small mammals – using two phyla provides a broader explanation – but that is an opinion about the type of ecology one thinks should be done. Another person might say the study would be more valuable if restricted to small mammals – it could use more intensive sampling and hypothesis testing – but that is an opinion about the importance of statistical inference.

4.2.6   Continuation of Process 1: Defining a research question

Axiom 2 in Box 4.3 defines the assumed importance of debris piles, for small mammals specified in Axiom 1. But Axiom 2 is still weak in that habitat is an integrative concept which can mean different things under different circumstances. Furthermore, as this study was being planned, some field ecologists suggested to Steel that small mammals would not be found in the debris piles on the cobble bars. So it was important for her to acknowledge the assumptions of use by Axiom 2, and the assumption of the existence of small mammals in riparian areas, which includes the areas where the debris piles exist. Axiom 2 is more specific, and more at risk, than Axioms 1 or 3.

Note that although Axiom 2 is uncertain, at least in some respects, it has two characteristics that distinguish it from a postulate. (1) It is an assumption essential for the research, but it will not be investigated directly (no data statement will be constructed). Instead it will be investigated indirectly through the study of its dependent postulate(s). (2) It is less specific than a postulate. A number of postulates may have to be examined before this axiom could be considered confirmed as false.

These three axioms are typical of field-oriented studies. They follow two assumptions of field ecology: that everything we see has its function in relation to the way that ecosystems work and that this general assumption defines a purpose for field ecologists, to explain ecological phenomena.

These axioms are weak because they do not place strong limits on the study. This is particularly so for the over-arching axiom (Axiom 1). One could go on trying and trying and trying to find some function for a phenomenon – in this case, debris piles – but, at every failure, consider something new. Paradoxically, an advantage of this type of axiom is that it can produce extensive exploratory research. Many relationships have been discovered in this way. However, the pervasive use of weak axioms in ecological research has had two contrasting influences:

(1)   Some scientists may make a forceful assertion of theory from initial field observations, i.e., they come to hold axioms that assert the existence of a phenomenon more strongly than is really justified, and tend not to acknowledge that their first investigations are exploratory. Axioms are applied beyond what they were developed for. This is discussed more in Section II.

(2)   Other scientists, recognizing the exploratory nature of much field investigation, assert that the most effective research is the experiment. The alternative to weak axioms and exploratory research is a detailed specification of a particular theory with strong axioms – and examples of this type of research are considered in Chapters 6 and 11. But this approach also has a weakness. In ecological sciences, one can develop

well-specified theories most effectively for precisely defined locations or circumstances. Some decisions must be made about treatments, measurements, and sites that may limit the generality of the conclusions. As evident in the experimental investigation of herbivory relative to competition (Chapter 2) certain decisions about treatments, measurements, and sites, limit the generality of the conclusions. Because a theory must be practically examined for particular circumstances, there is not (usually) a single, experimental test of a theory.

Steel listed six statements to be investigated (Box 4.2). The first two reiterate the general importance of the problem; Statements 3 and 4 emphasize that woody piles are important to birds and small mammals; Statements 5 and 6 point the way to a comparative study of different piles that may show different use. Yet none of these six statements can actually be investigated. Each uses terms that are too general. This is typical of the early stages of research planning. It is necessary to refine these statements to the point where they can be researched directly, so that the measurements made can be assessed to classify a given statement as true or false, or most usually in ecology, as probably true or probably false.

Statement 1 of Box 4.2 has already been transformed into Axiom 1 (Box 4.3). As the over-arching axiom it will not be investigated directly but can be assessed as a result of the whole research program.

Statement 2 is typical of statements written in the early stages of planning ecological research. It proposes connections between this system and others, i.e., the CWD piles in the riparian zone, and the streams and uplands. These connections are not specified, which is why the statement cannot be investigated directly. The word *integrate* – another holding concept – is used but not defined. The types of relationship considered are that logs produced in the forest, transported by rivers and streams, come to reside in the riparian zone as debris piles and then are visited by birds and animals from the uplands. Statement 2 is of a general type that occurs frequently in proposals with origins in field observations. It exemplifies the philosophy that everything is connected, or at least that there are many mutual influences between environment and organisms, and between organisms, and the ecologist's task is to define them. For Steel's research, a first investigation of the use of CWD piles, it is not necessary to consider Statement 2. Indeed, for practical purposes, it is necessary to exclude considerations of from where either the animals or the logs come. That is, the study must be bounded in its scope, although it may become important to investigate bounding decisions in later research.

Statements 3 and 4 both use the word *critical* without defining it. Whether the use of debris piles is critical will depend upon the actual use and assessing that against the use of other habitats. Defining the use of the CWD piles is the

> **Box 4.4. Steel's first written translation of statements to be investigated into postulates**
>
> Postulate 1:   Debris piles in the study area can be classified as old piles or new piles and as piles in the forest or piles on a cobble bar. Differences exist in debris piles that are distributed over the study area.
>
> Postulate 2:   Small mammals use debris piles in riparian areas for denning, feeding, and shelter.
>
> Postulate 2A: Small mammal use of old debris piles is different from small mammal use of new debris piles.
>
> Postulate 2B: Small mammal use of old debris piles that are at the forest edge is different from small mammal use of debris piles that are on a cobble bar.

first task, specified as Postulate 2 (Box 4.4). (At this stage in Steel's planning, birds had been dropped from the study.)

Statements 5 and 6 use the word suitability, which again implies that a particular assessment will be made. When the focus changes to making comparisons (of use between piles of different types or in different locations), these statements are translated directly into Postulates 2A and 2B. At this stage of its development, the whole study depends upon there being differences between piles of woody debris. So the very first task is to explore just what differences do exist (Postulate 1). There is circularity, of course: you would not do the survey, and you would not know what to survey, unless you thought there were differences.

At this stage, considerable progress has been made in developing the research plan. The investigator criticized her own initial statements and made them more precise. The sets of axioms and postulates produced define the study objectives. Defining concepts is the next task, particularly using the scientific literature. This process of concept definition can call both axioms and postulates into question and can be used to see whether they follow logically.

Not surprisingly there can be considerable resistance to defining concepts. Definition is hard work. Moreover, scientists may deliberately leave some concepts vague and make an approximate, or broadly based, measurement. What is important is to define what is known, even if that is not as complete as you would like. Definition is particularly important where, as in this case, concepts established in one type of study are extrapolated for use in a different situation. Of course, scientists debate differences in the definition of concepts. Refining concepts is an important way that science advances. However,

---

Box 4.5. Concepts (C) used in Postulate 1

Postulate 1:   Debris piles in the study area can be classified as old piles or new piles and as piles in the forest or piles on a cobble bar.

C_1 (*debris piles*): The following criteria define a debris pile in my study.

> The pile must be at least 3 m × 1 m.
> The pile must be composed of more than one piece of wood.
> The pile must have been deposited by the river.
> The pile must be within 50 m of the active channel.

C_2 (*study area*): The study area is a one-mile (1.6 km) stretch of the south bank of the north fork of the Skykomish River.

C_3 (*old debris piles*) and C_4 (*new debris piles*): I will define these as concepts by considering all debris piles deposited in the 1990 flood as new and all debris piles deposited before the 1990 flood as old. Any of the following criteria will categorize the piles as new.

> Presence of green branches inside the pile.
> Lack of dead leaves on top of the pile.
> Lack of seedlings growing on top of the pile.

C_5 (*piles in the forest*) and C6 (*piles on a cobble bar*): These concepts will be distinguished because piles in the forest will be on soil or next to a coniferous tree other than a cedar tree. Piles on a cobble bar will be on a cobble substrate and next to only deciduous trees.

In order to gather the data, I will map the entire study area using a compass and tape measure. I will tag each pile with a five-digit code and collect the following information about the piles: age, substrate (forest or cobble bar), length, width, distance to channel. These facts will allow me to choose a stratified random sample of piles for trapping. These facts will also allow me to hold size of the study area within manageable limits.

---

concepts can be refined only when they have been examined through use in postulates, or as the basis of a study in axioms.

The concepts used in Postulates 1 and 2 (including 2A and 2B), respectively, are described in Boxes 4.5 and 4.6. Postulate 1 is really two postulates: one about new and old piles, and one about the forest and cobble bar. Both are concerned with uncodified knowledge. They examine whether a particular habitat type can be classified – though without examining whether that

---

Box 4.6. Concepts used in Postulate 2

Postulate 2:   Small mammals use debris piles in riparian areas for denning, feeding, and shelter.

Postulate 2A: Small mammal use of new debris piles is different from small mammal use of old debris piles.

Postulate 2B: Small mammal use of debris piles that are at the forest edge is different from small mammal use of debris piles that are on a cobble bar

C_7 (*small mammals*): At this point in the study, I am unable to define small mammals. I will define them as whatever is caught in my traps. I expect to catch deer mice (*Peromyscus maniculatus*), forest deer mice (*Peromyscus oreas*), Townsend meadow mice (*Microtus townsendii*), Oregon creeping voles (*Microtus oregoni*), shrew moles (*Neurotrichus gibbsii*), jumping mice (*Zapus princeps*), wandering shrews (*Sorex vagrans*), marsh shrews (*Sorex bendirii*), Trowbridge shrews (*Sorex trowbridgii*), and, perhaps, Townsend chipmunks (*Eutamias townsendii*), Townsend moles (*Scapanus townsendii*), or long-tailed meadow mice (*Microtus longidcaudus*).

C_8 (*riparian area*): Riparian areas will be defined as the area in which woody debris has been deposited by the river up to 50 m from the active channel.

C_9 (*use*, implies *denning, feeding, and/or shelter*): Although this is a complex concept in the theory of animal use of debris piles, it is made a simple concept by the measurement selected. Any small mammal caught in a trap will be considered to be denning, feeding, sheltering, or otherwise using the debris pile. Future studies will need to refine this measurement but for my survey this is a sufficient way of measuring the concept.

In order to investigate Postulate 2, I will place two pitfall traps and two Sherman live traps at each debris pile selected above. These traps will be baited and monitored for six consecutive days. Animals will be marked to identify that they have been captured but will not be individually tagged.

---

classification is important. They are exploratory and prerequisite to the intended study.

The first concept, C_1 is established as a concept from research by the very first exploratory investigation. C_2 is established in the same way. Both C_3 and C_4 and C_5 and C_6, are concepts by imagination.

4.3 Process 4: Ensuring the proposed research is technically feasible and can be completed with available resources

There are two points to note about Postulate 1 (Fig. 4.5) and its concepts:

(1) It defines a classification of the area that determines the investigation of mammal use.
(2) It requires exploratory investigation of type and distribution of debris piles.

It is important to realize that classifying debris piles into old and new and as in the coniferous forest or on the cobble bar is a postulate. Classification is thought to reflect a real difference in debris pile use, but one not established yet. So the classification, although containing much that is empirical (note particularly the definition of a pile itself), determines the theory of mammal use of debris piles that Steel was developing. If the classification was incomplete or inappropriate, it would not be a basis for the subsequent investigation of mammal use, e.g., if the species of timber in the pile were actually more important than pile location or age.

This is typical of field investigations. Decisions about the location and distribution of studies must be made that may seem practical, but their effect on the outcome of the research can be major though difficult to determine in advance. It can be important to make some preliminary investigations, or pilot studies, to establish both the technical precision and breadth of application of concepts such as C_1 (*debris piles*). For example, are these measurements precise and accurate? Does classifying piles into C_3 (*old debris piles*) and C_4 (*new debris piles*) on the basis of the three indirect measurements of pile age produce a clear division into two ages of piles? How many piles show one or two, but not three, of the criteria?[1] The difficulty may be directly with the choice of measurements, e.g., they may not represent the aging process, or even with the concept that piles of different ages actually do exist (lack of precision), or with the technical procedure for measurement, e.g., measurement cannot be made accurately. If there is not a clear division in age between piles, then perhaps another approach needs to be taken. Breadth of application of the concept could be examined by seeing whether the classification, made on the basis of three measurements, holds on other sites.

The fundamental causal relationship to be investigated in the study is described by Postulate 2. Its subdivision into component postulates, and analysis of its concepts into *parts* and *kinds* with different functions, are the key aspects of the research. There are questions of definition about the concepts used in Postulate 2 (Box 4.6). Although the concept of *use* of piles

---

[1] This commentary was written before Steel analyzed her results. After analysis, she did find that it would have been valuable to have analyzed pile age more precisely. This is a natural part of the scientific process.

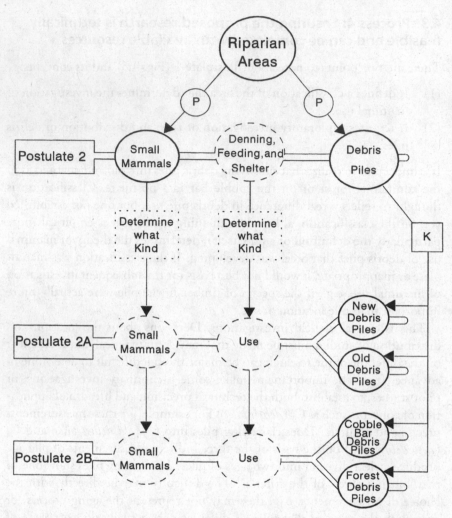

Fig. 4.1. Diagrammatic representation, using the symbols defined in Fig. 3.3, of the network of theory to be investigated by Postulate 2. The results of Postulate 1 were to establish that there were different *kinds* of debris piles (shown down the right-hand side of the figure). Notice that investigation will proceed horizontally, by investigating postulates, but that there will be a result on the vertical structure by defining and classifying the *part* and *kind* relationships between concepts.

includes a range of animal functions, the first measurement is simply one of animal presence in the piles. Any trapping device must be used with a sampling system and, as often happens in ecological research, both the number of traps and the length of time for which they could be set were determined by the available equipment and observer effort. The important question is how to distribute this most effectively.

Steel's task was to define the kinds of small mammals she would find. C_8 is an integrative concept, from research, that is used here as a classification concept. C_9 is a holding concept, for animal *use* of the piles but is obviously related to the general functional concepts of *denning, feeding,* and *shelter*. It is a concept by imagination that will quantify animal use. How the theory to be investigated by Postulate 2 may develop as a network is illustrated in Fig. 4.1.

Box 4.7. Vertebrate use of coarse woody debris piles on the north fork of the Skykomish River: biological, historical, economic, and political perspectives

Most streams in the temperate region of the Pacific North-west transport an enormous amount of coarse woody debris in the form of logs, sticks, and rootwads, which is deposited in large piles after flood events. These accumulations can be found on gravel bars in the channel, on the channel shore, or in the riparian forest as far up as the highest waterline. Debris piles may provide critical resources (forage areas, shelter) for small mammals. This study proposes to investigate the function of large accumulations of woody debris in riparian areas for providing habitat and forage for small mammals. Although an extensive literature exists describing the importance of downed logs and standing snags for providing increased habitat complexity in forest ecosystems (Spies and Cline 1988), large piles of debris, deposited within and along river channels, have received little investigation except as they provide structure for fish habitat (Hawkins *et al.* 1983, Angermeier and Karr 1984, Benke *et al.* 1985, Bryan 1985, Johnson *et al.* 1985, Dolloff 1986, Elliott 1986, Swales *et al.* 1986). The ecological role of these piles needs to be understood because management practices such as upland clear-cutting, riparian harvest, or stream cleaning may influence their size and/or distribution (James 1956, Marzolf 1978, Simpson *et al.* 1982).

I am proposing a two-part study design. In the first part of the study, I will gather exploratory data on the distribution of debris piles and on small mammal usage. The second part of the study will test whether certain characteristics of the piles, age, and location, make them more or less appealing to small mammals. I will test the effects of age and location of piles because these can be substantially affected by management practices through removal or movement of woody debris.

Axiom 1:   Debris piles are an important component of stream ecosystems.

The distribution of debris in streams of old-growth Douglas fir (*Pseudotsuga menziesii*) has been documented by Lienkaemper and Swanson (1986). They measured amounts of debris, ranging from 92 Mg ha^{-1} to 300 Mg ha^{-1}, and noted that amount of large debris decreased with increasing stream order. In the first- and second-order streams of the Pacific North-west, large woody debris commonly covers as much as 50 percent of the channel (Sedell *et al.* 1988). First-order blackwater streams in southeastern Virginia were found to have 8–13 debris dams per 100 m of stream and 9.22 Mg ha^{-1} to 33.56 Mg ha^{-1} (Smock *et al.* 1989). Historically, woody debris piles covered enormous areaas of small streams and large rivers. As an example, a driftwood jam on the Skagit River,

Washington, was reported to have been three-fourths of a mile (1.2 km) long and one-fourth of a mile (0.4 km) wide (Sedell *et al.* 1988). Current estimates of woody debris biomass and volume in aquatic ecosystems are extremely variable. These findings suggest that coarse woody debris is a fundamental component of many river ecosystems but that the amount of debris may vary within or between river corridors.

> *Axiom 2:*   *Debris piles in forested areas provide habitat and/or forage for small animals.*

Woody debris piles in close proximity to the channel not only increase the diversity of aquatic habitat for fishes but also add complexity to the available habitat for terrestrial life. Twenty North American species of small mammal are known to use coarse woody debris for denning, feeding, and reproduction (Harmon *et al.* 1986). Kaufman *et al.* (1983) determined that the white-footed mouse (*Peromyscus leucopus*) selects habitat according to vertical complexity, but use of large debris piles was not studied. In another study, several less commonly captured species of small mammals were collected only in riparian habitat. Doyle (1990) found that deer mice (*Peromyscus maniculatus*) and Townsend chipmunks (*Tamias townsendii*) were often located in microhabitats that contained relatively large amounts of woody debris. Doyle proposed that riparian habitats might be a population source for small mammals and upland habitats might be a population sink.

> *Axiom 3:*   *Small mammals use riparian areas.*

In coastal Oregon, there are at least 80 species of wildlife that are snag- or log-dependent and that also frequent riparian forests (Cline and Philips 1983). No studies have investigated whether certain species of small mammals are obligate users of riparian areas; however, data from a study by Doyle (1990) indicate that riparian environments in montane areas provide superior habitat for several species of small mammals.

Phase One

> *Postulage 1:*   *Debris piles in the study area can be classified as old piles or new piles and as piles in the forest or piles on a cobble bar.*

There is no research about debris piles on the north fork of the Skykomish. I am investigating this postulate on the basis of information from the work referenced under Axiom 1 and from observation.

C_1^{imag} (*Debris Piles*)[2]: Debris piles are now a *concept by imagination.*

---

[2] Recall the superscripts *imag, res,* and *meas* indicate the status of concepts respectively as by *imagination, from research* or *by measurement.*

For now, I have defined them to be any cluster of logs, deposited by the river, larger than 1 m in any direction. I may redefine this concept after the data exploration phase of the project.

C_2^{imag} (*Study area*): Study area is a *concept by imagination*. There is no research on which to base my definition. The study will examine the north fork of the Skykomish River because of its similarity to many rivers on the west side of the Cascade crest. Although results from this river segment can not be applied to all rivers in the region, it is reasonable to assume that trends on the north fork of Skykomish are comparable to trends on other mid-order reaches of nearby rivers. Because access to the river is limited and anthropogenic influence increases downstream, I have further refined the concept to include only the south shore from the Galena Bridge to Trouble Creek.

C_3^{imag} (*Old debris piles*): I will define old debris piles as those that existed before the flood of 1990. There is no literature that has defined this concept before. (Concept by imagination.)

C_4^{imag} (*New debris piles*): I will define new debris piles to be those that were deposited in the 1990 flood. I will use three main criteria for categorizing a pile as new (at this point, the piles become concepts by measurement).

(a) Presence of green branches inside the pile.
(b) Lack of dead leaves on top of the pile.
(c) Lack of any seedlings growing on top of the pile.

Postulate 2: *Small mammals use debris piles in riparian areas for denning, feeding, and/or shelter.*

I will document the small mammal communities that use the piles. Documentation of pile use will not determine whether the piles provide critical resources but, in determining which animals make use of the pile, we can begin to understand which resources the piles provide. References can be found under Axioms 2 and 3; however, little research has investigated this question before. The lack of literature is the reason for Phase One of my study, the aim of which is data exploration.

C_7^{imag} (*Small mammals*): Small mammals are a *concept by imagination* because I don't know what I will capture. This concept will be refined after Phase One.

Debris piles: Debris piles are a concept by imagination defined above.

C_8^{res} (*Riparian areas*): Riparian areas have been defined in the literature in different ways for different reasons. According to the forest practices regulations, a riparian management zone is 25 m wide on either side of a stream that has fish in it (or of a river with or without fish). The definition of the word is "pertaining to the bank of a stream". I have defined it to mean as far away from the channel as debris has been deposited. In some areas, this distance is 30 cm, and in others, it is hundreds of meters. Riparian area is a concept from research.

C_9^{imag} (*Small mammal use*): Denning, feeding, and/or shelter are concepts from research and have been referenced in the literature extensively, but I will not be measuring these things. Instead, I will use mammal presence as an index of resources provides. For example, if there are many deer mice in a pile, I will assume that they are denning, feeding, or getting shelter from the pile. The corresponding concept by measurement will be trapped mice. Later the exploratory date set is completed, more research will be needed to investigate this link. At this stage, I must first determine which animals are using which type of piles; then, I can ask specific questions about what the animals use the piles for.

Phase Two

> Postulate 1: *Small mammal use of old debris piles is different from small mammal use of new debris piles.*

I will compare vertebrate use of piles that were deposited in the 1990 flood with use of piles that existed before the flood to examine the importance of pile age in providing food and habitat resources to the vertebrate communities. No research has investigated this question before. It will become an imporotant question because management may influence flood frequency and, therefore, the average age of piles.

> *Small mammal use*: Small mammal use will be a *concept by imagination*. I will still have to use presence as an index of denning, feeding, and/or shelter.

> Postulate 2: *Small mammal use of piles that are at the forest edge is different from small mammal use of piles that are on a cobble bar.*

I will investigate the importance of pile location because riparian harvesting will influence the number of trees surrounding a pile. No research has investigated this question before.

> *Small mammal use*: Small mammal use will be defined as in Postulate 1.

C_{10}^{imag} (*Piles at the forest edge*): A pile at the forest edge is a concept by imagination that I will define as any pile that is on soil or next to a coniferous tree.

C_{11}^{imag} (*Piles on a cobble bar*): A pile on a gravel bar is a concept by imagination that I will define as any pile that is on a cobble substrate.

The policy context

The north fork of the Skykomish River is a state scenic river. Its status as a scenic river mandates that there be no further development on its banks and that there be no disruption of the river corridor. Legilsation to create the Scenic River Program was primarily initiated by recreational river users who wanted to preserve pristine and beautiful rivers for both ecological and recreational reasons. However, debris piles that overhang a bank or initiated by recreational river users who wanted to preserve pristine and beautiful rivers for both ecological and recreational reasons. However, debris piles that overhang a bank or are in the path of the current during high water may ensnare both boats and people. For rafters, kayakers, and canoeists, debris piles present a deadly threat. Whitewater rafting is a large industry on the Skykomish River, but after each flood, shifting logjams threaten the ability of rafting companies to provide safe trips.

The State Scenic River Program will soon be forced to decide whether debris piles are protected under the legislation. If they provide an important ecological function, they will probably be protected. The whitewater industry will demand that some piles be removed to provide safe passage. My research will assist policy-makers in deciding whether debris piles shoule be removed, and, if so, which type of pile can be removed with the least disturbance to the riverine, riparian, and forest ecosystems.

The management context

Stream rehabilitation is a current focus of both forestry and fisheries managers. Some rehabilitation work requires that woody debris be put back in the streams. How should the debris be placed? No studies have tried to answer questions about optimal debris distribution, and all studies about stream rehabilitation have focused on fish habitat. My research will investigate the importance of river-deposited debris in providing resources for the terrestrial communities. Perhaps stream rehabilitation can simultaneously enhance fish production and riparian function.

Potential bias

There are many sociological considerations in my research. First, I must consider whether research in this area is conducted by an "open" or a "closed" group of scientists. Although the original work on coarse woody debris was conducted by a fairly small group of closely associated researchers at the H. J. Andrews Experimental Forest, the topic is of such broad importance that researchers from distant areas of the country have addressed the issue (Angermeier and Karr 1984; Smock *et al.* 1989). Despite all the research, no one has reported that coarse woody debris is not important. Perhaps the unanimous conclusions reflect the universal importance of coarse woody debris or perhaps the conclusions reflect a unifying paradigm in which these scientists exist.

If I were to find that debris piles are not important, the scientific community would be surprised. The answer would probably not be acceptable until the project had been repeated with the same conclusion. I am certainly expected to find that some small mammals live in the piles, but I do not think that this expectation could bias my results. I am not expecting a particular result for the second part of my project in which I compare cobble vs. forest piles and old vs. new piles.

With respect to the policy implications of my project, I think that there is much potential for bias. However, I am both a whitewater raft guide and the researcher, so I think that I can keep a balanced perspective. I do not think that my research has been designed so that only certain answers are possible. The implications of my project will depend on whether or not I catch many small mammals. Once I set the traps, it will be out of my control. I do think it will be very important to keep the possible ramifications of my research results in mind when I publicize my findings.

I have tried to tailor my project towards providing results that are useful to management. In order to be most helpful in meeting management goals, I would need to set up experiments in stream rehabilitation. At this stage in the project, experiments are not possible. Future research may focus on providing strict guidelines for stream rehabilitation.

References

Angermeier, P. L. and Karr, J. R. (1984). Relationships between woody debris and fish habitat in a small warmwater stream. *Transactions of the American Fisheries Society*, **113**, 716–726.

Benke, A. C., Henry, R. L., Gillespie, D. M. and Hunter, R. L. (1985). Imporotance of snag habitat for animal production in southeastern streams. *Fisheries*, **10**, 3–13.

Bryan, M. D. (1985). Changes 30 years after logging in large woody debris, and its use by salmonids. In *Riparian Ecosystems and their Management: Reconciling Conflicting Uses*, pp. 329–334. USDA Forest Service General Technical Report RM-120.

Cline, S. P. and Philips, C. A. (1983). Coarse woody debris and debris-dependent wildlife in logged and natural riparian zone forests – a western Oregon example. USDA Forest Service General Technical Report RM-99.

Dolloff, C. A. (1986). Effects of stream-cleaning on juvenile coho salmon and Dolly Varden in southeast Alaska. *Transactions of the American Fisheries Society*, **115**, 743–755.

Doyle, A. T. (1990). Use of riparian and

upland habitat by small mammals. *Journal of Mammmmology*, **71**, 14–23.

Elliott, S. T. (1986). Reduction of a Dolly Varden population and macrobenthos after removal of logging debris. *Transactions of the American Fisheries Society*, **115**, 392–400.

Harmon, M. E., Franklin, J. F., Swanson, F. J., Sollins, P., Gregory, S. V., Lattin, J. O., Anderson, J. R., Lienkaemper, G. W., Cromack, K. and Cummins, K. W. (1986). Ecology of coarse woody debris in temperate ecosystems. Advances in Ecological Research, **15**, 133–302.

Hawkins, C. P., Murphy, M. L., Anderson, N. H. and Wilzbach, M. A. (1983). Density of fish and salamanders in relation to riparian canopy and physical habitat in streams of the northwestern United States. *Canadian Journal of Fish and Aquatuatic Science*, **40**, 1173–1185.

James, G. A. (1956). *The Physical Effects of Logging on Salmon Streams in Southeast Alaska.* USDA Forest Service Report, Alaska Forest Research Center-5.

Johnson, R. R., Ziebell, C. D., Parton, D. R., Folliott, P. F. and Hamre, R. H. (1985). Changes 30 years after logging in larger woody debris and its use by salmonids. In *Riparian Ecosystems and their Management: Reconciling Conflicting Uses*, pp. 329–334. USDA Forest Service General, Technical Report RM-120.

Kaufman, D. W., Peterson, S. K., Fristik, R. and Kaufman, G. A. (1983). Effect of microhabitat features on habitat use by *Peromyscus leucopus. American Midland Naturalist*, **110**, 177–185.

Lienkaemper, G. W. and Swanson, F. J. (1986). Dynamics of large woody debris in streams in old-growth Douglas-fir forests. *Canadian Journal of Forest Research*, **17**, 151–156.

Marzolf, G. R. (1978). The potential effects of clearing and snagging on stream ecosystems. *FWS/OBS-78/14, Fish and Wildlife Service – National Stream Alteration Project.*

Sedell, J. R., Bisson, P. A., Swanson, F. J. and Gregory, S. V. (1988). What we know about large trees that fall into rivers. In *From the Forest to the Sea: A Story of Fallen Trees*, ed. C. Maser, R. F. Tarrent, J. M. Trappe and J. F. Franklin, pp. 47–79. USDA Forest Service General Technical Report PNW-229.

Simpson, P. W., Newman, J. R., Keirn, M. A., Matter, R. M. and Guthrie, P. A. (1982). *Manual of Stream Channelization Impacts on Fish and Wildlife.* FWS/OBS-82/24, Fish and Wildlife Service Report.

Smock, L. A., Metzler, G. M. and Gladden, J. E. (1989). Role of debris dams in the structure and functioning of low-gradient headwater streams. *Ecology*, **70**, 764–775.

Spies, T. A. and Cline, S. P. (1988). Coarse woody debris in forests and plantations of coastal Oregon. In *From the Forest to the Sea: A Story of Fallen Trees*, ed. C. Maser, R. F. Tarrent, J. M. Trappe, and J. F. Franklin, pp. 5–24. USDA Forest Service General Technical Report PNW-229.

Swales, S., Lauzier, R. B. and Levings, C. D. (1986). Winter habitat preferences of juvenile salmonids in two interior rivers in British Columbia. *Canadian Journal of Zoology*, **64**, 1506–1514.

4.4 Process 3: Ensuring the proposed research has relevance to prior scientific knowledge

At this stage in project development, it was important for Steel to examine the proposed axioms and postulates in relation to the scientific literature. Of course, both the literature and practical considerations had guided the development of the work this far, but a more thorough analysis was required now that the plan was developed to this stage.

In the analysis of the literature, and in the discussion of potential bias (Box 4.7), two theoretical ideas underlying the work become apparent. First, structural complexity in a habitat increases animal *use*. Some literature indicates that in montane regions riparian environments provide habitat superior to non-riparian. The integrative concept of structural complexity is wider than can be applied in this study at a practical level and is not part of the logic of the investigation in detail, i.e., that different types of structure within the riparian zone might have different value as habitat. However, the first investigations into classifying woody debris piles are an attempt to define the structure of this particular environment.

Second, there is the general theory that CWD is important. Steel notes (Table 4.7) that if she finds this not to be so, other scientists studying CWD would not accept her result, at least in having general application. Axiom 1 is the over-arching axiom of this work, one that has to be assumed because it specifies that the problem exists in general terms. Over-arching axioms may be challenged, but as Steel recognized they are rarely completely rejected by one study.

Concepts refined through research involving field investigations or experiments, and particularly those in published work, are concepts from research. Though already subjected to challenge – and therefore more objective – they are not infallible. A concept from research may be applied in a different situation to where the work was done that established it. Such an extension is a form of induction (see Chapter 7); at that point, a concept from research may revert to a concept by imagination for that particular new investigation.

An important part of defining the status of a concept is by reviewing the scientific literature. The crucial questions are:

Can I use this concept as defined by previous research?	If so, this is a concept from research
Must I specifically investigate the concept?	If so, this is a concept by imagination

Measurements may be devised for both concepts from research and concepts by imagination. In both cases, the measurements should themselves be thought of as concepts, i.e., concepts by measurement, because measurements almost certainly do not correspond exactly to a concept from research or by imagination. It is not unusual for there to be a number of different ways to measure something, and for there to be considerable debate about what the different measurements tell. For example, different types of small mammal trap catch different types of small mammals or have different catch efficiencies.

4.5 Process 2: Applying creativity to develop new research ideas

Creativity is already apparent at many points in the construction of this research plan. Steel recognized that the storm and flood event of 1990 produced substantial deposits of woody debris and provided a study opportunity. Her theory that piles of debris on different land and forest types may have different animal use defined a contrast to be used in a comparative study. To consider this a possibility, Steel had to be sufficiently aware from the scientific literature of how small mammals might use such piles. The extension of that body of knowledge to consider piles of different ages, and on different types of land, involved a use of imagination through the construction of new concepts.

Another creative process is the use of analogy. As previously noted, Steel proposed to enlarge the already developed theory (published literature) indicating the importance of woody debris within rivers and streams by extending it to debris piles on their land margins (her project). The importance suggested for these debris piles is analogous to, but not the same as, that in rivers and streams. Research about the importance of the physical structure of forests in providing habitat, particularly for birds, was used. The theory is that trees provide cover and different niches for animals: they *structure* the environment. Considering debris piles in this way is again an analogy, i.e., that piles do the same types of thing as trees, providing places where mammals may den or shelter.

There are two important points to remember about creativity. It is not the exclusive property of a few people, and it does not necessarily arise as a bolt out of the blue. Improving definitions, refining measurements, and defining how your study relates to wider problems are all creative processes. These tasks continue throughout the duration of the project.

4.6 Process 5: Determining how conclusions can be drawn

Construction of the data statement is a crucial stage in which you translate your ideas (postulates) into a concrete, measurable form. If it is possible to construct one, the hypothesis becomes a restricted version of the postulate. Whether you use the acceptance or rejection of a hypothesis to accept or reject the postulate must take into account the decisions about measurements and the statistics used.

Constructing data statements for postulates was illustrated in Chapter 2. That example involved plant survival, which was easy to measure. Here the concepts are more complex – which is more typical of ecological research. Consider, for instance, Postulate 2A.

Postulate 2A

| Small mammal | use of | new debris piles | is different from that of | old debris piles |

Table 4.1. *Questions about concept definition for Postulate 2A*

Small mammal	use of	new debris piles	is different from that of	old debris piles
Should species of small mammals be distinguished? May there be differences in species "*use*" of piles? One type of small mammal trap will not capture all types of small mammal equally effectively. What trap type should be used?	*Use* may include denning, feeding, and nesting. May there be a difference in "*use*" at different times of year? "*Use*" can only be measured as occurrence in a pile	If we are to sample use of new and old piles, then there must be sufficient piles of each age to measure. Confounding factors, i.e., could influence small mammal use – for example, if all new piles are closer to the river edge than old piles. The mixture of differently aged piles over the area of investigation should be random – or we must expect to account for any effect of a non-random distribution		

These concepts must be translated from their general form used here into the particular form that applies where measurements are made. This is done in two stages: defining the concepts, and deciding about measurements.

Data statement: Part One

Specifying the type of investigation to be used.

A comparison of small mammals trapped in new and old coarse woody debris piles.

Data statement: Part Two

Specifying the conditions of the investigation and the measurement details. General questions to ask about concepts by measurement are given in Chapters 3 and 5. Specific questions about the concepts of this postulate, sampling design, and measurement decisions are given in Tables 4.1 and 4.2.

Notice the many choices. One, to sample in November, was influenced by logistics and by other research being planned. Another, the use of visual assessment for classifying old and new piles, depended upon previous exploratory research.

Table 4.2 *Decisions about measurements and the structure of a survey*

Small mammal	use of	new debris piles	is different from that of	old debris piles
All animals were recorded by species. Preliminary studies showed considerable variation in small mammal capture between sample times, even for the same pile. Steel decided to make one sample of 38 piles – a large number for one investigator to handle. This meant leaving traps open for 24 hours rather than merely overnight, knowing that animals may not survive the trapping. Consequently, this study was left until November, the end of the season, when animal death would not affect subsequent investigations. That possible seasonal differences are not accounted for in this study limits interpretation. Three trap types were used simultaneously at each pile to capture as many different species as possible		New and old piles were distinguished by three visual assessment characters. All debris piles within 25 m of the active channel and along 1609 m of the river were mapped, classified, and randomly sampled. No statistical tests were made for spatial aggregation of piles by different ages – but there were no apparent patterns		

Data statement: Part Three
Specifying any statistical hypotheses to be used and calculations to be made.

A hypothesis was constructed:

> *There is no significant difference between the number of small mammals trapped at old or new piles.*

Piles were stratified by age (old or new), location (forest or cobble bar), and size (large or small) and then randomly sampled within these strata. By the time of this investigation it was known from investigations of other hypotheses that numbers of animals trapped had a very skewed distribution.

Data were analyzed by ranking piles according to the numbers of animals caught, then calculating whether more old or new piles were in the top ranks (Wilcoxon rank sum test, Zar 1996). For this type of data, where numbers of animals caught can be very variable, calculating by ranking is more effective than calculating means and variances and testing for differences between means.

When using statistical inference to test hypotheses, many ecologists look for probabilities of 0.05, i.e., 1 chance in 20 that the event could have occurred as a result of random effects. On this basis the hypothesis that there is no significant difference between total numbers of small mammals caught in old or new piles can be rejected with a probability of only 0.056. That is, if there were actually no difference between the types of pile, there would be 1 chance in 17.9 of finding this rank ordering of the piles. Because this does not meet the usually accepted standards for rejecting a hypothesis of no difference, should the investigation be stopped at this point?

No – because the whole investigation is really exploratory (Chapter 8). We need to look one layer deeper into what might be causing this tendency towards different *use*, if anything. And indeed further analysis showed differences between genera. More *Peromyscus* ($P = 0.024$, only 1 chance in 41.7 that this could be a random effect) were caught in the new piles, but there was no difference in the numbers of *Sorex* and *Microtus*. More of both species of *Peromyscus* present at the site were caught in new piles; the difference was significant for *P. oreas* but not *P. maniculatus*. New piles had more species richness of small mammals than old piles ($P = 0.078$, 1 chance in 11.5 that this could be a random effect).

Scientific inference

What scientific inference can be made? The use of statistical inference indicates a number of things about the postulate – but interpretation must be qualified by the decisions made in developing measurements and establishing a procedure for statistical inference.

We can say:

Peromyscus *use new debris piles more than they do old debris piles.*

Significance was achieved for *P. oreas* but not *P. maniculatus*, although it did show greater numbers in new rather than old piles. To investigate whether there is a clear distinction between the two species requires further information – more sampling over extended periods with the knowledge of the differences in biology between species. If we are prepared to accept a probability of 0.056, we might be tempted to say:

Small mammals use new debris piles more than they do old debris piles.

But that would be an incorrect scientific inference, since we have found out that there are differences between the species. Some showed preference and some not, and the P value obtained was due to the proportionally large numbers of *Peromyscus* measured. This single proposition must be redefined into two:

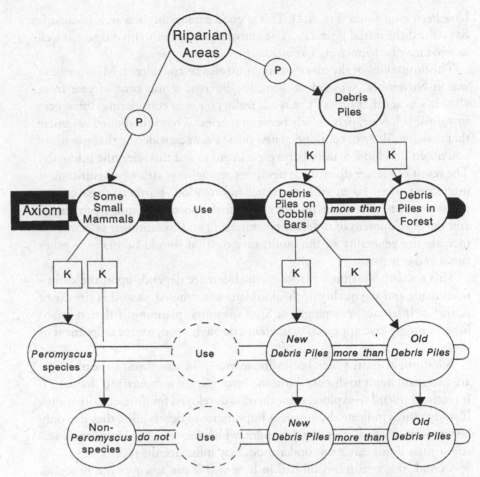

Fig. 4.2. The developing network of scientific knowledge about small mammal use of coarse woody debris (CWD) piles. The advance comes from refinement of concepts into *parts* and *kinds*. Information from Steel's project – that more small mammals were caught in CWD piles on cobble bars than in forests – is incorporated. However, it would require further study to discern whether this reflects differences in proportional *use* relative to other habitat types or differences in overall animal density, discussed in Chapter 7. Initial results about Postulate 2 are inconclusive; it also requires further research that possibly depends on a further classification of *Peromyscus* species according to their habitat preferences. Within the cobble bars, research is needed to define what properties different types of CWD pile have; this would allow restructuring of the theory (network) and so allow more progress.

Peromyscus *use new debris piles more than they do old debris piles.*
Other species of small mammals do not use new debris piles more than they do old debris piles.

However, given the numbers of animals caught and the restricted nature of the investigation (one sample, one season, one site), these two propositions are best considered as postulates for further investigation and not axioms that have been confirmed (Fig. 4.2). This is good exploratory research because it has refined the initial postulate. The knowledge of what to investigate as well as what may be happening has increased.

The limitations of the investigation must also be considered. Measurement was in November. Was animal behavior different at this time of year from what it was at other times? The result really points to considering differences in animal behavior, particularly between species. What is it about *Peromyscus* that makes it likely to be found at new piles? From considering that question, you might develop a manipulative experiment to test this scientific inference. The result of greater diversity at new piles, though not strongly significant, is interesting when taken with the other information. Further investigations really would require more precisely formulated postulates and hypotheses – and the development of different measurement and/or sampling systems. To increase the generality of the result, investigation should be made at other times and other sites.

This research illustrates that scientific inference depends upon the definitions made and the quality of the postulates constructed, as well as the use of statistical inference. A strength of Steel's research planning is that it established a progressive approach to inference through a progressive refinement of ideas.

What might limit general scientific inference in this investigation? There are two main items to discuss. First, the investigation is *inductive* (Chapter 7). It reaches forward to explore, using theories developed for different situations. If study results indicate the apparent importance of debris piles, they can only be tentative. Factors other than the piles, which may not have been considered, or their significance not understood, may influence the results.

Second, this research is limited in how well it can uncover the processes that control small mammals. It is bounded in both the generality and detail of what it can find out.

The generality of the conclusions is limited because the investigation was restricted to one site and one year. The technical requirements for conducting a duplicate investigation were not available to Steel. How many sites should be studied before a result is accepted? The answer is not simply to repeat the same research at more and more sites, although some repetition would be important, but to develop a theory that can be subject to rigorous investigation.

The detail of the conclusions is limited by how concepts were defined. Pile use was measured by trapping. However, this is not a direct measurement of real use, i.e., what the animals do inside the piles, and that may indicate why the piles are important. This exemplifies aggregation in constructing a concept, i.e., *use* was defined in a way that could be measured. As a result the concept by measurement bounded the sophistication of the study – something that happens frequently in ecological research.

These two issues, bounding the generality of an investigation a by studying a limited system (e.g., one site, one year) and bounding the detail of an investigation by use of a particular measurement system, are common problems in ecology and there consequences must be investigated through subsequent research.

4.7   Steel's comments on the planning process after completing her Master's thesis

After her thesis (Steel 1993) was complete Steel made the following comments on the planning process.

(1)   She conducted no pilot study of trapping and wished that she had. Even trapping small mammals for a week would have given her valuable information to gauge the scope of the work, particularly estimation of the variance of numbers caught in traps (see Chapter 8 for a discussion of power). From this, calculations can be made of sample sizes needed.

(2)   The written plan was valuable when she started to write her thesis. Additional objectives and techniques were added to the study, particularly a survey of bird use of the woody debris piles and measurement of pile temperature regimens. However, it was useful to have the overall objectives to which to react. Steel wished she had kept more notes of her ideas during the research process.

(3)   The process of having to define debris piles during the planning stage proved invaluable. The classification made was the source of some important findings, particularly differences in both the species and type of use close to the forest and cobble bar.

(4)   Steel used reference sites at some distance from the woody debris piles but on the same cobble bar as a contrast for animal use between piles and no piles. She wished that she had compared other complete sites, perhaps sacrificing some of the intensity of her bird observations. Reference sites on the same cobble bar provide only a limited contrast. For example, small mammals are known to have a wide foraging area, so those caught at the reference sites on the cobble bar may actually have been using piles there as a denning base. When community-level questions of this type arise, then communities must be compared.

(5)   Steel had asked for comments on her proposal from a number of scientists, in addition to the original three faculty members who were also members of her guidance committee. However, at the beginning she had been reluctant to ask for criticism from specialists in small mammal research. This was because her interests were oriented towards the woody debris piles, not the small mammals, except as a possible indicator of pile importance, and she did not intend to make small

mammals the focus of her research. Subsequently, as she realized the importance of understanding small mammal biology in interpreting her research, she wished she had asked for that criticism earlier.

This type of reluctance is not uncommon. Why do we have it? (And most of us do!) I suggest three reasons. First, at the start of any research, we may be diffident about parts, or maybe all, of our investigation. Second, we cannot be sure of receiving a hospitable reception or constructive criticism from all scientists. (This was not the case for Steel.) Unfortunately, some scientists are caught up in protecting their position in their own field or asserting themselves, and you may only encounter this a few times before you come to expect it. Third, we have to put a great deal of personal effort into research – frequently, a person's life and thoughts are bound up in it – yet we still have to play with the ideas, toss them around, pinch and poke at them, and scrutinize them. Normally, people do not like that done with their life but it is important that it happens with research.

In my opinion, critical analysis is the most important process of the scientific method. It is essential to overcome feelings that stop you from seeking criticism and to learn to distinguish between a tough but genuine critique, which you need, and some sort of power play, which you do not.

4.8 Further reading

Carey and Johnson (1995) discuss small mammal ecology in forests. Nakamura and Swanson (1994) define the distribution of coarse woody debris in a riparian zone and propose factors that controlling it. The research that developed from this plan is reported by Steel *et al.* (1999).

5 How theories develop and how to use them

Summary

Theories are generalizations. The theoretical constructions used must explain what is true in common between particular examples, and must be sufficiently detailed to explain what may be different between them, and how commonalities and differences may be resolved. Theories contain questions. For some theories the question(s) are explicit and represent what the theory is designed to explain. For others the questions are implicit and relate to the amount and type of generalization, given the particular choice of methods and examples used by researchers in constructing the theory. Theories continually change, as exceptions are found to their generalizations and as implicit questions about method and study options are exposed.

The development of a theory about vegetation and climate changes in central Alaska during the late Quaternary illustrates the following approaches: disconfirmation of a simple postulate, exploratory analysis in constructing new postulates, research to refine those postulates, and synthesis with knowledge from different types of investigation that give confirmatory support.

The use of a theory developed for one situation and applied to another requires particular care. A practical example, in which a theory about mate selection in one species of fish provided the starting point for investigating other species, illustrates the need to specify axioms very carefully when using theories.

Theories may develop in a progressive or regressive way. Development is progressive if the generality of a theory increases. It may be regressive if the theory becomes unable to incorporate the conclusions from some investigations without its originator making *ad hoc* adjustments that invoke special exceptions for particular circumstances.

5.1 Introduction

We use theory when we analyze an ecological problem, no matter how practical or field oriented the problem may be. In research planning, conceptual and propositional analysis defines the content of the theory, and specifies its axioms and concepts from research. But it is important to go further and to determine the scope and stage of maturity of the theory – how extensive and effective its predictions are and how well supported it is. As theories develop, they go through stages in which different types of investigation may play a major part. If you can recognize the stage of development of a theory, it may help you to understand in which logical direction the research may turn next.

1. Theories contain generalizations

Constructing a theory is an attempt to establish some generality, a set of statements that will explain a wider set of circumstances than the immediate scope of the measurements, experiments, or investigations used in constructing it. Generality has three components:

(1)   From what is studied to what may be claimed, e.g., that the particular causes investigated are representative of a wide class of circumstances.
(2)   That what is measured (concepts by measurement) do actually represent concepts from research and concepts by imagination.
(3)   In the construction of the theory itself – that the theoretical constructions used are sufficiently detailed that they explain what is true in common, what may be different in particular examples, and how commonalities and differences may be resolved.

Much scientific activity is designed to examine the extent to which these three components of generality hold. As soon as we use a theory in a new situation, we attempt to extend at least one of these components. This was apparent in Ashley Steel's use of the theory about coarse woody debris (Chapters 3 and 4). A theory had been constructed through work within rivers and streams; she applied it to land alongside rivers using different concept definitions. Simply by using the theory, Steel attempted to extend its generality. But if we want to accept that the theory applies both within rivers and streams and along riverbanks, then we have a larger set of qualifying conditions to specify, and that needs to be done by defining concepts.

Each set of axioms of a theory rests upon (1) particular data, (2) the conditions chosen in collecting those data, and (3) the logic used in accepting or rejecting particular propositions. The theory is defined explicitly on its axioms, which is the part we usually present in written accounts, but it also depends on their data statements, which describe the processes of reasoning and investigation used. To be rigorous when specifying a theory, we should also specify these qualifying conditions.

All theories propose explanations about things that can not be completely measured. But an important distinction must be made between theories that could be directly confirmed or refuted by observation or experiment – if only it were possible to make the necessary measurements with sufficient precision and accuracy (Chapter 7) – and those theories that use concepts that can not be directly measured except through particular instances. For example, some ecological theories seek to explain concepts such as diversity and stability. Although we may develop measurements to make relative assessments of the type "This community is more diverse than that one," a theory of what controls diversity must be built up from studies of many communities. This problem is discussed in Chapter 10.

The examples in this chapter illustrate how generality of theory is constantly challenged by research into particular instances. Where a generality holds, the theory builds. Frequently in ecology a theory also becomes more complex to account for some particular feature of the current investigation. Where a generality no longer holds, substantial reorganization is required, although the objective of the theory may remain the same. The constant battle between what we theorize as occurring in general, through the construction of axioms, and what we find in new particular instances is one of the most important features of ecology. Differences of opinion about whether thought should originate with axioms or observation is expressed as differences in philosophy (see Chapter 9) about the nature of ecological knowledge (Section II). For the researcher, it is important to ask, "Is the theory I am using at a stage where its generality is building through successive confirmations, or is a realignment due to resolve observed differences?" The answer may determine the type of reasoning used in the research (Chapter 7).

2. Theories contain questions

Questions are implicit in all theories, and a researcher must discover what these questions are. Two types of question can be asked about every theory: (1) a question about its right to exist, i.e., is it really justified by the research so far? And (2) a question about how effective and extensive its generalities are, i.e., is it really effective in providing explanations about the subject? The particular cases and data used in establishing the axioms of a theory always have methodological or technological uncertainties. Choices always have to be made about measurements, sampling, and general lines of approach; these choices imply questions.

Frequently, the choices made about instruments, methodology, or types of example used become standardized among investigators. For example, ecophysiologists may choose a limited number of plant or animal species to investigate – and they may use standardized ways of preparing tissues for physiological analysis. This limited range of investigation and/or methodological standardization may minimize or remove some recognized difficulty in the investigation. Standardization can be expected in science, because it speeds progress and allows researchers to compare results, but it may limit how extensive the explanations and predictions of the theory are.

Theories, even good ones, are established on the basis of limited sets of investigations. There is always the question of whether a theory will apply in another related, but different, circumstance.

3. Theories change continuously

Every investigation has some effect on a theory. If the investigation simply assumes that the theory is correct and finds nothing to contradict that assumption, then the investigation is confirmatory. The use of the theory then establishes further support for it, even though that may not have been the direct objective of the work. The efficiency of different types of investigation, whether attempting to confirm or to challenge a postulate, is discussed in Chapter 6. Confirmatory work, if taken from a situation not previously investigated, extends the basis for a theory. Repeated use of theory in practical application can entrench belief in the theory's effectiveness, even though some of its axioms may define only a limited scope of application. C. A. S. Hall (1988) gives critical examples of this in the use of predictive models in fisheries management.

Theories may gain in power as their generality is extended by work in new situations or may lose power if counter-examples are found. Often, theories change by the increase in number of axioms or change in their content. No theory is safe from change, and all theories contain the seeds of their own destruction, either in unanswered questions about their right to exist or in questions about the effectiveness of their generality. Sometimes the demise of a theory can be dramatic – when contradictory evidence challenges the right of the theory to exist. Often, theories simply seem to disappear from active use or investigation because they are no longer effective. Either the generality for which scientists are striving for in developing a theory is no longer needed for some current questions, or it becomes apparent that the hoped-for generality cannot be reached.

The first example in this chapter illustrates development of a theory of vegetation history and climate change in central Alaska during the late Quaternary when a simple postulate was tested and found to be wrong. Two aspects of theory development are presented: (1) use of exploratory analysis when a postulate is rejected but there is then no explicitly defined logical alternative to follow, and (2) incorporation of very different types of information into a theory. A theory was first developed using pollen analysis of lake sediment cores; then information about individual tree-species ecology and from global circulation models was incorporated. There can be no absolute test of this more complex theory. Its strength is that its predictions are broad – they are both ecological and meteorological – and they must be coherent between the two types of knowledge.

It is quite usual for graduate students to take a theory and apply it in what seems to be a related field. The second example in this chapter illustrates some problems that can be encountered when doing this. In this example a theory,

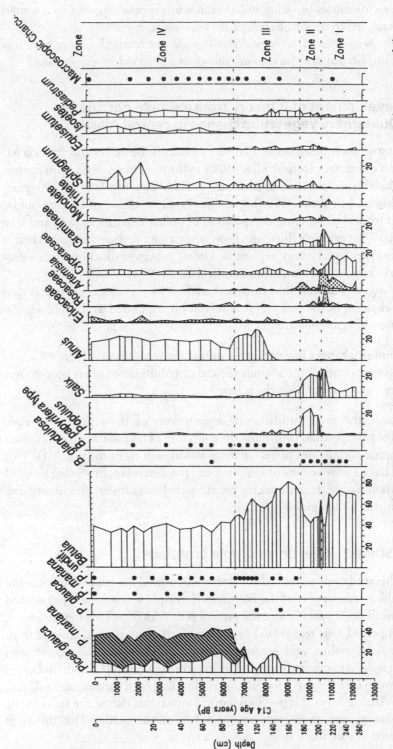

Fig. 5.1. A pollen percentage diagram from Wien Lake, central Alaska. Stippled curves represent 5 × exaggeration, solid dots represent the presence of plant macrofossils for *Picea* and seeds for *Betula*, and macroscopic charcoal (> 0.425 mm). Note the increase in *Picea mariana* pollen from 6500 BP (before present). (From Hu *et al.* 1993, with permission.)

how behavior causes breeding isolation in a fish species, applied to a related fish species, led to reexamination of the theories' axioms.

Finally properties of ecological theories are summarized. These have important implications for how to use theory in a practical investigation.

5.2 Development of a theory from a simple postulate: Late-Quaternary vegetation change in central Alaska

This description is based on an analysis by Linda B. Brubaker and Patricia M. Anderson of research from the late 1970s to the present on vegetation change in Alaska during the late Quaternary period, i.e., the last 20 000 years (Anderson and Brubaker 1994). For this period, evidence of change can be deduced from the changing composition of pollen deposits in peat and lake sediments. The example illustrates how, when a simple theory is researched, a more complex theory may replace it. Unless otherwise documented, references may be found in Anderson and Brubaker (1994).

The basis of research into historical analysis of vegetation change is that climate changed in the past and that vegetation responded to that change. So the over-arching axioms are:

(1) Climate changes have occurred during the past 20 000 years.
(2) Climate controls the abundance and distributions of plant species over large temporal and spatial scales.

These state the most fundamental assumptions of the study. The basic method of investigation is through the collection of lake sediment cores, their stratigraphic dating, and pollen analysis within each stratum (Fig. 5.1). This method has its own axioms of sampling and measurement, including those of radiocarbon dating, techniques for preparing pollen samples, and identifying and counting pollen grains (Box 5.1).

5.2.1 Stage 1: Rejecting a simple postulate

When this study was started, the literature supported the general assumption that in the northern part of the Northern Hemisphere a warming period had commenced 6000 years before the present (i.e., 6 ka BP). This hypsithermal, or warm period, was recognized in eastern North America and Europe from palynological studies, and assumptions were made about the response of different species to climate change. Pollen records in central Canada indicated that between 6 ka BP and 3 ka BP, the tree line had been further north than at present. Arising from this previous work, four axioms define the basis of the initial theory, and there was one postulate of the study. The theory is represented as a network in Fig. 5.2.

Box 5.1. Axioms of sampling and measurement for an investigation of Alaskan vegetation history by stratigraphic analysis of vegetation remains in lake sediment cores

(1) Cores taken from lake sediments with a square rod piston sampler (Wright 1967) contain a continuous record of the pollen rain at that point

Single cores are analyzed from each lake and attempts made to sample within the central region. Cores are taken in 1 m segments, the hole is cased, and error in positioning each segment is < 1 cm. The entire core length varies depending on sedimentation rate and deposit age but is often 3–10 m. This sampler is strong, lightweight, and easily used by two people.

(2) The pollen within a core is a record of the species surrounding the site

Pollen in a core is a biased record of plants in the vicinity of a lake (say, 20 km radius.) Pollen is produced in different amounts by different species, and has different dispersal characteristics, e.g., wind versus insect dispersal. This bias has been studied. Two hundred and seventy-five mud surface cores have been taken from lakes to investigate relationships between current surrounding vegetation assemblages and pollen deposition. Some guidelines have been produced, e.g., 10 percent spruce generally corresponds to the tree line, > 20 percent spruce is closed boreal forest, 30 percent birch and 30 percent sedge are low shrub tundra. A difficulty with this calibration system is that some past assemblages are not the same as current ones.

(3) The levels within a lake sediment core can be dated by the radiocarbon technique

A proportion of the carbon atoms in naturally occurring CO_2 is of the radioactive isotope ^{14}C rather than the stable ^{12}C form. Natural processes in the atmosphere maintain this proportion. Once carbon is fixed into organic material, the balance between ^{14}C and ^{12}C changes gradually due to radioactive decay of ^{14}C. Decay rate is assumed a physical constant, so sediment age can be estimated by measuring the $^{14}C/^{12}C$ ratio. Dating precision is often 50–150 years depending on the amount of carbon in a sample.

Whereas radiocarbon dating may be precise, the measurement is not accurate for two reasons. (a) The ^{14}C proportion in the atmosphere has changed over time. There has been systematic variation, but this has been studied, e.g., by measuring ratios in known tree-ring sequences thousands of years old. (b) Organic carbon can be eroded from surrounding vegetation and deposited on the lake bottom, increasing the apparent age. New techniques of accelerated mass spectrometry allow small amounts of carbon to be dated, e.g., pollen grains collected from individual levels, so greater accuracy may be obtained.

(4) Three hundred pollen grains from each level of a single core are an effective sample

The taxonomic level of identification is often just to the genus, and not all pollen grains may be identified. In this work, *Picea mariana* and *Picea glauca* proportions are determined by measuring pollen dimensions (pollen size of *P. glauca* < *P. mariana*) and comparing with modern populations by maximum likelihood techniques. Near-shore cores are used to look for macrofossils, e.g., seeds and leaves, identifiable to species.

Axiom 1:	Climate change is uniform on continental and global spatial scales.
Axiom 2:	Summer temperature controls the location of the tree line.
Axiom 3:	In central Canada, the tree line was north of its present location 6 ka BP to 3 ka BP.
Axiom 4:	In north central Alaska, the tree line is defined by the presence of Picea species.
Postulate 1:	Picea species in north central Alaska moved north of their present location from 6 ka BP to 3 ka BP.

To test the postulate, lake sediment cores were collected and analyzed (Anderson and Brubaker 1994). Initially, seven lakes were sampled across the present tree line, running along the Brooks Range in northern Alaska and turning south at the westward end of the mountains. There was no evidence of *Picea* (spruce) pollen, above expected background levels usually found in tundra, in cores collected just north (10–20 km) of the present tree line, or to the west where the present tree line turns south. In cores collected just south of the tree line, there was no evidence of an increase in tree species around 6 ka BP followed by a decrease around 3 ka BP. So the postulate should be rejected.

Having rejected the postulate, should you also reject Axiom 1? Postulate 1 was based on the axiom but, of course, the axiom itself was based upon indications from a number of studies over a much larger area. The critical evidence for rejection is based on only five samples. Could there be anything special about the Alaskan situation? For example, might the Brooks Range be

Fig. 5.2.   Network at Stage 1 of the investigation into vegetation and climate change in Alaska. Axioms are solid bars; the postulate is an open bar. Concepts are connected vertically; summer temperature is a part of climate change, both central Canada and north central Alaska are kinds of continental scale that have the same properties. The connection that the limit of *Picea* defines, and so is a part of the tree line, is defined by an axiom. Refer to Fig. 3.3 for symbol definitions and related text-representing theories as networks.

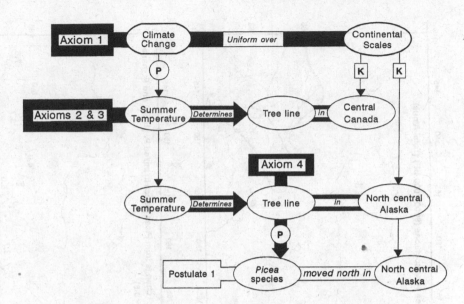

a large-scale physical barrier to northward migration? This is not thought likely. There was no evidence of changes in the pollen record within the forested area that would have probably occurred in a warming period. But before rejecting an axiom outright, the weight of evidence used to develop it needs to be balanced by a similar weight of evidence to reject it. What happens is that Axiom 1 can be replaced by a more complex statement – rather than merely being rejected.

5.2.2   Stage 2: Exploring for spatial and temporal changes

Once the pollen data disproved the postulate that the tree line in Alaska moved north beyond its present range from 6 ka BP, the research refocused to explore the pattern of variation between tree species. If there was no general northward movement, was there any movement at all? Other lake sediment cores were collected from throughout Alaska. Taken together, the complete record provided much information on other spatial and temporal patterns of vegetation change. The primary technique was to make maps of equal frequency of the pollen of a species at successive times through history (Fig. 5.3).

Exploratory analysis investigates a simple postulate of the type "there is a pattern to be found, and this pattern will be informative about the process being investigated". It is an essential phase in scientific investigation – but a difficult one. Patterns may be found that are not relevant to the study, and this can only be determined by subsequent investigations. Patterns may be missed because the wrong type or scale of data is investigated. Some scientists feel uncomfortable about exploratory analyses because they seem to lack rigor.

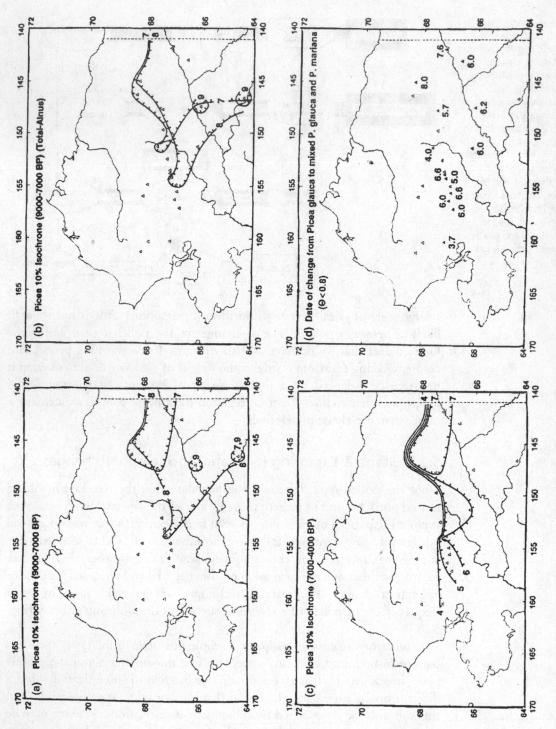

Fig. 5.3. Maps showing lines that mark 10 percent occurrence (isochrones) of *Picea* pollen with dates marked on the appropriate isochrone of each map: (a) 9–7 ka BP; (b) 9–7 ka BP, recalculated with *Alnus* pollen not included as part of total pollen; (c) 7–4 ka BP; (d) estimated dates of charge from *Picea glauca* to mixed *P. glauca* and *P. mariana*. Θ is the proportion of *Picea glauca* in the *Picea* pollen and Θ ≥ 0.8 is taken to indicate that *Picea glauca* was the exclusive *Picea* species in the vicinity of the lake where the core was taken. (From Anderson and Brubaker 1994, with permission.)

This is unfortunate because such analyses are an essential part of ecological research, albeit a part that subsequently has to be balanced with a rigorous approach to postulate testing.

What emerged from the exploratory analysis was that, although tree pollen did not increase beyond the present tree line from 6 ka BP, there were marked changes in the spatial distributions of tree-species pollen within the area of Alaska that had been occupied by trees.

About 6 ka BP, *P. mariana* replaced *P. glauca* in eastern and central Alaska, and both species invaded the western lowlands. *Picea mariana* became dominant around 6 ka BP and has remained so from then until the present (Fig. 5.1). The westerly expansion of *P. glauca* was thought to be limited to riparian zones and south-facing sites with well-drained soils. As these species spread, birch shrub tundra disappeared as the dominant vegetation type in west central Alaska.

Why should *Picea* species spread from east to west through Alaska around 6 ka BP? Why were these species not in western Alaska, although previously existing at other places just as far north? Why did *Picea* species not extend the tree line further north during this period as they did in central Canada?

5.2.3   Stage 3: Introducing axioms from tree ecology

The exploratory stage established some axioms of species distribution, which became the bases of the next version of the theory:

Axioms of species distribution
Axiom 1:   *At 6 ka BP,* P. mariana *replaced* P. glauca *in east and central Alaska.*
Axiom 2:   *At 6 ka BP, both* P. mariana *and* P. glauca *invaded western Alaska.*
Axiom 3:   *From 6 ka BP, shrub tundra disappeared as the dominant vegetation type in western Alaska.*

At this stage the theory can offer only a possible explanation of climate change when these axioms of species distribution are combined with axioms about environmental preference of species:

Axioms of species environmental preferences
Axiom 4:   Picea glauca *occurs on warm, well-drained soils, typically on stabilized flood plains, levees, and south-facing slopes.*
Axiom 5:   *Within the same broad geographic region,* P. mariana *will replace* P. glauca *where there are relatively lower soil temperatures and wetter soils.*

> *Axiom 6:*   *Birch shrubs prefer cold, dry climates, and, when found in association with species of Cyperaceae, indicate a tundra-type climate.*

These axioms are based on ecological study of current species distributions. Notice that Axioms 4 and 5 both refer to soil conditions. To use them in interpreting climate requires a general axiom:

> *Axiom 7:*   *Climatic conditions determine conditions of soil moisture and temperature.*

Now the axioms of species distribution can be combined with the axioms of ecological preference to form postulates about climate change. The decrease in *P. glauca* and the spread of *P. mariana* are taken to be a response to cooler, wetter conditions in central Alaska from 6 ka BP. The replacement of birch tundra by *Picea* spreading from the east also indicates the development of a moister, less continental climate in central Alaska. New postulates emerge: there was a climate change – but it represents a change in both temperature and moisture rather than simply a change in temperature, and for central Alaska the temperature change was a cooling not a warming.

This theory of climate change in Alaska now has two features. First, it is based upon interpretations of tree species' ecological preferences, which have been stated as axioms and used to interpret particular patterns of species distribution. Second, the emphasis of climate change has become east–west and related to a combination of moisture and temperature influences, rather than north–south and related to an overall warming. There are now two new postulates developed from the exploratory investigation combined with the axioms of distribution and ecological preferences of individual tree species.

> *Postulate 1:*   *From 6 ka BP, there was change to a colder, moister climate in central Alaska south of the Brooks Range.*
>
> *Postulate 2:*   *From 6 ka BP to 3 ka BP, western Alaska became wetter.*

In this account, Stage Three is presented as distinct and following Stage Two. Practically, for the scientists involved, there was less distinction in time – although it was important for them to maintain the logical distinction.

5.2.4   Stage 4: Increasing the precision of the theory

What should be done next? Can the two new postulates be tested? Not yet, and unlike Postulate 1 at Stage One, they cannot be tested by pollen analysis of cores; indeed there may be no simple test. These new postulates are radically different from the axiom of the hypsithermal – so this is a major challenge to the accepted theory. These postulates suggest a new pattern, not just a small revision such as, say, a slower but similar change. The radical difference between Stage One and Stage Three means that it is essential to refine the postulates and reexamine the new axioms and their supporting data statements. Refinement of postulates is a further stage of the exploratory work, in which a procedure for testing must be developed.

Two questions motivate refinement:

(1)   Are the data and techniques used in the exploratory analysis adequate? The exploratory analysis depends upon the numbers of cores and the interpolation system used to map the species spread. The number of cores can be increased and a range of different interpolation methods between the coring sites tested to refine the maps of species spread. The number and spatial distribution of lakes limit just how many lake sediment cores can be obtained.

(2)   Are the axioms of tree ecology really correct? They may be the best we can do from present knowledge, at least as far as our understanding of ecology is concerned. But there may be features other than response to soil temperature or moisture that influence rate of species spread. For example, restrictions in seed distribution could limit the rate of spread. More work could be done to investigate the present ecology of the species. Can their preferences be distinguished with greater precision so that a more precise map of temperature and moisture change could be constructed? This type of work would also require understanding more about the nature of vegetation changes as a whole, particularly the possible influence of soil changes that might affect species distribution and the role of such disturbance agents as fire. Anderson and Brubaker (1994) describe recent work along these lines.

Axioms 1 through 3 are a new foundation (which is why they are axioms), from which further, more detailed investigations could be made. For example, consider two confirmatory postulates:

> Picea mariana *always replaces* P. glauca *in central and eastern Alaska (though not in western Alaska since they moved at the same time).*
> *Where* P. mariana *increases then* P. glauca *decreases.*

Finding conditions where these do not happen would challenge these axioms – although such results would be more likely to lead to their modification rather than complete overthrow. Other types of postulates could be proposed for the various species movements implied by the theory.

5.2.5   Stage 5: Working towards explanations that are coherent with meteorological theories

Frequently, to increase the effectiveness of an ecological theory, information must be sought outside current lines of investigation. The new theory cannot be tested in a comprehensive way by a postulate test, as was possible for the old theory at Stage One, but it must be reconciled as part of a more complete explanation with parallel information from the study of global climate change. This information is about lake-level fluctuations, marine plankton depositions, and general circulation models (GCMs) (COHMAP Members 1988). These models predict variation in temperatures and precipitation across the earth's surface and have been used to simulate climates since the last ice age, i.e., from 18 ka BP. The task is to ensure that the new theory is *coherent* with other information on climate change. Coherence is discussed in general terms here – principles and methods for arriving at *explanatory coherence* will be discussed in Sections II and IV.

Simulation with GCMs, and the correlation of their predicted effects with changed lake levels, plankton, and pollen stratigraphy, suggest that climate change is complex. For example, the jet stream in the Northern Hemisphere, which marks the boundary between cool northern and warm southern air masses, undulates round the north pole; these undulations change over time. A meteorological theory is that, because of change in the tilt of the earth's axis, summer solar radiation levels were greater and winter levels less in northern latitudes (7–10 percent more in summer and less in winter than at present). It is predicted that this effect peaked at 9 ka BP and decreased from then. In Alaska, the warming effects of higher summer insolation were modified by the intensified eastern Pacific subtropical high-pressure zone, and possibly by sea ice. Some of these effects may have been a response to disappearance of the land connection between Russia and Alaska and the formation of the Bering Sea. These climate changes may account for the northward expansion of the tree line in north-western Canada (9 ka BP), a response to warming, but no observed advance in Alaska because the warmer air was diverted around Alaska.

5.2.6 Assessment of theory development

The theory for the vegetation history of central Alaska has been developed through five stages.

(1) Rejecting the simple postulate of a similarity in climate change between central Canada and Alaska based upon the movement of the tree line over time.

(2) Exploring the elements of the new theory for spatial and temporal change based upon the assumption that patterns would be found that could be explained.

(3) Integrating tree ecology information with the patterns of tree distribution to produce a comprehensive theory, but one that was untestable by a single investigation and could be improved only within the terms used for its discovery.

(4) Increasing the precision of the new theory by refining the postulate.

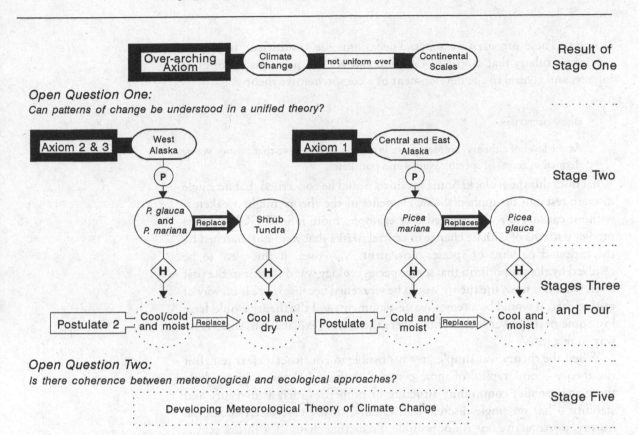

Fig. 5.4. Developing theory network to explain climate and vegetation change in Alaska. Axioms about the two regions of Alaska are represented as unconnected, and neither is directly connected to the over-arching axiom, although both are clearly relevant to it. Further work, *Open Question One*, is required to provide a unified understanding of exactly how climate change may have caused vegetation change and how trees may have responded. This is likely to require refinement of general circulation models and increased understanding of the pollen record and tree species movements across landscapes, *Open Question Two*. (Refer to Fig. 3.3 for symbol definitions.)

(5) Comparing the theory for vegetation changes in Alaska with information from other types of data that may be indicators of climatic change and predictions from GCM models.

The development of this theory has been presented as five logical and sequential stages; however, in practice these stages overlapped. For example, work on GCMs proceeded in parallel with the palynological investigations, but vegetation-history scientists only gradually accepted that the accuracy of GCM predictions made them relevant to this work.

The theory (Fig. 5.4), which attempts a description of climate change for Alaska south of the Brooks Range, now contains generalizations and questions. It has changed and is likely to change in the future. The original axiom, of uniform climate change in Northern America, is now replaced by a more complex description. An important area of research, to investigate whether GCM predictions and vegetation change theory is coherent, would require improving the spatial and temporal resolution of both approaches. But there seem likely to be limits to both of these possible improvements.

It is now an objective to understand the changes across Alaska as part of a complete theory (How are changes in east and central Alaska linked to those in west Alaska?) and to reconcile meteorological and vegetation history changes. These are *open questions*. The axioms are insufficient to specify precise postulates that can be investigated and yet the questions are of great interest and central to the development of a comprehensive theory.

OPEN QUESTION

> At the level of a theory, an *open question* defines an objective that cannot yet be defined in terms of specific axioms and postulates.

What does this theory lack? Some scientists would be concerned that no single discrete test can be applied. Some elements of the theory might weaken it without causing it to be rejected. For example, more refined GCMs might predict patterns of climate change in central Alaska that were not matched by the expected patterns of species movement. Moreover, if this were to be resolved by the proposition that some species ecology was different in the past from what it is now, the theory would be weakened because there is no way of testing that. It would be a regressive development, and the theory would have lost some of its power – even if this were the correct explanation, there is no way to test it.

When the theory was simple, it was possible to construct a clear test. But the theory – now typical of more complex ecological theories, particularly those that predict community structure or properties such as diversity and stability – has no single discrete test. Clearly, in the case of an ecological history, a retroactive test is not possible. Predicting change for future centuries and millennia does not provide a test – even though it may be an intriguing exercise. As is discussed in Section II, tests of large-scale ecological theories have particular difficulties.

5.3 Practical application of a theory: Hybridization in fish species

When faced with a practical problem not previously researched, we naturally try to use what we consider to be the closest available theory. But just how precisely might that theory describe the new system? This will depend upon the properties of the theory, not only the logic specified by its axioms but also the questions contained in the theory. We may also gain insight by assessing the stage of development of the theory we intend to use.

In the following example, previous research on mate choice in sockeye salmon had defined some behavioral processes that maintained two forms of the species, the larger anadromous[1] variety and smaller resident kokanee

---

[1] Anadromous: a fish species that is hatched and has its early development in fresh water, migrates to the sea, and then returns to breed in the same fresh water. Some species have anadromous individuals, and individuals who remain resident in fresh water and go through all stages without visiting the ocean. In sockeye salmon, the non-anadromous form is smaller and known as kokanee.

variety. The hope was that this theory, which described behavioral control of mate selection, could be applied to other salmonid species, and particularly to analyzing the practical problem of hybridization between cutthroat and rainbow trout.

Box 5.2 is a research proposal made by Denise Hawkins at the start of her doctoral research. It contrasts with that presented by Steel, which started with the emphasis on field observations; Hawkins' starts with the emphasis on theory definition. My comments are boxed insets, headed as Scientific Method Notes. The introduction of the theory about mate selection follows the specification of the main problem. This order of presentation is typical in this type of research problem – but it can lead to a lack of appreciation of the issues involved in applying a theory to a new problem.

Box 5.2. Analysis of hybridization of cutthroat trout (*Oncorhynchus clarki*): behavioral mechanisms and hybrid survival by D. Hawkins, School of Fisheries, University of Washington

Introduction

Cutthroat trout (*Oncorhynchus clarki*) have the most widespread distribution pattern of all the present forms of western North American trout. This widespread distribution, in addition to life history characteristics of cutthroat trout such as homing to their natal stream to spawn (Campton and Utter 1987) that leads to many separated populations, has produced a highly variable species; the variability exhibited within a single drainage is sometimes as high as that exhibited between separate basins (Behnke 1972).

Steelhead trout coexist with the cutthroat throughout much of their native distribution. This is the only example of overlapping geographical ranges in western North American trout. Although these two species are able to live in sympatry without losing their species integrity, cutthroat trout are highly susceptible to the introductions of non-native trouts (Behnke 1972). For example, hybrid swarms are frequently found when rainbow trout (*Oncorhynchus mykiss*) are introduced into interior waters where cutthroat trout are the native inhabitants (Campton 1981, Forbes and Allendorf 1991).

Coastal cutthroat trout (*O. clarki clarki*) have historically been a valuable sport fish in the Puget Sound area, and there has been interest expressed in preserving this fishery (Johnston and Mercer 1976). However, past attempts to enhance this fishery have failed for a variety of reasons (Johnston and Mercer 1976), and the number of cutthroat trout is decreasing due to the species' sensitivity to habitat deterioration and competition from introduced species (Trotter 1987). At the same time, rainbow trout have been extensively introduced into many areas through stocking programs. This introduction of hatchery-origin rainbows has resulted in hybridization with the cutthroat, even in streams where the cutthroat and steelhead were living in sympatry without hybridization. This hybridization has resulted in the reduction of genetically pure strains of cutthroat (Behnke 1972); in a species as diverse as the cutthroat, the loss of a particular stock represents a significant loss of diversity.

Hybridization between coastal cutthroat and steelhead was not thought to be a common occurrence and had not been formally documented (Campton and Utter 1985). However, Campton (1981) and Campton and Utter (1985) found a significant number of hybrids in 2 out of 23 streams studied, suggesting that hybridization is not an uncommon event. However, since the fry could not be typed, it was not clear whether the hybrids were produced by cutthroat–native steelhead crosses or by cutthroat crosses with introduced rainbows or steelheads (Campton 1981). It would seem that the hybrids were cutthroat–hatchery crosses, since the introduced fish were the perturbing factor in a stable system.

The ability of cutthroat and steelhead to coexist and yet maintain their species integrity has generally been attributed to differences in spawning time and habitat preferences (Campton and Utter 1985). However, cutthroat and steelhead redds have been seen interspersed in some streams, and both species will spawn on the same day if adults of both species are present (Campton 1981). These findings would indicate that factors other than simply joint occurrence in the stream, possibly behavioral or density dependence, contribute to the presence or absence of hybridization between cutthroat and hatchery rainbow trout.

Campton and Utter (1985) also found an absence of hybrids over one year of age. This may indicate possible postzygotic isolating mechanisms, which could be manifest in many ways, though absence of one-year hybrids could be due to migration. Although the complete interfertility of these two species has been assumed, it has never been fully tested (Campton and Utter 1985). Therefore, it is not known whether the fitness of the hybrids is greater than, or equal to, or less than that of the pure lines, or whether the hybrids can backcross with the cutthroat parental stock.

Objectives

To determine whether premating spawning behavior mechanisms are responsible for maintaining the species integrities of coexisting cutthroat and steelhead populations.

In order to determine whether postzygotic mechanisms are responsible for maintaining genetic differences by selection against hybrid progeny, the second section of the study will examine the fitness of the hybrids in comparison to the cutthroat parental stock. Swimming performance, feeding aggression, egg-to-fry survival, smoltification, egg size, and alevin development will be used as indexes of fitness.

> *Scientific method notes.* These axioms and postulates about fitness of hybrids were part of the complete research proposal but are not presented here.

Over-arching axiom:	Populations interact when part of the same ecosystem.
Species specific over-arching axiom:	Cutthroat trout, rainbow trout, and steelhead trout interact through interbreeding under certain circumstances.
Axiom 1:	Introgression is deleterious to a species' survival.

C_1^{res}:   *Introgression* is the introduction of foreign genes into an adapted gene pool or loss of genetic integrity.

C_2^{res}:   A *species* is defined as an interbreeding group of animals that produce viable offspring.

C_3^{res}:   *Survival* is defined as the continued existence of the species as an evolutionarily distinct unit.

Axiom 2:   Loss of genetic diversity is deleterious to a species' survival.

C_4^{res}:   *Genetic diversity* is the available gene pool of a species.

Axiom 3:   Hybridization between two species leads to the loss of genetic integrity of the species and, therefore, to the loss of genetic diversity of the species.

C_5^{res}:   *Hybridization* is the interbreeding of two species.

Axiom 4:   Spawning behaviors, including mate preferences and nest site selection, are heritable traits.

C_6^{res}:   A *heritable trait* is one that is passed on to the offspring

Postulate 4.1a: Differences in mate preference exist in the spawning behavior of cutthroat trout and steelhead trout.

Postulate 4.1b: Differences in mate preference exist in nest site selection behavior of cutthroat trout and steelhead trout.

Postulate 4.1c: These differences are responsible for maintaining

genetic isolation under circumstances of sympatric existence.

C_7^{res}:   *Genetic isolation* is the lack of interbreeding between species.

C_8^{res}:   *Sympatric existence* is when two species coexist without geographic isolation.

C_9^{imag}:   *Mate preference.*

C_{10}^{imag}:   *Nest site selection.*

The crucial aspect of this postulate is:

C_9 (cutthroat) $\neq C_9$ (steelhead trout)
and C_{10} (cutthroat) $\neq C_{10}$ (steelhead trout)

Postulate 4.2a: Differences exist in mate selection between wild
and hatchery reared steelhead trout.

Postulate 4.2b: Differences exist in nest site selection between
wild and hatchery reared steelhead trout.

Postulate 4.2c: Differences exist in nest site selection between
wild and hatchery reared steelhead trout.

C_{11}^{imag}:   Premating isolating mechanisms are behaviors that maintain genetic isolation
by precluding mating.

Axiom 5:   Spawning behavior can be changed by density of fish on
the spawning grounds.

Postulate 5.1:   Density of fish on the spawning grounds affects the
premating isolating mechanisms.

Analysis of Postulate 4.1a: Differences in mate preference exist in the spawning behavior of cutthroat trout and steelhead trout

Males

Previous work on male mate choice in fishes has indicated that
males prefer the largest available female (Hanson and Smith 1967;
Sargent *et al.* 1986). However, there is now evidence that male
choice is dependent on relative size and form of females (Schroder
1981, Foote 1988, Foote and Larkin 1988). For example in anad-
romous/non-anadromous sockeye the smaller non-anadromous
kokanee prefer the smaller kokanee females to the larger sockeye
females (Foote 1989). Cutthroat and steelhead trout spawning
behavior is typical of other stream-spawning salmonids (Pauley *et al.*

1989); therefore, the experimental design outlined by Foote and Larkin (1988) is appropriate for the study of assortative mating between cutthroat trout and steelhead trout.

Scientific method notes. This is a crucial step in the investigation. The two concepts C_9 (*mate preference*) and C_{10} (*nest site selection*) are concepts by research from sockeye, where they have been experimented with, but become concepts by imagination, C_9^{imag}, C_{10}^{imag}, when applied to cutthroat and steelhead trout. The assumption is made that there will be processes of mate preference and nest site selection that will act in the breeding process, and these will differ for cutthroat and steelhead.

In order to examine cutthroat trout and steelhead trout male mate preference for steelhead and cutthroat females, the following experimental competitive situations will be established: (A) no competitor present; and (B) conspecific competitor paired with like female, i.e., cutthroat trout male paired with cutthroat trout female along with lone steelhead trout female when testing preference of added male cutthroat trout. This is a first attempt to examine Postulate 5.1 by extending the experiments used to examine Postulate 4.1a.

Scientific method notes. What should the complete set of tests be? As written, the two conditions to be examined – (A) mating with no competition present and (B) competition effect of conspecific male – will both be conducted with one female cutthroat trout and one female steelhead trout.

To minimize affects of female ripeness, only territorial females will be used. Three females of both species will be placed in an experimental arena on the night before male testing and allowed to establish territories overnight. Just before males are added, one female of each species will be selected based on defense of a nest site and the other females will be removed. The male will then be added and courtship displays and time spent oriented to the female, within 1 m distance, will be recorded. An experiment will consist of two five-minute observation periods separated by a seven-minute rest.

Scientific method notes. These experimental conditions are based upon work with similar species. This seems sensible. But there can be no guarantee that these conditions will be appropriate. When you apply a given theory in a new situation, the limitations used in establishing it are implicit – in this case, a particular experimental and measurement technique. What would it be best to do?

The conditions used by Foote were almost certainly based upon his understanding and observation of the natural history of sockeye and kokanee. It would seem appropriate to obtain natural history observations of cutthroat and steelhead before making experiments.

If male mate preference in cutthroat trout and steelhead trout follows the patterns exhibited by sockeye salmon, it would be expected that cutthroat trout and steelhead trout males would prefer conspecific females in the absence of competition regardless of size of the respective females. However, in the presence of competition, it would be expected that male cutthroat trout would show a greater tendency to court steelhead trout females than steelhead trout males would show to court cutthroat trout females owing to relative size of the two species.

Females

It has been demonstrated (Schroder 1981, Foote and Larkin 1988, Foote 1989) that female salmonids indicate mate preference by the delaying of spawning when courted by unacceptable males. If mate preference in female cutthroat trout for cutthroat trout males is apparent it would indicate one possible pre-mating isolating mechanism. Foote (1989) examined kokanee females and found that when attended by less desirable males the kokanee deposited fewer eggs as measured by a smaller weight loss. This technique should also be applicable to cutthroat trout females as the spawning behavior of cutthroat trout has been shown to be similar to that of other salmonids (Pauley *et al.* 1989).

In order to examine cutthroat trout female mate preference the following experiment will be carried out. Ten cutthroat trout females will be tagged, weighed and placed in each of two experimental arenas and allowed to establish nest sites overnight. In the morning, ten steelhead trout males will be added to one arena and ten cutthroat trout males will be added to the other arena. After 18 hours the females will be reweighed and all fish released. It would be expected that the females attended by the steelhead trout males will lose less weight than those attended by cutthroat trout males.

Analysis of Postulate 4.1b Differences in mate preference exist in the spawning behavior of cutthroat trout and steelhead trout

Foote (1990) demonstrated that large and small kokanee females established nest sites in the same general region in the absence of competition indicating similar nest site requirements. However, cutthroat trout and steelhead trout females have been shown to prefer slightly different types of nest site. Cutthroat trout generally spawn in smaller tributary streams whereas steelhead trout tend to choose larger main-stem rivers (Trotter 1987). This difference in spawning site selection would suggest that in areas where the two species spawn together the cutthroat trout would nest in shallower water than did the steelhead trout, which would result in isolating the two species.

Preferred spawning sites will be determined as described by Foote (1990). Five pairs of cutthroat trout and steelhead trout will be separately allowed to establish nest sites in an experimental arena over a ten-hour period. The location of each pair of the first species will be mapped after the ten hours and then replaced with the five pairs of the second species. This will be repeated three times.

If the above speculated differences are found in cutthroat trout and steelhead trout mate preference and nest site selection, this would indicate the involvement of these two behaviors in the isolation of sympatric populations of the two species.

Scientific method notes. For nest site selection experiments, it is already known that cutthroat trout and steelhead trout females have different preferences. Consequently, differences in the environment favored for nest site and differences in mating behavior may both be part of the isolating process. This was not the situation in the kokanee–sockeye experiments.

Although an attempt is being made to establish experiments that test hypotheses, the underlying purpose really is to define two concepts, *mate preference* and *nest site selection*. It is typical of much research to redefine concepts, particularly, as in this case, where concepts are borrowed from similar but not identical situations. For this work, it is a strength that a number of different types of experiments will be attempted. The implication, of course, is that no single experiment would be adequate to satisfy Postulate 4.1, and breaking it down into a series of sub-postulates, while helpful to planning, would still leave the need to integrate all the results.

References

Behnke, R. J. (1972). The systematics of salmonid fishes of recently glaciated lakes. *Journal of the Fisheries Research Board of Canada*, **29**, 639–671.

Campton, D. E. 1981. Genetic structure of sea-run cutthroat trout (*Salmo clarki clarki*) populations in the Puget Sound area. Unpublished M.S. thesis, University of Washington.

Campton, D. E. and Utter, F. M. (1985). Natural hybridization between steelhead trout (*Salmo gairdneri*) and coastal cutthroat trout (*Salmo clarki clarki*) in two Puget Sound streams. *Canadian Journal of Fish and Aquatic Science*, **42**, 110–119.

Campton, D. E. and Utter, F. M. (1987). Genetic structure of anadromous cutthroat trout (*Salmo clarki clarki*) populations in the Puget Sound area: evidence for restricted gene flow. *Canadian Journal of Fish and Aquatic Science*, **44**, 573–582.

Forbes, S. H., and Allendorf, F. W. (1991). Associations between mitochondrial and nuclear genotypes in cutthroat trout hybrid swarms. *Evolution*, **45**, 1332–1349.

Foote, C. J. (1988). Male mate choice dependent on male size in salmon. *Behavior*, **106**, 63–80.

Foote, C. J. (1989). Female mate preference in Pacific salmon. *Animal Behavior*, **38**, 721–723.

Foote, C. J. (1990). An experimental comparison of male and female spawning territoriality in a pacific salmon. *Behavior*, **115**, 283–314.

Foote, C. J. and Larkin, P. A. (1988). The role of male choice in the assortative mating of anadromous and non-anadromous sockeye salmon (*Oncorhynchus nerka*). *Behavior*, **106**, 43–62.

Hanson, A. J., and Smith, H. D. (1967). Mate selection in a population of sockeye salmon (*Oncorhynchus nerka*) of mixed age groups. *Journal of the Fisheries Research Board of Canada*, **24**, 1955–1977.

Johnston, J. M. and Mercer, S. P. (1976). *Sea-run Cutthroat in Saltwater Pens: Broodstock Development and Extended Juvenile Rearing (with a Life History Compendium)*. Fisheries Research Report AFS-57, Washington State Game Department.

Pauley, G., Oshima, K., Bowers, K. and Thomas, G. (1989). *Sea-run Cutthroat Trout*. U.S. Fish and Wildlife Service Biological Report, **82**(11.86), U.S. Army Corps of Engineers TR EL-82-4.

Sargent, R. C., Gross, M. R. and van den Berghe, E. P. (1986). Male mate choice in fishes. *Animal Behavior*, **34**, 111–122.

Schroder, S. L. (1981). The role of sexual selection in determining the overall mating patterns and mate choice in chum salmon. Unpublished Ph.D. thesis, University of Washington.

Trotter, P. C. (1987). *Cutthroat Native Trout of the West*. Colorado: Associated University Press.

Scientific method notes. On starting her field studies, Hawkins found she needed to establish some basic information about the cutthroat population. She needed an exploratory investigation before setting the conditions of her experiments. Both anadromous and resident cutthroat existed, and, although anadromous fish are known to be larger, the size differences were not sufficient for her to distinguish the two types. It was important to examine these differences. In some species, there are set differences in the proportion of anadromous type – if this were so in cutthroat, it could be important to understanding and defining a hybridization system. There was also some suggestion that the cutthroat may be resident in headwaters, whereas the anadromous fish may be resident downstream. So fish may be isolated physically by the structure of the river.

Practical problems were also encountered. Cutthroat return in smaller numbers than sockeye, the species on which this work had been based, so there was less experimental material. Furthermore, Washington State regulations limit the transfer of fish between rivers, so transporting fish for experimentation was not possible. This may result in a more intensive study at one experimental river – Big Beef Creek, owned by the University of Washington.

5.4 Development, properties, and use of ecological theories

When starting research, it is essential to analyze the theoretical basis of the subject critically by a detailed analysis of the literature. Axioms and their component concepts can be defined, and the weak or uncertain points in the theory identified and related to the question under investigation. But, to decide on the best strategy for investigation, one must go one step further. The stage of development and properties of the theory must be analyzed. Is it a mature theory, or one in the early stages of development? Does the theory contain open questions? Has the theory been borrowed from a related, but different, application? Is the theory potentially powerful in what it may explain – but not based on many examples? This analysis is rarely made, perhaps because of a preoccupation with understanding of the content of a theory. Yet, as the examples in this chapter illustrate, the structural characteristics of a theory, as well as its content, may give the researcher important indications how to proceed.

For example, the theory that Hawkins intended to use to study isolation through breeding systems in trout species was based upon detailed study of one species – sockeye. However mature that theory may have been for sockeye itself, it was not mature for all trout species. Hawkins found she was unable simply to repeat the experiments conducted on sockeye with other species.

Theories are attempts to provide general descriptions of ecological processes. Most usually they are built upon a number of different types of investigation, particularly as they develop over time. Conclusions from experiments, observations, and different types of measurement may all be integrated in a theory and these different pieces of information typically are linked by sets of assumptions that may not be directly testable. As a theory becomes more complex, it is less likely to be overthrown by a single experiment or observation.

Theories may develop in a *progressive* or *regressive* way.

PROGRESSIVE THEORY DEVELOPMENT

The development of a theory is *progressive* if its generality increases. This may manifest as:

(1) increases in the number of circumstances in which the theory holds,
(2) increased evidence that what is measured represents concepts from research; and
(3) evidence that the theoretical constructions used are sufficiently detailed that they do explain what is true in common and what may be different in particular examples

The theory of uneven spatial and temporal invasion of species into north central Alaska is a progressive development from the theory it replaced of uniform northward spread across the Northern Hemisphere. But notice that the theory for central Alaska is a regional variant of the as yet incomplete theory of Northern Hemisphere climate change for the period 6 ka BP to 3 ka BP.

REGRESSIVE THEORY DEVELOPMENT

Development of a theory may be *regressive* if the theory becomes unable to incorporate the conclusions from some investigations without making *ad hoc* adjustments that invoke special exceptions for particular circumstances.

If, on investigation of further lake sediment cores, the theory of plant invasion in central Alaska could be modified to account for some but not all of the new data, then some data would remain unexplained. If the theory is excused from having to explain this data and, particularly, if special circumstances outside of those typically used in this type of work, i.e., *ad hoc*, are invoked, the theory has become regressive. *Ad hoc* explanations might be:

(1) The lake of a newly cored site had had a very different type of hydrological development, which disrupted sediment formation – without any additional supporting evidence for this apart from the core itself, e.g., that the core was taken next to an upwelling spring on the bottom of the lake.
(2) There were unique problems with the measurements that altered the proportion of pollen grains found or the dating – again without independent ancillary information to suggest this, e.g., that at this site pollen of a particular species decomposed, or was eaten, or some such explanation.
(3) Unique events are added to the theory that apply only to that core site, e.g., that the site shows none of a particular species because of unique meteorological events there.

Theories are likely to go through a regressive stage, where *ad hoc* explanations are used because a few anomalous results may not be sufficient to overthrow a complex theory. Yet as the number of anomalous results increases, then a reason why the whole theory is wrong may become more apparent.

To use an ecological theory properly, its implicit questions must be discovered. What challenges its right to exist? What challenges its effectiveness and the extensiveness of its generalities? One thing is certain. Something always triggers these challenges; there is no perfect theory.

5.5 Further reading

Philosophers of science have extensively researched the process of theory development, and what governs its effectiveness. Three of their own theories about that, those due to Kuhn (1970), Lakatos (1970) and Shapere (1977) are discussed in Chapter 11. However, there are many types of theory, some focusing on the scientific details, e.g., Thagard (1992), and some concentrating more on the effects of science as a social process, e.g., Hull (1988).

Pickett *et al.* (1994) discuss the structure, content and origin of ecological theories and give a classification based on structure and purpose as well as content.

Further analysis of the vegetation history of south-west Alaska is discussed by Brubaker *et al.* (1999). Further information on the cutthroat trout hybridization problem is in Hawkins (1997) and Hawkins and Quinn (1996) and Hawkins and Foote (1998).

6 The art of measurement and experiment

Summary

Making measurements and conducting experiments both require that the investigator make choices about what and how to measure and about what treatments to apply and what controls and replication to use. The development and use of measurements, and the design of ecological experiments, are not routine tasks with single correct solutions.

Four principles of measurement for new concepts are presented:

(1) There must be a postulate under investigation so that the purpose of measurement is clear and precision required can be defined.
(2) Each new measurement must be specified in a data statement so that the accuracy required can be assessed.
(3) More than one measurement of a new concept should be investigated because different measurements may inform in different ways.
(4) The variability of a measurement should be investigated explicitly.

Two principal types of experimentation are distinguished; response-level experimentation and analytical experimentation. Response-level experiments are usually designed to investigate the magnitude of an effect when the type of response is already known. Most of the statistical techniques of experimental design and analysis have been developed for response-level experiments.

Analytical experimentation – the focus in this chapter – investigates how ecological systems function. To make an experiment, choices have to be made to investigate particular postulates and develop specific measurements and design appropriate control procedures, treatment controls and replication.

6.1 Introduction

At some point research moves from observation and general conjectures to detailed analyses. Then measurements stimulate the development of postulates and determine whether new postulates will be accepted. Scientists devote large amounts of time and money to developing measurement systems. Indeed, whole new fields of discovery have been opened up by development of a measurement technique. For example, satellite imaging for problems in landscape ecology (Haines-Young *et al.* 1993), radiotelemetry for animal ecology (Amlaner and Macdonald 1980), and radioactive isotopes for studying movement and patterns of distribution of elements (Schutz and Ward Whicker 1982) were each measurement techniques that were researched, have become standardized, and now advance ecological knowledge.

For many scientists, the experiment is their principal research method. Some consider it the essential method that sets science apart as the process that develops objective knowledge. For them the experiment is a neutral arbiter: treatments are set and results observed and reported in an unbiased way – whether for or against the postulate that generated the experiment.

If measurements and experimentation occupy such important positions in science, why should this chapter be titled "The art of measurement and experiment"? What can be an *art* about them? The answer is that all measurements and experiments require choices about what and how to measure, what treatments to include in an experiment, or how to determine the balance between numbers of replicates and treatments. Making measurements can also require a skill that develops with practice. The process of making these choices and developing skills cannot be standardized so that each scientist comes to the identical decision or has the same technical skill. Granted, some branches of ecological science have standardized certain measurement or experimental techniques that they use as the basis for further advance. Such standards may be the product of long experience or custom. Nevertheless, whatever the reason and current practice, at some point someone or some group made choices about what was considered most important, and those choices continue to be followed.

Measurement and experimental science have principles and techniques of methodology that I shall describe, but neither measurement nor experiments are completely neutral arbiters of a postulate. When devising experiments to investigate possible effects of behavioral processes in the breeding of trout species, Denise Hawkins was faced with many choices (Chapter 5). What should the combination of exposures be between males and females of different species? What is a reasonable length of time to observe whether mating will take place? And so on. Where possible, experimental conditions should be based on the species natural history, but some conditions will certainly run counter to it. For example, to ensure encounters between the required fish, and to observe the fish as this happens, the creatures must be enclosed in some way. What type of enclosure should be chosen? Will enclosure focus the animals on mating more or less than would otherwise be expected? Can the effects of enclosure themselves be evaluated? Hawkins' answers to these questions cannot be guided entirely by the species natural history – they may come from previous experiments or simply be guesses that may have to be investigated subsequently.

The choices you make in developing a measurement or designing an experiment set the stage for what you can find out. Herein research ecologists face a particular problem. Not only can the variability of ecological systems pose severe problems for developing a measurement or standardized experimental procedure, it may be precisely this variability that we need to investigate. Ecological science can be advanced by making different measures of what is initially thought of as a single quantity, quality, or process but what is eventually realized to be a complex concept that needs comprehensive definition.

This chapter presents and illustrates principles of measurement and techniques of experiment for ecological research, showing where in the process, and how, the individual scientists' subjectivity and creativity are important. It also illustrates that there can be a natural succession in methodology from developing increasing efficiency, accuracy and precision of measurement, through experimenting with parts or simplified representations of ecological systems, to attempting to understand increasingly complex systems.

6.2   Principles of measurement for new concepts

Three properties of measurements need to be considered: effectiveness, accuracy, and precision (Fig. 6.1). Effectiveness is how completely a concept by measurement represents a concept from research or a concept by imagination. For example, Ashley Steel had to choose a measurement that would represent animal *use* of piles of coarse woody debris (CWD) (Chapter 4). She decided to trap animals in piles – this could be replicated (which influence accuracy and precision) – and it was effective in measuring what was needed and nothing else. Other measurements, e.g., counts of small mammals' nests or dens within piles, would be an incomplete measure of *use* – the target measurement would not cover the concept. Trapping throughout the area where piles were, not focusing specifically on measurements within the piles, would be too large a target (Fig. 6.1).

Of course trapping in the piles allowed her only to record that animals were present, not what the animals did, e.g., nest, feed, or den. So, while the measurement was effective for her postulate it had limited effectiveness in answering the motivating ecological question, i.e., the causal relationship between piles and the presence of small mammals. However the postulate – *small mammals use CWD piles* – was necessary at that stage of the investigation because, at its outset, some scientists did not think small mammals would use the piles for any purpose. Furthermore, with no understanding of where or when small mammals occurred, devising a sampling procedure for nests and dens would be difficult and the postulate – *small mammals use CWD piles* – could be falsely rejected if none were found.

For a particular measurement, accuracy is how well it represents a concept

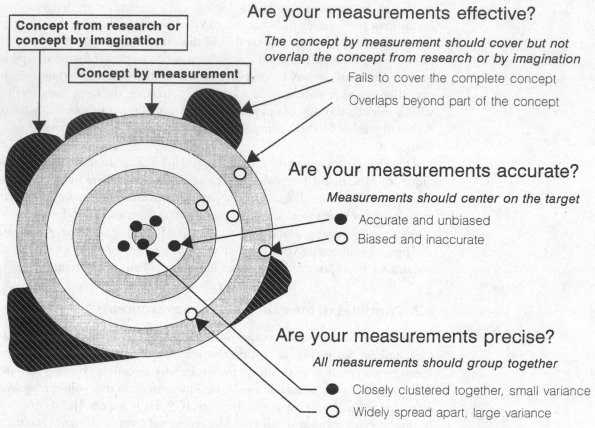

Are your measurements effective?

The concept by measurement should cover but not overlap the concept from research or by imagination

Fails to cover the complete concept

Overlaps beyond part of the concept

Concept from research or concept by imagination

Concept by measurement

Are your measurements accurate?

Measurements should center on the target

● Accurate and unbiased

○ Biased and inaccurate

Are your measurements precise?

All measurements should group together

● Closely clustered together, small variance

○ Widely spread apart, large variance

Fig. 6.1. The three qualities of a measurement are effectiveness, accuracy, and precision.

in an unbiased way. For example, individual trap types do not capture species in the same way and Steel used different types of trapping to capture different species. Together these different trap types gave a more accurate measure of use than a single trap type. A single trap type that does not catch all small mammals would give a biased measure of use. Accuracy could have been investigated further by exploring the comparative effectiveness of the trap types in parallel investigations and so document their relative bias. You must judge what level of accuracy is necessary to test a given postulate. Yet forming that judgement can be particularly difficult, since at any stage there is usually some further investigation that could be made to increase accuracy, or at least to define it.

Constancy or repeatability of values or quantities under the same conditions is used to judge precision. In the case of trapping small mammals, precision is controlled by sampling procedure and intensity. The measurement required is the number of small mammals using piles of a particular type. So, in this case, the number of traps used, and how they are distributed between piles of the same type, will influence the precision of the estimate.

Too few traps are likely to lead to variable estimates. Precision can be difficult to obtain, e.g., different species of small mammal, may become either trap shy, or alternatively, repeatedly attracted to traps for the bait. Both behaviors may cause variation in numbers caught over time.

Four principles of measurement for new concepts should be followed. Certainly concepts by imagination are new concepts but some recently promoted concepts from research may also be considered as new as far as their measurement is concerned.

PRINCIPLES OF MEASUREMENT FOR NEW CONCEPTS

(1) *There must be a postulate under investigation.* Measurements do not exist in isolation from the scientific research they serve. Extensive research can be devoted to developing and testing a new measurement – but the efficiency, accuracy and precision required are determined by the scientific objectives defined by the postulate.

(2) *Each new measurement must be specified in a data statement.* It is essential to establish the relationships among the data that are required to test a postulate and how measurements are to be used to collect the data, and what that use can achieve. This may require exploratory analysis.

(3) *More than one measurement of a new concept should be investigated.* The philosophy of multiple measurement recognizes that a concept by measurement is not equivalent to a concept by imagination or from research. Instances where a single measurement is sufficiently well understood that multiple measurements need not be considered have usually been established by consistent use. In research using new concepts it can be important to use different measures.

(4) *The accuracy and precision of a measurement should be investigated explicitly.* A measurement should not be assumed to have both the required accuracy and precision. Although some measurements are very precise some may be more exploratory and generally informative.

The list of four axioms of sampling and measurement for stratigraphic pollen analysis of lake sediment cores (recall Box 5.1) exemplifies measurement standardization. Such standardization allows results to be compared and used together in an interpretation of results from different investigations. For example, this is particularly important where pollen data from a number of cores taken by different investigators are used to construct past climate changes over a large region. But standardization does not ensure that the measurement procedure will not change. As a science advances, it is usual for both the measurement technique and its associated sampling system to be overhauled and accuracy increased. For example, with the advent of instruments that can measure the $^{14}C/^{12}C$ ratio on smaller amounts of material, lake sediments can be sampled differently. Previously, samples for radiocarbon dating have been taken from a mixture of all the material in a stratum.

However, older organic material, such as peat from the land surrounding the lake, may have washed in and been deposited on the lake bottom, giving a stratum an older apparent age than when the pollen in it was actually deposited. The ability to analyze small amounts of material means that identifiable individual plant parts (macrofossils) or collections of pollen grains, unlikely to have been a part of a previous peat formation, can be collected together and provide sufficient material for dating. In this case, refined measurement and sampling technique improves the effectiveness (the correct thing is measured) and accuracy (the bias due to washed-in deposits and the variability of that between lakes are removed).

Sometimes we may be lulled into inaccuracy and imprecision in scientific measurement because of everyday associations with concepts. For example, the common language concept of *age* is associated with time, which we measure with a clock or calendar but the concept *age of CWD* pile is associated with its function as a habitat. Time measured by calendar or clock may be an imprecise measure of that function. The important characteristic may be whether the pile still supports sufficient fungal growth to provide food for small mammals or insects, or whether the structure of the pile has remained sufficiently intact to act as a shelter. These properties may change in irregular ways over calendar time. Other times, we may forget that the jump from natural history observations to detailed scientific understanding is more than it seems. For example, watching different species of fish spawn in a river does not reveal everything that may influence spawning.

6.3 Experimental analysis of ecological systems

The basis of much successful science is in explaining the difference between contrasting situations. Sometimes natural contrasts occur, and examples where these have been used in ecology will be discussed in Chapter 11. In experiments a deliberate attempt is made to establish differences, through applying specific treatments, and to observe their effects relative to a situation with no treatment.

EXPERIMENT

An experiment involves deliberate action(s) by an investigator, *treatment(s)*, imposed upon an ecological system in such a way that the system *response* to the treatment can be observed and/or measured by contrasting it with a condition of no treatment, the *control(s)*. Controls must be designed so that changes not due to the intended effects of the treatment may be accounted for. *Treatment(s)* and *control(s)* may be observed at the same time on different samples (a *synchronic contrast*) or the treatment may be applied at a particular time to create a before (control) and after (treatment) (a *diachronic contrast*).

Defining experiments as deliberate actions sets them apart from observations, whether of the effects of a naturally occurring change, e.g., species removal due to a selective epidemic disease, or of a management action, e.g., removal of a fish predator from a fishery. The essential feature of experiments is that treatment(s) are designed with some purpose that is related directly to current scientific questions and by replication and choice of controls, the type and magnitude of response can be calculated. Experiments are not automatically of more value than observations of naturally occurring or chance differences. There can be experiments that investigate uninteresting scientific questions and there can be experiments where the artificial nature of the treatments gives the experiment little interpretive value.

The challenge lies in constructing treatments and developing measurements that will inform about how an ecological system functions.

SYSTEM

A system has component parts that interact to achieve or maintain a particular property in a variable environment. Sometimes this property is called *homeostasis*, which maintains the constancy or directed purpose of a function, status, or process when there is an external disturbance, e.g., thermoregulation in animals. Sometimes the property is called *dynamic*, which emphasizes the change within the system; e.g., the way that animal behavior may change in different environments and maintain important life functions.

CLOSED SYSTEM

In a *closed system* all the components that influence the systems function can be defined.

An automobile with a full tank of fuel can be considered as a closed system. Its input is fuel rate to the engine and its outputs can be defined in terms of motion, fuel spent, and exhaust produced. We may be interested in analyzing the homeostasis or dynamics of a closed system. For example, the cruise control of an automobile is a mechanism designed to give a constant speed (homeostasis). The dynamics include the change in speed per unit change in fuel supplied. These concepts of homeostasis and dynamics could be used to explain why one automobile obtains greater mileage per unit of fuel consumed than another. This could be investigated experimentally by manipulating inputs and observing outputs under different conditions.

OPEN SYSTEM

An *open system* exchanges matter and/or energy across its boundaries and/or is influenced by external stimuli. The components of an open system, and/or their functioning, may change in response to an external event or stimulus.

These definitions of systems are perhaps less reflective of true properties of particular systems than about our supposed knowledge of them. The term *open system* is sometimes used in ecology to emphasize that many influences may act on ecological systems and that identifying the components and their properties can be a continuing task. However, an automobile can also be considered as an open system in the sense that wind speed and/or weather conditions such as snow or ice on the road may affect fuel consumption.

In the example of mating of trout species, the system has all the influences that may effect a successful hatch. The environment is variable both in river habitat, e.g., bottom substrate and water flow rates, and in presence or absence of fish of the same or different sex or species. We may consider this as an open system and include in the analysis the dynamic behavioral and physiological characteristics that define the response of individual fish to varying circumstances. The proposed experiment (Chapter 5) involves enclosing different combinations of fish and observing the dynamic response but to be effective an experiment requires controlling other factors or measuring their possible effects.

The essential strength of the experiment is that it contrasts situations with limited differences. We try to consider each experiment as a closed system. In the analysis of ecological systems it is important to investigate whether the simplifications or restrictions involved in constructing the experiment influence its results by constraining the system in a way that influences the response. How trout species mate may be influenced by many factors: sexual maturity of the individual, presence of a suitable substrate and water conditions, presence of other fish of the same and the opposite sex and of different species. Will the exposure of fish in an enclosure inform about this system?

EXPERIMENTAL ANALYSIS OF ECOLOGICAL SYSTEMS

In the *experimental analysis of ecological systems* treatments are designed to discover the functional relationships between organisms and/or between organisms and their environment. Because the nature of the response is under investigation attention must be paid to precisely how the treatment causes any effects. Generally, measurement of a number of response variables may be required to detect and interpret this.

In analytical experiments with ecological systems there should be interest not only in the difference between treatment and control but in the effects of restraining or "closing" the system so that the contrast can be observed. Measuring the response of more than one variable should be considered, e.g., the whole sequence of the trout mating procedure is of interest and should be observed. Analytical experiments should proceed with single or few treatments at a time, and so choices must be made about the sequence of possible treatments if a comprehensive experimental analysis is envisaged. Measurements of auxilliary variables that are not expected to differ between treatment and control should also be considered as a check on the functioning of the experimental system.

The large body of statistical theory and techniques called experimental design, and particularly the analysis of variance (ANOVA) (Fisher 1966, Zar 1996), is powerful but it can not be taken for granted that it can be used in an any experiment. Development of ANOVA was motivated by response-level experimentation.

RESPONSE-LEVEL EXPERIMENT

In *response-level experimentation*, treatments are designed to discover the magnitude of an already established or likely effect. Examples are experiments to find out how much the growth of a crop or an individual plant may respond to a fertilizer, how much certain animals may grow in response to a new diet, or what concentration of a pollutant kills 50 percent of test organisms.

The general methodological principles of response-level experimentation are:

(1) Few (usually one) measures of response, e.g., yield of a crop receiving a fertilizer, growth rate of an animal given a new diet, mortality of an organism exposed to a pollutant. Additional measures usually inform about the nature of the response, e.g., quality of the crop, fat content of the animal.

(2) Multiple treatments in the same experiment, e.g., different quantities of the same type of fertilizer, diets that differ in a few components, or different concentrations of the same pollutant.

(3) Extensive replication designed to ensure that quantitative differences between treatments could be estimated, e.g., difference in yield or growth rate or percentage killed.

Analysis of response level experiments using ANOVA makes assumptions about the statistical properties of the data and, particularly, homogeneity of variance. Usually these assumptions are sufficiently closely met in agricultural or health science trials that the techniques can be applied effectively. ANOVA uses a null hypothesis (see Chapter 8) that there is no difference between treatments. Generally, and in contrast to analytical experimentation, there is certainty in response-level experimentation about the type of result that will be obtained though not the magnitude of the effect.

Response-level experimentation is used in ecology and for practical problems in forestry and fisheries. In ecology, the difficulty is most often in achieving both an adequate range of treatments and their replication.

Frequently, the investigator must reduce the problem to one where an artificial environment can be constructed (see Semlitsch 1993). In forestry, response-level experiments are often carried out with seedlings. For example, Elliot and White (1994) investigated the response to N and P, and nutrient use efficiency, with *Pinus resinosa* seedlings using a split-plot design that incorporated effects of shading. Morris *et al.* (1993) used the basic structure of a response-level experiment to investigate the effects of weed competition on young plants; they combined the regular analysis with measures of water relations and light interception to obtain a process-based interpretation. Response-level experiments with whole ecological systems are more difficult, not only because their spatial heterogeneity makes adequate replication of treatments difficult to achieve, but also because their response can be complex rather than "more" or "less". Underwood (1997) gives examples of experiments and conditions where ANOVA has been used in ecological analysis.

There are similarities between response-level and analytical experimentation: treatments are designed, treated is compared with untreated through the use of controls in the experimental design, and treatments are replicated. But there can be substantial differences in measurements, controls and statistics (Table 6.1).

Analytical experiments examine how systems work where alternative postulates about the system must be examined and usually the development in measurement of responses and experimental procedures are closely related. There are seven steps to follow in designing an analytical experiment and these are described in the rest of the chapter by application to an example.

Component steps in designing analytical experimentation

(1) Making a conceptual analysis of the problem.
(2) Constructing multiple postulates.
(3) Choosing a postulate to study.
(4) Developing measurements.
(5) Defining the experimental conditions.
(6) Designing treatment application, replication and controls.
(7) Investigating ancillary processes to aid interpretation.

Although these steps are listed in a logical sequence, some may have to be used repeatedly in a planning process. For example, choosing a postulate may depend upon pinpointing or developing effective measurements. If measurement is not possible then you may have to choose another postulate. Similarly, if controls and replication cannot be constructed, you may need different experimental conditions or a different type of experiment, see 6.5.

Table 6.1. *Summary of similarities and differences between response-level experimentation and analytical experiments in ecology*

	Response-level experimentation	Analytical experiments in ecology
Typical purpose	Quantify an effect strongly suspected to occur, e.g., crop yield in response to fertilizer, organism response to differences in a diet or to pollution	Investigate how a system functions, e.g., response of a community to removal or addition of an organism, response of an organism to a structural or chemical change in its habitat
Design	Frequently there are a number of different levels of each treatment. The purpose of replication is to estimate quantitative differences between treatments against a background of homogeneous variation	Treatments are frequently simple contrasts, e.g., ambient CO_2 concentration against twice ambient CO_2 concentration in much climate change response research, enclosure against non-enclosure in top-down research (Chapter 11). The purpose of replication is to gain in certainty in the nature of the response. However, owing to technical difficulty in the treatment, there may be few replicates
Measurement	Usually a single measure of response suffices	Many components of system behavior should be measured so that characteristics of the response can be defined
Type and purpose of controls	A control is included, or a standardized treatment given, that acts as a baseline against which quantitative comparisons can be made	Controls are difficult to achieve. The effect of the treatment must be investigated in detail, and other effects it may have, apart from the postulated ones, must be analyzed rather than merely assumed to be presented in the control

continued

Table 6.1. (*cont.*)

	Response-level experimentation	Analytical experiments in ecology
Methods of analysis	Linear models, i.e., ANOVA or multivariate ANOVA (MANOVA), regression	Frequently response is complex. Each component of the response should be analyzed individually for each replicate and then groups of replicates compared. Where multiple measurements are made, and they are each considered a response, as opposed to interdependent indications of a system response, then MANOVA should be used (Scheiner 1993)

6.4 Planning an analytical experiment: An example – control of photosynthesis rate of *Pinus strobus* trees

This is an example where actual field measurements (Maier and Teskey 1992), made possible by a technical advance in measurement, have led to the need for analytical experimentation. That experiment has not yet been conducted – but the planning required for it, laid out here, illustrates the principles of measurement for new concepts and the design steps for analytical experiments in ecology. This example points up difficulties faced by ecologists, both in measurement and in experimental design, when they investigate ecological systems.

6.4.1 Results from an improved measurement technique

The most frequently used field measurement of photosynthesis rate involves:

(1) Enclosing foliage in a transparent cuvette so it can be illuminated.
(2) Passing a measured air stream through the cuvette.
(3) Using an infrared gas analyzer to measure the difference in CO_2 concentration of the air stream before and after it is passed through the cuvette.
(4) Calculating photosynthesis rate as the difference in CO_2 concentration times the flow rate.

These four steps summarize a technically complex process (Long and Hällgren 1985).

Early infrared gas analyzers required laboratory standard alternating current electricity. They had long response times and did not always give stable outputs. Cuvettes had to be in place around the foliage for numbers of minutes to ensure a precise measurement, and during this time the cuvette, and more importantly the air and leaf inside it, heated up if illumination was high. So the measurement was inaccurate. To counter this overheating, thermostatically controlled cooling systems were developed for cuvettes. Gas flow rates were also difficult to measure, particularly at small flows. Usually, large amounts of foliage were enclosed to give measurable CO_2 concentration differences at larger gas flows. There are similar technical considerations in measuring plant water status and atmospheric humidity that must be taken into account in the interpretation of photosynthesis measurements. These technical considerations played a large part in restricting the first comprehensive studies of tree photosynthesis to laboratory measurements with seedlings raised in a glasshouse or laboratory, or foliage cut from a tree and brought into the laboratory (e.g., Ludlow and Jarvis 1971). Scientists spent much of their time constructing and maintaining the equipment and developing the technical skill to use it.

Development of fast response infrared gas analyzers and electronic controls that improved their stability enabled development of portable photosynthesis measurement systems that could be used in the field. Small cuvettes are placed over foliage for a short period and do not require elaborate temperature control. These improvements increased measurement accuracy and precision and confidence in the results grew. This technical advance opened up new areas of study, particularly direct measurement of photosynthesis in the field rather than in the laboratory. Tree ecologists have been particularly interested in investigating the control of photosynthesis of mature trees.

6.4.2   Observing an anomaly

Maier and Teskey (1992) measured water status and gas exchange in the upper canopy of two 31 m tall *Pinus strobus*, eastern white pine, at Coweeta, North Carolina. Photosynthesis was measured *in situ* using a portable measurement system of the type described previously. Tree water status was measured by cutting a needle and inserting it into a pressure chamber with only the cut end exposed through a pressure-tight orifice. Pressure in the chamber was increased until water was forced out of the exposed end of the needle (Ritchie and Hinckley 1975). The pressure required, i.e., the xylem pressure potential measured in mega-pascals (MPa), increases as needles contain less moisture and the needle has decreasing water potential. Measurements made just before dawn (pre-dawn xylem pressure potential, PXPP) indicate the extent that the tree has recovered from the previous day's transpiration, and this recovery depends upon soil moisture. Gas exchange,

Pinus strobus

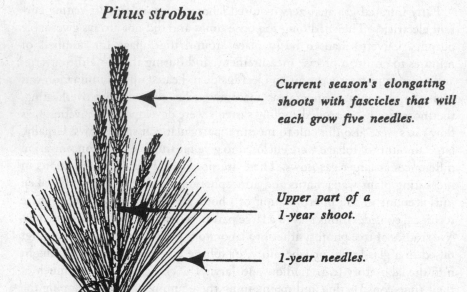

Current season's elongating
shoots with fascicles that will
each grow five needles.

Upper part of a
1-year shoot.

1-year needles.

Fig. 6.2. Morphology of the developing shoot of *Pinus strobus* in early summer.

both CO_2 uptake and water vapor loss (transpiration), were measured repeatedly throughout the day on the same one or two fascicles from each age class of foliage (Fig. 6.2).

Needles were arranged in a small cuvette to minimize self-shading, and CO_2 concentration in the cuvette was stable after 30 to 90 seconds. At the end of the day's measurements the fascicles were cut and their weight and area measured. The ability to use small amounts of tissue in the cuvette enabled needles to be taken from just a few branches used repeatedly throughout the year, reducing a possible source of variation. Air humidity was recorded as absolute humidity deficit (AHD); the weight of water needed to raise the air to 100 percent saturation, measured in grams per meter cubed ($g\,m^{-3}$).

Maier and Teskey intended to determine the parameters of the photosynthesis equations for different ages of foliage, i.e., to quantify how radiation, tree water potential, and air humidity produced specific levels of photosynthesis. This type of investigation can be time consuming because the environment of large trees cannot be controlled and many measurements must be made to ensure that the required range of variation in the physical environment is sampled. However, for two days, 21 June and 20 July 1986, the environmental conditions were remarkably similar, yet the rates of net photosynthesis were considerably higher on 21 June than on 20 July for both the new (current season) and one-year foliage (Fig. 6.3). These results are exceptional – it is unusual to have two days of almost identical weather

Fig. 6.3. Comparison of net photosynthesis rates in *Pinus strobus* foliage in the field conditions on 21 June (*open symbols*) and 20 July (*closed symbols*): (a) environmental conditions, temperature (*diamond symbols*), AHD, absolute humidity deficit (*square symbols*); PAR, photosynthetically active radiation (*triangular symbols*); and (b) rates of net photosynthesis (P_{net}) for current year (*square symbols*) and one-year foliage (*triangular symbols*) on 21 June (*open symbols*) and 20 July (*closed symbols*). (Redrawn from Maier and Teskey 1992, with permission.)

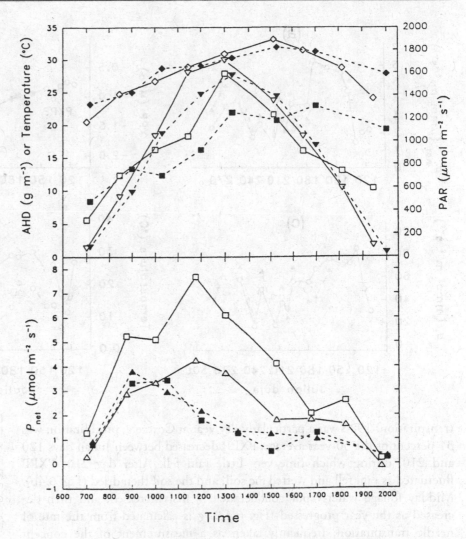

conditions, separated by a month, and where photosynthesis rates have been measured continuously over each day so that the differences between the days are clear.

When observing what seems to be an anomaly, you must check it against other results to see whether the anomaly has been produced by the system being investigated rather than some artifact of measurement. In this example, evidence from successive midday rates for one-year foliage indicates that this decline is part of a trend (Fig. 6.4a). Midday photosynthesis rate, P_{net}, the value at saturating light intensity, increased to a maximum around Julian day 170 and then fell quickly. (Julian days are numbered continuously from 1 January.) Why should there be such a difference in photosynthesis rates between 21 June (Julian day 172) and 20 July (Julian day 201)? Could the decline be related to changing foliage water potential or rate of

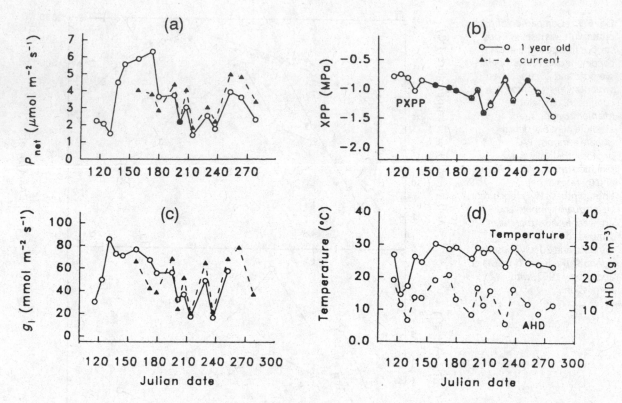

transpiration? 1986 was a particularly dry year at Coweeta; precipitation was 31 percent of the 50-year average. PXPP decreased between Julian days 120 and 210, during which time very little rain fell. After day 210, PXPP fluctuated as rain fell and wetted the soil, and the soil then dried (Fig. 6.4b). Midday foliage water-vapor conductance, g_l first increased, and then decreased as the year progressed (Fig. 6.4c). g_l is calculated from the rate of needle transpiration, frequently taken as a measurement of the concept stomatal opening, and measured by the increase in humidity of the air flowing through the cuvette. Although PXPP and g_l both changed throughout the year, the patterns of change differ from those of P_{net}.

6.4.3 Making a conceptual analysis of the problem

Anomalies capture attention because they present a contrast between what is observed and what is expected. They stimulate ideas about making experiments if only to try to repeat the apparent anomaly. But how should such an anomaly be investigated? You must start with what you know, or at least think you know. In this example the physiological processes of photosynthesis have been well researched, at least in the laboratory. Figure 6.5 shows the theory as a network of axioms, although not the numerical relationships, which are

Fig. 6.4. Seasonal trends in the canopy of 31-year *Pinus strobus*: (a) midday net photosynthesis, P_{net}; (b) xylem pressure potential, XPP; (c) midday foliage water-vapor conductance, g_l; and (d) air temperature and absolute humidity deficit, AHD. Each point for (a), (b), and (c) is the mean of 12–16 measurements from two trees made between 1030 and 1400 hours, except for pre-dawn xylem pressure potential (PXPP), which was measured 30 minutes before sunrise. (Redrawn from Maier and Teskey 1992, with permission.)

Fig. 6.5. Representation of the axioms of a theory as a network, marked by shaded connections between concepts, for the control of photosynthesis rate in *Pinus strobus* foliage.

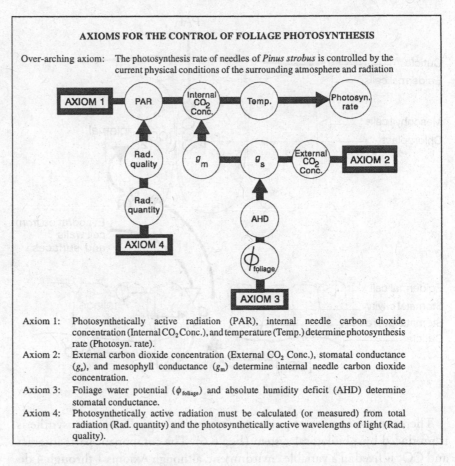

AXIOMS FOR THE CONTROL OF FOLIAGE PHOTOSYNTHESIS

Over-arching axiom: The photosynthesis rate of needles of *Pinus strobus* is controlled by the current physical conditions of the surrounding atmosphere and radiation

Axiom 1: Photosynthetically active radiation (PAR), internal needle carbon dioxide concentration (Internal CO_2 Conc.), and temperature (Temp.) determine photosynthesis rate (Photosyn. rate).

Axiom 2: External carbon dioxide concentration (External CO_2 Conc.), stomatal conductance (g_s), and mesophyll conductance (g_m) determine internal needle carbon dioxide concentration.

Axiom 3: Foliage water potential ($\phi_{foliage}$) and absolute humidity deficit (AHD) determine stomatal conductance.

Axiom 4: Photosynthetically active radiation must be calculated (or measured) from total radiation (Rad. quantity) and the photosynthetically active wavelengths of light (Rad. quality).

known to depend on species and plant condition. Figure 6.6 represents foliage photosynthesis as a system.

Axiom 1 states the over-arching axiom in more precise conceptual form, which is further defined by Axioms 2 through 4. Axiom 2 defines internal foliage CO_2 concentration of a leaf in terms of its external CO_2 concentration and two conductances that control its rate of movement to the site of photosynthesis, stomatal conductance, g_s, and mesophyll conductance, g_m.

Axiom 3 specifies that foliage water potential and atmospheric humidity deficit control stomatal conductance. The precise mechanism of stomatal control is an active subject of investigation and still debated in the scientific literature (e.g., Aphalo and Jarvis 1991, 1993). It is typical of a scientific investigation that you may have to use a theory, some parts of which are still uncertain.

Axiom 4 specifies that the quantity used in calculating radiation must account for both radiation quantity and quality, i.e., wavelength. There is a generally curvilinear response of photosynthesis to temperature, with a maximum at 15–25 °C depending on species.

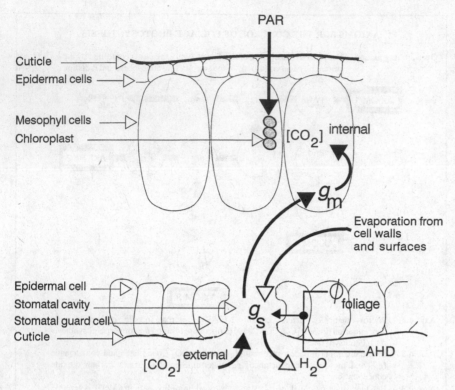

Fig. 6.6. Diagrammatic representation of the foliage photosynthesis system, described by Axioms 1 through 4 (Fig. 6.5), in foliage cells. The control of g_s by $\phi_{foliage}$ and absolute humidity deficit (AHD) is represented as taking place through the stomatal guard cells, although that is not described in the axioms. Solid-head arrows represent CO_2 movement to the site of photosynthesis and photosynthetically active radiation (PAR), H_2O movement out from the leaf by open-head arrows.

There are two important points to note at this stage. First, photosynthesis is produced by a biological system (Fig. 6.6). The component parts interact and CO_2 is fixed in a variable environment, although Axioms 1 through 4 do not define any homeostasis. Furthermore, one part of that system, the stomata, is itself a system regulated by plant water status and/or water vapor concentration in the atmosphere. Details of the axioms describing the control of stomata are not included in Fig. 6.5. It is typical that descriptions of ecological systems include features that are only partly, or incompletely, described, and an implicit assumption is made that they will operate within previously experienced limits and their fluctuation will not dominate system response. When such features are found to have an important effect ecologists sometimes refer to ecological systems as open. The problem of lack of complete specification of the theory is discussed in Chapter 11. In the example in this chapter it might be possible to consider a comprehensive investigation of photosynthesis and tree water status but practical limits of cost and investigator time may make that difficult to implement – so again choices have to be made.

Second, the axioms of the photosynthesis system have not been elucidated using *Pinus strobus*. That is, the theory has been extended to this species via these axioms. It is *assumed* that the axioms in Fig. 6.5 will hold for *P. strobus*, although the values of the component relationships in each axiom are not known.

6.4.4 Constructing multiple postulates

Imagination is required in constructing postulates that might explain the observed anomaly. This is not simply a random process, although ideas do sometimes seem to "come out of the blue" (see Chapter 14). Imagination can be directed particularly where, as in this case, an anomalous result is being investigated. Even though the anomaly is not understood – that is the object of the research – the conditions under which the anomaly occurred may indicate the postulates that might be developed. Frequently, postulates extend a concept in some way, and often a concept linking the system under investigation with another system is developed.

The object is to develop a plan for research. Three postulates (Fig. 6.7) about the process of control of photosynthesis are advanced that might explain the anomaly. Postulate 1 creates an analogy with crop plants and suggests within-tree control of photosynthesis, introducing the idea that growth requirements set a demand for photosynthate (the products of photosynthesis). This postulate identifies some homeostasis in the photosynthesis system that is not under environmental control. Postulate 2 continues the line of reasoning of Axioms 1 through 4, that photosynthesis is controlled by the physical environment and that greater understanding of tree water relations will resolve the anomaly – recall the incomplete specification of stomatal control in the theory (Fig. 6.6). Postulate 3, like Postulate 1, suggests control from within the plant but is not specific. It is included because some researchers who have found different parameter values for Axioms 1 through 4 in field studies call this *acclimation*, a change in photosynthesis rate that follows the seasonal change from winter to summer and back again.

- Postulate 1. *Growth rate determines photosynthesis rate.* One difference observed between 21 June and 20 July was that the needles of the new shoots were actively elongating in June whereas they were almost fully developed by the July measurement – so the question is asked "Does growth rate influence photosynthesis rate?" This postulate is analogous to the sink control of photosynthesis in some annual plants (see, e.g., Herold 1980). A *sink* is a part of a plant utilizing photosynthate; a source produces it. Photosynthetic rates in a number of plant species have been manipulated by controlling the growth rate of sinks for carbon (see, e.g., Moss 1962, Herold and McNeil 1979, Kirschbaum and Farquahar 1984, Ho 1988). The physiological theory of sink control of photosynthesis involves some form of end-product inhibition (Keener *et al.* 1979) in which the cyclic regeneration of phosphorylated compounds within the plant might become limited by reduced growth.

So, a slower growth rate might reduce photosynthesis rate (Fig. 6.8). The strength of this postulate is: (1) it draws from a well-researched mechanism for the control of photosynthesis within the growing season found in plants other than trees, (2) growing shoots and needles are a substantial sink for carbohydrate in conifers (e.g. Schneider and Schmitz 1989, Hanson and Beck 1994). The concept of sink control would be extended from these other species and conditions, to growth of *P. strobus* shoots on mature trees.

- Postulate 2. *Dry mid-summer weather decreases shoot water potential and increases AHD, which reduces photosynthesis rate.* PXPP for 21 June and 20 July were both greater than −1.0 MPa. This suggests that, on both dates, water deficits that developed during the previous days were replenished overnight. However, between the two days there is a continuing downward trend in midday foliage conductance (from 69 to 33 mmol m^{-2} s^{-1}), a measure of reduced transpiration and, by implication, reduced stomatal opening – and so a possible limit to photosynthesis rate. There was also a small decrease in midday xylem pressure (−0.97 to −1.06 MPa); indicating that shoots had less internal moisture. The actual value for 20 July occurs at a small peak of increase – due to recent rain – on a more general decline. Photosynthesis is known to decrease with decreasing stomatal conductance, but the relationship is complex; uncertainties remain about diurnal patterns, the factor(s) that trigger stomatal closure, and whether an optimum g_s minimizes water loss for given CO_2 gain (for a short review, see Baldocchi 1994). The question is whether the values of midday foliage conductance measured here were sufficient to cause the reduction in photosynthesis.

- Postulate 3. *The photosynthesis system acclimates to a changing physical environment.* Photosynthesis rate of conifer foliage is known to acclimate to changing conditions over time. Foliage taken in winter cannot be restored to summer rates by short-term warming (Pisek and Winkler 1958), and recovery from winter depression is related to increasing ambient temperature (Lundmark *et al.* 1988): this is termed acclimation. Decrease in photosynthetic potential during winter has been associated with the process of cold adaptation (Strand and Öquist 1988), and the pattern of photosynthesis during winter has been used as an indicator of dormancy (Bolhàr-Nordenkampf and Lechner 1988). Postulating changes to the photosynthesis system seems reasonable as a parallel development to the physiological process of dormancy (analogy). The strength of this postulate is that photosynthesis rate is already known to acclimate seasonally (summer–winter). Here we propose to extend the concept of *acclimation* (and the theory) to within a season (summer) as well as from one season to another, although what may induce the acclimation is not specified.

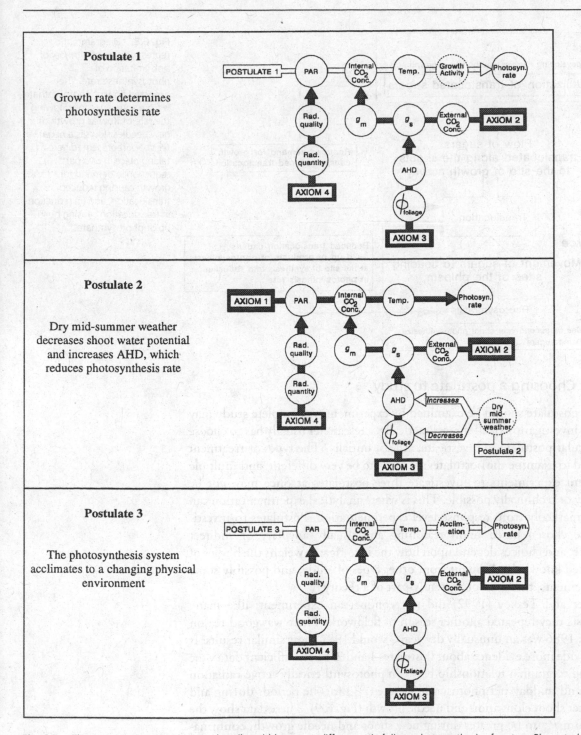

Fig. 6.7.   Three competing postulates to describe within-season differences in foliage photosynthesis of mature *Pinus strobus*. Dashed circles represent *concepts by imagination*; solid circles represent *concepts from research*. For abbreviations, see Fig. 6.5.

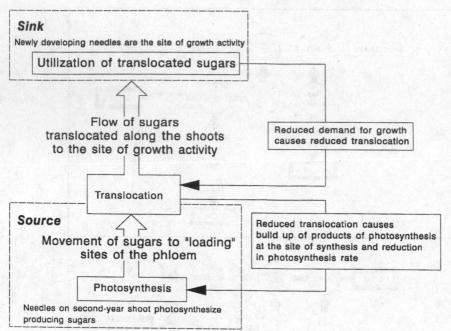

Fig. 6.8. Diagrammatic representation of a theory of sink limitation of photosynthesis rate. The expected flow of photosynthate is from source to sink. Slowing the rate of flow as growth of new needles slows is a *negative feedback* represented here as taking place in two parts: (i) reduction in demand for growth causing reduced translocation, and (ii) reduction in translocation causing build up of photosynthate.

6.4.5 Choosing a postulate to study

Which postulate should be examined by experiment? A complete study may require investigation of all three. Practically, a researcher usually has to choose a particular postulate to investigate, at least initially. The types of treatment required to examine the postulates are likely to be very different, and multiple treatment experiments to investigate three postulates at once may not be logically or technically possible. This is where analytical experimentation can depart markedly from response-level experimentation. Postulates to investigate are chosen based upon reasoning and usually ancillary or indirect evidence, and choices depend upon how the investigator weighs the bodies of associated knowledge brought from other types of study, and possibly some measurements, that are not presently part of a theory.

Maier and Teskey (1992) did not conduct an experiment; like many ecologists they repeated another season of fieldwork. There was good reason for this: 1986 was an unusually dry year – would 1987 show similar results? It did provide more evidence about Postulates 1 and 2. First, sufficient data were obtained to graph a relationship between photosynthetically active radiation (PAR) and midday net photosynthesis rate (P_{net}) for the periods during and then after shoot elongation and needle growth (Fig. 6.9). These data show the maximum P_{net} to be greater during new shoot and needle growth, confirmatory evidence that growth rate of the new shoot and needles may influence photosynthesis rate.

Fig. 6.9. The relationship between net photosynthesis (P_{net}) and photosynthetically active radiation (PAR) during periods of high (Julian days 110–175) and low (days 175–290) growth rate. Each point is the mean P_{net} within a PAR interval. Means were calculated in 50 μmol increments for PAR above 100 μmol m^{-2} s^{-1}, and 20 μmol increments for PAR below 100 μmol m^{-2} s^{-1}. Relationships were derived from combined 1986 and 1987 data sets. (Redrawn from Maier and Teskey 1992, with permission.)

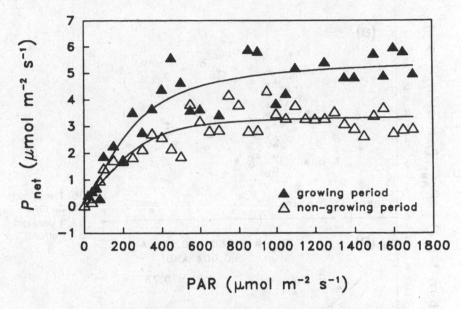

Second, pooled results from the two years indicated that, for PXPP greater than -1.0 MPa and at saturating PAR, P_{net} showed no relationship with AHD; but for PXPP less than -1.0 MPa, i.e., with less water in the plant, then P_{net} decreased as AHD decreased (Fig. 6.10). So, -1.0 MPa PXPP appears to be a threshold value. However, the information about these possible relationships appears to be limited by the measurements being made. Perhaps PXPP is not an effective measurement of within-day water stress? It represents something about day-to-day changes in tree water status, but it does not measure what happens between pre-dawn and the time of measurement (i.e., the target measurement, PXPP, does not completely overlap the concept that should be measured, Fig. 6.1). Direct measurement of xylem pressure potential at the time of photosynthesis would probably be more effective and the effects of AHD on water uptake and loss must be determined. This illustrates a general principle: measurements designed to analyze the cause of a response must be made close in time and space to the measured response variable. In this case such measurements would have added extra work to the investigators daily schedule.

There now seems sufficient evidence that Postulates 1 and 2 should be investigated. Subjectively I would choose to suspend interest in Postulate 3: (a) because of the increasing confirmatory evidence for 1 and 2 and, (b) because the examples of acclimation of photosynthesis most known about are the summer–winter transition and acclimation of photosynthesis to shade (Leverenz 1988). The latter is not relevant because branches were at the top of the tree canopy and there does not seem to be a parallel circumstance in this situation. Only summer growing conditions in a warm environment are being considered, and developing Postulate 3 for investigation would require

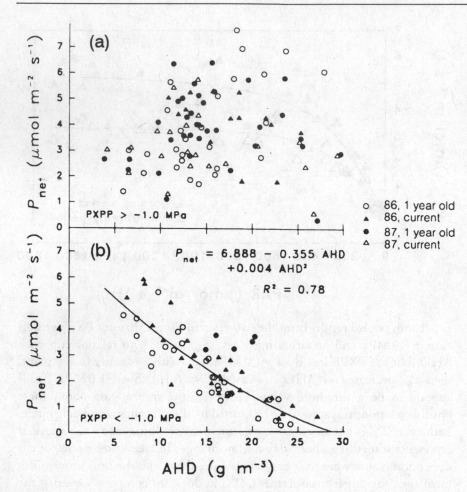

Fig. 6.10. Response of net photosynthesis (P_{net}) to absolute humidity deficit (AHD), (a) when pre-dawn xylem pressure potential (PXPP) > -1.0 MPa, and (b) when PXPP < -1.0 MPa. In all cases photosynthetically active radiation (PAR) $> 800\,\mu mol\,m^{-2}\,s^{-1}$. (Redrawn from Maier and Teskey 1992, with permission.)

developing both a concept of imagination for acclimation under this rather different situation and an equivalent concept by measurement for within a warm period.

Additional field measurements suggest that photosynthesis may be limited by both decreasing growth rate and increasing ADH, combined with decreasing plant water potential. But from the field measurements there is no estimate of the independent effects of these processes. Experiments are required to gain these independent estimates, to separate the possible effects of Postulates 1 and 2, and to determine the possible magnitudes of their effects on photosynthesis rates under different conditions.

Remember the objective is to design a research plan. What should an investigator do when faced with two postulates that may not be mutually exclusive? In this example the choice is to investigate Postulate 1, but to design an investigation that takes into account that Postulate 2 may be important.

Most important would be to develop measurements and sampling procedures for water relations and status of the tissues.

Recall Postulate 1:

Growth rate determines photosynthesis rate.

In the previous discussion a theory based on source–sink relationships was used to describe why this postulate might have caused the anomaly. This postulate could be examined in two ways. First, the concept by imagination, *growth rate*, could be measured directly using the additional postulate:

Postulate 1.1: *Needle growth rate plus shoot growth rate defines growth rate.*

But we could go further than this and also assume that a change in *growth rate* (as it is now defined) produces a change in the concentration of available products of photosynthesis, i.e., non-structural carbohydrate concentration (soluble sugars and starch).

Postulate 1.2: *Growth rate determines non-structural carbohydrate concentrations.*

Postulate 1.2 is bolder than Postulate 1.1: it assumes much more – not only that *growth rate* of needles and shoot controls photosynthesis rate but also that it does so by affecting the concentration of non-structural carbohydrates. In this example the bolder postulate identifies a causal process (Chapter 7), i.e., how a change in *growth rate* would influence photosynthesis rate.

BOLD POSTULATE

A *bold postulate* has a high information content, and its confirmation, or rejection, advances the theory substantially.

Whether or not it is an effective strategy to investigate a bold postulate can depend upon the measurements required. A bold postulate certainly is at greater risk of being falsified and, in that sense, it has greater value (Popper 1972).

FALSIFICATION

A postulate is *falsified* when its predictions are shown by empirical evidence not to be true.

But in this example there may be greater measurement and sampling difficulties with the bolder postulate and so greater risk of rejecting the postulate when it is actually true. For this reason it is important to leave the general statement of Postulate 1 in place and to consider different ways to test it. Postulate 1 defines the idea, but needs some detailed exploration before it can be rejected or accepted. A decision about Postulate 1 should not be made until a number of different specific descriptions and measurements of growth rate have been tested and exhausted.

The following sections describe the procedure for developing a data statement for an experiment to investigate Postulate 1, with the knowledge that there is a competing postulate in Postulate 2. (A complete data statement cannot be produced because that requires knowing where and when the experiment would be conducted and, almost certainly, exploratory research into measurements.)

6.4.6   Defining the experimental conditions

Data statement: Part One. Define the scientific procedure to be used in investigating the postulate

(1)   Growth rate will be manipulated experimentally; this will be the treatment. Photosynthesis rate will be measured; this will be the response.
(2)   All trees will be watered in order to remove possible effects of PXPP and ADH in reducing photosynthetic rate.

Note that this is not an experiment to examine both Postulates 1 and 2 at the same time, i.e., there is no condition where trees will not be watered and receive either the experimental control on manipulated growth rate, or its control. Nor is the intent merely to water trees and observe what happens for an additional year. That might provide confirmatory evidence for Postulate 1, if the same type of decrease in photosynthesis rate was observed as growth rate declined. But such confirmatory evidence lacks the power of an experimental approach that deliberately attempts to falsify the postulate through direct manipulation.

How might growth rate be manipulated? The introduction of *growth rate* as a condition that may influence *photosynthetic rate* is made as an *implication*. In the terms of propositional logic growth rate determines photosynthetic rate. So

If growth rate is high, then photosynthesis rate is high.

The postulated situation is that, if conditions of light, CO_2, water, and temperature are equal, then:

	Julian days 110–175	Julian days 176–290
Assumed growth rate	High	Low
Observed photosynthesis rate	High	Low

To falsify the postulate, we must seek to construct conditions where we observe low growth rate during days 110–175 but high photosynthesis rate, or high growth rate during days 176–290 but low photosynthesis rate. However, the only experimental opportunity is to decrease the growth rate of needles during days 110–175 and observe the effect on photosynthesis rate. *Pinus strobus* has determinate shoot growth, i.e., the numbers of needles it produces along a shoot are determined as the overwintering bud forms during the previous summer. By day 212 (the end of July), growth of needles and their supporting shoot is complete – the possibility of increasing growth rate after day 212, at least for needles and shoot length, is non-existent.

We could remove newly growing shoots during Julian days 110–175 (complete excision) or trim off the developing needles (partial excision). Excision would be cheap and simple, but it damages the system. In particular, measurement of photosynthesis of one-year shoots after the period of shoot growth could not be compared under uniform conditions with that of a shoot without excision.

A more complex but possibly more informative experiment would be to cool the growing needles on the newly forming shoot during days 110–175, to slow their growth. This requires introducing an axiom that temperature affects growth (Lanner 1964, Wiegolaski 1966). An enclosure could be placed around the newly growing shoot and cool air injected. Ambient temperatures when photosynthesis was measured (Fig. 6.3) were in the range 20–30 °C, so an effective change might be to reduce temperature markedly, to, say, 10 °C. If such a decrease in temperature decreased needle growth rate of the currently forming shoot, but no decrease in photosynthesis rate of the second-year shoot, then Postulate 1.1 would be falsified.

Note that photosynthesis rate of the newly forming foliage should also be measured since photosynthesis rates of both the new and the previous year's foliage were found to decline during days 176–290 (not possible if the new shoot has been excised). However, the effect of reduced temperature on the developing foliage may act through a direct effect on photosynthesis, not just through an effect on growth rate.

Should a treatment be planned to cool the one-year foliage but not the new foliage, and measure photosynthesis rates? What would that tell us? Needles along the one-year shoot do not grow, nor does that shoot elongate. (The stem of the shoot makes radial growth, but that tends to occur at a different time of year.) So this treatment has no relevance for the implication

growth rate $\Rightarrow$ *photosynthesis rate.*

If we did find that the growth rate of new shoots and foliage was reduced under this treatment that would imply

photosynthesis rate (one-year shoot) $\Rightarrow$ *growth activity (new shoots and foliage)*

This implication is about the carbon economy of growing needles, not about the control of photosynthesis, which is what we are considering. In a complete theory for the physiology of trees these are interrelated, but in this analysis they are separate.

Under the proposed treatments, non-structural carbohydrate concentrations could also be measured and Postulate 1.2 thereby investigated at the same time as Postulate 1.1. If both were falsified, then that clearly is stronger evidence for rejecting Postulate 1 as a whole. However, the additional time and cost may prohibit both Postulates 1.1 and 1.2 being investigated at the same time and a choice may have to be made.

The first two principles of measurements for new concepts are met: the postulates and data statements are specified. The value of the third principle – using more than one measurement for a new concept – can be seen. It would allow a bold postulate (1.2) to be examined along with a more cautious one (Postulate 1.1). However, to satisfy the fourth principle of measurement, that accuracy and precision of a measurement should be investigated explicitly, both of these measurements should be investigated before they are used together in an experiment. Can the growth rate of whole shoots be measured by measuring sample needles or some other procedure? How many samples are needed to estimate a mean value of non-structural carbohydrates? Without a reliable measurement, the experiment might be valueless.

6.4.7　Developing a measurement

Data statement: Part Two. Specify the measurement for each concept of the postulate

Postulate 1 uses C^{imag} (*growth rate*) as a general concept. But a procedure must be developed for measuring it. There are two components to the growth of a new shoot: the growth of the shoot apex itself, which can be measured by extension in length, and the growth of the foliage. Foliage growth rate in *P. strobus* might be difficult to measure. Five needles grow from each fascicle positioned along the new shoot (Fig. 6.1), and these tend to grow at different rates, lower fascicles starting and finishing earlier. To investigate this postulate it would have to be decided how many needles to measure, and at what positions along the shoot. An overall estimate of growth rate, integrating change due to both shoot extension and needle growth, should be obtained to examine whether it correlates with changes in photosynthesis rates. Using an integrated measure, based on weight of new tissue formed rather than shoot length or volume, may correspond more closely to the process to which it is postulated to be linked, i.e., CO_2 uptake.

Postulate 1.2 suggests an alternative approach. It proposes that non-structural carbohydrate content depends upon growth rates. In the case of *P. strobus* shoot growth, the theory of sink control is used to imply that concentrations of non-structural carbohydrates would be greater in July when growth is slower. Non-structural carbohydrates would have to be measured, which would almost certainly involve taking repeated destructive samples from shoots. So both a measurement and sampling system would have to be specified.

It would also be necessary to define measurements of *photosynthesis rate*. Because we cannot rely on experiencing two days of identical weather conditions in future years, the most appropriate technique might be to use P_{net}, the maximum photosynthesis rate under saturating radiation – the measurement that Maier and Teskey (1992) used in their comparisons (Figs. 6.3, 6.4, 6.9). P_{net} could be measured either when the foliage was illuminated directly by the sun or by providing ancillary radiation to the foliage in the photosynthetically active wavelengths. Measurements at two different radiation levels both normally thought to be saturating could be made to ensure that saturation had been reached. (This would probably be determined in an exploratory investigation and an appropriate value determined for routine use.) Other environmental conditions would still have to be measured and their influence on the measured value of P_{net} calculated. Before *Data Statement: Part Two* could be completed, exploratory investigations would be needed to determine the feasibility of these measurements.

6.4.8 Designing treatment application, replication, and controls

Data statement: Part Three. Specifies the requirements of the data for any statistical test to be applied

This is primarily concerned with the technical process of assessment, and in an analytical experiment, replication and controls determine the assessment procedure.

EXPERIMENTAL UNIT

The *experimental unit* is the system to which the treatment is applied, and for which there is a corresponding control.

In this example, the one-year and new shoot and foliage together are the experimental unit, not just the new shoot and foliage that are cooled or excised. The complete experimental unit must be replicated.

MEASUREMENT UNIT

The *measurement unit* defines the response to the treatment. The measurement unit and experimental unit are not always the same.

In this example, the measurement units are photosynthesis rate of the one-year foliage, growth rate of the new shoot and foliage and/or non-structural carbohydrate concentrations. It is important to maintain the distinction between the experimental and measurement units because it is the experimental unit that must be replicated.

TREATMENT

In analytical experiments, a *treatment* is designed to manipulate a system and produce an observable or measurable effect.

REPLICATE

In analytical experiments, *replicates* are systems assumed to have identical properties and receiving identical treatments, either as treatments or controls. The purpose of replication is to estimate variation in response to a treatment due to unknown differences in the system itself and/or the application of the treatment.

Although everything should be done to make application of the treatment the same between replicates, there may still be small differences between them. For example, while it may be decided to experiment with only one type of shoot, say the leading shoot of the main axis of the tree, there may still be differences between trees due to such things as tree height and/or relative position in the canopy.

CONTROLLED ANALYTICAL EXPERIMENT

In a *controlled analytical experiment*, the performance of the treated system is compared with that of a system not treated, i.e. the *treatment control*.

In the proposed experiment the treatment is to reduce growth rate of newly forming shoots and foliage (or stop it through complete excision). The measurable effect is the comparison of photosynthesis rates between one-year shoots with and without cooling the new shoot and foliage (or with and without excision). Growth rate of the new shoot and its foliage and photosynthesis rate of one-year foliage must also be measured in the treatment control. Unless the growth rates differ, the experiment is not successful. The comparison is the requirement of the experiment.

In this ecological experiment, as with most others, there are many possible sources of ecological, biological, and experimental variation whose effects on the system could be studied. For example, there are different types of shoot on a tree: shoots on side-branches of different orders, shoots with cones, or the leading shoot of the tree. If you think that differences in shoot type may affect the system, i.e., how the experimental unit functions, then such differences could become the foci of new experimental investigations, but not replicates for this experiment. You would stratify this variation.

STRATIFICATION

Experimental units may be divisible into groups for which you expect to see treatment response and variation to behave similarly within groups but dissimilarly between groups. Identifying the potential sources of variation and sampling from within these groups is the process of *stratification*.

First, you must choose a stratum (or strata if you want to look at more than one axis of variation) within which you can replicate. Your choice would be aided by exploratory analysis. For example, in this experiment we might choose to investigate just one type of shoot, say the topmost side-branch on a tree where radiation levels are high. Second, you must accept that the restricted choice of biological variation examined restricts the way the postulate is examined and limits the conclusions that can be drawn.

Experimental design requires independence (for a definition, see Chapter 8) between the experimental units. Failure to recognize this requirement is one of the biggest problems facing the ecologist who intends to do field experiments. The result is all too frequently pseudoreplication, the assumption of independence when none exists. Fortunately, Hurlbert (1984) has dealt with this topic directly and presents examples of how this problem may arise in ecology.

PSEUDOREPLICATION

Pseudoreplication occurs where more than one measurement is made within an experimental unit and yet the measurements are considered as separate (independent) indicators of the experimental treatment(s) or the control(s).

Imagine in this experiment if it was considered technically easier to make a chamber that cools one whole branch that had many {one-year shoot and new shoot} units rather than a series of small chambers each cooling a {one-year shoot and new shoot}. If measurements were made on a number of {one-year shoot and new shoot} units within the cooled whole-branch chamber, and these were treated as separate indicators of the treatments that would be *pseudoreplication*; many measurement units would be contained within a large experimental unit. This is wrong. The purpose of replication is to sample for effects of variance in the application of the treatment and variation in the biological materials. If many {one-year shoot and new shoot} units were measured in one cooled branch chamber they would all experience the same variance that may occur in that particular branch chamber.

Pseudoreplication is, unfortunately, extremely common in ecological research. The temptations are easy to understand, particularly with experiments. For example, we may wish to investigate the effects of deer browsing on tree seedling regeneration in a forest, or what happens to invertebrate fauna on the exclusion of crabs from an area of the seabed. In both cases, building and maintaining the exclusion is expensive. Ecologists want to know about the interactions between animals, and between animals and their environment; this leads them to make many measurements on a few plots. But statistical analysis of samples from within the exclusion area can not be used to make general inference about the whole forest or whole seabed. Replication of the exclusions would be necessary to get an estimate of the random effects due to the exclusion process and spatial variability on the forest or sea floor. This may lead to making fewer measurements on a larger number of plots. However, it may be important to make pilot (exploratory) experiments with a few treatments before attempting to establish many replicates. This strategy is discussed in detail in Chapter 8.

For the branch-cooling experiment the implication of the requirement for replication may seem to be having many cooling systems. That may not be possible and an alternative strategy may have to be adopted involving more detailed study, and more measurements of the proposed effect, using few systems. For example, it may be possible to study the details of the effects of one cooling over time and compare it with a paired shoot that is not cooled. The significance of any effect would be interpreted in the details of the measurements, e.g., rate of change of photosynthesis rate in relation to rate and/or duration of cooling. If such experiments gave results over a few days then a number of paired comparisons could be made, but these repetitions would not be replicates in the sense required for use of ANOVA. The power of the experiment would come from the detailed measurements and measuring pairs of shoots in contrasts. Using the excision treatment, rather than cooling, may enable replication and a standard use of ANOVA. However, experimental manipulation, by definition, interferes with a system, and it is frequently difficult to isolate the intended treatment effects from ancillary effects due to the experimental techniques.

In this experiment, two types of control must be considered in addition to the treatment control.

1. Control against ambient variation

In this case water deficit effects may be a source of variation that should be eliminated (i.e., the possible effects of Postulate 2). It was proposed to irrigate trees so as to maintain PXPP greater than $-1.0\,\text{MPa}$. Only if Postulate 2 is investigated at the same time as Postulate 1 should irrigation be a treatment. For response-level experimentation, investigating two postulates at once can allow you to estimate interaction effects, i.e., one effect in the presence of another. But in analytical experimentation, examining two postulates and their possible interactions can create a large and difficult experiment, not the least because new measurement techniques are so frequently used.

CONTROL PROCEDURE

A *control procedure* allows effects other than those being studied to be removed from influencing the system.

Of course, the effectiveness of any control procedure should be examined by measurement (see 6.4.9) and even by repeating the experiment using a different control procedure.

The use of control procedures is one of the most difficult issues in ecological experimentation because, although they are designed to reduce ambient variation, exactly how this is done, and the effects on the system, are not part of the experiment unless they are part of investigating ancillary processes, i.e., the treatment *and* control are *both* made within the control procedure. Laboratory, aquarium and greenhouse experiments can be considered as extensive control procedures, e.g., by homogenizing the environment and/or restricting investigation to particular biological material. But their relevance for ecology is repeatedly questioned because the effects of such simplifications may not be studied. In this planned experiment, watering (the control procedure) is explicitly designed to remove one possible set of ancillary effects yet, even then, its actual effects may not be known, may be unexpected and certainly need to be investigated.

2. Control treatments to estimate non-treatment effects of the experimental procedure

Consider the treatment of cooling the growing foliage and shoot. The effects of enclosing the current-year shoot without cooling should be investigated because the enclosure itself may interfere with other conditions, e.g., water relations, radiation. The problem with this type of control is that it is already known that simple enclosure will raise shoot temperature. So enclosure alone is an inadequate measure of the enclosure effect. One possible control would be enclosure with an attempt to keep temperature close to, rather than cooler than, ambient.

The excision treatment, although it may appear more direct and less complicated than cooling, may have unintended effects. For example, excision of one shoot may result in enhanced growth of other adjacent shoots. This can not be controlled for – additional measurements may be needed to examine for its occurrence.

TREATMENT CONTROL

Inclusion of a control for non-intended treatment effects allows the designed effects of the treatment to be estimated more precisely.

6.4.9 Investigating ancillary processes to aid interpretation and assessment

The two main purposes of investigating ancillary processes are monitoring whether important assumptions are holding and looking for signs that alternative postulates may be important. In this experiment, monitoring soil moisture and shoot water potential would be essential to examine the effectiveness of the control procedure. The need to investigate ancillary processes, and particularly those that monitor the effectiveness of control procedures, or the possible operation of a competing postulate, is a major difference from response-level experimentation.

6.5 Whole-system analytical experiments

The photosynthesis experiment is a study of the functioning of one part of an ecological system – photosynthesis is a component of tree growth. Two approaches have been taken to study whole systems rather than analyze them component by component, *large-scale perturbations* and *synthetic constructions*.

LARGE-SCALE PERTURBATION EXPERIMENT:

Large-scale perturbation experiments attempt analytical experimentation to large ecological systems, e.g., a whole river or forest. These are diachronic contrasts but the large scale limits, or may completely remove, the possibility of replication and or the construction of controls. This complicates assessment.

Large-scale manipulations of ecological systems, e.g., fertilizer application to a tundra river (Peterson *et al.* 1993) or catchment liming (Dalziel *et al.* 1994), application of iron to the ocean to test for its effect on productivity (Kolber *et al.* 1994, Martin *et al.* 1994), may be repeated but are unlikely to be replicated. Replication requires systems or a set of conditions sufficiently identical that choice between treatments and controls could be assigned randomly. Such perturbations may become increasingly important to examine large-scale effects (Matson and Carpenter 1990). Perturbation analysis has been a regularly used technique of biological control systems analysis (Milsum 1966). The value of such manipulation in ecology depends upon the process studies accompanying the manipulation, i.e., that a theory must be specified and its components measured in detail during the diachronic contrast imposed by the treatment.

SYNTHETIC CONSTRUCTION

Synthetic constructions seek to design and construct all or part of an ecological system. The synthesized system is usually closed. It sometimes comprises organisms that represent primary producers, herbivores, decomposers, or predator—prey combinations, it may be exposed to treatments or constructed under different conditions designed to investigate the ecological relationships and overall properties of the ecological system.

It is valuable to distinguish this type of research, e.g., the construction of microcosms or artificial communities, from analytical experimentation because synthetic contrustructions are attempts to construct models of natural systems to see whether particular characteristics are reproduced. In this sense they are not inquiries that use a limited contrast between treatment and control. Rather, they are attempts to reproduce the working of a theory, e.g., a synthetic microcosm may have primary producers, herbivores, predators, and decomposers. The most important consideration is how well the synthetic construction represents some real ecological system. The level of inquiry is broader in scope than for analytical experimentation, where the system is manipulated but left largely intact.

6.6 Discussion

A recurring difficulty in analytical experiments with ecological systems is the many possible interactions between system components. What standpoint should the experimenter adopt? Clearly ecological systems are not closed – different factors may come to control an ecological process under different circumstances. But, from the perspective of the researcher, to adopt the position that ecological systems are completely open may not be helpful. If you think that the range of possible effects cannot be defined then it may seem pointless to strive for effective, accurate and precise measurement and attempt rigorous experiments with well-defined controls. The most appropriate position may be to consider ecological systems as *partly open systems* where many possible factors may become important, although observation at any one time may show the effects of only a few. Postulates 1 and 2 propose a more complicated description of the photosynthesis system than Axioms 1 through 4: in this example (Fig. 6.11), both postulates introduce negative feedback but Postulate 1 is more explicit in its proposed mechanism.

The original question of the example in this chapter was to explain the anomaly "Why were photosynthesis rates different on days with almost identical meteorological conditions?" This requires two stages:

Stage 1: Recognizing that a particular negative feedback could have occurred.

Stage 2: Identifying that it did occur.

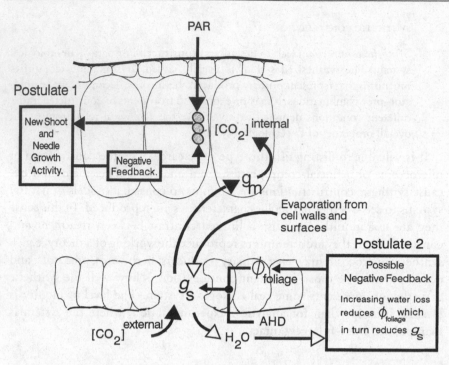

Fig. 6.11. Postulates 1 and 2 increase the complexity of the system studied (recall Fig. 6.6). For both postulates, two questions must be asked: "Can this mechanism occur?" and "Does the mechanism explain a particular ecological observation?" Note that in this example the process of Postulate 2 has not been completely specified. For abbreviations, see Fig. 6.6.

Even though Stage 1 may be accomplished (Postulate 1 was confirmed) it could not be shown retrospectively that this explained the anomaly, i.e., Stage 2. The question about the anomaly should be rephrased as "If (either) Postulate 1 (or 2) can occur then under what conditions does either of them control photosynthesis rate?" In the study of partly open ecological systems we constantly ask these two types of question: "What might the structure of the system be?" and "When do different parts have an influence?" Answering this second question defines the domain of influence of particular parts.

It is not easy to falsify postulates. If you fail to see contrary evidence it can be difficult to ensure that you have created the proper environment for the experiment to succeed, even if the postulate is not false! For example, when a shoot and its needles are cooled it might take some time before first-year needle growth rate slows down sufficiently to affect the photosynthesis rate of second-year shoots. The measurement would have to account for this, without the investigator first knowing what the lag is! Similarly in Postulate 1 the particular treatment implies an exclusive physiological connection in terms of carbon economy between a one-year shoot and the new shoot and foliage it supports. But this may only be part of the relevant system. For example, one-year shoots may supply a number of new shoots and their foliage. If some of that new growth were not exposed to experimental cooling it might create a sufficient sink demand that photosynthesis rate would not drop even though the postulate may be true, and that is why measurement and experiment are an art because understanding is incomplete.

Understanding the system is the essential foundation to the art of experimentation and an inductive guess must be made. Of course, experimental investigation should explore ahead of present knowledge but for practical reasons it should not be too bold, particularly if you are attempting a falsification. McIntosh (1985) relates that some scientists believed marine biology was retarded by the early introduction of laboratory experimental methods before investigations had revealed what might be appropriate to investigate through experiments. Many experiments can be planned but they may not be informative.

Hairston (1989) extensively reviewed published ecological experiments and concluded that individual experiments should be small in scope. This seems particularly good advice for analytical experimentation – where an experiment is designed to investigate the functioning of a system. The power of an experiment lies in the logical weight it gives to an argument, based upon contrasts, and it is essential that competing postulates should be constructed. This in turn can require much exploratory work both to design and test measurements and to appreciate whether a proposed treatment will have the intended effect.

The art of measurement is in ensuring that the measure represents the essence of the concept, and investigation with multiple measurements may be needed to investigate this. It is important not to let a single measurement, which may have accuracy and precision but may not be effective, come to represent a concept to the point where the measurement is an implicit definition of the concept.

The art of analytical experimentation is finding a system that is sufficiently simple to manipulate, but sufficiently complex to represent the essential elements being investigated. It is not to find the simplest possible system. To investigate Postulate 1, the simplest possible system to manipulate might be a young tree growing in a glasshouse. Although such an experiment may be much easier to conduct, it would be closer to a synthetic construction than an analytical experiment – and it would be essential to show that this system (the young tree in the glasshouse) had the same properties as the tree in the forest. Experiments are not neutral arbiters of truth or falseness of a postulate. The onus is on the investigator to show, not merely assume, the relevance of the experiment.

6.7 Further reading

Mead (1988) gives a comprehensive account of the principles of experimental design using many examples from applied biology and Underwood (1997) describes applications in ecology and use of analysis of variance. Eberhardt and Thomas (1991) provide an extremely valuable survey and classification of the types of environmental field study used by ecologists and the problems associated with experiments. *Design and Analysis of Ecological Experiments* (Scheiner and Gurevitch 1993) contains valuable articles illustrating techniques that can be used with experiments when some of the assumptions of ANOVA are not met, or where there may be unusual design features. Chapter 1 of that book, which outlines basic considerations, is recommended if you are planning experiments, and Chapter 5 illustrates the procedure where multiple response variables are measured and MANOVA should be used.

Jaech (1985) describes statistical methods, and includes computer programs, for analyzing measurement errors where two methods (or observers) repeatedly measure the same thing, or things, and a series of paired measurements or observations are obtained.

The criterion that postulates should be bold is one of the principal tenets of Popper's philosophy for the scientific method. His reasons for specifying this criterion are discussed in Popper (1972), particularly the importance of bold postulates in analyzing complete theories.

There has been widespread use of *synthetic constructions* in ecology to analyze questions that are not amenable to field experiment. For example, Gonzalez *et al.* (1998) and Drake *et al.* (1993) have constructed fragmented systems to study the potential effects of patches and habitat corridors in organism survival and conservation. Aquatic systems have been used to study the species balance necessary to achieve continued survival, e.g., Kersting (1997), and the necessary design components to make aquatic mesocosms work has been the subject of considerable scrutiny – particularly maintenance of physical and chemical characteristics, e.g., Sanford (1997), Williams and Egge (1998). The effects of enclosures on the ecological system being investigated is not confined to physiological studies, as in the cuvette effect mentioned in this chapter. Some indications of the types of effects that may occur can sometimes be obtained where enclosures of different sizes have been used, e.g., Jacobsen *et al.* (1997), Quinn and Keough (1993), or where enclosure effects are compared with non-enclosures, e.g., Sarnelle (1997).

7 Methods of reasoning in research

Summary

There is no single method of reasoning that scientists can, or do, follow. We reason in two general ways: deductively, when we use the logic of a theory to make a deduction that we then investigate. If we find the deduction correct then we add additional evidence to the theory – inductively, to extend a theory to explain more, where we consider an idea that applies in one situation will also apply in another.

Deduction follows the rules of propositional logic. In research we can use those rules with axioms and postulates to make predictions that we test. Four rules are particularly valuable:

(1) A conjunction is true if, and only if, all of a series of linked propositions are true.
(2) A disjunction is true if any of a series of linked propositions is true.
(3) Some deductions require that concepts be equivalent, and that must be established by research.
(4) The implication statement has the form, "if p, then q" and sequences of such statements are used to establish chains of scientific reasoning.

While the rules of propositional logic are the base of much scientific reasoning, their use is constrained in ecology by the probabilistic nature of many concepts and the problems of investigating systems with multiple causality.

Much reasoning that ecological scientists use to construct theories is inductive. We argue "If such an event/process/condition (A) occurs, then we will observe a particular result (B)." We investigate to see whether B actually occurs, and if it does then we may conclude that A is true. However, B may not be the result of A but of some other factors overlooked in the investigation. This problem is called the *fallacy of affirming the consequent*. We may repeat the investigation in other situations, but these are confirmatory investigations and do not rectify the reasoning.

Scientists depend upon inductive reasoning for much research but the above-mentioned fallacy has led to modifications being suggested for how inductive reasoning can applied. One suggestion is to attempt falsification;

another is to base reasoning on causal processes and in this way add strength to an argument. Suggestions have also been made that reasoning should proceed by use of contrasts, and that multiple postulates for a phenomenon should always be considered. These different approaches are discussed in this chapter.

7.1 Introduction

Theories contain different types of knowledge and how certain we consider them to be varies as research progresses. Previous chapters have shown how this can be represented, and have illustrated use of different types of research technique applied with different methods of reasoning. Is there a way to determine which is the most effective method of reasoning for particular types of question?

There is no single method of reasoning that scientists can, or do, follow. We reason in two general ways:

(1) Deductively, when we use the logic of a theory to make a deduction that we then investigate. If we find the deduction to be correct then we add additional evidence to the theory. If we find it to be wrong then we have to revise the theory. The first postulate of northward movement of tree species in Alaska during 6 ka BP to 9 ka BP is an example of deduction.

(2) Inductively, to extend a theory to explain more, where we consider an idea that applies in one situation will also apply in another. Both Ashley Steel and Denise Hawkins used this method of reasoning, taking a theory developed in one circumstance and using it as the basis for analysis of another. This process uses the internal consistency of the theory, but extends beyond it. However, success in finding additional cases is taken as further support for the theory.

Most reasoning in scientific research is inductive. Induction introduces a subjective element and although we may find apparent confirmation through an investigation we can never be certain that the inductive reasoning is correct. Some other factor, or factors, may really explain the observations or results obtained.

Dependence upon, but difficulty with, induction has led to modifications being suggested for how inductive reasoning can best be used. One approach suggests falsification – that instead of seeking confirmation of inductive ideas we should carry out investigations attempting to falsify them and that if an idea withstands an attempted falsification it has more merit than one that has only been confirmed. While this is valuable, a difficulty is that falsification is not a practical approach for all propositions.

Another approach suggests that reasoning should be based on causal processes and in this way adds strength to an argument. The basis for investigating small mammal use of debris piles is that these animals must den, feed and reproduce; these are essential causal processes for the animals to live. However, a difficulty in using causal reasoning is in knowing exactly what the underlying controlling causes are in any situation. Finding these is frequently an important part of theory we need to develop and in ecology we usually want to know how causes operate.

Two further approaches suggest how reasoning should proceed. One is through the use of contrasts – that we should attempt to advance by isolating comparisons where a process or condition does and does not occur. This is an important method and of course is the underlying method of reasoning of experiments where a treatment is compared with a control. However, it is not always possible to find appropriate contrasts. Another approach acknowledges the difficulty of making inference and advocates always holding multiple postulates for a phenomenon, as in the photosynthesis example given in Chapter 6, to avoid favoring one unduly and becoming so attached to it that no alternative is even considered. While this may seem sound advice there is the practical difficulty that usually few postulates can be investigated and understood comprehensively by one scientist.

There is no single method of reasoning that can be universally approved and followed. These different methods are described in this chapter, their strengths and difficulties outlined, and how they can best be used is discussed.

7.2 Principles of propositional logic

Deduction uses propositional logic – a large and complex subject (for texts, see Quine 1972 and LeBlanc and Wisdom 1976) whose methods include Venn diagrams, ways of representing complex arguments in truth tables, consistency trees, and Boolean schemata.

DEDUCTION

Deduction is the application of the rules of propositional logic in an argument. Propositional logic assigns statements as true or false and the conclusions made do not claim more information than is already contained in the statement of the theory.

Typical deductive reasoning is:

(1) All Xs are Ys.
(2) All Ys are Zs.
(3) Therefore, all Xs are Zs.

Note that the conclusion restates information already contained in the theory (1 and 2), although deduction can involve complex reasoning, particularly the use of the implication statement (see later in this section of this chapter).

Mahoney (1976) lists definitions and assumptions of propositional logic. These numbered rules are given here, slightly modified, and with discussion on their implications for use in ecological analysis.

(1) PROPOSITION

A proposition asserts a relationship between a set of concepts. A proposition can be classified true or false.

Before you can logically analyze a theory, you have to do two things. First, you must define the terms with which the logic is to be argued. In this book, these are defined as concepts, which are classified in Chapter 3 according to their type and status. Second, you must list out the component statements as propositions. These two things are done in a conceptual and propositional analysis.

(2) LAW OF THE EXCLUDED MIDDLE

All propositions are true or false. No proposition is both true and false, and no proposition is neither true nor false.

A frequent aim in research is to obtain a set of propositions that are classified as true or false. At the start of research, some propositions (the axioms) are assumed to be true or false, and these are used in deductive reasoning. If they can not be classified in that way then deductive reasoning can not be used.

(3) CONJUNCTION

The conjunction of propositions is expressed as "and"; a conjunction is true if and only if all of its constituent propositions are true.

Consider axioms of research into climate change and tree-species distribution in Alaska (Chapter 5):

(1) *Climate change is uniform on continental and global spatial scales.*
(2) *Summer temperature controls the location of the tree line.*
(3) *Summer temperature is a component of climate.*
(4) *In central Canada, the tree line was north of its present location 6 ka BP to 3 ka BP.*

If all of these were true it would be reasonable to conclude that the tree line would have been north of its present location elsewhere than central Canada. Through research we examine the consequence of this deduction. If it is found to be false then only one of the propositions needs to have been untrue. Confirming a proposition that is deduced from a conjunction can show a great deal.

(4) DISJUNCTION

The disjunction of propositions is expressed as "or"; a disjunction is true if any of its propositions are true.

Consider Postulate 2 of the coarse woody debris study (Chapter 4):

Small mammals use debris piles in riparian areas for denning, feeding, and shelter.

In practice, what was tested was the presence or absence of small mammals (there was also the axiom that each activity may lead to capture in traps). When the postulate was confirmed then any one of the following could have been true:

Small mammals use debris piles in riparian areas for denning,
or, *Small mammals use debris piles in riparian areas for feeding,*
or, *Small mammals use debris piles in riparian areas for shelter.*

If the postulate had not been confirmed then all of the component members of the disjunction would be untrue. This example is an open disjunction because there may be propositions other than those included here that are possible uses, i.e., these three have not been shown to be the exclusive uses. Falsifying a postulate deduced from a disjunction can show a great deal.

(5) NEGATION

The negation or denial of a proposition always has a truth value opposite to that of the proposition: in cases of double negation, the original truth-value is maintained.

(6) EQUIVALENCE RELATIONS AND SYMMETRY

An element of a proposition is said to be equivalent to itself (the identity relationship): equivalence is represented by the symbol $\equiv$. A proposition is said to be symmetrical when its truth-value is unchanged by reversing the order of its elements, i.e., $\{p$ is related to $q\} \equiv \{q$ is related to $p\}$, such as in $p + q \equiv q + p$. Equivalent terms may be substituted in a proposition without changing its truth-value, i.e., if $r \equiv p$, then $\{r$ is related to $q\} \equiv \{p$ is related to $q\}$.

Equivalence and symmetry are properties that have to be established through research. The insistence (Chapters 3 and 6) on explicitly distinguishing concepts by measurement from concepts from research or concepts by imagination is to ensure that the need to establish equivalence is recognized. Assuming equivalence can lead to faulty deductions. Sometimes, where reasonable equivalence can not be established, the meaning of a concept is allowed to change to be equivalent to a measurement – but then the logic of the argument also must change. For example, it is important to keep constantly in mind that field measurements of photosynthesis are not completely effective measurements (Chapters 3 and 6), and neither is trapping of small mammals a completely effective measurement of the use of coarse woody debris piles. In neither case should the measurement be used as an unqualified equivalence to the concept from research.

(7) IMPLICATION

A compound proposition which asserts that an antecedent condition entails (requires or necessarily implies) a consequent condition is called an implication (symbolized as $p \Rightarrow q$, or "if p, then q," where p is the antecedent and q the consequent).

Symmetry cannot be assumed simply because an implication is assumed to be true, i.e., $p \Rightarrow q$ does not mean $q \Rightarrow p$. For example, deciding after research that needle growth rate determines current photosynthesis rate does not imply that current photosynthesis rate determines needle growth rate. Needle growth may also be supplied from storage carbohydrate, which may compensate for a decrease in current photosynthesis rate.

(8) TRUTH STATUS OF IMPLICATIONS

An implication, e.g., $p \Rightarrow q$, is false only if the antecedent is true and the consequent false (i.e., observe p and $not\text{-}q$); otherwise it is true. From this, it follows that:

(a)   a true antecedent occurs only with a true consequent, if p is true then q must be true;

(b)   a false consequent always indicates a false antecedent, $not\text{-}q \Rightarrow not\text{-}p$;

(c)   a false antecedent has no necessary bearing on the truth-value of the consequent, if $not\text{-}p$ we can deduce nothing about q; and

(d)   a true consequent has no necessary bearing on the truth-value of the antecedent, if q we can deduce nothing about p.

There are two valid forms of implication and two invalid forms. Consider the following:

Proposition: If the premise (p) is true, then the conclusion (q) is true.
Example: If Rex is a dog, then Rex is a mammal.

This simple example can be represented in Venn diagram form (Fig. 7.1).

The valid forms of reasoning are confirmatory, where the premise is true and so the conclusion is also judged to be true and disconfirmatory where the conclusion is false and it is deduced that the proposition must be false.

Fig. 7.1. Venn diagram representing the simple proposition:

If Rex is a dog, then Rex is a mammal.

The square, U, is all possible events. If we establish, as an initial condition, that all possible animals will be considered, then U is all possible animals. The large circle represents all mammals, and the proposition:

Mammals are one among a number of types of animals.

Everything outside the larger circle is all animals that are not mammals. The small circle represents all dogs.

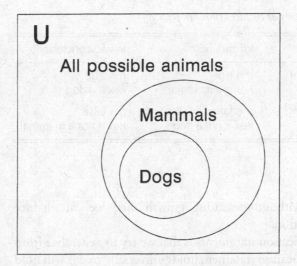

Table 7.1. *Two valid forms of implication for $p \Rightarrow q$*

		Information	Valid conclusion
(1)	Confirmatory reasoning	p is true	q is true
	Example	Rex is a dog	Rex is a mammal
(2)	Disconfirmatory reasoning	q is false	p is false
	Example	Rex is not a mammal	Rex is not a dog

The two invalid forms of reasoning are affirming the consequent and negating the antecedent. If you affirm the consequent you forget that the conclusion may be satisfied by conditions other than merely the proposition you are using, simply because Rex is a mammal does not mean that he is a dog. If you deny the antecedent you assume that because the first part of the proposition is found to be false the conclusion is false: because Rex is not a dog does not mean that he is not a mammal.

In the example from Chapter 6 in which shoot and needle growth rate may influence photosynthesis rate, a change from high to low growth rate is the antecedent condition, a change from high to low photosynthesis rate the consequent condition. As long as we continue to observe the two changes then the implication remains true. It is only falsified if we observe that a change from high-to-low growth rate is not followed by a change from high-to-low photosynthesis rate.

If we observe no change in growth rate (false antecedent) we can deduce nothing about whether photosynthesis rate might or might not change if growth rate did change. Similarly, if we observe a change in photosynthesis

Table 7.2. *Two invalid forms of implication for $p \Rightarrow q$*

		Information	Invalid conclusion
(1)	Affirming the consequent	q is true.	p is true
	Example	Rex is a mammal	Rex is a dog
(2)	Disconfirmatory reasoning	p is false	q is false
	Example	Rex is not a dog	Rex is not a mammal

rate from high-to-low (without measuring growth rate) we can deduce nothing about the implication.

The problem with implication statements is that we try to generalize from them. We say, "If the implication statement holds universally true it will hold in these cases. It holds in these cases. Therefore it holds universally." This is the fallacy of affirming the consequent. In the photosynthesis example, if we consider the implication statement to be universally true then we may want to change our theory of what controls photosynthesis in *Pinus strobus* and assume that the reasoning that gave rise to the postulate is true (Table 7.3). This is inductive reasoning and, as Mahoney (1976) says: "Induction involves a . . . conjectural leap to a conclusion whose content exceeds the information contained in its premise."

INDUCTION

An inference is inductive if it passes from particular statements such as accounts of the results of observations or experiments, to universal statements such as theories. (After Popper 1968.) In practice if a series of instances of a phenomenon is found then induction is the process of stating some aspect of generality from that. This may be based on empirical evidence, i.e., inference from repeated occurrences of an observation, or based on the hypothetico-deductive method, i.e., investigation of a conjecture based on an existing theory.

But don't scientists use this type of reasoning all the time? Yes! Inductive reasoning is the basis for most research. Nevertheless it can come as something of a surprise that a fundamental method used in science is based on a logical fallacy. This is why science is a challenge and why critical analysis is so important – not only of what we think we may have found out but also of the concepts, axioms and postulates that we have used as the basis for discovery.

The more detail with which we investigate a biological or ecological process, the more certain we may feel about our explanations, and the more inclined we may be to elevate a postulate to the status of an axiom. However, this is only subjective confidence in that the propositions have escaped

Table 7.3. *Logical statements of scientific investigation defining the fallacy of affirming the consequent*

Logic	Scientific activity
If A, then B	Formation of new postulate (B) from axioms of the theory and subjective ideas about the cause of the observed problem (A)
B is the case	Examination of postulate in relation to data and eventual confirmation
Therefore, A is the case (This is a fallacy)	An assumption that the reasoning that gave rise to B is correct

falsification. We never prove the postulate through confirmatory investigations. In practice, big errors of falsely turning a postulate into an axiom are likely to be exposed quickly. More insidious, however, is that the limitations on postulates may not be detected. A postulate may be true some of the time but not all of the time. Inductive reasoning has no way of detecting this other than through developing additional contrasting postulates that may also happen to be true some of the time.

Inductive reasoning can also be used, for example, where items are classified:

(1) X_1 is a Y; X_2 is a Y; X_3 is a Y; . . . X_n is a Y.
(2) Therefore, all Xs are Ys.

This type of reasoning can be seen in the project investigating animal use of woody debris (Chapter 4). The over-arching axiom of that study was:

Debris piles are an important component of stream ecosystems.

With the more specific postulate:

Small mammals use debris piles in riparian areas for denning, feeding, and shelter.

Concluding that woody debris was important within rivers and streams in determining the habitat for fish stimulated the study of woody debris along riverbanks. The pattern of reasoning is "Woody debris in rivers and streams (X_1) provides an important habitat (Y_1) for animals," extending to "Woody debris along river banks (X_2) provides an important habitat (Y_2) for animals." Of course, both of these are part of the general argument that "Woody debris (X_n) is an important habitat (Y_n) for animals," which is an extension of the over-arching axiom. This is inductive reasoning, and such propositions, even when classified as axioms, remain uncertain.

There are four approaches to the thorny problem of induction.

(1) Attempt to falsify a proposition. If the attempt fails, then the certainty about the proposition may increase. We would wish, however, to judge the increase in certainty according to the type of falsification used. For practical reasons falsification can be difficult to apply in ecology.

(2) Advance by using contrasts where the conditions that produce a phenomenon can be isolated from the conditions that do not produce it. This is the basis of experiments, where treatments are compared with controls, but the method is not restricted to experiments.

(3) Seek causal and organizational reasons that connect postulates to, or define, fundamental biological, chemical or physical processes. However, the basis for causal or organizational reasons may themselves require much ecological research.

(4) Construct multiple competing postulates. In this way we seek to check the temptation to be too ready to confirm a particular postulate.

These four approaches are discussed through the latter sections of this chapter.

(9) CONTRADICTION

The conjunction or implication of any proposition with its own negation is false.

(10) TRANSITIVITY

If the consequent of a true implication is itself the antecedent of another true implication, then it is true that the antecedent of the first necessarily entails the consequent of the second (i.e., if $p \Rightarrow q$ and $q \Rightarrow r$, then $p \Rightarrow r$).

For example, adding the proposition *If an animal is a mammal (q) then it is a vertebrate (r)* to the argument represented by Fig. 7.1 leads to *Rex (p) is a vertebrate (r)*. A circle containing the circle representing mammals would represent vertebrates in Fig. 7.1. In ecological research, conclusions made from transitivity are those most vulnerable to the problems associated with a researcher assigning a true or false classification to a concept rather than a probability.

7.3 The use of propositional logic in ecological research

Propositional logic is fundamental to scientific reasoning – it is what we aim to use when we generalize, when concepts have absolute definitions and our interest is to explore the consequences of a theory. The difficulties must be explored because it has been claimed by some that this is the reasoning all scientific research should use. The use and potential pitfalls of the four most important rules are summarized in Table 7.4.

Table 7.4 *Four important rules of propositional logic, their principal uses, and some potential pitfalls in using them*

Rule	Use	Potential Pitfalls
Conjunction (Rule 3) A conjunction is true if, and only if, all of its constituent propositions are true	This is the principal rule used in searching for inconsistencies within an established theory consisting of a number of propositions. Predictions made must be tested	(1) Inadequate definition of concepts (2) Lack of equivalence between concepts used in the linked series of propositions
Disjunction (Rule 4) A disjunction is true if any of its propositions are true	It is useful to examine existing propositions to see whether they are a disjunction, i.e., that they do contain alternatives. This procedure is particularly valuable for new propositions and theories. It is important to try to close a disjunction if it is to be used in making deductions	Recognizing a disjunction depends upon precise definitions of concepts. In particular, the use of holding concepts can protect a proposition from being recognized as a disjunction
Equivalence (Rule 6) Equivalent terms may be substituted in a proposition without changing its truth value	The procedure is to ask the question: Is this {concept or proposition} equivalent to that {concept or proposition}? This is particularly valuable for examining the relationship between a measurement and the concept it represents, and for examining whether new supposed examples or cases belong to a theory	The pitfall is in not asking questions of equivalence!
Implication (Rule 7) An antecedent condition requires or necessarily implies a consequent condition	This is the principal rule used for examining a single proposition, both its own implication and its role in the whole theory, i.e., in compound implications	(1) Fallacies of affirming the consequent or denying the antecedent (2) Failure to use probability rather than deterministic reasoning when necessary

Platt (1964) is a forceful and frequently quoted proponent of this approach – he gives it the catchy title *strong inference*, sometimes used in textbooks dealing with ecological subjects (e.g., Trudgill 1988, Hairston 1989, and Haefner 1996). However, Platt writes as a biochemist. For him questions were a series of alternatives about a biochemical process. The aim was to develop what he termed a "logical tree" of argument, "If A then B, if B then C; if X then Y, if Y and C then D . . ." i.e., connected series of implication statements (Transitivity, Rule 10 of propositional logic).

Platt (1964) argues that the conclusion from an experiment must always exclude some postulate. So we can say, "We now know it can't be that," or in his terms, "any conclusion that is not an exclusion is insecure and must be rechecked". He emphasizes that you should write down alternatives, do crucial experiments, and focus on excluding a hypothesis,[1] referring to this type of reasoning and argument as "a conditional inductive tree." However, the process Platt actually describes is one of deduction, and he does not explain the inductive part of his approach, i.e., how you decide what to consider next in developing an investigation.

Platt proposes the use of small and "elegant" experiments. He attributes a quotation to Levinthal that "You must study the simplest system you think has the properties you are interested in." Platt's view is that surveys, taxonomy, equipment design, systematic measurements and their tables, and theoretical computations "all have their proper and honored place, provided they are parts of a chain of precision . . . of how nature works." He considers the important issues of science as qualitative, not quantitative – equations and measurements are useful, but the most convincing arguments are made without quantification. He defines the use of logic as capturing an argument within a coarse but strong box, in contrast to that of mathematics. "The mathematical box is a beautiful way of wrapping up a problem, but it will not hold the phenomena unless they have been caught in a logical box to begin with."

Platt argues that his experimental science could produce answers in the form of alternatives and considers this stronger than other types of science. For Platt the objective in experimentation is to define the relationship between events; use of the term causal relationship should be restricted to where an antecedent is both necessary and sufficient for a consequent.

But can we use this approach in ecology? Platt's science was one where extensive exclusive procedures were necessary to perform controlled experiments. Biochemists and molecular biologists working when Platt wrote were intent on analyzing the biochemical processes within cells. They gained

---

[1] Platt uses the word "hypothesis" in the general sense of theory on some occasions, and sometimes in the precise sense as a postulate that can be examined and rejected or confirmed.

extensive experience with a few experimental systems (e.g., rat liver, eryth-rocytes, and strains of *Chlorella, Escherichia coli*), and most of their research was *in vitro* (following Levinthal's injunction to "study the simplest system with the properties of interest"). Typically, enzymes or enzyme sequences were prepared from living cells, usually from a cell culture. The questions asked were how these worked in sequences in the metabolic process (discove-ring transitivity), or what set of conditions was necessary and sufficient for a process (specifying conjunctions and logical deductions from them). Experi-ments frequently consisted of providing different substrates to the cell culture, or enzyme extractions. But "studying the simplest system . . ." came with a penalty of interpreting the subject of analysis, i.e., enzymatic and cellular processes, separated from its context, i.e., the whole cell and the interactions between cell and organism.

Inference based on logical outcome is clearly a powerful form of reasoning. But its use depends upon particular conditions being met, particularly that probabilities are removed and that the subject being studied does have a complete determinism, i.e., a complete chain of necessary and sufficient conditions can be worked out. But it is not applicable to all science and it is not a superior form of reasoning, as Platt (1964) suggests. Indeed Platt's paper is really the assertion of a particular philosophy (Chapter 9).

First, it makes an assumption about the prime role of experimentation. Experimental results sometimes should not be considered paramount because they depend upon both the assumptions made in establishing the experiment and sufficient prior knowledge about the subject that makes the experiment reasonable. Hafner and Presswood (1965) analyze the development of some scientific theories in physics where, they suggest, "strong inference" is an idealization. They note that its application requires the existence of an already developed tree of hypotheses. If a theory has not reached such a stage, however, strong inference will not work (i.e., exploratory analysis is required first, see Chapter 8).

Logical outcome analysis frequently requires experimentation, but the questions that controlled experiments can answer are limited. It is not possible to construct an experiment to answer every ecologically interesting question. Even when experiments are constructed, it may not be possible to design a control for every influence other than those that are the purpose of the experiment, which means that experimental treatments may be less penetrating than the investigator had hoped.

Second, concepts in biology and ecology do not always have uniform definition and application across the whole subject and they are often prob-abilistic (Sattler 1986). Working to define concepts and examining whether they can have uniform application is an important part of ecological research (Section II). Defining the probability of a concept is not only important for the definition, but for the logic that must be used in research, particularly

LOGICAL ARGUMENT

IF **A** ENTAILS **B** AND **B** ENTAILS **C** THEN **A** ENTAILS **C**

PROBABILISTIC ARGUMENT

PROBABILITY (**R** GIVEN **P**) — PROBABILITY (**Q**, GIVEN **P**) X

PROBABILITY (**R**, GIVEN **P** AND **Q**) +

PROBABILITY (NOT- **Q**, GIVEN **P**) X

PROBABILITY (**R**, GIVEN **P** AND NOT- **Q**)

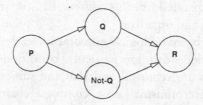

Fig. 7.2. Transitivity (Rule 10 of propositional logic) is a property of logical entailment. If being a whale (A) entails being a mammal (B) and being a mammal entails having hair (C), then being a whale entails having hair. The probabilistic analogue is more complex. The probability that a human (P) will get lung cancer (R) equals the probability that the human (P) is a smoker (Q) times the probability that a human smoker (P and Q) gets lung cancer plus the probability that (P) is a non-smoker (not-Q) times the probability that a human non-smoker (P and not-Q) gets lung cancer. The argument that you can not prove smoking causes lung cancer is a fallacious application of logic. Calculating probabilities is the only appropriate procedure for reasoning in this case.

when you consider the implications. In this the probabilistic argument is more complex than the logical one (Fig. 7.2) but clearly offers a more effective reasoning process where concepts exist as probabilities. Salmon (1973) discusses the importance of recognizing that, in the probabilistic argument, confirmation is incremental.

Attempts to assert that the logical argument of entailment should be used rather than the probabilistic one do occur in ecology and resource management. For example, statements such as "One cannot prove that acid rain or tropospheric ozone damages trees" are fallacious: these problems, like the relationship between smoking and cancer (Fig. 7.2), are about probabilities (whether they be low or high) and must be left as such.

Third, these difficulties with transitivity are really a symptom of attempting to analyze complex systems with simple logic. Whether a system will produce a particular condition may require specifying many conjunctions, disjunctions, and conditions. The restricted use of the term "causality" proposed by Platt – the occurrence of A must be both necessary and sufficient for B – is often not appropriate for ecology. In ecology some results can be produced by more than one effect and certain effects operate on some occasions but not others, and we need to determine when those are. For example, in the study of breeding isolation in trout species (Chapter 5), different processes may operate – those related to behavior, geography, habitat selection, or timing of breeding cycle. For a comprehensive understanding we must know when, and why, any of these may occur, i.e., their domain of influence (Chapter 10), as well as the details of any individual process.

Seeking to construct a transitive sequence of implication statements may not be an appropriate approach for analyzing dynamic systems where links of influences are cyclical with recursive networks, not linear and exclusively conditional. For example, what we may encounter in ecology are relationships of the type: A influences B, B influences C, B influences D, and C influences A. So C has an effect on D through an effect on A's influence on B. Such indirect influences may appear important sometimes and unimportant at other times. It is this type of conditional effect that plant water status and atmospheric humidity deficit may have had on mid-summer photosynthesis rates (Chapter 6).

There are two particularly important dangers in viewing logical analysis and the methods in Table 7.2 as premier methods of reasoning for scientific inference in ecology:

(1)   In order to classify a postulate as true or false, you may encourage yourself to attribute a close identity between a measurement and a concept when that is not justified.

(2)   Classifying what really should be probabilities as true or false may encourage premature or false generalization, e.g., Fig. 7.2.

7.4 The hypothetico-deductive method and use of falsification in scientific reasoning

HYPOTHETICO-DEDUCTIVE METHOD

The method of creating scientific theory from which the results already obtained could have been deduced and which also entail new predictions that can be verified or refuted by observation or experiment. (After Flew 1984.)

The hypothetico-deductive (H-D) method is wider in scope than empirical induction because it seeks support from the deductive consequences of existing theory – but it is essentially an inductive method and predictions should not be viewed as pieces of evidence completely independent of the theory used to produce them. Testing the first postulate of northward movement of tree species in Alaska is an example of the H-D method because everything was logically consistent from within the existing theory. Neither Steel's nor Hawkins' work followed the H-D method. In both cases their induction extended beyond the content of existing theories.

When you make a prediction based on deduction there are three things to consider: the evidence you will seek (measurement, observation, result of experiment), the postulate giving rise to the prediction, and the theory that gave rise to the postulate. Predictions are not independent tests of postulates because any measurement, observation or experiment is carried out against a particular theoretical background. Both what can be measured, and how it

can be measured, become integral parts of theories, e.g., the axioms of
sampling and measurement for lake coring and pollen counting (Box 5.1). So
a prediction is not just a test of a postulate but also a test of these auxiliary
axioms – some of which may have considerable theoretical content.

Although some scientists refer to the H-D method as **the** method of science
there is little support that it actually describes what scientists do, or should do.
In particular the inductive nature of the H-D method, despite its name, has
drawn attention. Can this be improved by insisting that attempts must be
made to falsify postulates?

Popper's (1968, 1972) case for bold conjectures and attempted refutations
attracted the notice of many scientists. Popper (1968) proposes that a
statement is scientific if, and only if, it is testable and identified testability
with falsifiability but not with verifiability or probability. If an attempt is
made to falsify a postulate, and the falsification fails, then the postulate is
corroborated, although it may still be found false in the future. The position is
that you can disprove things but you can never prove them. It is this point of
Popper's philosophy that is sometimes used as a talisman in the scientific
culture.

Attempts to falsify postulates, rather than to seek their confirmation, are
clearly likely to expose a theory to failure or change, or to make it more
certain. Popper (1968) constructs the following set of methodological rules
based on this strategy. I include comments about their application to circumstances in ecology:

(1)

    (a)   Admit into science only propositions that can be falsified.
          Comment: There are of course difficulties with this.

               Woody debris piles are important to small mammal use of cobble bars

          may not be falsifiable practically, as a complete postulate, but it is
          still the proposition of purpose. However, related propositions
          can be considered, such as:

               Postulate 2B. *Small mammal use of old debris piles at the forest edge*
               is different from small mammal use of debris piles that are on a cobble
               bar.

          This is falsifiable, in the sense that a direct investigation could be
          made which may show that this postulate is not true. While this
          postulate would be acceptable to a falsificationist, its value can be
          determined only in the context of other postulates, some of which
          may not be directly falsifiable.

    (b)   Once a postulate has been tested, i.e., attempt(s) have been made
          to falsify it and have failed, then it may not be removed from a

theory. But it may be replaced by another postulate which is more testable.

Comment: For example, once investigated, Postulate 2B can not be removed merely because some new postulate is thought of – say, that differences in ground vegetation explain the differences in small mammal numbers between the forest and cobble bar. Postulates must be constructed that would investigate some possible effect of ground vegetation, taking into account the result of differences between piles.

(2) Have no other rules that protect a proposition against falsification.

(a) If there is a conflict between a theory and what is observed, the theory can be modified only by increasing its falsifiability, i.e., there should be no *ad hoc* postulates or axioms added to the theory.

Comment: For example, suppose use of debris piles in the forest was found to be greater than that on cobble bars. Then Postulate 2B should not be saved by invoking a special event (such as a dramatic and unusual increase in birds of prey that ate the small mammals on the cobble bar but not in the forest) without investigating whether it occurred and was indeed the responsible factor.

(b) Changes in concept definition must be regarded as changes to the theory, and rule 2(a) should apply to these definitions.

Comment: For example, *use* is defined in the CWD investigation by trapping animals. A change to a more specific concept, say, breeding, denning, or feeding, is a change to the postulate. So if the postulate with *use* has been falsified, it should not be saved by explaining that there are nevertheless differences in breeding, denning, or feeding despite the falsification of *use*. New postulates must be constructed.

(c) The reliability of a scientist whose results challenge a postulate or theory cannot be an excuse for rejecting those results. Counter-investigations must be performed.

Comment: Ideally this should be the approach taken. However, it is important to appreciate the bias that scientists may have both in what they decide to do and how they may interpret results (see Section III).

(3) In principle, the testing of theories is without end.

These are valuable rules, although they are not sufficient for the practicing scientist. A major problem is Popper's core idea that the task is to increase the falsifiability of theories. In some cases, practicing scientists balance the heuristic of over-arching axioms, which can not be falsified directly but give a driving purpose to a research effort, with detailed arguments about possible falsification of the component postulates (see the discussion of Lakatos' (1970) methodology of scientific research programs, Chapter 11). So defining postulates that can be falsified can work – but only within a framework that is generally inductive. Moreover, as described later in this chapter, frequently in the systems that ecologists have to describe, the logical status of a postulate – whether it is necessary, sufficient, or both sufficient and necessary – can not be studied simply by considering whether or not a postulate is falsifiable.

Popper was arguing in general terms without necessarily considering the practical difficulties in performing meaningful experiments or of making effective, accurate and precise measurements. His objective was to illustrate that repeatedly seeking confirmatory evidence in a straightforward inductive way is not an adequate procedure and he wished to provide an alternative.

For some postulates there may be large technical difficulties in conducting experiments that may falsify a postulate. For example, prior to starting her investigation Steel found some scientists skeptical that any small mammals used the CWD piles on the cobble bar. Her research showed that they did use the piles. Imagine, for the sake of argument that we replace the word "important" that Steel used with "necessary" as Platt urges. To falsify the postulate:

> *Woody debris piles are necessary to small mammal use of cobble bars.*

would require removing piles, over some minimum size of area, so that animals would not stray in from surrounding areas, and replicating such a treatment. Certainly the word *necessary* makes the postulate bolder and more testable. If an attempt to refute this postulate fails, i.e., no small mammals are found from where the CWD piles have been removed, then the postulate has been corroborated. It withstood a serious challenge and is much stronger. But such an experiment would be expensive, and most possibly not permitted by the agencies responsible for the land. In practice, as discussed in later sections of this chapter, this postulate should be refined and more specific postulates derived that can result in falsification. For example, there is already the observation that CWD piles are frequented more by some species than others, and tests must take such information into account. There are no naturally occurring cobble bars along this river that do not have CWD piles.

So, while falsification is an important quality that a postulate may have, we should not require it of all postulates. We do have to use some positive inductive support. Further, Lipton (1991) points out that falsification does not occur exclusively through the refutation of postulates or theories after comparison with data. Scientists may reject a theory as false because, while the evidence may not contradict it, it fails to explain something important. In this cases a positive judgement is made about whether the explanatory failure is more likely to be due to incompleteness and/or error, and this judgement depends upon inductive reasoning.

The H-D method is not rescued by insisting upon falsification. It has important weaknesses, with or without an emphasis on falsification. These are:

(1) In the H-D method there need be no empirical constraint on the formulation of postulates. The H-D method emphasizes the origin of postulates from existing theory or metaphysical argument. Examples appear in the literature where investigators have sought confirming instances for such postulates – but demonstrating that something can apparently occur lacks power in reasoning unless conditions when it does not occur are also studied. Answering questions of the type "Why P?" (which can be produced from the H-D method) rather than "Why P not Q?" builds incomplete knowledge. This is discussed further in 7.7.

(2) Because the H-D method gives priority to theory as a guide to observation and experiment the origins of the theory may be neglected and its relevance to a particular question is ignored. In this way attempts may be made to use a theory that is not appropriate for answering a particular question.

(3) The H-D method does not say where a postulate should come from – any speculation from within a theory is acceptable. Of course the postulates must be logically consistent with previous data – but there are likely to be alternatives and the H-D method does not inform about how to judge between them. By selecting postulates that are not bold then it may be possible to obtain confirmatory evidence that is not particularly telling. Selecting a bold postulate and scouring for a single confirming instance can lead to misleading conclusions if generality is then claimed. Both these extremes happen, although usually there is consideration by scientists of what is a "good" postulate.

(4) Scientists are supposed to check observable consequences – but this is too restrictive, since there may be relevant data that are not entailed by the theoretical system but are relevant to the question and how to construct a postulate.

7.5 An exercise in choosing between postulates expected to be true and postulates expected to be false

Induction is our natural form of reasoning. We observe the world, seek patterns, devise explanations, and extend those explanations. We do not proceed in our everyday living by attempting to falsify the basis of our everyday actions. Popper introduced the idea that postulates should be falsifiable. An additional idea is that we might actually choose to investigate a postulate that we think is false.

The value for scientific research of choosing a postulate that we expect to be false has to be learned. One practical exercise – the Eleusis game (after Romesburg 1979) – demonstrates the value.

A deck of 52 normal, shuffled, playing cards is used. A sequence of cards is laid down, in a horizontal row, face up and one at a time by a postulator, who attempts to discover a rule known only to one observer, the target-holder. The target-holder, who decides upon the rule in advance of any cards being laid down, and responds to each card with "true" or "false" according to whether or not it follows the rule. The analogy is that each card laid down represents a scientific inquiry, the postulator is a scientist, and the target-holder knows the true nature of how the world works. The postulator is usually one of a group discussing what the target rule may be, developing postulates about it, and considering whether a particular card should be laid down in the sequence. The postulator may choose to discard cards that would not produce a required test. The postulator may ask the target-holder only once what the rule is, so the idea of the game is that the group working with the postulator should be certain of the rule before the postulator asks. During the game the target holder should say nothing more than "true" or "false" when a card is laid in the sequence.

Suppose the target-holder produced a simple rule, that of alternating colors (where ♥ = hearts, ♣ = clubs, ♦ = diamonds, and ♠ = spades; A = ace, K = king, Q = queen, and J = jack). Then the following sequence of cards happened to be turned over – 10♥ (initiates the sequence), 5♣ (the target-holder would declare this true, because black followed red), J♣ (false, because black followed black), 9♥ (true), 3♦ (false), 4♥ (false), K♠ (true), 2♥ (true), A♠ (true) – and laid out in the following way:

10♥	5♣	9♥	K♠	2♥	A♠
	J♣	3♦			
		4♥			

The cards in the top line (10♥, 5♣, 9♥, K♠, 2♥, A♠) follow the rule. The cards in each column (J♣ and then 3♦, 4♥) are those that failed to follow the rule. They were incorrect when laid down in the horizontal line after the card at the top of the column. This leaves a visible record. To ensure progress in a reasonable time, the types of rule the target-holder should select include:

> Adjacent cards must be separated by a difference of 1.
> Adjacent cards must be separated by a difference of < 3.

Odd and even cards must alternate.

Cards must alternate either in terms of parity (odd vs. even) or color (black vs. red) or both.

Usually, at some point in the game, the group is faced with a mixture of true and false trials and, on this basis, will develop one or more postulates. They can then decide whether to lay down a card that they think should confirm their postulate, or one that will attempt to falsify it. For example, while the above sequence certainly follows the rule of alternating colors it also follows another rule, that black cards must separate hearts. Given the sequence that would be a reasonable postulate. It could be confirmed (falsely) by laying down a heart after A♠, or a (successful) attempt to disconfirm it could be made by laying down a diamond.

The game can be played in different ways. For example, one group may follow a confirmatory strategy choosing trials that are always expected to be true, one group attempts falsification choosing trials expected to be false, whenever a postulate is made, and one group uses a mixture of the two strategies.

The game usually illustrates how difficult it can be to formulate a postulate from a series of trials. Participants frequently forget, or ignore the complete evidence obtained. It also illustrates that there can be difficult choices to make over whether to attempt a confirmation or disconfirmation, but that usually at least one attempted disconfirmation is required somewhere in the sequence. Most people who play the game find it particularly valuable to have a turn as target-holder when they can listen to the discussion among the group about what card to lay next.

7.6 How to decide whether to attempt confirmation or falsification

Klayman and Ha (1987) have made a penetrating analysis of the circumstances where attempts to use postulates that are expected to turn out false are appropriate. They review instances from different types of research, discuss why scientists most frequently test postulates by examining tests that they expect to come out true, and reference many instances of research into the scientific method indicating that scientists seek to confirm a bias rather than falsify a bold conjecture. However, much of this research has been based upon analysis of a "rule discovery" game simpler than the Eleusis game in the previous section. Subjects are presented with the sequence 2, 4, and 6, which, they are told, follow the rule. They then can choose a sequence of any three numbers, ask whether it follows the rule, and so attempt to discover the rule by a series of such tests. The rule that many investigators have wanted their subjects to discover is any ascending numbers (Klayman and Ha 1987), but the initial sequence the subjects have been given (2, 4, 6) and told is true also contains an implication that numbers must be even and/or ascend by 2s. Falsification, applying a test that is expected to come out false, is the strategy to eliminate these possibilities of evenness or intervals of 2. However, Klayman and Ha point out that this is only one type of problem; other types of problem exist for which falsification may not be the best strategy. The important conclusion is that the more efficient strategy, whether to use a test that is expected to come out true, or one that is expected to come out false, depends upon the relationship between the postulate (your idea about the truth) and the actual "truth" (what they call the target). At first, this conclusion may seem to be of little value as the target is unknown – which is precisely why there is a postulate in the first place! The value of the approach is that it makes you think critically about a postulate and the stage of uncertainty of the whole investigation.

Four types of relationships are considered between postulate (P) and target (T):

(1)   The postulate is true, i.e., it is contained within the target, but there may be other postulates that also are true.
(2)   The postulate is true under some circumstances and false under others, and not all true circumstances are explained by the postulate.
(3)   The postulate is true under some circumstances and false under others, and all true circumstances are explained by the postulate.
(4)   The postulate is never true.

I illustrate how these four relationships can arise: first, by considering different examples of the rule discovery game and, second, by examining possible scenarios from the practical examples given in this book.

Let us consider the most frequently used instance where the sequence 2, 4, 6 has been presented by the target-holder and classified as true. Psychological research has shown that the chosen postulate for the rule (target, T) is most likely to be numbers *ascending by intervals of 2*. Now the actual target is *any three ascending numbers* so that the postulate is contained within the target (Fig. 7.3). Under these circumstances, the postulate proposes a + Ptest (one expected to come out true): another three numbers ascending at intervals of 2, e.g., 8, 10, and 12. The target-holder's answer "yes," which indicates that P conforms to T, is an ambiguous verification of P: there can be other circumstances that are true and that the postulator does not know exist. A − Ptest (one expected to come out false) could give two possible results. For a sequence such as 8, 6, 2, deliberately chosen not to conform to P (i.e., ($\bar{P}$), the target-holder's expected answer, "no" is another ambiguous verification that P represents T. The postulate is conclusively falsified only when a − Ptest is submitted that is contained in T but not P, e.g., 2, 6, 8, in which case the

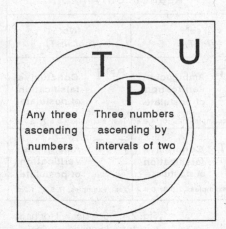

TYPE OF TEST PERFORMED	RESULT OBTAINED	
	"Yes" (in T)	"No" (not in T)
+Ptest Test we expect to come out true	$P \cap T$ Ambiguous verification of postulate Test examples: 2,4,6; 8,10,12	$P \cap \overline{T}$ Impossible
−Ptest Test we expect to come out false	$\overline{P} \cap T$ Conclusive falsification of postulate Test examples: 2,6,8; 2,3,4	$\overline{P} \cap \overline{T}$ Ambiguous verification of postulate Test examples: 8,6,2; 2,4,2

Fig. 7.3. Left: Venn diagram representing the situation where a postulated rule, P, is contained within the target rule, T. In this diagram P happens to be correct, but other rules may also be correct so the target has not yet been discovered. U represents all possible instances. Right: Outcomes of attempting tests we expect to come out true (+Ptests), where sets following the P rule are tried, and tests we expect to come out false (−Ptests), where sets are deliberately constructed that do not follow the postulated rule. Note that while only a −Ptest (i.e., testing something contrary to the postulate) can provide conclusive falsification that P is sufficient, not all such tests do that. The symbols P ∩ T mean that the postulate P is within the target, T; ∩ means the "intersection of". $\overline{P}$ (i.e., anything that does not follow the rule of the postulate) is the complement of P; $\overline{T}$ is the complement of T. Some rules that do not follow the postulate may be in T and some in $\overline{T}$. (Redrawn from Klayman and Ha 1987, with permission.)

target-holder confirms that the sequence follows the rule when the postulator expects it not to.

Now consider another hidden target: any three even numbers. Again 2, 4, 6 is the initial given card sequence classified by the target-holder as true, and the P of "three numbers ascending by intervals of 2" is chosen. In this case, P and T overlap (Fig. 7.4). The postulate is conclusively falsified by a + Ptest (e.g., 3, 5, and 7), which follows the postulate but is false to the target, or by a − Ptest (e.g., 2, 6, and 8), which is false to the postulate but true to the target. The important point is that a + Ptest conclusively falsifies the postulate in this instance.

What we can draw from this work is that the logical relationship between the target (i.e., the actual truth about the science you are investigating) and the type of postulate you propose can determine the type of investigations that will be most effective. For example, if, on the one hand, I felt that growth rate always influences photosynthesis but is likely to be one of a number of factors (other than light, water, and temperature), such as acclimation, that does so, then I should attempt falsification (i.e., a − Ptest) as in Fig. 7.3. If, on the other hand, I felt that my postulate

high growth rate ⇒ high photosynthesis rate

is only sometimes true, I could attempt a + Ptest, as in Fig. 7.4. In this case, P is not the complete description of T (i.e., P is falsified) if growth rate is low and photosynthesis rate high ($\overline{P} \cap T$), or if growth rate is high and photosynthesis rate low ($P \cap \overline{T}$).

Two other cases are of interest. If the postulate is too broad (Fig. 7.5a), i.e., it surrounds T and includes other possibilities, then only a + Ptest will

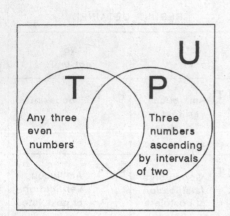

TYPE OF TEST PERFORMED	RESULT OBTAINED	
	"Yes" (in T)	"No" (not in T)
+Ptest Test we expect to come out true	$P \cap T$ Ambiguous verification of postulate	$P \cap \overline{T}$ Conclusive falsification of postulate
	Test examples: 2,4,6; 4,6,8	Test examples: 3,5,7; 5,7,9
-Ptest Test we expect to come out false	$\overline{P} \cap T$ Conclusive falsification of postulate	$\overline{P} \cap \overline{T}$ Ambiguous verification of postulate
	Test examples: 2,6,8; 6,4,2	Test examples: 7,5,3; 1,2,3

Fig. 7.4. See Fig. 7.3 for basic structural description. In this instance, conclusive falsification, as well as ambiguous verification, can be obtained from either a + Ptest or a – Ptest. (Redrawn from Klayman and Ha 1987, with permission.)

provide conclusive falsification. For example, if the target is *three consecutive even numbers* and the postulate is *numbers ascending by 2s*, only a + Ptest (e.g., three numbers such as 3, 5, 7) will provide conclusive falsification. Consider, in the photosynthesis example, if both low growth rates and high non-structural carbohydrate concentrations were necessary to produce a low photosynthesis rate. Then one might cool growing tissue but not reduce the photosynthesis rate. Recall that the bolder postulate (Fig. 6.7) suggested that the effect of reducing growth rate operates through an effect on non-structural carbohydrate concentrations. The effectiveness of experiments that reduce the temperature of the growing shoot depend upon close coupling between growth rate of that new shoot and foliage and photosynthesis of the one-year shoot that supports it and is having its photosynthesis rate measured. If other unchilled shoots drew more photosynthate, which compensated for the reduction from the chilled shoot, then photosynthesis rate might not be reduced. In that case, monitoring non-structural carbohydrates would be important.

+ Ptest and − Ptest are also balanced in their effectiveness if P is completely removed from T (Fig. 7.5b). This could apply to the situation where you guess a postulate without having any test sequences confirmed as true. The postulate is completely wrong, and you can obtain conclusive falsification from either a + Ptest or a − Ptest. Since you may also obtain ambiguous verification from a − Ptest, + Ptests might be more efficient at the exploratory stage. This may be a common situation in the early phases of research − when postulates are not informed by investigations of the actual situation but by guesses based on other situations. − Ptests comes into their own when P is closer to T.

This form of analysis can be applied to the problem of evaluating the role of CWD piles in the riparian zone (Chapter 4). Steel found that, on cobble

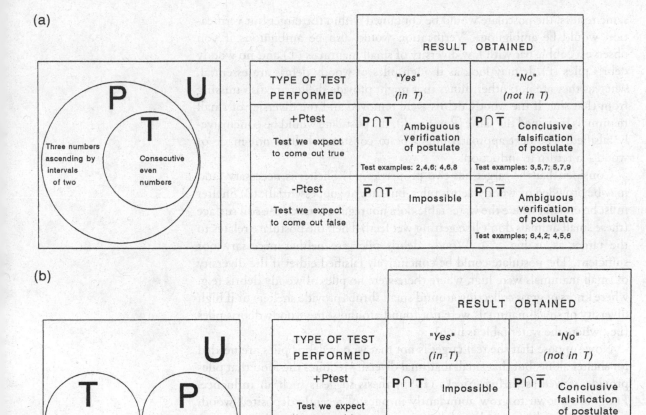

Fig. 7.5. See Fig. 7.3 for basic structural description. (a) If the target is more constrained than the postulated rule, then conclusive falsification can be obtained only from a + Ptest. A – Ptest is not as potentially informative. (b) This instance cannot apply to the situation where 2, 4, 6 were initially given as a true sequence and a postulate based on that. The postulate has no overlap with the target. (Redrawn from Klayman and Ha 1987, with permission.)

bars, diversity of small mammals in traps close to piles of woody debris was greater than that in reference areas away from the woody debris. At this point, a postulate could be constructed that *woody debris piles provide necessary shelter for a diverse small mammal population on cobble bars.* We can now consider three different targets, and how a strategy of + Ptests or − Ptests might vary according to the relationship between postulate and target.

If the target, i.e., the "real" truth, is that shelter is needed but can be anything, not necessarily CWD piles, then the postulate would be contained within the target, as in Fig. 7.3. Woody debris is sufficient but not necessary. If woody piles were added to the site and small mammals found to be associated with them, or if Steel's study were repeated at another site with the same results, the postulate would be contained within the target but verification would be ambiguous. Verification would also be ambiguous if you observe a cobble bar with low diversity of small mammals ($\bar{T}$) and no woody debris piles ($\bar{P}$). It may look as though piles of woody debris are essential, whereas they are not: other things that might provide shelter are also missing from that site. If the woody debris were removed and the diversity of small mammals remained the same ($\bar{P}$ not in T), the postulate would be conclusively falsified. The next approach might be to construct shelters not made of wood – a return to induction.

Consider that the target has two properties. (1) Shelter is necessary (and may be provided by woody debris piles, but they are not essential). (2) Shelter must be on land where the water table does not regularly reach the soil surface (these small animals don't like getting wet feet!) Then the postulate relates to the target as in Fig. 7.4. Woody debris piles are neither necessary nor sufficient. The postulate could be conclusively falsified either if the diversity of small mammals were high where there were no piles of woody debris (e.g. where knots of grass collecting around small shrubs provide shelter) or if high diversity of small mammals were not found around some woody debris piles (i.e., where the water table is high).

Now suppose that the real target is not that woody debris piles are needed for shelter alone, but that small mammal diversity requires the food that piles provide as well as shelter. Steel's (1993) thesis suggests such an influence. Fungi are known to grow abundantly in piles of recently deposited wood, numbers of small mammals and diversity were somewhat greater in new piles, and the literature indicates that a number of the small mammal species found consume fungi in considerable amounts. In this case, a postulate *that woody debris piles provide shelter* is contained within the target, as in Fig. 7.5a. Piles of woody debris are necessary but not sufficient – because not all piles may provide the required fungi used as food. In terms of implication statements: all CWD piles $\Rightarrow$ shelter, new CWD piles $\Rightarrow$ food, shelter + food $\Rightarrow$ high diversity of small mammals. Conclusive falsification of the simpler postulate could be obtained from a + Ptest by providing shelter alone.

This type of reasoning in examining the relation of a postulate to possible theories is useful in stimulating the development of multiple postulates (see 7.9). An important device is to apply the words *necessary* and *sufficient*. Ask yourself to find the circumstances under which your postulate(s) may be necessary but not sufficient (what else might be important that is being missed?), sufficient but not necessary (what else would do instead?), or neither necessary nor sufficient.

In addition to the logical relationship between a postulate and target, the relative probability of events within a universe of observations can influence the effectiveness of using + Ptests or − Ptests. Klayman and Ha (1987) give formulas for calculating this.

7.7 Using contrasts

Contrastive questions consist of a fact to be explained, e.g., diversity of some tropical lowland forests, and a foil that helps to explain it, i.e., other lowland tropical forests that are not so diverse. Different foils may be used to explain, and so allow inferences about, different aspects of the fact. The choice of foil is governed by what aspect of the fact needs to be studied, and foils may be chosen that help to determine which of the competing explanatory hypotheses is correct. Not all contrasts make sensible contrastive questions.

CONTRASTIVE QUESTION

A contrastive question asks "Why this and not that?". It seeks to compare what needs to be understood, the fact, with something, the foil, that is similar in most respects but different in an important component. Construction of good contrastive questions requires an understanding of background knowledge (axioms) and a postulate of the difference between fact and foil.

The contrast in photosynthesis rates between late June and late July gives rise to the question "Why does the low photosynthesis rate (P, the fact to be explained) occur in late July rather than a high photosynthesis rate (Q, the foil producing the contrast)?" It assumes that only P or Q are possible and simply seeks a causal explanation of the difference, not to describe a complete theory for photosynthesis rates in either case, although the research may contribute to wider understanding. Of course the strategy of using contrastive questions first requires the existence of a contrast. One of the most important features of the photosynthesis research was to confirm and define the difference between rates at the two times (Fig. 6.9) after the first observation (Fig. 6.3). Background knowledge about theory may strongly influence where a contrast may be looked for – but need not dictate it.

There is less focus in non-contrastive questions. For example "What controls photosynthesis rate of large trees?" is a non-contrastive question of the type "What causes P?". The tendency is to respond to such questions by falling back on related knowledge about the subject. In this case what is known about photosynthesis of tree seedlings might be used, and then an attempt made to use the H-D method to postulate how this situation (mature trees) is different from the body of theory. But this approach tends to dictate the type of explanation in terms of existing theory about something that may be quite different and so provide no direction about how an explanation might be found.

Seeking explanations for a series of contrastive questions can advance knowledge. Contrastive questions give a focus to research, which is particularly important when incomplete specification is a recurrent (and expected) problem, as it is in much of ecology. When the contrasts are close then incomplete specification in the theory is likely to apply to both parts of the contrast, i.e., fact and foil, equally.

7.8 Causality

Attempting to evaluate a theory simply by falsifying a postulate is unlikely to be sufficient. Stage One of the Alaskan Quaternary vegetation and climate change problem (Chapter 5) is a case where it was sufficient, but in subsequent stages of the research and in other examples in this book theories are more complex. Emphasizing falsification may attribute insufficient importance to the process links within a theory. For example, if growth rate of new needles and shoot, and photosynthesis rate, were both reduced in June by chilling the new needles and shoot, i.e., the postulate was confirmed rather than rejected, what should you do next? Attempt to seek a further falsification in another way? It might be much more informative to find out just what amount of temperature reduction produced what amount of reduction in photosynthesis rate, i.e., explore the link between the two processes, and in so doing return to inductive reasoning in an attempt to define and then explain the cause.

CAUSATION

If C and D are two actual events or occurrences such that D would not have occurred without C, then C is the cause of D. If C, D, E is a finite sequence of particular events such that D depends causally on C, and E on D, then this sequence is a causal chain. One event is a *cause* of another if, and only if, there is a causal chain leading from the first to the second. (After Lewis 1986.)

What should count as a cause in ecology needs definition and the definition must be neither too strong nor too weak. A requirement that something (say C) should be both necessary and sufficient for some event (say D) to occur is too strong a requirement. It imposes an unjustified requirement on what can be called knowledge in ecology. There can be causation without causal dependence, i.e., an event other than C may also produce D. So, a bird may require nesting sites in order to inhabit an area. We may say that trees that provide nesting sites are part of the cause for the bird inhabiting the area. But the birds are not causally dependent on trees if things other than trees may be used as nesting sites, e.g., buildings in urban areas.

Causal explanations can be too weak. The "big bang" theory of the origin of the universe may be part of the causal history of everything (Lewis 1986, Lipton 1991) but it explains no problems in ecology. We have to ask whether the cause provides a good explanation of a particular phenomenon. Because we can define causal differences only in relation to particular contrastive questions, then defining what will be accepted as a cause can be part of both problem definition and the research itself.

Ecologists frequently work by observing correlation and considering why they exist. There are three problems in interpreting causes from correlation analysis or some other observation of regularity.

(1)   *The problem of effect rather than cause.* Rather than C being the cause of D it may actually be an effect of D. This is not a problem when there is a clear time sequence in the occurrence of first C and then D. If time sequences can not be observed then circumstances must be found where C occurs without D. Of course, causal loops do occur, i.e., where C and D have a mutual causality, but they need to be described as such and one event should not be given priority in a causal chain. This is a problem in the photosynthesis example (Chapter 6). Does a lower shoot growth rate cause reduced photosynthesis rate – or is it an effect of it? An experiment is required to determine the sequence.

(2)   *The problem of multiple effects.* This is where C causes first E, then F, but E does not cause F. However, we may regularly observe E then F and be tempted to attribute a causal relationship. Isolating C as the cause of F requires scientific analysis beyond simply observing a correlation. For example, it could be that a disease may have caused both a reduction in growth rate and in photosynthesis rate but the two are not causally related.

(3)   *The problem of alternate causes.* Lewis (1986) refers this to as *preemption*. "Suppose C_1 occurs and causes E; and that C_2 also occurs and does not cause E, but would have cause E if C_1 had been absent. Thus C_2 is a potential alternate cause of E but is pre-empted by the actual cause C_1" (Lewis 1986). This is a problem in ecology where it is frequently difficult to conclude that a particular cause is the only possible one.

Following postulation of causality after observing correlation, or apparent regularity, then further research may be required to eliminate these problems.

Causes in ecology are frequently developed in two parts. The first is how some condition arose, and this most often describes a spatial or temporal organization. The second describes the process itself, a biological, physical, or chemical transfer of matter, stimulus or genetic information. The development of the study of piles of CWD is a beautiful example of how an organizational explanation (classification of kinds of piles) may open the way to defining a cause for the differences between them (e.g., which types provide shelter and which food for particular species of small mammal) and so of the functioning of the whole system of piles of CWD on cobble bars and adjoining forest.

CAUSAL REASONS IN ECOLOGY

When we say B causes C we mean there is some direct transfer of energy or mass, or a direct stimulus that produces a response, or a genetic property that influences or determines something. Causal reasons are produced in response to questions that ask why something comes about.

Ecologists may not be able to define particular causes precisely – yet the assumption that they do exist can drive research. Steel made the first assumptions that small mammals den, nest, and feed and that those requirements would provide causal explanations for the distributions that she found (Chapter 4). Similarly, end-product inhibition of photosynthesis is the causal reason underlying the postulate that reduced growth rate may result in reduced photosynthesis rate (Chapter 6).

ORGANIZATIONAL REASONS IN ECOLOGY

These begin with the observation of a structure or pattern, and questions are asked about how that may either determine a causal relationship or result from one. Research into an organization–cause relationship (frequently referred to as structure–function in ecology) tends to involve successive improvement in the depth or detail of the explanation. This improvement is usually symmetric, i.e., involving improvement in scientific understanding of both organization and cause.

The influence of CWD on animals, within streams and rivers and potentially on cobble bars and adjoining forest, is organizational. It provides structure through which causal reasons operate. Further examples are given in Chapters 10 and 11.

Using contrasts is essential for establishing causal and organizational explanations. In causal explanations the difference condition, between fact and foil, identifies the cause.

DIFFERENCE CONDITION FOR CONTRASTS

To explain why P occurred rather than Q we must cite a causal difference between P occurring and Q not occurring consisting of a cause of P and the absence of a corresponding event in the case of Q not occurring. (Adapted from Lipton 1991.)

Note that to explain why one thing happens (say P), rather than another (say Q), a positive difference for P must be cited and not just a reason why Q does not occur. This is a stronger requirement than just citing a difference between P and Q. The reason why contrastive questions are so advantageous is that the question limits and focuses what theory is needed for explanation of the difference.

Contrastive analysis can be used when organizational causes are analyzed. Typically, ecological organization is recognized as needing considerable research in definition, whether as population or community structure. Where organization is complex, then it can be difficult to define its influence in a way that allows experiments. The first investigation into the effects of CWD on animal populations (Chapter 4) was to seek an organization, i.e., old piles and young piles, piles on cobble bars and piles in the forest, and a real value of that classification was that it defined a series of contrasts.

Ideally there should be only one difference between fact and foil, the difference condition. The central requirement for a contrastive question is that fact and foil have a similar history, against which differences stand out. When the histories are too different and there are many differences then isolation of a cause or causes becomes difficult.

Use of causal reasoning may prompt three important types of discussion. The first is associated with the problem of infinite regress, i.e., asking "Why?" to every answer given and when a failure occurs, then denying the whole causal chain. This is the question of what should be accepted as the final cause and whether one can actually be determined. For example, assuming Postulate 1 (Chapter 6),

> *Growth rate determines photosynthesis rate.*

was confirmed using the method of measuring growth rates concurrently with measuring photosynthesis rates, and Postulates 2 and 3

> *Dry mid-summer weather decreases shoot water potential and increases AHD, which reduces photosynthesis rate.*
> *The photosynthesis system acclimates to a changing physical environment.*

were falsified, would it be sufficient to say that the July depression of photosynthetic rate was caused by a reduction in growth rate? Or would it be essential to investigate the actual mechanism of reduction, i.e., to extend the causal chain? (Possibly following a scheme such as that shown in Fig. 6.8.)

The example in Chapter 6 does seek to provide a positive difference *for* P (low photosynthesis rate in July) and not just a reason why Q (high photosynthesis rate in July) did not occur. A process in the form of an implied causal chain is proposed. Low growth rates *cause* an increase in metabolite concentrations in tissues that *causes* a reduction in photosynthesis rate. But to answer the questions about high and low photosynthesis rate we do not have to go further yet into the mechanism of photosynthesis. Causal and/or organizational reasons are not overthrown by infinite regress.

A cause must be stated relative to defined interests, the particular question asked, and the definitions made. These set limits to what is required and to the value of the information (so we would not attribute every reduction in photosynthesis rate to reduced growth rate).

A second type of discussion focuses on the difference between causality and causal dependence. Because we typically deal with adaptive and acclimating organisms there may not be a universal cause for an observation. Research questions we ask can not always specify the controlling factors of the system that must be in place to determine causality, let alone causal dependence. In ecology an experiment may not be possible, and causal explanations remain provisional. Grubb (1992) gives an excellent illustration of how different particular explanations of a phenomenon can arise in different environments and that a unified explanation, requiring understanding of evolutionary processes in those different environments, can take some considerable time to emerge.

A third discussion can occur about the sensitivity of a causal chain to surrounding circumstances. For example, if some links in a causal chain are influenced by events other than those in the chain itself.

Given these difficulties a single contrastive investigation may not provide a satisfactorily complete explanation of a phenomenon. A contrast does provide what Lipton (1991) terms *causal triangulation*, i.e., the fact and the foil give two points of view according to the conditions, and these can be used to triangulate to the position of the cause of interest in the causal history of what needs to be explained. Successive contrastive investigations should improve the triangulation by using different facts and foils. So, for example, there was a series of contrasts in the CWD example with the possibility for more, and the potential for a series of contrasts in the photosynthesis example. A series of contrasts can be used to construct a knowledge network that provides an explanation, which is discussed in Section II.

7.9 A strategy for constructing theory using multiple working postulates

The fallacy of affirming the consequent, the weakness of falsification as a comprehensive system for scientific reasoning, and the difficulties in constructing causal explanations, lie at the heart of the difficulties associated with the scientific method. We can not devolve our responsibilities as scientists onto a sure-fire scientific method. We depend on our critical abilities – which is why I place so much emphasis on critical analysis in Chapters 1 and 2. One way to promote critical analysis in your thinking is to have multiple postulates about the same phenomenon. Since we cannot be sure of our propositions, we should examine different ones, as a community of scientists or as individuals.

In 1890 Chamberlin (reprinted as Chamberlin 1965) advanced this philosophy as the method of "multiple working hypotheses". He used the term "hypothesis" in its general sense of an idea to be investigated. I continue this discussion of his idea using "postulate" rather than "hypothesis" for consistency within this book and because it is absolutely essential to distinguish between the two levels of argument, the general idea and the detailed test of it. He compared the attitudes of working with multiple postulates to those of working with a single postulate or a ruling theory, and pleaded with scientists to keep their minds open. He considered that adopting a single postulate could lead to an unconscious process of selection and neglect of argument and debate in favor of that postulate. With just a single postulate a scientist could become parental toward it – nourish and favor that postulate so that its loss would become unthinkable.

Chamberlin suggested that multiple postulates are particularly valuable where complex explanations can be anticipated, e.g., the geological processes that led to the formation of the Great Lakes. This example is of particular interest because in geology, as frequently in ecology, we are intent on developing a theory about a system in which multiple processes may be operating. In the photosynthesis example (Chapter 6), multiple postulates were proposed to explain a particular result. Postulates positing control of photosynthesis rate by the growth process, changes in tree water status, or acclimation of the photosynthesis apparatus to changes in conditions, are each possible explanations of the observed phenomena and they are not necessarily mutually exclusive. In the CWD example (Chapter 4), the alternative to the postulate that CWD are "important" is that they are not. This is, in effect, a null postulate.

Chamberlin suggested that the process of examining multiple postulates could lead to vacillation. If it is appropriate to examine multiple postulates, then a complete specification should include ways of deciding among them. Of course, in some circumstances, particularly in analyzing complex systems, different responses can be obtained for the same treatment as the system itself changes. At this point, then you are adopting the approach of analyzing a system, something that can respond in different ways under different circumstances. The experimental analysis of systems is not easy. You can never be sure what a fundamental control is, or even whether there is a fundamental control in any sense. This limits application of the traditional experimental method (for further discussion, see Chapter 10; for discussion of the need to apply the principle of multiple working postulates to foraging theory, see Ward 1993).

7.10 Discussion

We can not depend on any single method of reasoning. From the strategic aspect it seems likely that theories pass through phases when different types of investigation are important. Postulates may change from broad ones needing the type of confirmatory research that is typical of exploratory investigations to those that clearly might be falsified. We can expect to see a balance between inductive research and testing the consequences of deduction where generalizations have been possible. It is not possible to insist on using only postulates that can be found to be untrue. Nor is it advisable always to use postulates expected to be true, or expected to be false. However, it is important to test theories widely and to research situations where you may not actually expect a particular result. Causal and organizational processes are the foundations of how we reason in ecology, but they need substantial research to define what they are and how they apply in particular situations.

Constructing rival postulates rigorously is one of the most difficult aspects of research, but it is the way in which we can advance the growth of objective knowledge. Sometimes rival postulates are explicit and exciting; sometimes they may require the diligent and routine investigation of a series of alternatives. Most usually in ecological research, we have to use statistical analysis for this (Chapter 8) because our concepts and measurements of them are rarely absolutes.

The concept of a causal explanation, introduced in this chapter is further developed in Section II, where it is described in the wider context of what makes an effective explanation and how to construct an explanation in ecology when complex integrative concepts are used.

7.11 Further reading

Nola (1987) has reviewed Popper's theory of the scientific method and some philosophical difficulties with Popper's approach of bold conjectures and Grünbaum (1976) describes attempted refutations.

In this chapter I have discussed the importance of contrasts. What is sometimes referred to in biology as the comparative method uses contrasts looking at differences between similar things. Givinish (1987) gives a detailed review of application to an ecological problem. Cole *et al.* (1991) present cases studies and discussions of methods in comparative analysis of ecosystems.

Klayman and Ha (1989) and Slowiaczeek *et al.* (1992) describe experimental analysis of hypothesis testing and cognitive processes. Sloep (1993) presents an example of how ecologists may attempt to use falsification as persuasive argument in a dispute rather than making a detailed methodological analysis. He concludes that this was an unfortunate state of affairs and that contributing reasons were: "a touch of ignorance on the part of some scientists as regards the intricacies of falsificationism, and a touch of conceit on the part of some philosophers as regards their claims for practical usefulness of falsificationism".

8 Assessment of postulates

Summary

Most frequently in ecology statistical analysis rather than propositional logic is used in assessing postulates. However, a clear distinction must be made between calculating a statistic for some observed data, e.g., a correlation coefficient, where there was no formal hypothesis prior to the investigation, and making a hypothesis test. Postulate assessment through hypothesis testing should be an explicit and acknowledged procedure and must be made in the context of what the postulate means for the whole theory. Premature testing must be avoided and exploratory analysis should be used to refine a postulate and the hypotheses that can be constructed for it up to the stage where testing is appropriate.

Exploratory analysis is used to clarify the meaning of the postulate. The scope of the postulate must be determined and its quantitative and relational concepts defined. Preliminary measurements may be used, sometimes in conjunction with further conceptual analysis. Exploratory analysis must be used to investigate the possibility of confounding, where there are hidden or mixed relationships in data.

To use the theory of statistical inference the investigation and data obtained must satisfy the assumptions that statistical theory makes. An example is given of the elementary probability law, based on the normal distribution, combined with the concepts of observation, universe, sample, independence of observations, random sample, statistic, and degrees of freedom. Each of these concepts is defined in this chapter, the possibility of bias occurring is discussed, and the type of exploratory analysis they require is described.

Testing statistical hypotheses is a powerful technique in scientific investigations but its limits must be clearly understood, both in the types of problem that can be investigated and how the consequences of a test can be used. There are limits to the types of problem that can be investigated. A null hypothesis can not always be constructed in scientific investigations, and sometimes where it can be constructed it cannot be used effectively. Statistical hypothesis testing is not a neutral arbiter for a postulate. An investigator *chooses* the number of observations to be made, and this determines the power of the test.

It may be more appropriate to calculate confidence intervals and use them in a considered scientific inference rather than use reject-or-accept hypothesis testing to a particular P value.

8.1 Introduction

For some ecologists, it is important to be "testing hypotheses". This emphasis may be a reaction to the time when ecology was largely a descriptive science with little precision, and the feeling, perhaps justified, that theories have been constructed on little evidence and a subjective approach (McIntosh 1980). Hypothesis testing has appeared to some to offer a necessary and welcome certainty. However, attempting to proceed only by formal hypothesis testing can itself cause scientists to formulate hypotheses prematurely and attempt to test them with inadequate quality and numbers of measurements. Statisticians being consulted by beginning ecology researchers see many cases of premature testing that run the risk of not finding differences in a test when one really exists (Type II error, see 8.3).

An important goal of scientific investigation can be to test a hypothesis and in ecology this usually involves statistics. One view is that it is quite straightforward (Fig. 8.1). This is naive testing. If a statistical test is presented in a dissertation or thesis, two types of question are likely to be asked about it and the testing procedure in Fig. 8.1 does not help to answer either of them.

> Question One: Have you interpreted the relationship or condition you have tested in the correct way – or is your interpretation confounded?

CONFOUNDING

> Confounding is when alternative explanations can be given for a study result. While A may be interpreted as influencing B there is an alternative explanation that the design or measurements of the study did not eliminate.

For example, another factor, C, may influence both A and B. Or sometimes when no effect is found between A and B it may be because either or both of them were not measured properly, e.g., A influences part but not all of B. The ways in which confounding can be minimized is discussed in 8.3.

> Question Two: Was your study designed effectively – or was it biased?

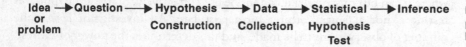

Idea → Question → Hypothesis → Data → Statistical → Inference
or   Construction Collection Hypothesis
problem   Test

Fig. 8.1. A naive view of the procedure for statistical hypothesis testing.

BIAS

Faults in study design, or in the misapplication of a statistical test, lead to incorrect estimation of a quantity or a misrepresentation of the relationship between A and B.

For example, incorrect measurement or inadequate sampling of a quantity, or a relationship between A and B, may result in a faulty conclusion, as may comparison of measured samples in a statistical test when the measurements do not actually meet the requirements of the test, e.g., of independence or normality. Both of these situations are discussed in 8.4.

Statistical inference is important because it enables us to make, or work towards, causal-statistical explanations. It overcomes some of the problems of scientific method based upon inductive and deductive logic that limit the range of problems to which they can be applied. It does this by adopting the notion of a causal process as an objective for scientific discovery (Chapter 7) and by recognizing probability and providing a theory that deals with it. As a method for producing scientific explanations these two things, the objective of reaching a causal understanding and the statistical theory, go hand-in-hand. Generally problems of confounding are difficulties in making causal interpretations, whereas problems of bias are difficulties in the application of statistical theory. Statistics courses, and the volumes of statistical methods in academic libraries, concentrate upon the technicalities of defining and applying the theory, but for the scientist the essential requirement is to define the causal process. In this chapter the case is made that unless attention is focused on constructing the underlying causal interpretation then application of statistical inference can be misleading. Making statistical inference is just one component of the more comprehensive process of making scientific inference (Fig. 8.2). Scientific inference is defined in the introduction to Section II.

STATISTICAL INFERENCE

To make a statistical inference:

(1) a sampling or measurement protocol must be designed,
(2) a measurement must be selected,
(3) a statistical hypothesis must be constructed which may be focused on investigation or experiment, and
(4) the requirement of the statistical theory used in testing the statistical hypothesis must be satisfied.

Recall from Chapter 3 that a distinction is made between a postulate, which is a new or unexplored proposition, and a hypothesis, which is a statement that will be tested by investigation. This chapter is about how you

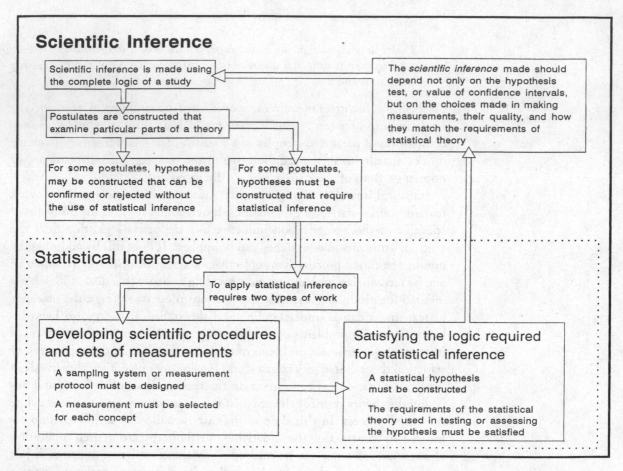

Scientific Inference

Scientific inference is made using the complete logic of a study

The *scientific inference* made should depend not only on the hypothesis test, or value of confidence intervals, but on the choices made in making measurements, their quality, and how they match the requirements of statistical theory

Postulates are constructed that examine particular parts of a theory

For some postulates, hypotheses may be constructed that can be confirmed or rejected without the use of statistical inference

For some postulates, hypotheses must be constructed that require statistical inference

Statistical Inference

To apply statistical inference requires two types of work

Developing scientific procedures and sets of measurements

A sampling system or measurement protocol must be designed

A measurement must be selected for each concept

Satisfying the logic required for statistical inference

A statistical hypothesis must be constructed

The requirements of the statistical theory used in testing or assessing the hypothesis must be satisfied

Fig. 8.2. Use of statistical inference requires that you can construct a scientific procedure and a set of measurements for use in statistical assessment.

can produce a hypothesis for a postulate and why refining a postulate through exploratory analysis is so important, and is a distinct and essential piece of research.

8.2 Refining postulates using exploratory analysis

Statistical inference is designed for making assessments in the presence of chance effects. It works properly only if you first remove confounding and bias. To do that it is essential that you make an exploratory analysis of a postulate. The approach is largely quantitative, and the field or laboratory work you need to do can best be determined by considering the principles of exploratory data analysis.

Tukey (1977) states the overriding principle of exploratory data analysis as: "It is important to understand what you can do **BEFORE** you learn to measure how well you seem to have done it". (Tukey's capitalized emphasis.) Tukey (1980) takes great pains to distinguish between exploratory and what he refers to as confirmatory analysis, i.e., using statistical inference in hypothesis testing. Exploratory analysis is not just a play with some data. It is a process to develop precision and accuracy in the scientific argument. It can result in postponing – or even abandoning a statistical test altogether. But this is an argument for, rather than against, rigor. Tukey implies that to go straight to statistical testing without exploratory data analysis neglects the following three crucial questions:

(1) *How are scientific questions generated?* Tukey (1980) answers that these are generated mainly by what he terms quasi-theoretical insights and exploring past data. A new problem may have origins in previous data in which some pattern or feature was not explained. Generally in ecology, research questions seem to arise from considerations wider than observing irregularities in previous data, but exploratory analysis can have a dual role with propositional and conceptual analysis in generating and formulating questions and particularly in making them precise.

(2) *How are designs for new investigations guided?* Tukey (1980) answers that these are usually guided by the best qualitative and semi-quantitative information available. Investigations do not spring into existence completely formed. The last stage in propositional and conceptual analysis is the development of the data statement, and this requires exploratory analysis to investigate the quantitative background surrounding the scientific question. Only then can a hypothesis test that has a well-defined meaning be constructed and the numbers of observations required to make the hypothesis test be calculated. Analysis of previous research, and possibly some specially planned new investigations, can be valuable in ensuring effective design.

(3) *How do we proceed in analysis?* Tukey (1980) answers that we proceed by exploring the data – before, during, and after analysis – for hints, ideas, and sometimes a few conclusions with probability attached to them. This may seem an extreme view almost relegating conclusions to rare events! But the philosophy Tukey advances is to be inquisitive about your data and not to put your trust in significance tests and then relax when you have selected one. Exploratory analysis should be continued when the results of an investigation are obtained: The results may initiate the next research investigation by illuminating the unexpected.

A typical example was the analysis of small mammal use of old and new CWD piles (Chapter 4). A statistical test for the hypothesis of no difference in use of piles of different ages was applied to total number of small mammals caught. The null hypothesis of no difference in use could be rejected with a probability of only 0.056 of being a random event (1 case in 17.9 that the result could have occurred by chance) whereas biologists and ecologists usually choose a probability of 0.05 (1 case in 20 that the result could occur by chance). Recall that the purpose of testing a statistical hypothesis is to examine the effects of random variation, i.e., chance occurrence. Further analysis of the data – an exploration – revealed possible confounding, in that there may have been a difference between species in their use of piles, some being affected by differences in pile age while others were not. This analysis should be treated as exploratory research, and a new postulate(s) constructed that can be tested without a confounded effect. The analysis is invaluable in showing that the structure of the postulate has to be reworked and further propositional and conceptual analysis is required. It has contributed something to scientific inference – but not confirmed anything with a specified probability.

Loehle (1987) describes this stage of the scientific method as theory maturation. New postulates always arise from previous theory, although it may be rudimentary or very general. The function of exploratory analysis is frequently to establish a causal basis for some phenomenon so that further wok can provide precision within a theory and progress towards researching testable hypotheses that can be interpreted without confounding and bias.

EXPLORATORY ANALYSIS

Exploratory analysis is used to refine a postulate so that a testable hypothesis can be constructed for it or it is reconstituted into postulates for which hypothesis tests can be developed. There are two processes (Fig. 8.3):

(i)   Developing a scientific procedure and set of measurements.

This involves research to define parts (1) and (2) of the data statement:

   (1)   Define the scientific procedure to be used in investigating a postulate,
   (2)   Specify the measurements to be made for each concept of a postulate.

The result is that you define the alternatives to be examined in a hypothesis, and how it might illuminate the importance of a postulate in a complete theory. You establish the formal conditions of the test and this requires construction of a null hypothesis.

(ii)   Satisfying the logic required for statistical inference.

This involves research to specify part (3) of the data statement:

   (3)   Specify the requirements of the data for any statistical test to be applied.

The result of this process is that you ensure that the mathematical assumptions on which the test is based are met. For example, for a *t*-test this requires:

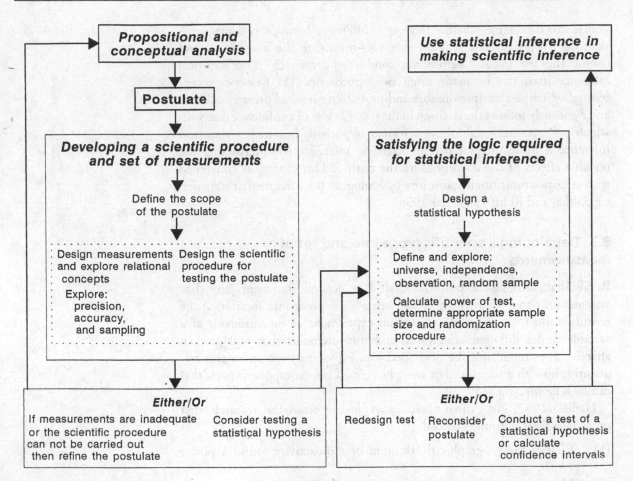

Fig. 8.3. The processes involved in using exploratory analysis to refine a postulate so that a statistical test of a hypothesis can be made.

defining the universe of observations,
ensuring independence between samples,
ensuring that the measured variable has, or can be transformed into, a normal distribution,
estimating the variance of the measure and calculating the sample size necessary to detect a specified difference in means.

Although using statistical inference may be a stage you wish to reach, much can be learnt from the exploratory process itself. Notice that exploratory analysis does not mean measuring everything you think possible about a phenomenon and then hoping to sort out some patterns during an analysis. Exploratory analysis is a very directed procedure.

A clear distinction must be made between, on the one hand, statistics that might be applied in an exploratory analysis, such as the median and boxplot of a previously unmeasured variable, a correlation coefficient, or multiple regression equations, and, on the other hand, a rigorous hypothesis test, perhaps to determine whether there is a difference between two means. A rigorous hypothesis test requires that you formulate the alternatives and choose the test statistic before you conduct the research. The statistical inference itself can be made using two approaches. (1) Reject-or-accept testing, which sets sample number and probability level as prior conditions and rigorously follows the decision of the test. (2) Use of confidence intervals, which still requires establishing a formal hypothesis but makes statistical inference a more integral part of scientific inference by interpreting the possible effects of the conditions on the testing. This chapter is concerned with statistics most often learnt first by ecologists (i.e., frequentist statistics, e.g., Sokal and Rohlf 1981, Zar 1996).

8.3 Developing a scientific procedure and set of measurements

Recall that, although a postulate itself is a general statement, any data statement written for it must be very specific. A procedure for testing the postulate must be designed, perhaps an experiment or measurement of a variable under different conditions, and measurements that will give an effective assessment must be developed. But exploratory measurements frequently show that the postulate must be refined into component parts that can each be investigated.

Loehle (1987) gives three questions, typical in ecological research, that must have their meaning clarified:

(1) What is the geographic distribution of radioactivity round a power plant?
(2) What is the breeding season of animal species X?
(3) What severity of drought will kill plant species Z?

Translating question (1) into a postulate starts propositional and conceptual analysis:

> *Power plants affect the surrounding distribution of radioactivity.*

which contains the assumption that the principal interest is with nuclear power plants. Questions (2) and (3) each appear to be concerned with unknown relationships, e.g.,

> *The breeding season of animal species X is from time t_1 to time t_2.*
> *Plant species Z is killed by a drought of severity s.*

where t_1, t_2, and s must be discovered.

These postulates need considerable exploratory analysis before any investigation could be made of them.

Defining the scope of the postulate   Frequently in research, new ideas are being explored, the scope of a theory is being increased, and exploratory analysis is necessary to investigate just what the scope of a postulate is. In the example about radioactivity surrounding a power station, there may be motivating questions that need to be defined and made explicit. Does the interest lie in the fate of radioactivity, e.g., input and loss from the area and radioactivity decay rates, or in its potential biological and human health effects? Additional information is needed to clarify the question. For example the concept surrounding should be defined, both in terms of distance from the power station (1, 5, 10 km, . . .) and, possibly, amount of data available, or that should be collected, within that area. Here conceptual and exploratory analyses are used together.

In the other examples, is the breeding season of species X to be considered for only one place or throughout the species ecological range? Determining what needs to be measured may establish just what can be investigated. Is drought to be considered at all stages of development (seedling through mature) for plant species Z? And is drought to be considered singly or in combination with other factors? Here, initial observations may have to be carefully scrutinized to generate ideas on what is most important.

Defining quantitative concepts You must be sure that concepts by measurement represent both concepts from research and concepts by imagination in a way that you understand. The examples in previous chapters illustrated that measurements, particularly those of new concepts, can be biased, inaccurate, or incomplete representations in some way or another, and you have to define in what way. Before making a hypothesis test you must be able to make measurements that allow you to detect an effect!

For the power station example: what measurements of radioactivity are to be considered, e.g., airborne or soilborne, at specific locations or distances, directions and local positions, and specific times. This definition requires specification of concepts by measurement and investigating techniques to determine measurement accuracy and precision and what sampling procedure should be used. For the other examples: for animal species X will *breeding season* be defined as attempts to mate, or successful production of offspring, or complete development of offspring to individuals that are independent of their parents? For plant species Z, will *drought* be defined in relation to both atmospheric and edaphic conditions? Will environmental indices be used or an attempt made to calculate the water balance of the soil–plant–atmosphere system? There are no right and wrong answers to these questions, but the answers chosen determine the precise nature of the postulates that can be investigated.

Defining relational concepts An important task in science is to define the relationship between quantitative concepts. In his textbook, Tukey (1977) describes a series of techniques defining these relational concepts. He emphasizes graphical display of different types of data and, particularly, how to organize the displays so that patterns and anomalies can be detected (see also 8.6). He describes smoothing techniques, data transformations and calculating and examining residuals from fits. The objectives are "treating the data to look one layer deeper" and "simplifying description". Tukey suggests:

(1)   To say that we looked one layer deeper and found nothing (e.g., to examine a frequency distribution and find a simple shape, or to fit a function to a data set, and find no pattern in the residuals) is a definitive step forward – though not as far as to say that we looked deeper and found some relationship that might be explained by a new postulate (e.g., to find an unusual but consistent frequency distribution or a distinct pattern in residuals after fitting a function). In the animal species X example, if you observed the starting dates for mating displays by individual animals, you can plot a histogram (numbers starting against successive time intervals) rather than simply calculate a mean, and then examine what the shape of the histogram might suggest about how mating display starts.

(2)   To say "if we change our point of view in the following way . . . things are simpler" is always a gain. For example, consider that species X is a migrating bird. To find that time of arrival at a site for a migrant is highly correlated with start time of its mating display may simplify your view of the mating processes although it adds a factor {*arrival time*}. However, more may be gained if we do not have to change our point of view and some additional things are found to be simple. For example, to find that the probability of successful mating is highly correlated with starting time of mating display may simplify your view of the factors that control both initiation time and length of breeding season. Similarly, if you make field observations of the soil water potential at which plant species Z wilts and you find it correlated with soil type then that is a gain – but you have changed your view by adding a complicating factor {*soil type*}. However, if you find that some soil characteristic, say {*soil organic content*}, correlates with soil water potential, which correlates with wilting, then you have simplified by reducing the complexity that needs to be considered. Note that correlations are mentioned. These give indication; they are used as guide points in an inductive process to refine a postulate and, in these examples, could establish the set of implication statements that must then be tested.

Before you make an investigation or conduct an experiment from which you wish to make statistical inference you must have a sufficiently accurate prior mathematical description of the primary relationships on which you are going to make the hypothesis test. Where a description contains bias or is confounded it may not be possible to resolve differences and the hypothesis test could be used to reject the postulate prematurely.

From Chapter 4 the first investigation of Postulate 2A,

> *Small mammal use of old debris piles is different from small mammal use of new debris piles.*

was made for all small mammals that could be caught in three trap types at one sample time. As mentioned previously this turned out to be a valuable exploratory investigation because it suggested that there were different kinds of small mammal as far as use of pile types was concerned. However, to present that single investigation as a hypothesis test is premature because the data, number of animals caught, was confounded. Some species seemed to behave differently from others. A further reduction is needed in the conceptual analysis to remove that confounding before statistical inference about a hypothesis is made.

Remember that in statistical hypotheses we always use the null hypothesis so that Postulate 2A becomes

> *Null Hypothesis:*   *there is no difference in the use made of old and new debris piles by small mammals.*

Two types of error can be made:

Type I error :   A true null hypothesis is rejected.
If the null hypothesis is true, e.g., there really is no difference in use of old and new piles, nevertheless to conclude there was a difference.

Type II error: A false null hypothesis is accepted.
If the null hypothesis is actually false, e.g., there really is a difference in use of old and new piles, nevertheless to accept the null hypothesis and say there was no difference.

Having developed a postulate so that a sharper image can be obtained there remains the problem of chance effects. There are influences, measured by the variance, that are not related to the process or relationship you are trying to investigate – but nevertheless influence your capacity to measure whether it is there or not. The number of measurements you make determines your ability to make statistical inference about a relationship in the presence of chance effects. This is discussed as power of test.

Typically the probability of making a Type I error is called α and the probability of making a Type II error β.

CONFIDENCE IN THE TEST OF A STATISTICAL HYPOTHESIS

The *confidence in the test of a statistical hypothesis* is $1 - \alpha$. By convention α, the specified significance level is usually set at 0.05 for ecological investigations.

POWER OF A TEST

The *power of a test* is $1 - \beta$; the probability a hypothesis will be concluded as false if it is false.

For a given sample size n, the value of α is inversely related to the value of β, i.e., lower probabilities of committing Type I errors are associated with higher probabilities of committing Type II. Normally, ecologists seem most concerned that they will commit a Type I error, rejecting a null hypothesis when it is actually true, and thereby accepting some feature as real when it is not. But in the first stages of an investigation, if the image is blurred, and variance may be large and sample sizes tend to be small, there is less chance of Type I and greater chance of Type II error. It is difficult to detect genuine patterns when the techniques of measurement and investigation are immature and sampling may be sparse relative to what is actually required to test a postulate. It is too little appreciated that attempting rigorous statistical inference too soon results in a loss of understanding.

Statisticians, acting as consultants to beginning researchers in ecology, see more cases where Type II errors seem to have occurred than they do Type I. This is where there are suggestions of differences, or relationships in some data, but they are not "significant" in terms of a statistical hypothesis test. Consequently the investigator comes to a statistical consultant hoping to "rescue" the research (achieve some statistical significance with $\alpha = 0.05$) and that some novel statistical analysis will help. Frequently the root cause is lack of exploratory analysis so that postulates are murky and unfocused and measurements are too few. The reasons for these problems are partly ecological and partly social. Ecology should be developed as the most precise of sciences because it seems that both causal and chance variation determine what we investigate – yet techniques for measuring and sampling, and conducting experiments, are difficult to develop. Socially we may expect, or feel pressure, to make rapid advances (Questions 26, 27, and 28 of Table 1.1) but can proceed too quickly from exciting but broad questions, through inadequate postulate analysis, lack of precise and accurate measurements, to unsatisfactory answers. The social aspects of this will be discussed more in Section III.

The result of investigation into Postulate 2A can be used in a positive way as an exploratory investigation that results in a reduction into:

Postulate 2A(1):   Peromyscus *use new debris piles more than they old debris piles.*

Postulate 2A(2): *Other species of small mammals do not use new debris piles more than they do old debris piles.*

There are three values in clarifying a postulate, in this example by reducing it to components.

(1) The more focused a postulate becomes, the easier it is to make measurements. Given the different trapping systems required for different small mammal species, this would be true for both of the new postulates. The test of Postulate 2A used rank ordering and compared which type of pile had more individuals in higher ranks. Use of this statistic was necessary, given the distribution of numbers of animals caught and that results from different trap types were pooled. Reducing the postulate may allow a more targeted measurement and so a more quantitative estimate of the differences. It would be a gain to be able to estimate mean use rather than just saying that there were differences.

(2) Statistical hypothesis testing can be most effective when a postulate contains no hidden relationships (confounding) – then all variation can be considered as *chance* and can be represented in the statistical theory of random variation. In Postulate 2A(1) there may still be concern whether *Peromyscus* species should be separated because they behave differently, and that may require further exploratory analysis. Tukey (1980) notes that "The solidest confirmatory analyses we have are based upon randomization theory – the best way to ensure the applicability of randomization theory is to randomize, carefully and appropriately." Randomization, defined later in this chapter, is the essential requirement in design of investigations to test hypotheses. Effective randomization ensures that all factors that are not the subject of your investigation have an equal effect on the measurements made. In the case of trapping small mammals this may include the chance of a small mammal encountering and entering a trap once it is placed within a pile, and where repeated sampling is used, the chance of individual animals becoming either trap-shy or trap-happy. If the two *Peromyscus* species differed in these, or similar respects, then the postulates should be reduced to consider them separately – even if no biological or ecological difference in their use of debris piles is thought to exist. You must separate the effects of random factors from those that are the subject of investigation. This is an active process where you determine beforehand the potential influences on your results against which you are randomizing.

(3) If there is a causal relationship, i.e. a real Target to discover, then the more focused the postulate, the more likely the cause is to be understood (Chapter 7). In Postulates 2A(1) and 2A(2) you can now look for differences between the biology of *Peromyscus* and other species to understand what the important differences might be in use of old and

new piles. Considering Postulates 2A(1) and 2A(2) gives a more focused approach than does the broader Postulate 2A.

The requirement for statistical inference is objective criteria for rejecting or accepting the null hypothesis. For example Postulate 2A(1) translates to the null hypothesis

There is no difference in the use made of old and new piles by Peromyscus.

A formal statistical hypothesis test of this could be made (see 8.4) because the same measurement would be used in both sides of the contrast. Compared with the investigation of Postulate 2A, the measurements and/or sampling would most likely be redesigned to improve precision and accuracy for this particular case based on the investigation already made.

Statistical hypothesis testing requires a measurement that can be used consistently. For example, Steel (1993) found indications that *Microtus* species were more frequent at debris piles on the cobble bar than in the forest and at large rather than small debris piles on the cobble bar. One measurement (trapped *Microtus* in a debris pile) could be used to analyze these apparent habitat preferences by constructing a series of alternative postulates for which statistical hypotheses could be developed. For example if characteristics of the habitat surrounding debris piles were considered as possibly important:

Postulate: Microtus *use debris piles surrounded by herbaceous vegetation more than they do debris piles not surrounded by herbaceous vegetation.*

Postulates in a null form can be constructed for different cases in preparation for constructing null hypotheses:

There is no difference in the use made by Microtus *of debris piles surrounded by herbaceous vegetation and not surrounded by herbaceous vegetation in the forest.*
There is no difference in the use made by Microtus *of debris piles surrounded by herbaceous vegetation and not surrounded by herbaceous vegetation on the cobble bar.*

Since the mean number of *Microtus* found on the cobble bar was different from that in the forest, and the concept of ground vegetation is likely to be different in the two cases, the two hypotheses must be considered separately. The single measurement (trapped *Microtus* in a debris pile) could be used further to explore differences in size of debris piles (e.g., no difference in use for piles greater or less than a critical size).

Not all postulates present the possibility for constructing null hypotheses. This is particularly so for postulates without a single measurement that can be used to pick one's way through a set of possibilities (as with the *Microtus* example). For example, once Steel had demonstrated some importance for

CWD piles, a number of postulates could be constructed about that importance. Particularly whether woody debris piles are sufficient, necessary, or necessary and sufficient for small mammal use of cobble bars (Chapter 7). These questions of sufficiency and necessity stimulate a more detailed examination of the general postulate – *that woody debris gives a physical structure in riparian zones that provides habitat for animals.* The difficulty, as so often in ecology, is in establishing an effective null hypothesis for more general postulates.

Can an overall assessment be made of this general postulate using statistical inference? Consider trying to construct the data statement. A null hypothesis for the postulate could be:

> *There is no difference in use made by small mammals of cobble bars with, and without, debris piles.*

This illustrates one type of limit to the application of the causal-statistical theory of scientific explanation in ecology. Such an experiment would have to be conducted with whole cobble bars as the experimental unit, and so with cobble bars as replicates. However, removal of debris piles from whole cobble bars is not a practical experiment. It is destructive and would not be tolerated by the river management agency. It may not even be a particularly useful experiment if the treatment itself (removing piles) caused other effects owing to disruption of vegetation and possible damage to the cobble bars likely to influence small mammals (i.e., it would probably be a confounded treatment). In this research problem the causal-statistical approach is best applied to refining the concept of "use," and examining it in the context of small mammal biology. With that approach, direct hypothesis testing may be both useful (because it may inform about the details of the system) and possible (because contrasts may exist in type of use, either between types of pile and/or species, that could be manipulated experimentally).

Harvey *et al.* (1983) analyze developments in a number of ecological theories and demonstrate that hypothesis testing is not a straightforward and easy task. To test for the existence of patterns using a null hypothesis, you have to establish a null model. All models, including the null model, make assumptions and these assumptions, have to be understood. Strong (1980) describes ecology as a subject that deals with uniqueness rather than sameness and, because of this great variation "needs the most explicit sorts of null hypotheses". The null hypotheses suggested in this discussion of small mammal use of debris piles are explicit because the postulates have been reduced to the condition that they can be tested with single measurements.

Null hypotheses are not frequently used in ecology, and Strong (1980) suggests that most research is either phenomenological or corroborative, i.e., generally it does not attempt more than demonstrating correlation or association. A difficulty with tests of ecological hypotheses is that they are usually weaker than those that can be applied in sciences that are predominately experimental. Hypotheses can be resolved only in terms of probabilities, and chains of implication statements based on such hypotheses cannot be extended very far (Chapter 7).

Quinn and Dunham (1983) define two broad classes of problems[1] that make posing ecological questions in terms of testable alternatives difficult:

(1)   It may not be possible to construct meaningful hypotheses in patterns of multiple causes. An example of this is attempting to construct hypotheses of species interactions that can be investigated experimentally by species exclusions. Multiple interactions among species may make the results of such experiments unpredictable and not interpretable in terms of hypothesis-testing logic. When there are strong interactions the behavior of multivariate processes can not be inferred from combinations of strictly univariate tests. This is the problem of trying to examine the functioning of systems through examining components piece by piece, as required in hypothesis testing. The investigation of small mammal use of CWD piles has not reached the stage of being described as a system of complex interactions, and successive refinement and testing of statistical hypotheses is a valuable approach at this stage.

(2)   The force that statistical hypothesis testing provides comes through rejecting a null hypothesis. However, it may not be possible to construct null hypotheses because we cannot specify what may exist in the absence of a particular factor. For example, in a feeding experiment with an animal to examine the efficacy of a particular diet, we cannot construct a null hypothesis of no difference in animal growth between a treatment diet and no food at all. In this circumstance, a standard diet must be developed as a reference, and any treatments assessed relative to performance on that reference diet, i.e., tests of the type that there is no significant difference between a particular and a reference diet. The use that can be made of such an approach depends upon defining an animal's performance on the reference diet itself. What happens in this type of analysis is that the investigator constrains the type of causal explanations to be sought in order to be able to use statistical theory. In questions of factors that control community structure or, for example, analysis of migratory patterns, there is no effective null hypothesis and there may be no possibility of developing a "standard," e.g., a "standard" migration.

Strong (1980) describes a series of ecological theories where formulation of null hypotheses had been essential to progress but had been achieved with

---

[1]   Quinn and Dunham (1983) list three classes in all, separating the first class of problem given here into two.

difficulty. He points to the requirement for multiple independent tests and the use of experiment. In some cases, problems in constructing null hypotheses may arise because propositional and conceptual analysis has not proceeded far enough. For example, migration might be analyzed in terms of a number of component processes, which in turn may be explored experimentally. However, some problems of developing a standard or a null may be intractable. Quinn and Dunham (1983) state a belief that rigid insistence on "hypothesis testing" can distract from advancing understanding by focusing on distinct alternatives where in fact there is a range of possible outcomes. This difficulty with hypothesis specification and experimental analysis is symptomatic of a wider problem, the need to consider the integrative concepts of ecology (Section II) and to make a more comprehensive definition of what constitutes a *scientific explanation*.

8.4 Satisfying the logic required for statistical inference

Statistical science provides us with mathematical models for the unexplainable, random variation we encounter in Nature and that enable us to distinguish it from the variation due to causal and organizational features that we want to understand. One of the objectives of statistical research is to develop new theories and mathematical models for distinguishing between random and non-random variation for new types of investigation.

Elementary probability theory is the first probability theory encountered by most scientists in their training and is still the most widely used. The simplicity and efficiency of elementary probability theory are beguiling because, although its concepts and axioms are themselves straightforward, they are demanding of the real world and how we can conduct our research. The theory is based upon a law, a mathematical formulation about the distribution of chance effects.

ELEMENTARY PROBABILITY LAW

The *elementary probability law* can be expressed by an integral of the form

$$P = \int_a^b p(x)\mathrm{d}x$$

where P is the probability that an observation occurs between a and b. $p(x)$ is the distribution function of the observations. This function may take various forms, but the one most commonly considered is the normal distribution function, for which $p(x)$ has the form

$$p(x) = \frac{1}{\sqrt{2\pi\sigma^2}}e^{-\frac{(x-\mu)^2}{2\sigma^2}}$$

which when plotted has a bell-shaped curve that approaches the axis asymptotically in both directions. (*Asymptotically* means that the curve gets closer and
closer to the axis but never actually reaches it so that values further from the
mean are less and less likely but never impossible.) The values μ, the "true"
mean, and σ^2, the "true" variance are defined in terms of $p(x)$:

$$\mu = \int_{-\infty}^{\infty} xp(x)\mathrm{d}x$$

$$\sigma^2 = \int_{-\infty}^{\infty} x^2 p(x)\mathrm{d}x - \mu^2.$$

There are two things to keep in mind. First, this is a very useful law and
mathematical description because test statistics have been derived from it that
allow you to calculate the probability that samples are from the same
population, i.e., they are used to examine a null hypothesis of no difference.
The law and mathematical description is the foundation of a theory, a
"formalization of Nature" describing variation due to chance effects, i.e., it
defines the limits within which a mean can be approached. When using
probabilities nothing is "known" in absolute terms: for a quantity, this means
describing *upper and lower bounds* and a *probability* that the value lies within
those bounds. These two concepts of *bounds* and *probability* go hand in hand.
The definitions that follow apply to data that are continuous (not categories
or classes) and have (approximately at least) a normal distribution function.

Second, the method of calculating upper and lower bounds for estimating
the mean is set within a formal theory. Exploratory analysis should be used to
examine whether the theory can actually be applied to the case of interest. The
following definitions describe the principal concepts in the theory. Required
exploratory analysis follows the relevant definition where appropriate. If the
theory does not apply, then strictly the method of calculation can not be used,
although the theory itself contains the concept of robustness, which defines
the tolerance to exactly how much the assumptions of the theory can be
transgressed and yet the main results still apply.

OBSERVATIONS

Observations are numbers from single measurements.

SAMPLE

A *sample* is a number of observations from a universe. A sample can be defined
by various statistics, e.g., mean, variance, median. The sample can then be
represented as a point in n-dimensional space, where n is the number of
statistics used, e.g., successive samples of small mammals from a repeatedly
randomized trapping line would have different means and variances. Each time
that a sample is taken it occupies a different point in n-dimensional space, i.e.,

the sample can be represented by the different statistics you calculate from it.

UNIVERSE

A *universe* is the set of all observations that could possibly be obtained that follow the probability law.

Immediately there arises a potential problem in applying the elementary probability law. In the equation defining μ the distribution of observations is defined as a continuum between $-\infty$ and ∞. Yet, practically, few variables we investigate extend across this complete range, i.e., their universe is not precisely represented by the mathematical definition. The question is: "If this equation can not be strictly applied, is the use of this elementary probability theory jeopardized?". If the distribution function of a sample remains bell shaped, even though not extending from $-\infty$ to ∞ around the mean, then for practical purposes it is justified. You can still calculate the mean and variance of, say, leaf length or fish weight, even though there is no probability of encountering a leaf or fish of infinitesimal or infinite size.

Required exploratory analysis for universe. It is important to define the universe of observations. For example, particular care must be taken about sampling when estimating the size or weight of animals or plants. Measurements must not exclude some part of the universe or overestimate other parts. In Steel's investigation (Chapter 4), different types of trap preferentially captured different species of small mammals. The important requirement is to define such a bias[2] and correct it where possible (e.g., by changing the measurement) or change the postulate to reflect the actual population you can investigate. There is no foolproof way of ensuring that you are sampling from the universe you have defined. You should examine data for, e.g., sex ratios different from what you would expect, or truncated frequency distributions, which may indicate that you are failing to sample for either the large or the small end of the universe. The principle to follow is to question every measurement process, not trust to luck! Estimating animal populations is particularly complex and has resulted in the development of a whole statistical methodology discussed by Skalski and Robson (1992). If a measurement of the frequency distribution appears non-normal (for tests of departure from normality, see Zar 1996), mathematical transformations may be required.

INDEPENDENCE

Two events, x_1 and x_2, are *independent* if, when p_1 is the probability that x_1 lies

---

[2] For the χ^2 test a bias can be introduced when sample size is small (Zar 1996). this is a different aspect of bias, due to the formulation of the statistic, and can be corrected for mathematically.

within an interval I_1 and p_2 is the probability that x_2 lies within an interval I_2, then the probability of the *joint* event (x_1 in I_1 and x_2 in I_2) is p_1 times p_2.

The important criteria are as follows. If one event or measurement is related in some way to the next one, and this relationship influences the joint probability, then the observations are *dependent*. Where some form of mutual influence does occur, the statistics calculated might not represent the population as a whole (they would be confounded), which means that a different approach must be taken. For example, the competition process may influence tree size along a planted row in the forest. Successive measurements of tree size along the row are not independent. If x_{large} is a large tree in the top size quartile, then there may be a greater probability that it has neighbors x_{small}, i.e., trees in the bottom size quartile. In this case the probability of the joint event along a row (x_{small}, x_{large}, x_{small}) is greater than simply the product of the probabilities (twice the probability of x_{small} times the probability of x_{large}) based on the size frequency distributions of all trees. Independence is an important assumption that is frequently transgressed in ecological work. Much research may be required to discover dependencies. Perhaps the two most common occurrences that lack independence are successive samples in time and close neighbors in space.

> *Required exploratory analysis for independence.* Ecologists frequently investigate situations where spatial heterogeneity is encountered, or where an event at one time influences a subsequent one. There may be a distinctive influence upon many individuals of a subset of the population. For example, a year in which an insect population increases to epidemic levels is likely to influence subsequent years. When samples are collected at different times or in different locations the data must be examined to see whether there are patterns.

The technique used to achieve independence is randomization. But it is essential to know what must be randomized for – and exploratory analysis may be necessary to determine this. Section II makes the case that it is frequently the spatial and temporal dependencies between organisms and their environment that we seek to understand; we aim for organizational causes in ecology. Then it is important to study the dependencies, not assume that they can be treated as chance effects, and so try to remove their apparent influence through randomization. That does not mean that statistics where independence must be assumed are of no use at all, far from it. It does mean they must be used with care.

RANDOM SAMPLE

A *sample* is drawn at *random* to ensure there is no bias influencing what is to be measured. A formal definition of *random sample* has the following necessary and sufficient conditions:

(1) The observations of the sample are from the same *universe* in the sense that they obey the same probability law.
(2) The observations are *independent* draws. (Recall definition of "independence".)

Consider using the test statistic *mean* to compare samples of the same species of a fish but living in two different situations, A and B, because it has been suggested that A is polluted whereas B is not and pollution is postulated to influence growth of individual fish. Other circumstances that might influence fish size must be the same between the situations. For example, predation may differ between A and B, which could alter the frequency distribution of fish size. One approach then might be to stratify the samples by age of animals, so that comparisons between A and B are made between strata within the same *universe* (compare A and B for one-year fish, then two-year, . . . separately). Then the assumption would be that within each stratum the concepts of this statistical theory hold. "Randomness" implies consistency of factors influencing the measurement other than those being tested for.

When considering a random sample, the assumption is that accumulation of chance effects influencing the measurement behaves in a very particular way – in this chapter the elementary probability law is considered. Mathematically, this is approximately defined by the frequency distribution of the observations. The assumption of randomness is really a demand for regularity in Nature in the way that all influences not of interest will act upon what is measured. Randomness is a "presupposition" about Nature that allows us to consider that we can describe it quite well. It can be difficult to achieve postulates where this can be assumed so, and even more difficult to be sure it has been achieved!

Notice that a random sample is defined in terms of its properties. Do not assume that use of random number tables is a sufficient guarantee of achieving a random sample. It is how they are used that is important, particularly to define the correct universe. For example, if size is measured, then individuals should not be taken preferentially in a way that might affect the sizes obtained. If a feeding trial is proposed, e.g., to see whether males and females may respond differently to a particular diet, then initial size differences between them may require that they should be treated as separate classes before randomization, e.g., hypothesis tests may first have to consider differences in male and female responses separately.

The most important question concerning the application of a given probability law is what constitutes a pertinent set of observations. The concepts of *universe* and *randomness* are intertwined with that of *independence*. All three are involved in defining what may be considered as a sample. These concepts are the foundation of elementary probability theory and do not exist distinct from each other. If, for example, through an exploratory analysis it is found necessary to challenge that individual values in a sample are independent, then the sample will not have been random.

SAMPLE STATISTIC

A *sample statistic* is any mathematical function calculated from the observations.

For example, for observations, x_1, x_2, x_3, . . ., x_n, the sample mean, $\bar{x}$, is defined as

$$\bar{x} = \frac{\sum\limits_{i=1}^{n} x_i}{n}$$

The *sample variance*, s^2, an estimate of the unknown universe variance, is defined as

$$s^2 = \frac{\sum\limits_{i=1}^{n} (x_i - \bar{x})}{n-1}$$

There are many other sample statistics (e.g., median, skewness, kurtosis) but the unique importance of the mean, $\bar{x}$, and variance, s^2, is that they are estimates of μ and σ^2, respectively the "true" mean and variance of the universe (often referred to as population parameters).

A *confidence interval* or *interval estimate* of μ can be calculated from the sample statistics s and $\bar{x}$, the number of observations in the sample, and a critical value of the *t*-statistic selected for a chosen probability, in this case a 95 percent probability that the true mean will lie within these bounds:

$$\mu = \bar{x} \pm t_{0.025} \frac{s}{\sqrt{n}}$$

Recall that the distribution of the universe, and so of the sample, is assumed to be normal. For small sample sizes the *t* distribution is more acceptable as an approximation than the normal distribution itself (the *t* distribution becomes flatter and with larger tails as the degrees of freedom are fewer, and so is a more conservative model than the normal distribution). A value of *t* is read from the table of critical values depending on the probability required and degrees of freedom available. For a confidence interval around the mean, since there is a possibility that $\bar{x}$ may be greater or less than μ, then a value for $t_{0.025}$ is used to give a positive and negative bound.

Fig. 8.4. Width of the 95 percent confidence interval $2t_{0.025}s/(\sqrt{n})$ for increasing values of n, and with $s = 1$.

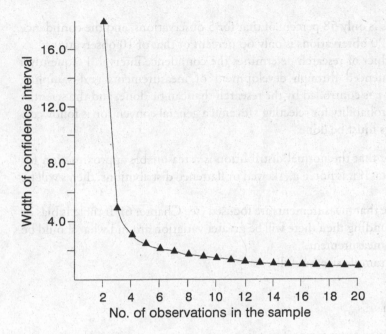

DEGREES OF FREEDOM

The larger a sample is, the more general confidence we can have that statistics calculated from it represent values for the whole universe of observations. But if calculating one statistic uses another statistic that has already been calculated, then there is repeated use of the same information. To account for this "double use," we impose a restriction on the second statistic. If n is the sample size, and k the number of independent restrictions, then the *degrees of freedom* are defined as $n - k$. The sample mean has no restrictions placed on it, because you simply add all the values together and divide by the number of values (i.e., no other statistic is used in calculating the mean). But the sample variance has $n - 1$ degrees of freedom, because the mean is used in calculating it.

For illustration of the principle: consider a sample of just two observations with actual values of 9 and 5. Since $\bar{x} = 7$ the residuals (used to calculate the variance) are $+2$ and -2. The second residual merely has to be the negative value of the first. Whereas the first residual can be considered free, the second is strictly determined: so there is only one degree of freedom. Even where (as in most cases!) the numbers are not so nicely arranged as this, the principle remains the same.

As n increases, the confidence interval is reduced by reducing both the critical value of t and the term $s/\sqrt{n}$. The confidence interval decreases markedly for every additional observation when numbers of observations are small (Fig. 8.4). In Fig. 8.4 the width of confidence interval for 10 observations is only 58 percent of that for 5 observations, and the confidence interval for 20 observations is only 66 percent of that of 10 observations.

The conduct of research determines the confidence interval. Frequently s can be influenced through development of measurements and sampling procedures, n is controlled by the research that can be done, and the scientist selects the probability for selecting t (even if a general convention is followed). Three things must be done:

(1) Ensure that the normal distribution is a reasonable approximation for the data. If it is not, e.g., skewed or flattened distributions, then s will be large.
(2) Ensure that measurements are focused (see Chapter 6). If there is bias or confounding then there will be greater variation around what should be a true measurement.
(3) Select sample size carefully.

ROBUSTNESS

A statistical test is robust if its validity is not seriously threatened by moderate deviations from the underlying mathematical assumptions. (After Zar 1996.)

This is not a precise definition because how well a test tolerates abuse depends upon the test, and the type of abuse and the size of the sample and probabilities being considered. Zar (1996) discusses the robustness of the most frequently used statistics. The most frequently used antidote to transgressing assumptions is transforming the data. But note that this concept of robustness applies to deviations from the mathematical assumptions of the statistical procedures, e.g., normality for a t-test, homogeneity of variance for analysis of variance. It does not apply to the requirements for defining universe, sample, independence, and randomness. Obtaining the "given set of observations" is the most complex part of an investigation, involving (1) achieving a clear understanding of the elements of probability theory being used and (2) designing the investigation to produce adequate observations.

8.4.1 Constructing and assessing a statistical hypothesis

There are two issues. First how to construct hypotheses, and, second, what type of assessment should be made of them – a formal test, or calculation of confidence intervals.

Statistical hypotheses should be inclusive itemizations of the conditions of the test. For example, which of the following pairs of hypotheses (H) is correct?

H_1:   Use of old and new debris piles by *Peromyscus* is the same.

H_2:   Use of old and new debris piles by *Peromyscus* is different.

Or

H_1:   Use of old and new debris piles by *Peromyscus* is the same.

H_2:   Use of new debris piles by *Peromyscus* exceeds the use of old piles by an amount k.

Neither of these simple formulations is sufficiently complete to make H_1 and H_2 alternatives that can be tested for no difference (use of a null hypothesis). Information about the character of the observations must be included. And this information cannot be derived from the probability theory, which is general and makes no assertions that apply only to one or another specific set of observations. The observations themselves are simply groups of numbers and do not contain information about the kind of universe from which they were drawn, or how they were drawn. This information must be included in the way in which the hypotheses are structured (after Churchman 1948). Consider the following set of propositions

(1)   The observations are divided into two groups, use of old debris piles (O) and use of new debris piles (N) on cobble bars not in forest.

(2)   The observations in O constitute a random sample, as do those in N.

This requires that any necessary stratification has been made and random samples are drawn from groups that experience the same causal influences, apart from the one being tested, and are subject to the same chance effects.

(3)   The random sample O comes from some normal universe as does the random sample N.

(4)   The mean of the O universe, $\bar{O}$, is the same as the mean of the N universe, $\bar{N}$.

(5)   The variance of the O universe is the same as the variance of the N universe.

One formulation about the use of debris piles is the following pair of alternative hypotheses where H_0 represents the null hypothesis and H_1 the alternative:

Case One: H_0: accepts 1, 2, 3, 4, and 5.
            H_1: accepts 1, 2, 3, and 5, but rejects 4.

If we add to the list of propositions:

(4*)

The mean of the N universe exceeds the mean of the O universe by at least an amount $k(k > 0)$.

Then another formulation would be:

Case Two: H_0: accepts 1, 2, 3, 4, and 5.
            H_1: accepts 1, 2, 3, 4*, and 5.

allowing the possibility that $\bar{O} \neq \bar{N}$ (where $\neq$ means "not equal to") and $O < \bar{N} + k$. H_0 and H_1 do not exhaust the possibilities, and technically a third hypothesis would have to be included, i.e., to include possible differences in variance between O and N.

If we want to ask only if there is some difference between old and new piles then we would have:

Case Three: H_0: accepts 1, 2, 3, 4, and 5.
              H_1: accepts 1, 2, 3, but rejects either 4 or 5 or both.

This example illustrates the care that must be taken in formally constructing statistical hypotheses. The propositions that are the same in the alternatives are the *propositions of the method*, and the remaining propositions are the *basis of the inquiry*. The conditions imposed on the alternatives define a precisely formulated question. But this precision depends upon statistical theory and so you must show the meaning, and relevance, of the principal components of the theory, i.e., "observation," "random sample," etc, where you use the theory. The term "hypothesis construction and testing" is reserved for the situation when it is conducted within a logical and statistical framework. Other branches of probability theory are used in different applications, and they too have their own implications for formal hypothesis construction and testing.

Consider an instance of Case One where the null hypothesis and alternative have been stated. To test a hypothesis, both sample size and a probability for rejection of H_0 must be selected prior to the test. Because the variances are assumed equal (Proposition (5)) then a pooled estimation is calculated

$$S^2_{pooled} = \frac{1}{(n_1 + n_2 - 2)} \left[\sum_{i=1}^{n_1} (x_{1i} - \bar{x})^2 + \sum_{i=1}^{n_2} (x_{2i} - \bar{x}_2)^2 \right]$$

Now the *t*-test statistic can be calculated:

$$t = \frac{\bar{x}_1 - \bar{x}_2}{s_{pooled} \sqrt{\frac{1}{n_1} + \frac{1}{n_2}}}$$

So *t* is a measure of the difference between the mean values of the two samples relative to the sample variance. If the calculated *t* exceeds the critical value for *t* read from the tables, for the selected α for the degrees of freedom, $n_1 + n_2 - 2$, then H_0 must be rejected.

While $1 - \alpha$, the confidence in the test, is chosen by the investigator, the power of test, $1 - \beta$ must be calculated. Recall that power of test is the

Table 8.1. *Factors that influence the power of the test and their effects*

Influence	Effect
The actual difference between μ_1 and μ_2	Power increases as $|\mu_1 - \mu_2|$ increases
The number of observations, n, made in each sample	Power increases as n increases
The variance of the observations, s^2, the estimate of σ^2	Power increases as s^2 decreases
The chosen value of α	Power increases as α increases (e.g., for particular values of $|\mu_1 - \mu_2|$, n, and σ^2, there is more power of test when α is chosen as 0.05 than as 0.01. The less confidence in the test of a postulate you specify – the greater power the test has).

probability that a null hypothesis will be rejected as false if it is false. Four things influence power (Table 8.1).

For a *t*-test, the power of test, $1 - \beta$, is expressed as the probability that a difference in means will be detected for a given variance. Tables are published that show critical values of the calculated standardized deviation between the means, i.e., the same value calculated for the *t*-statistic, (though referred to as *d* when used in power calculations) for particular values of degrees of freedom and particular values of power (Table 10 in Pearson and Hartley 1976). To illustrate how the power of test changes, consider a two-sided test with $\alpha = 0.05$ and 8 degrees of freedom (d.f.) (Fig. 8.5a) (equivalent to a *t*-test for difference in means with $n_1 = n_2 = 5$). If the value of $d = 1.0$ then the power of test is 0.15 – there is only a 15 percent chance the null hypothesis being rejected as false if it is false. The test has very low power. It is more likely that a true difference in means, if one existed, would not be detected than that it would be. A value of $d = 4$ indicates a test with power of 0.95, i.e., a 95 percent chance of rejecting the null hypothesis if it should be rejected. Increasing that number of observations in a sample, e.g., so there are 16 d.f. (Fig. 8.5a), increases the power of test.

The power calculation is particularly valuable in the early phases of a study and can be used to estimate the numbers of observations that should be made in a sample. First make an exploratory investigation to determine the variance; then the number of observations that must be made in a sample to detect particular differences in the means can be calculated (Fig. 8.5b). So, if you wish to detect a difference in means of 6.0 and you calculate s to be 10.0 then $d = 6.0/10 = 0.6$ (indicated by the arrow on Fig. 8.5b). Then to have a power of test of 0.5, n_1 must $= n_2 = 16$, while to obtain a power of 0.9 then $n1$ must $= n_2 = 48$. As d increases, i.e., you would be satisfied with merely being

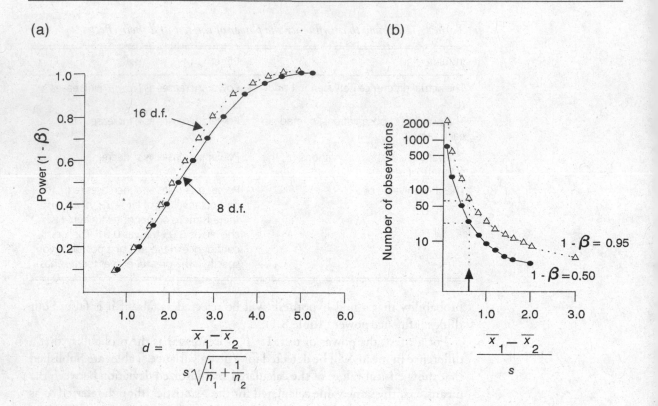

able to detect bigger differences between the means, then the required sample size for particular power decreases.

This problem of small sample size is most critical to ecologists, although the converse problem also exists. By taking very large sample sizes then very small differences in $|\mu_1 - \mu_2|$ are detected as "significant". For example, suppose you were looking for taxonomic criteria to separate two species. You might find some measurement for which there were real, but very small, differences between them. In a research study you could determine the difference by sampling large numbers of both species (you increase the power of the test). But small differences, even though you can establish their significance, may still not be particularly useful as a practical taxonomic criterion.

The point is that in selecting sample size you have a major influence on the result of the test. This is one reason why many scientists do not follow the classic reject-or-accept approach of formal statistical testing. For them it is more appropriate to use confidence intervals along with specification of sample numbers and probability used in selecting the critical t-value. Although statistical theory provides a rationale for rejecting the null hypothesis there is no formal reason for accepting it – we can only say, "H_0 is not

Fig. 8.5. (a) Curves for power of test for t-test, two-sided with $\alpha = 0.05$. (b) Curves of number of observations (natural logarithm scale) required to attain two levels of power ($\alpha = 0.05$, one-sided test).

rejected". This is an important caution. A single test of significance, at say $\alpha = 0.05$, is a promising start but not sufficient to discredit the null hypothesis convincingly. R. A. Fisher maintained that discrediting the null hypothesis convincingly requires repeated tests of significance:

> "In relation to a test of significance, we may say that a phenomenon is experimentally demonstrable when we know how to conduct an experiment which will rarely fail to give us a statistically significant result." (Fisher 1966.)

If the variance of a measurement is known, then the degree of difference between means that samples of different sizes will detect can be estimated. Remember, when deciding on the sample size the scientist determines the magnitude of the difference that is detectable. In this sense a statistical test, e.g., t-test, is not a neutral arbiter of the differences in means. Many researchers take it for granted that they will use a value of α of 0.05 in confidence in the test of a hypothesis – but that value is set only by convention.

8.4.2 Completing the data statement

The third part of the data statement should *specify the requirements of the data for any statistical test to be applied.* The examples given in this chapter have been centered on a possible use of the t-test, the assumptions of normality and equal variances between the samples, and the concepts defined above. Where other statistical procedures are used, e.g., analysis of variance, or chi-squared, it is important to realize that they too have assumptions and that exploratory analysis may be needed to see if the assumptions are met.

8.5 Discussion

There are three frequent complaints about the use made of statistical hypothesis testing, two particular values in using it, and two cautions must be given of conditions when it may not be appropriate.

The complaints   These must be aired first because they come from responsible applied statisticians and, unfortunately, they are all too frequently valid complaints. They are:

(1)   *Inappropriate construction of hypotheses.* Scientists apply test statistics, and quote P values in their results, without the type of reduction and exploratory analysis described in this chapter and consequently the hypothesis test is not sharply focused (Poole 1987, Greenland 1990, Brennan and Croft 1994). Although these papers use health science examples, their message applies equally to ecology.

(2) *Overdependence on limited statistical inference.* Most hypotheses are constructed in a particular way from a postulate and a single hypothesis test can not be taken as unequivocal evidence about a postulate. Favoring reject-or-accept testing over confidence intervals ignores these subjective elements required in setting up a hypothesis test, e.g., choice of number of observations in a sample, selection of critical value for the test (Altman 1982). Despair at scientists' failure to recognize this has led some applied statisticians to urge that the focus should be on exploratory analysis (see e.g. Greenland 1990). The result in any single test of significance is strictly provisional and conditional upon additional evidence about how measurements were made and other details of the investigation.

(3) Writing about ecology and its application to resource management Ehrlich and Daily (1993) suggest: "There is unfortunately, *an uninformed scientific culture that attempts to exclude from science any results not carrying an arbitrarily high level of statistical 'significance'*, any recommendations based on first principles, or any science that is descriptive." (Italicized emphasis added.)

Hypothesis testing should not be used without exhaustive examination of the conditions under which the test is made. And once a hypothesis test is made it should not be treated as absolute evidence about the postulate. Assessing a statistical hypothesis, whether you use the reject-or-accept approach, or confidence intervals, should be an exhaustive process. It requires detailed analysis of what you are investigating and refinement and reduction of general postulates to specific component postulates so that confounding is actively minimized and possible bias is excluded. Then a statistical inference can be attempted.

The values But is it worth it? Why not stop after the first detection of a difference in means, even though one might not be certain how effective randomization may have been? Or make inference from a correlation coefficient, even though not certain whether a causal relation exists? Or calculate a multiple regression formula, pick out the most significant terms, and make scientific inference on that? There are two related reasons for carrying through research to hypothesis testing.

(1) Initial observations of differences in means, or of correlation, or regression parameters, are usually only hints about what is going on. It can be argued that progressing with new investigations would inevitably lead to failure if the hints were wrong and that it would be quicker to proceed in that way than pursue investigation of the current relationship to the stage of a comprehensive assessment. The counter-argument is that, if science is developed on the basis of hints and it does go wrong,

the scientist may not be able to tell why. Was it the new postulate that was wrong? Or was the postulate turned into an axiom, perhaps too hastily? The network analogy for the growth of scientific theories is useful here. Progressing with no rigorous assessment is like constructing a large net with weak fibers. Rigorous assessment requires that you spend considerable effort isolating individual fibers (i.e., reducing the problem to its testable components,) deciding which are strong (i.e., doing accept-or-reject tests or calculating precise confidence intervals) and so constructing a smaller but sturdier network. For many people this sturdy approach is essential and they consider that something is not truly known and understood until an experiment or investigation can be constructed that will rarely fail to give a statistically significant result or acceptable confidence intervals. Although we frequently talk about using "statistics" the full name for this particular methodology of science is the causal-statistical approach. Statistical theory deals only with chance effects – and its strength is enabling the definition of the causal process. Continuing along on the basis of hints (even though buoyed up with calculations of P values or r^2 coefficients) is an abuse of the causal-statistical methodology.

(2) Attempting a hypothesis test, whether on the basis of reject-or-accept or using confidence intervals requires precision in specifying differences or similarities. The detailed work of exploratory analysis that established the test tells much about the measurement. However, it is most important to remember that statistical inference is an instrument of scientific investigation (Fig. 8.2). It does not replace a balanced scientific inference, and it is certainly not an independent arbitration. The researcher, by choosing both α and n, determines the power of the test.

The cautions   There are two reasons for not depending exclusively on the use of rigorous test procedures for statistical inference.

(1) A null hypothesis can be constructed only for certain types of postulate, usually those about how the details of a system works. Broad questions generally do not lend themselves to hypothesis testing. Inference about broad questions requires scientific explanations (see Section II) that may use information from statistical hypothesis testing but cannot depend entirely on it.

(2) Applying the causal-statistical approach may require constraining a system in order to define a consistent causal relationship. This may cause difficulties during the early stages of examining a postulate, where there are hints about a relationship but it is not clear precisely the conditions under which it exists. Or it may assume that the conditions under which the assessment made were representative. This could be a

difficulty if you are using experimental techniques. This issue is associated with reduction and refinement of postulates.

8.6 Further reading

In exploratory analysis, extensive use should be made of graphical representations of data in developing quantitative descriptions (Cleveland 1985). Two books, Tufte (1983) and Cleveland (1993) are useful in illustrating a range of different types of techniques for different types of data. Among computer applications S-plus (Statistical Sciences, Inc. 1993) is based on an exploratory data analysis approach that provides graphics and statistical analyses for interrogation of data (Becker *et al.* 1988) and is available for most popular workstation and portable computer operating systems. Magnusson (1997, see Table 1) in discussing the use of statistics made by ecologists gives an excellent list of the concepts that a student should understand before starting to collect data for a thesis. Johnstone (1987) discusses Fisher's approach to tests of significance.

Calculation of the power of the test can be made for statistics other than the *t*-test (Dixon and Massey 1957, Kraemer and Thiemann 1987) and Odeh and Fox (1975) provide tables and examples for sample size choice for *F*-tests. Computer programs can also be obtained for making sample size calculations: from Professor Barry Brown, University of Texas (*bwb@odin.mda.uth.tmc.edu*).

There has been extensive research in statistics to develop techniques for situations where the assumptions described in this chapter cannot be met. An introduction to some of the most useful for ecologists is given in a Special Feature of *Ecology* (Matson *et al.* 1993). The role of statistical analysis in development of causal reasoning is described by Holland (1986), particularly the argument "no causation without manipulation" and the paper is followed by discussions contributed by a number of statisticians. Schmidt (1996) details the important problems associated with overreliance on significance testing and recommends changes in how statistics should be taught.

9 Individual philosophies and their methods

Summary

A philosophy is a reasoned point of view of how to approach some task, or life itself. We all have our philosophies, and they do influence our approach to science. Understanding this can help us to appreciate that there will always be other points of view, which form a natural and important criticism of our own approach. In this chapter examples are described of research following particular philosophies and the counter-philosophies are also discussed.

Presuppositional philosophies assume that a particular methodology is correct, or that knowledge has a certain structure. The major philosophical positions are each opposed by a philosophy that makes contrary presuppositions. So, empiricism emphasizes the importance of data and initiating research through direct observation, while rationalism emphasizes that research should start with theory. Reductionism takes it that an understanding of complex systems can be made in terms of their less complex systems, while holism asserts that understanding cannot be complete if based on the study of components of a system. There are also fundamental differences in the approach to science itself and whether there can be objective standards that research can follow. Relativism argues that universal standards do not exist, while criticism, as a philosophy, attempts to achieve them through analysis of methods as well as results.

Philosophies can be understood by studying them in a historical and developmental perspective. As new types of scientific problem have been encountered, developments in the scientific method have been closely linked to developments in philosophy. Some methodological problems recur in scientific investigations and so previous debates remain relevant.

9.1 Introduction

There is no single correct method of science. The chapters so far have illustrated that the individual scientist must make choices about measurements and experiments, deciding on a confirmatory or falsifying strategy, seeking causal explanations, and developing statistical tests for a postulate.

What influences these choices? Popper (1972) states:

> "We all have our philosophies, whether or not we are aware of this fact, and our philosophies are not worth very much. But the impact of our philosophies upon our actions and our life is often devastating."

In this chapter, I illustrate some philosophies people bring to their science and how these influence both objectives and methods. Philosophy, as a discipline, has objectives different from those of science. It states theses that seek to be rational justifications (Althusser 1990). In science we attempt to be right about the real world; assessment is part of method and change in scientific view a result of work. In philosophy we attempt to be correct about our approach to understanding; the method is to define both the circumstances when a philosophy is valuable and the methods it uses, and successful philosophies find consistent application over time.

Scientists do have a dominant philosophy. Most scientists are scientific realists – at least to some degree.

SCIENTIFIC REALISM

Scientific realism has three tenets:

(1) There is a real world.
(2) Scientific methods find out about the real world.
(3) Science aims to give us, in its theories, a literally true story of what the world is like; and acceptance of a scientific theory involves the belief that it is true. (After van Frassen 1980.)

The scientific realist takes the position that predictively successful scientific theories are (approximately) true, since their truth is the best explanation of their success. The success of science would be miraculous if its theories were not largely true. This is an inductive explanation extending the apparent truth of what has been found out to infer the overall effectiveness of methods, and the existence of truth. It is not a scientific argument, but a philosophical argument about science. The scientific realist argues from the past observational success of a theory to its truth, not just from past observational success simply to future success. This argument is circular in a similar way to induction, although that may not concern scientific realists (Lipton 1991, pages 160–168). There are other suggestions for the appropriate philosophy for scientists. For example, that the scientific method provides a mechanism that eliminates falsehoods. This does not deny that scientific theories are true, it simply does not confirm that (for a discussion, see Lipton 1991).

While most scientists hold at least some element of scientific realism there are considerable differences between scientists, in how much they believe should be considered true, how much evidence and of what type is needed for such a belief, and what are the best methods to use. These differences in philosophy are recognizable in people's methodologies and what they emphasize as important. It is not uncommon for scientists to praise the philosophy they follow in coded expressions about methods or approaches to research. They may stress the need for "good data" (empiricism), or the importance of theory in ecology (rationalism), or the superiority of experiment (experimentalism). These are philosophical positions – attempts to be correct in the approach to understanding.

There are practical reasons why recognizing your own philosophy is valuable. First, a philosophy can bring with it a bias that can affect scientific progress. Most philosophies of the scientific method are what Shapere (1980, 1984) terms presuppositional. That is, they assume either that the nature of what is being investigated has certain features or that something about a particular method is correct. But there is always a counter-philosophy. So if you understand the bias, then you know places to look for valuable criticism, or how to develop a complementary approach that may enhance understanding.

Second, where there is not an obvious way to solve a major problem, scientists may argue about how to resolve it and then arguments are about philosophies and their associated methods. It is important to be able to recognize this and how your own philosophical position may determine your scientific opinion. For example, there has been a continuous debate since the beginning of ecology about whether ecological communities exist, and if so what their properties are (McIntosh 1985). This has led to philosophical debates about the value of using theoretical approaches (rationalism) as opposed to collecting and classifying information (empiricism) or making experiments (experimentalism). To appreciate the arguments and gain understanding of how they may be resolved it is important to suspend belief that your own philosophy is necessarily correct.

Among scientists, philosophical dispute is usually verbal, or implicit in what is written about other scientists' work. It is not within the bounds of generally accepted scientific discussion for me to write that I dislike your scientific problem, or that it is irrelevant, or that time spent on it retards the field. Such comments are clearly subjective. But if I think there are flaws in your method it is acceptable (and this is required of me if I am asked to review your manuscript submitted for publication) to say and write that I think your results are erroneous. However, this legitimate argument about method and results can also be used as a proxy argument for what is really a philosophical dispute. It is important to recognize when this occurs, particularly in your own thinking, so that genuine problems about methods can be distinguished from attack on, or defense of, a philosophy. One reason why some scientists may be reticent to admit to a philosophy (and thereby acknowledge that other philosophies exist) is that it would open them to criticism or lessen the impact of the criticisms they proffer.

From time to time, ecologists write papers directly in favor of some particular philosophy of method in ecology. For example, Murray (1992) argues in favor of applying Newton's rules of reasoning to biology and ecology; this was in response to Lawton's (1991) conversational review of method in ecology and how it might be improved by more respect and discussion between ecologists with theoretical and empirical approaches. Mentis (1988) argues that both the hypothetico-deductive and inductive approaches are valuable in ecology but acknowledges that some ecologists view hypothesis testing as the "acme of scientific practice". The number of papers by ecologists about scientific methodology is small compared with that about research itself, but the argument is often acerbic and infused with rich language – and this reflects the role of philosophies in making demarcations between approaches and in attempting to be correct.

Third, and perhaps most important, new philosophies evolve as scientific knowledge increases, new problems are met and new methods are developed. To understand your subject thoroughly, you need to appreciate the philosophies on which its methods are based, their strengths and weakness, and how they are changing. Laudan (1981a) considers that shifting scientific beliefs have chiefly been responsible for shifts in the philosophy of science and that development in its methodology has come as much from scientists as from philosophers.

Philosophies of scientific investigation are concerned with the fundamental questions:

> What can be known?   This is covered by metaphysics.

METAPHYSICS

> The process of analyzing the nature of reality and the fundamental structure of our thought about reality.

Dilworth (1996) draws the distinction between two aspects of metaphysics in science. The first refers to argument and discussion beyond the limit of empirical knowledge – and there has been a long-standing battle between empiricists and rationalists about whether such argument and discussion actually belongs in science. Speculative discussion can prepare the way for researching scientific problems – although clearly the more critical and analytical then the better the speculation may be. We use metaphysics for the task of speculating on problems that are not yet soluble, to coordinate knowledge and criticize assumptions. The second aspects of metaphysics are those facets of scientist's convictions or belief systems influencing the type of speculation that will be made. These are discussed in Chapter 16 as ideals.

How can it be known? This is covered by epistemology and there are two fundamental parts:

The *context of discovery*, dealing with constructing a heuristic, i.e., a way of finding things out.

The *context of justification*, dealing with processes of assessment, i.e., ways of testing how secure a result is.

Methods we may use can be judged by both these criteria and by how they apply to the overall objectives of the science. In this chapter, I look at philosophies of science related to the type of initial assumptions that we may make in approaching an ecological problem, first principles of methodology, and uncertainties that have arisen about the objectivity of scientific methodology. In each case, I am concerned both with problems of making discoveries and problems of assessment. The history of philosophy (as opposed to the history of science) is one of reworking ideas rather than completely supplanting them (for discussions, see Althusser 1990, Deleuze and Guattari 1994). For this reason, I give brief historical introductions to some of the major philosophies to show how their methods have been reworked and are still in practical use today because they accommodate newer versions of older problems. This chapter lays foundations for later discussion of a methodology for scientific research in ecology (Chapter 15) and an analysis of major criticisms of ecology (Chapter 16).

9.2 Initial assumptions

Before we start investigating, we usually assume that there is something to be found out. But our assumptions go further than that! Ecologists assume that there are processes of interaction between organisms, and between organisms and their environment, and that the ecologist's task is to discover these processes and details of operation. The precise nature of such assumptions is part of an ecologist's philosophy and determines what may be expected from investigations.

9.2.1 Teleology

Aristotle (384–323 BC) codified and developed the science of logic and produced an extensive natural science, particularly biology, that was still in use almost 2000 years after his death. It was this natural science that so stimulated criticism during the foundation of present Western science. Aristotle based his biology on direct observation, but he combined observation with classification and interpretation of similarities and differences. The syllogism, a central form of Aristotle's reasoning, has the following structure:

A major premise: plants are green.
A minor premise: grass is a plant.
A conclusion: grass is green.

The problem is that the major premise, which is taken for granted, is what actually must be proved. The syllogism can be used to provide some structure to argument – but does not aid discovery by itself.

Reasoning was important to Aristotle. He observed variation and similarity between organisms, used induction to draw conclusions, made deductions, and returned to the natural world to test them. But he did not experiment, and deductions were not exposed to stringent tests. Nevertheless, he made some remarkable observations and interpretations, e.g., that birds and reptiles are closely related in structure.

Aristotle was a teleologist.

TELEOLOGY

Teleology (Gr. *telos* end, *logos* a discourse) in biology is the interpretation of biological structure or function in terms of purpose.

Singer (1959) quotes Aristotle in translation from *On the Parts of Animals* in a passage that well illustrates his teleology:

"As every instrument and every bodily member serves some partial end, some special action, so the whole body must be destined to minister to some fuller, some completer, some greater sphere of action. Thus an instrument such as the saw is made for sawing, since sawing is a function, and not sawing for the saw. So, too, the body must be made for the soul and each part thereof for some separate function to which it is adapted."

Teleology is one way in which biologists and ecologists engage in speculation – it is a theory that something always has a purpose. Boylan (1986) distinguishes between monadic and systemic teleology. Monadic teleology contains a circularity – birds have wings so that they can fly, birds fly because they have wings – and makes no reference to scientific analysis. Boylan terms this monadic because a single entity is said to account for the effect it has, e.g., wings produce flying, simply because of their capacity to do so. Monadic teleology can be extended to describe variation in biology and imply purpose. "Eagles developed large wings so they can soar." "sparrows developed short wings so they can flit." Monadic teleology should be avoided because: (1) teleological statements are closed to logical development; (2) they are not true or false and they are not questions; (3) even in speculative discussion, teleological statements encourage further teleology. Monadic teleology is not harmless shorthand.

Most scientists would agree that this type of description is pointless.

However, as a form of speech, monadic teleology can be insidious. Sentences such as "Leaves have stomata so they can exchange CO_2 and water vapor with the atmosphere" are not uncommon – at least in conversations. Of course the result of much biological research is discovery of the function of various features and properties. It is reasonable to say, "The effect of stomata is that they control the exchange of CO_2 and water vapor between leaves and the atmosphere". This is a teleonomic statement, backed by research. Teleonomic statements tend to be more cumbersome than teleological ones and because of that teleology can creep into a conversation (for further discussion, see Loehle 1988).

Boylan describes systemic teleology in relation to inputs to and outputs from a system. As an example he interprets a well-known experiment on industrial melanism in moths. Two kinds of moth were found: peppered-winged, and black-winged, which is the mutant of the peppered. The species lived in an area of industrial pollution where the trees were black and the black-winged moths survived. When the trees were washed, the peppered-winged moths predominated, so predation was considered to be the "cause" of differential survival in that birds were considered less able to prey upon moths whose coloring was the same as that of the trees.

Boylan's view is that this is an example of "downward causation". That is, in terms of logic, systemic teleology is equivalent to inserting an implication (Chapter 7), so that,

$$(moth\ color + tree\ color) \Rightarrow survival$$

becomes

$$(moth\ color + tree\ color) \Rightarrow predation \Rightarrow survival.$$

As Boylan points out, other views about this relationship between moth color and survival are more cautious and consider that the mere fact of the effect being beneficial is not evidence of adaptation. An effect should only be called a function when chance can be ruled out.

In choosing his example of systemic teleology, Boylan selected an experimental situation where there was reason to offer some explanatory theory that might be the subject of further investigation and experiment. However, systemic teleology can also be applied far too loosely, most frequently under the disguise of "hypothesis generation" and in situations where there have been no prior investigations but simply casual observation that suggests to the observer a causal relationship. It is particularly dangerous when extended from organisms, where there may be some justification for teleonomy but not teleology, to ecosystems or communities where there is justification for neither. A statement such as "Ecosystems have structure so they can support high diversity" is closed to logical analysis (see discussion of propositions in Chapter 3) and attributes properties for a purpose to a theoretical construct.

The difficulty is that when studying biological systems we do need functional explanations. No understanding of vertebrate physiology would be complete without understanding of how the heart functions. But simply attributing a function to a feature is insufficient. A causal explanation is required along with an explanation of why that particular cause (or causal chain) arose. In biology the second part of the explanation is usually genetic and evolutionary. So for vertebrates we can give a causal explanation of how hearts function, including differences between vertebrates, and what the organ's role is within the body. The evolutionary parts of the complete explanation show the development of circulatory systems and their patterns of inheritance. In this case there is a sufficient (because it is based on causal reasons) and necessary (because of the genetic and evolutionary reasons that rule out alternatives) explanation.

9.2.2 Parsimony

Modern Western science began as part of a new questioning approach in society and developed into a direct challenge to Aristotle's science and its teleology. Peter Abelard (1079–1142), teaching in Paris, introduced a method of resolving theological questions by argument, by questioning everything (Flew 1984). His philosophy can be summarized as "constant questioning is the first key to wisdom. For through doubt we are led to enquiry, and by enquiry we discern the truth" (Muir 1976), and this approach became a foundation of early science.

Aristotle's written works became available through contact with Arab scholars, and an Aristotelean teleological approach to the natural world was adopted by much of the Catholic Church. Aristotle's philosophy had appeal because with his teleology came a plan for the universe – God's plan (Shapere 1974). However, its interpretation within the church was disputed; Aristotle considered the universe to be eternal, and so not created from nothing by God, and the Being responsible was not a personal God. Aristotle's philosophy also came under challenge in Paris during the fourteenth and fifteenth centuries (Blake *et al.* 1960, Montalenti 1974) from the Occamists, named after William of Ockham (d. 1349). Parsimony was a central feature of their approach.

PARSIMONY

No more causes should be proposed than are necessary to explain a phenomenon. Occam's Razor asserts the principle of parsimony in making propositions (Tornay 1938), *frustra fit plura quod potest fieri per pauciora*, . . . [it is vain to do through more what can be done with less].

Faced with the sometimes convoluted teleology of Aristotle, it was not surprising that the Occamists demanded simplification – and simplification was essential to perform the types of experiment that enabled them to make progress. Indeed, the victory of the Occamists' cause in the development of science was not that their theories were parsimonious and Aristotle's, and those developed in the same manner, were complex. The essential point was that their theories were tested and refined through measurement and experiment.

The Occamists based their approach to science upon measurements and the development of mathematics, and Blake *et al.* (1960) consider that modern mechanics had begun by 1375. Then, with a reliance on measurement (empiricism) and the formulation of mathematical rules (the development of laws, a rationalist approach), the Occamist philosophy contained the two important aspects of scientific philosophy that persisted until experimental approaches were explicitly formulated.

9.2.3 Holism and reductionism

Holism has been used to mean different things, Loehle (1988) gives four definitions, but in research holism is frequently characterized by claims that emergent properties exist. Thorpe (1974) gives Broad's (1925) definition.

EMERGENCE

"Emergence is the theory that the characteristic behavior of the whole could not, even in theory, be deduced from the most complete knowledge of the behavior of its components, taken separately or in other combinations, and of their proportions and arrangements in this whole." (Broad 1925.)

Emergence is also related to theories of hierarchical organization. Ayala (1974) describes complexity of organization as an outstanding characteristic of living matter and outlines a hierarchy of complexity through increasing levels from atoms and molecules, through cells, tissues, individual organisms, populations, communities, and ecosystems to the whole of life on earth. Partitioning things in this way into a hierarchical structure, where the constituents of each level can be considered as components of the next level of complexity, is an attempt to simplify the totality of things. It is a theory of how things may be organized. The concept of emergence depends upon this or some similar hierarchical organization being constructed, and that within levels there are at least some characteristic interactions unique to that level.

The question of whether emergence occurs depends on whether theories in a higher level (say, populations) can be shown to be special cases of theories in a lower level (say, organisms). If this can be done, then the higher level (populations) is said to be reduced to the lower (organisms), and there would be no emergent properties in the higher level. This is the theory of derivational reduction (Nagel 1961).

REDUCTIONISM

Any doctrine that claims to reduce the apparently more sophisticated and complex to the less so. (After Flew 1984.)

There are two conditions for a derivational reduction to be complete:

(1) Derivability – all theories of a higher level are logically deduced consequences of the theoretical constructs of the lower level. For example, everything in the theory of population dynamics can be explained by theories about organism behavior, e.g., birth and death, feeding, etc.
(2) Connectability – to achieve derivability, then concepts in the higher level can be reduced and redefined in terms of those in the lower level. For example, a population concept such as intrinsic rate of increase should be defined in terms of individual concepts of birth, death, and concepts that influence them.

The theory that there are emergent properties says that derivability and connectability are not complete. At various times in the history of their sciences, both biologists and ecologists have asserted the importance of emergent properties. This emphasized their subject's independent right to exist: that the biologist was not simply working out details of the consequence of physics and chemistry, nor the ecologist working out the consequences of biology. These assertions were important because, as with all ongoing research, it was not clear just what there was to be discovered. So, at times "emergent properties" have seemed almost mystical – concepts that defined a purpose for ecological science but could not themselves be defined with precision.

9.2.4 Teleology, parsimony, and reductionism in ecology

Shrader-Frechette (1986) describes the ecosystem approach as the dominant paradigm[1] for ecological research. She considers it to be unequivocally teleological. The roles of various species and organisms are interpreted in terms of their real or supposed contributions to the persistence of a larger entity, the ecosystem. Shrader-Frechette analyzes the use of teleology in ecology in the context of examining the extremes of holistic and reductionist approaches. She considers that ecological holists tend to make the following errors:

[1] Paradigm, to be discussed in detail in Chapter 11, refers both to a group of ideas that define an area of scientific investigation and to the scientists who have them.

(1) Failure to derive testable implications from theory and failure to use falsification.

(2) Tendency to claim that the analytical objectives of ecological reductionists are impossible to accomplish.

(3) Mistaken understanding of natural selection and adaptation, particularly the attribution of system properties to that theory.

On the other hand, she considers that ecological reductionists tend to make three inappropriate sorts of assumption:

(1) Only predictive science is real science. The tendency is to downplay theory and view it merely as models that do not or can not fit data, rather than as a conceptual framework.

(2) The operation of some ecological whole can be explained adequately by the operation of its parts.

(3) Any qualitative difference between organisms or communities counts against the presence of recurrent structures in ecology. She quotes Gleason: "since every community varies in structure . . . a precisely logical classification of communities is not possible."

Shrader-Frechette (1986) offers some postulates to guide ecologists in their study of allegedly teleological phenomena. First, she questions the expectation that good ecological science ought to be reductionist. "If biological phenomena are hierarchically organized, as indeed they seem to be, then perhaps they require special types of explanation." Reduction in biology is not complete without synthesis.

Second, more than one approach may be of value, which implies that there is more than one group of phenomena to be studied. One could say that autecology, the study of individual species, and synecology, the study of communities and complete ecological systems, are the divisions generally followed by reductionists and holists, respectively, although that is only an approximate division.

Third, "[g]iven the monumental problems requiring explanation, one does not have the luxury of any *a priori* assumptions that teleological accounts are in principle implausible" (Shrader-Frechette 1986). She considered that holistic ideas have a utility within which scientists can work toward falsification and prediction.

We do not have to accept either derivatational reduction or connectability or emergent properties as appropriate theories of knowledge (Maull 1977). The first requirement is to examine the types of interaction that may occur between different types of component of ecological systems. Levins and Lewontin (1980) criticize a type of hierarchical reductionism in ecology which "takes the form of regarding each species as a separate element existing in an environment that consists of the physical world and of other species. The interaction of the species and the environment is unidirectional: the species experiences, reacts to, and evolves in response to its environment. The reciprocal phenomenon, the reaction and evolution of the environment in response to the species, is put aside."

Reciprocal phenomena can be studied without invoking emergent properties. *Pedogenesis* is an example of such a reciprocal phenomenon. Soil conditions influence plant growth, and this has been studied extensively through experiments and laboratory analysis (see, e.g., Glass 1989 for nutrients, Kramer and Boyer 1995 for water). However, in the reciprocal influence, soil development is determined by such processes as root growth, organism death, decay, and the accumulation and downward transport of organic materials through the soil (e.g., for a review, see Ugolini and Edmonds 1983). The study of soil chronosequences below receding glaciers demonstrates marked changes in soil properties in parallel with vegetation succession. In south-east Alaska, Crocker and Major (1955) found a decrease in pH of the topmost soil horizon from 8.0 to 5.0 along a gradient from glacial material recently uncovered by melting of the ice to a *Picea sitchensis* forest on material uncovered for 70 years. Increases in soil organic carbon and soil nitrogen and a decrease in bulk density paralleled the changes in soil pH. Importantly, variation in the rate of change of pH in the first 30 years was related to the particular plant species that happened to invade the bare material. In this case, while investigation of effects of soil upon plants can be by direct experiment, the reciprocal phenomenon required an indirect investigation through chronosequences.

Pedogenesis is a property of ecosystems, not of plants and animals by themselves, nor physical and chemical properties by themselves. The particular biological, physical, and chemical conditions of an ecosystem both above and below ground determine the types and rates of change in the soil. Theories about pedogenesis that use these conditions are ecological theories, but there is no need to define pedogenesis, or any characteristic or consequence of it, as an emergent property. We can, and do, explain how pedogenesis occurs by using concepts about organic matter production and decomposition, physical and chemical weathering, root physiology and growth, microclimatology, and habitat preferences of individual plant and animal species. We can use such a theory to make predictions about future rates of change, say, of land recently uncovered by receding ice. The explanation for soil changes and plant succession depends on developing a synthesis of different types of knowledge into a theory that explains the phenomena but neither assumes, nor requires, that there is derivability and/or connectability in some hierarchical arrangement of the theories in those different fields (Maull 1977).

An attraction of reductionism is that it appears to offer the possibility for complete explanations of complex systems in terms of measurable components or inputs. Of course the question still remains, how is this to be done? (In practice, ecology, as a comprehensive subject, uses concepts that must be defined through construction of theory and can not be measured directly. See Chapter 10.) In ecology we see two types of reduction, *complete system reduction* and *partitioning reduction*. These attempt to achieve methodologically tractable problems in different ways.

COMPLETE SYSTEM REDUCTION

A *complete system reduction* asserts that the functioning of a system can be expressed as a simple relation or function between measurements of inputs to, and a few outputs from the system.

An example of a complete system reduction is the assertion that the primary production of plant communities is proportional to the radiation they absorb, i.e., we can explain a biological phenomenon in terms of the physical inputs to it: see Demetriades-Shah *et al.* (1992) for a brief historical review. The relationship can be formalized:

$$e = W/fS$$

where W is standing biomass, f is the fraction of radiant energy absorbed of sunlight (irradiance S), and e is termed the radiation use efficiency (Monteith 1994). Most investigations of this relationship have been for field crops, but see Jarvis and Leverenz (1982), who recommend it as a heuristic for an ecological understanding of forest growth. Many straight-line relationships between W and fS have been reported over the growth of annual crops: i.e., constant e, although e may vary between crops and circumstances of growth, e.g., differences in nutrition for the same crop (Arkebauer *et al.* 1994). This is not simply advanced as an empirical relationship with an undiscovered theoretical basis. It is considered by some scientists as "a fundamental description of plant growth which has enhanced understanding of both potential growth rates and reductions in those potentials owing to stressful conditions" (Arkebauer *et al.* 1994) and in this sense is a law (see Chapter 3).

The relationship is a reductionist approach, since explanation is sought in terms of lower level phenomena. A question is whether, or to what extent, the reduction is justified because its use may be a heuristic that provides insights and at least some types of assessment. The relationship and its use were criticized (Demetriades-Shah *et al.* 1992), defended (Monteith 1994, Arkebauer *et al.* 1994, Kinry 1994), and criticized again (Demetriades-Shah *et al.* 1994). These papers provide an unusual and interesting series because ultimately, in this case, the argument rests upon the philosophical justification for the type of reductionist explanation sought. There are direct implications for ecology because the method has been proposed for satellite-based estimations of fS for observed vegetation types.

Demetriades-Shah *et al.* (1992) criticize the logic of the formulation as follows:

(1) Its validity as an explanatory reduction. Anything that increases in size will intercept more radiation even though radiation may not have caused the size increase. They challenge the assumed causal basis. For them it remains an open question whether the increase in size was due to greater interception of radiation, or whether greater size caused greater interception of radiation.

(2) Its use in assessing crop performance. The use of cumulative values (W and fS), both of which depend upon the increasing foliage amount, guarantees a highly significant correlation. A sequence of pairs of random numbers, x_1, y_1; x_2, y_2; x_3, y_3; . . . will show no significant correlation, but when their progressively cumulative values are used x_1, y_1; $x_1 + x_2$, $y_1 + y_2$; $x_1 + x_2 + x_3$, $y_1 + y_2 + y_3$; . . . there is a significant correlation. It is not valid to calculate the correlation between successive values of W and fS measured as growth proceeds because successive values are not independent variables.

(3) Its value as a heuristic. The values of e may indicate the seasonal mean for a crop. But there is no measure of variation over the season and what may cause it.

The most direct defense is made by Arkebauer *et al.* (1994). They assert that the concept of e is valuable, and that "most (if not all) environmental stresses which are a 'critical limiting factor for growth' affect the plant through physiological pathways involving the photosynthetic use of radiation". They offer explanations of variation in e from different experiments based on interpretation of such factors as nutrient or water deficits. They also offer analyses of variation of e within seasons.

There are technical problems in both logic and assessment, but the main differences are in philosophy and whether, as Arkebauer *et al.* (1994) assert, it is reasonable to consider radiation as the main driving (causal) variable and to analyze other effects through their effect on that relationship. The alternative philosophy, as put forward by Demetriades-Shah *et al.* (1992), is that a reduction of this type is not appropriate:

"This form of analysis is intellectually appealing because it incorporates a well-known plant physiological response to light and appears to simplify a complicated situation. However, this simplification rolls all other (often more important) soil and environmental factors affecting the crop into a single term for radiation use efficiency. Energy may be the most fundamental resource from a physical point of view, but from the biological viewpoint it is no more important than water, nutrients, CO_2 or any other requirement essential for plant life."

So Demetriades-Shah *et al.* (1992) question both derivability and connectability – that production can not be explained by absorption of radiation alone. They go on to suggest that water use efficiency, nutrient use efficiency, etc., would be equally valid and that there would be conditions, e.g., low temperatures or arid regions, where *e* would not be informative. The great majority of the research into *e* has been done with annual crops and its value and application might be best defined as for expanding foliage canopies of annual crops in well-watered and nutrient-rich agricultural conditions. This specifies the domain of the relationship. If the problem of interest falls outside the domain where a reduction gives information then some other approach will have to be used. However, scientists using a particular reductionist approach may be unwilling to recognize, or define, the domain of application but rather to continue to maintain its universal application. This may be related to the values they hold, e.g., that investigation is not scientific unless there is an assessment procedure – giving priority in research to the context of justification (even though, as in the case of $e = W/fS$, there are legitimate doubts about the calculation). The alternative value is that an investigation is not scientific unless the explanation offered for phenomena (what causes crop growth) is complete – giving priority in research to the context of discovery. *Domains* are discussed in more detail in Chapter 10.

Even where simplicity of theory must give way to complexity, ecologists may still feel the need to justify this explicitly. A need for simple explanations, which are nevertheless complete, of the type offered by the light interception theory just described, has deep roots. Dunbar (1980) confesses "Occam's Razor has bothered me for years . . ." and goes on to assert that "the history of science is full of examples of bland disregard for Occam's Razor . . ." Foremost among the instances he quotes is Darwin's theory of natural selection. Grubb (1992) devotes his presidential address to the British Ecological Society to urging that simple theories be discarded when they are found wanting and be replaced by more adequate complex theories. He gives a comprehensive description of how three apparently competing simple theories of plant defense mechanisms can be resolved into a more complex one.

A more frequently found reductionist approach in ecology still assumes that explanation of an ecological system can be made in terms of its components, but recognizes that the interactions may be complex and seeks to build up knowledge through studying partitions of the system.

PARTITIONING REDUCTION

A *partitioning reduction* is made when a scientist considers that a system is complex and that explanation must be developed through studying partitions of the system. This can be motivated by conceptual difficulties, lack of resources, or development of techniques and skills that encourage a scientist to focus on some component.

Three important features of partitioning reduction must be remembered.

(1) Prior system analysis before reduction is implied, i.e., that the reduction has logic behind it and is not haphazard. For example, someone who studies root production, turnover, and nutrient uptake as an ecosystem process has made a partitioning reduction or at least implicitly agrees that these processes can be studied on their own without studying other ecosystem processes.

(2) Investigation of those parts of the reduction not selected for immediate study is postponed and synthesis with the results of the research may not be attempted.

(3) The connections between the selected partition and other partitions of the system may not be defined or studied. In root and nutrient studies, a complete system synthesis that examines their influence on other processes, e.g., on plant growth, may not be made. Reduction implies a choice about which piece or pieces should be studied from a complete analysis and a value judgement about that choice. The scientist says: "This bit is important and I will study it; other bits will (may) be studied later or by someone else". Of course, finishing study of the complete system is another choice and does not follow automatically after the first reduction has been pursued.

For example, in pursuit of a series of statistically testable postulates it might be useful to continue the reduction of the study of small mammal use of woody debris piles (Chapter 4). Such a study would most likely change focus from woody debris to small mammals (not necessarily a better or worse focus, but different). An explicit synthesis of the importance of woody debris on cobble bars might be postponed indefinitely while the details of small mammal biology are pursued. Two things encourage research to continue with a partitioning reduction: (1) requirements for statistical testing, and (2) incomplete methodology for making a synthesis that itself is testable (discussed further in Section II).

Teleology, parsimony, and reductionism have been debated for centuries without a resolution. This is because we assume causality in the nature of what we seek to find out and this provides us with a heuristic, while reduction provides us with a way of making assessments based on measurements and replication. As we encounter fundamentally new problems, such as in ecology, and those pose different types of question, we have to research our methodologies and find effective heuristics and assessment procedures.

9.3 First formalizations of methodology

By the early sixteenth century, the scientific method was established as a dialog between theory and experiment. Yet scientists of this time still felt the need to sweep away the complex ideas put forward by Aristotle because Aristotelean teaching was prevalent, having been adopted by the Church. More and more complex teleology had been developed to deal with observed features of the world not described in Aristotle's original writings. The requirement to start afresh with a completely new system of description was a huge task. It caused a demand for self-evident "starting points" that the scientist would not have to investigate. This demand was met in two different ways. The empiricist starts with data, the rationalist with axioms.

9.3.1 Empiricism

EMPIRICISM

The empiricist considers that experience and measurements are fundamental to understanding and determine the axioms we adopt. Nothing can be known unless its existence can be inferred from experience or measurement.

Naive empiricism is the extreme position that complete induction from experience and measurements is possible, i.e., using the method of empirical induction (Chapter 7). This philosophy received encouragement from Newton's research in mechanics, but subsequently philosophers realized that the very meaning of factual existence demands some *a priori* generalization, at least a minimum of theory. Some scientists place particular emphasis on statistical empiricism, the need to use statistical methodology when considering hypotheses. Statistics requires that we answer questions with data and that we judge those questions using rules about data, e.g., the normal distribution, randomization, etc., as given in Chapter 8.

Francis Bacon (1561–1626) is frequently considered to be the father of empiricism. Bacon, a product of the Protestant Reformation, worked in England when there was reaction against medieval Catholicism. From the age of 12, Bacon spent three years at Trinity College, Cambridge, and left with a confirmed hostility to the cult of Aristotle then dominant there. He resolved to set philosophy onto a more fertile path, to turn it from scholastic disputation to the illumination and increase of human good (Bowen 1963).

Bacon accepted Aristotle's scientific method of a continuous interaction between induction and deduction but was critical of the way it had been applied (Losee 1993). He proposed a system where knowledge can be obtained through direct observation. In one of a series of aphorisms he writes (Foster 1937):

> "Aphorism XXXVI
> One method of delivery alone remains to us, which is simply this: we must lead men to the particulars themselves, and their series and order, while men on their side must force themselves for awhile to lay their notions by and begin to familiarize themselves with facts."

Bacon's concern was to lay aside "notions," i.e., teleological theories, and to observe what he terms facts. In Aphorism XIV, he states "our only hope therefore lies in true induction". We should gather together all the facts we can and disregard previous theory. Bacon's proposed method of induction involves a gradual approach by which we

> "must analyze nature by proper rejections and exclusions; and then, after a sufficient number of negatives, come to a conclusion on the affirmative instances;"

He expounded a technique of arranging observations as tables of three types (Blake *et al.* 1960):

> tables of essence and presence, a search for commonalities,
> tables of deviation, or absence in proximity, a search for differences,
> tables of degrees of comparison.

When the three tables are constructed, then

> "induction itself must be set at work; for the problem is, upon review of the instances, all and each, to find a nature such as is always present or absent with the given nature and always increases and decreases with it."

Bacon's system is one of classification – and of course we use classification as an integral part of the scientific method, not just at the outset of research, but regularly when collating results to understand their meaning.

The focus of Bacon's method, and the contribution for which he is remembered, is that advance can be made through codifying knowledge rather than by propounding a teleological structure and that scientific inquiry can proceed in a gradual stepwise process of generalization from observation. He was aware that spurious correlation could exist and formulated a method of exclusion to remove them (Losee 1993). Bacon himself was not averse to experimentation and the development of hypotheses (Blake *et al.* 1960), writing that, "truth will sooner come out from error than from confusion". Bacon died as the result of conducting an experiment (Bowen 1963). While riding from London to Highgate he wanted to test how far cooling with snow might preserve flesh. He stopped at a cottage, bought a fowl, had it killed, and stuffed it with snow. Bacon considered the experiment a success, although no

Table 9.1. *Laudan's (1981b) distinction between the problems of Plebian and Aristocratic Induction. Laudan noted that "theory" in Aristocratic Induction must postulate one or more unobservable entities that could not be empirical generalizations*

Plebian problem of Induction	Given a universal empirical generalization and a certain number of positive instances of it, to what degree do the latter constitute evidence for the warranted assertion of the former? If we repeatedly measure the same relationship, when are we justified in considering it sufficiently confirmed to act as the basis for further investigation?
Aristocratic problem of Induction	Given a theory and a certain number of confirming instances of it, to what degree do the latter constitute evidence for the warranted assertion of the former? If we repeatedly use a theory successfully, when can we consider it to be confirmed?

mention was made of controls. Bacon was then seized with chills and weakness and died of his illness shortly afterward.

Recall that, in Chapter 7, the problem of induction was described as the fallacy of affirming the consequent, i.e., that confirming instances do not prove that a theory is true. But scientists use induction all the time, particularly in surveys and exploratory research. Because falsification is not possible in all circumstances more must be said about induction.

A distinction can be made between the method of empirical induction, termed as a philosophy Plebian Induction (Laudan 1981b), the type Bacon urges, and Aristocratic Induction (Table 9.1) the type Popper (1972) sought to combat with falsification, which are investigations using a more developed theory (Laudan 1981b) (Table 9.1).

Three conditions are necessary for Plebian Induction to hold:

(1) Natural concepts with little theoretical content are measured.
(2) The relationships expressed in the generalization are simple.
(3) The simple generalization holds in all the circumstances where it is looked for.

This form of induction is used in attempts to describe laws. Its principal difficulties are those of ensuring consistency in concept definition, even of simple concepts, and of developing a precise formulation and assessment for the simple generalization.

However, Aristocratic Induction is more demanding. Theories are complex. They contain not only those propositions that can be clearly recognized as axioms, and the concepts that are used, but also a series of ancillary axioms about measurements and the relation of other theories to the one under consideration.

9.3.2 Rationalism

Although both empiricism and rationalism developed in response to the same problem, one of excessive theorizing isolated from investigation, they took contrasting approaches.

RATIONALISM

There are four characteristics of *rationalism*:

(1) The belief that it is possible to obtain by reason alone a knowledge of the nature of what exists,
(2) the view that knowledge forms a single system, which
(3) is deductive in character, and,
(4) the belief that everything is explicable; that is, that everything can in principle be brought under a single system. (After Flew 1984.)

The term rationalism is also referred to a person who may ascend to only (2) and (4). For scientists, the important feature of rationalism is (1) – that at least some axioms exist as fundamental and are independent from particular pieces of experience.

Descartes' (1596–1650) philosophy is rationalist, although he wished to reduce the axioms with which he started to a minimum. His proposal was intellectually to doubt everything and then, from the minimum set of propositions, deduce the consequences. The introduction of deduction by formal reasoning distinguishes his approach from that of Bacon, who lived at the same time and observed the same problems. Seeking linkages and consequences became an integral part of the scientific method. Descartes was the first explicit proponent of a coherent, connected, reductionist science. We may now wish to challenge the particulars of his view, but at the time his stance was invaluable for future developments. Descartes himself considered this philosophy to be a very old one, similar to that of Democritus of Ancient Greece. The Democritans – the opponents of Aristotle – were mechanists who believed that the actions of living things resulted from the interactions of their component atoms (Singer 1959).

Singer states Descartes' four rules:

> "The first was, never to accept anything as true when I did not recognize it clearly to be so, that is to say, to carefully avoid precipitation and prejudice, and to include in my opinions nothing beyond that which should present itself so clearly and distinctly to my mind that I might have no occasion to doubt it.
> The second was, to divide each of the difficulties which I should examine

into as many portions as were possible, and as should be required for its
better solution.

The third was, to conduct my thoughts in order, by beginning with the
simplest objects, and those most easy to know, so as to mount little by
little, as if by steps, to the most complex knowledge, and even assuming
an order among those which do not naturally precede one another.

And the last was, to make everywhere enumerations so complete, and
surveys so wide so that I should be sure of omitting nothing."

Descartes' approach, particularly the first three rules, is an example of
metaphysics.

Detailed analysis, seeking logical connections and extensive examinations
are the strengths of the rationalists– method. However, there are two problems. The first is knowing how far to go in reducing an argument to its
supposed basics. A further problem is knowing whether one's deductions are
unique. One characteristic of rationalism, particularly from Descartes onwards, was a belief in the power of mathematics to define axioms. Some
strongly mathematical theories that are proposed for use in ecology do not
have their origins in ecological data, e.g., catastrophe theory and chaos theory.
This is rationalist science. It does not originate from measurement.

It was clear to Descartes that not everything could be deduced from first
principles and that conjectures were necessary but that these should be treated
as uncertain. He used the analogy of the world as a watch, with a visible face
but hidden internal construction: "for just as an industrious watch-maker
may make two watches which keep time equally well and without any
difference in their external appearance, yet [they are] without any similarity in
the composition of their wheels" (Descartes quoted in translation in Laudan
1981c). Laudan refers to this as the classic problem of *empirical underdetermination of theories*, that "indefinitely many mutually inconsistent micro-
structural hypotheses are all compatible with the visible effects". This problem lies at the heart of the difficulties with rationalism. The role of first
principles is to circumscribe the range of postulates by excluding some; a
sound postulate is compatible with the data and first principles, both of which
can be used to say which postulates are wrong – but not which are right.
Verification must not be confused with proof.

Arguments against the hypothetico-deductive methodology of the rationalists have a long and respectable history, and many remain as important as
when they were first proposed. Newton's laws of motion and his predictions
of planetary movements were successful. His self-professed method was
inductive. In contrast, many of Descartes predictions about the universe,
based on hypothesis, were shown to be wrong. Newton and his followers
argued this as a success for induction, and Newton called for a non-
conjectural science (Laudan 1981d) based on empirical induction not hypothesis.

9.3.3 Empiricism and rationalism in ecology

The apparent simplicity of Newton's approach remains attractive. Murray (1992) urges Newton's rules as a model for the scientific method of biology and ecology; he was responding to Lawton (1991), who suggests a major reason why ecologists fail to solve important questions is that theoretical and empirical ecologists lack respect for each other. As so often, when philosophies are being argued, Murray's indictment is strong:

> "I believe the reason for the failure of ecologists and other biologists to develop widely acceptable, explanatory, predictive theories, like those in physics, is far more fundamental. Biologist do not think like physicists."

Not content with that, Murray continued by quoting Dyson (1988) that biologists are "essentially diversifiers, scientists 'whose passion is to explore details' and 'who are happy if they leave the universe a little more complicated than they found it' whereas physicists are unifiers, scientists whose 'driving passion is to find general principles which will explain everything.'" Murray (1992) asserts "If biologists adopted Newton's four rules of reasoning, then, I submit, they would have unifying explanatory, and predictive theories . . ." (Table 9.2) and goes on to describe how that would apply to his particular research interests. Since this directly challenges a dominant philosophy in ecology and seems to urge reduction, parsimony, and induction, it deserves examination.

Could biologists and particularly ecologists adopt Newton's rules? Newton's rules have to be taken in their context. In Newton's *Principia* they are listed after his analysis of direct experiments on moving bodies conducted in the laboratory and before his description of the world. To a considerable extent they are Newton's justification for this inductive leap from laboratory to world. They define the experimental unit, which is any body, and provide justification that big and small bodies behave alike. The rules also assert that cause, which in Newton's case is gravity, is universal, i.e., the assumption of a universal domain. Whether these assertions about the experimental unit and the supposed applicability of the process being studied transfer to other subjects has to be examined as part of the research.

In ecology, two of the most important questions are "What is the experimental unit?" and "How universal is the process studied?" The experimental unit for Denise Hawkins (Chapter 5) is a fish preparing to spawn, and the property of interest is the process by which that fish determines a mate. Many factors may influence the mating process. To assert, *a priori*, that there is only one would be professing a philosophy that may not be useful for the purpose of the investigation. The problem with Newton's rules is not that they tell us how to proceed – but that they assert the nature of knowledge that we are to find out and in that, of course, they eventually proved limited even in physics.

Table 9.2. *Newton's rules of reasoning in philosophy and their interpretation*

	Newton's rule	Interpretation
I	We are to admit no more causes of natural things than such as are both true and sufficient to explain their appearances.	*Parsimony* This rule asserts the requirement that a proposition be both necessary and sufficient. It does not say how these properties are to be assessed
II	Therefore to the same natural effects we must, as far as possible, assign the same causes	*Assertion of a universal domain* Newton's examples for this rule were respiration in man and beast, that stones in Europe and America fell in the same way, the light of fire and the sun. He asserts that these phenomena have the same causes
III	The qualities of bodies, which admit neither intensification nor remission of degrees, and which are found to belong to all bodies within the reach of our experiments, are to be esteemed the universal qualities of all bodies whatsoever	*Principle of the justification of similarities through empirical experimentation* Newton was concerned with the physics of motion, which he analyzed through experiments. If experiments reveal the same principles among different bodies, then these are to be considered universal properties of all bodies. This rule asserts that a universal domain may be empirically justified, and Newton invokes the use of induction for this
IV	In experimental philosophy we are to look upon propositions inferred by general induction from phenomena as accurately or very nearly true, notwithstanding any contrary hypotheses that may be imagined, till such time as other phenomena occur, by which they may either be made more accurate, or liable to exceptions	*The general principle of induction* Newton comments that "the argument of induction may not be evaded by hypotheses". But note that he also implies that exceptions may occur and will alter the proposition

While Newton's rules I and IV might be applied in ecology, rules II and III could not. Rule II can be transgressed because there can be multiple causes for some effects – particularly for systems with components that interact in different ways. What determines the size that an animal or plant may reach? The answer is "many things". Any of a group of controlling factors, e.g., genetics, nutrition, or environmental conditions (such as temperature), may sometimes have a dominating control over growth. If we wish to determine how every factor operates and make use of that in ecological interpretation, we also must establish when and under what conditions it acts. We must define both the mode and domain of action of a series of factors to describe a complex system.

Quentette and Gerard (1993) offer a critique of Murray (1992) based on the assumption that evolution is a coordinating principle of life history and that unique events may be required to explain some phenomena. Even though evolution as a process may be deterministic, the evolution of an organism cannot be predicted precisely.

Blyth and MacLeod (1981a,b) give an example of an empirical study in applied ecology that illustrates important methodological problems with the inductive approach. Their objective was to develop predictive equations for the growth of plantations of Sitka spruce (*Picea sitchensis*) across north-east Scotland. The plantation technique was similar throughout the area and involved no intensive culture or addition of fertilizer, so the plantations could be considered to represent a large-scale bioassay of environmental factors that might influence forest growth. Because of this, the study is an interesting example of what an empirical approach using natural concepts, i.e., those that can be readily measured, such as rainfall, and temperature, can achieve in ecology.

The basic technique was to measure plantation growth in relation to age on 73 sites aggregated in three forest groups and to seek correlation with measured environmental factors. Growth was estimated (as a local yield class, i.e., mean annual timber increment over the time until harvest, (m of timber)^3 y^{-1}) from the height:age relation of the stand. Bimonthly rainfall and accumulated soil and air temperatures were measured for two years to characterize the sites, along with 6 physiographic, 13 soil, and 5 soil physicochemical variables. Stepwise regression equations were developed that selected variables in order of the size of their correlation with local yield class. The equations developed were tested by measuring 54 more sites. The work had a general postulate that the environmental variables selected influenced the measured growth. There was correlation between some of the environmental variables themselves; for example, summer air temperature decreased with elevation, and total phosphorus (P) and total nitrogen (N) were always strongly positively correlated with one another in the soil organic horizon.

Using stepwise multiple regression then increasing the number of predictor variables accounted for increasing amounts of the variance in forest growth (Fig. 9.1). There are three important points to note.

(1)   Predictions are more effective for calculations made for separations into individual forest groups than for the whole 73 sites taken together.
(2)   Prediction in one forest group (Perthshire) is more effective than that in the other two groups.

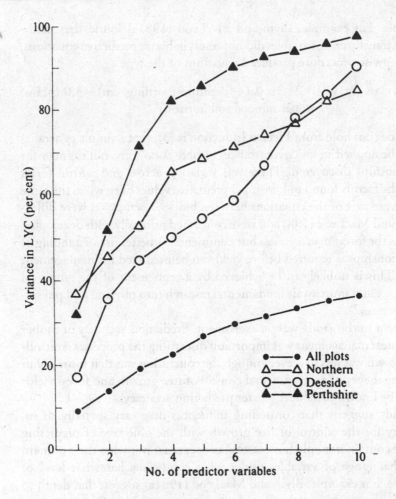

Fig. 9.1. Percentage of variance in local yield class (LYC) accounted for at successive stages in stepwise regression analysis, as increasing numbers of variables are included. Note that overall predictive efficiency is greater for individual forests considered separately than for all sites taken together. The equations for each case were different. (From Blyth and MacLeod 1981b, with permission.)

(3) The sequence of variables selected in the stepwise regression differed among the three groups.

Similar results were obtained when sites were grouped by elevation. Within each forest group, when sites within each of the forest blocks were considered separately, predictions were better, i.e., equations developed from initial sites explained more variation in the subsequent test sites. Particular variables appeared regularly as the first variable in satisfactory within-block equations, particularly depth to mottling in the soil and elevation. These results are typical of site survey studies in forestry (Carmean 1975). Generally, the study site must be divided into small uniform areas before clear associations can be detected.

Two of the three conditions necessary to complete a Plebian Induction are present: (1) The measurements were of natural concepts with little or no theoretical content and (2) the relationships expressed in the generalizations were simple. For example, Blyth and MacLeod (1981a) found that mathematically transforming variables did not result in better predictive equations, and the stepwise procedure produced equations of the type:

$$\text{local yield class } (m^3 y^{-1}) = 24.3 + 0.14 \text{ (depth to mottling, cm)} - 3.0 \text{ (pH of upper mineral soil horizon)}$$

What does not hold from Plebian Induction is (3), that a simple generalization can be applied in all circumstances – there is no universal domain for the relationships discovered. There was variation across and within forest blocks in the factors found to have most predictive value. Even when the same variables were part of the equations between blocks, coefficients were different. Blyth and MacLeod (1981a,b) were concerned primarily with developing predictions for forest management but comment: "A better understanding of these relationships is required before yield can be predicted for management purposes. This is unlikely to be achieved by a refinement of the empirical approach . . . and must await fundamental research into physical and physiological processes."

The accent in this study was on assessment. Prediction accuracy of timber yield for forest management was important; discerning the processes controlling tree growth was less so – even though the conclusion was that more about process (the theory of environmental control of tree growth and forest yield) needed to be known to achieve greater prediction accuracy.

The study suggests that controlling influences may vary spatially. If so, then theory for the control of tree growth with the objective of predicting yield must contain not only how natural concepts act but what their domain is, i.e., what range of variables under what conditions has what level of importance. For example, Blyth and MacLeod (1981a) suggest that depth to mottling is positively correlated with stone content. This factor is more important on wetter sites, indicating that drainage in these areas is important. In the terms available from this study, the domain of "depth to mottling" is wetter sites. Domains may be of different scales; for example, in an ancillary study Blyth and MacLeod (1981b) found a small-scale (0.01 ha) spatial variation in soil N and P that correlated with growth, but its effect could not be detected at the scale of investigation in the main study.

Where theory seems imprecise or imperfectly argued or lacks universality, as in much of ecology, an empirical approach has a strong attraction. However, it can have two weaknesses.

(1) Empiricists may not acknowledge the axioms that dictate how investigations are conducted and then assume that the patterns seen represent a fundamental truth rather than that they may have been influenced, at least in part, by the preconceptions that governed the investigation. For example, Blyth and MacLeod (1981a,b) made choices about the vari-

ables they used. Even though those choices were of simple variables with little direct theoretical content, e.g., elevation and aspect, they were determined by their initial ideas (theory) of what influences tree growth. For example, elevation and aspect may influence temperature, rainfall, and evapotranspiration and have consequent effects on soil properties

(2)   Surveys and classifications can proceed interminably and be self-perpetuating because no theory may emerge to give them coherence.

We depend on theory where we can not measure all that we study: first, to investigate the full details of functional relationships and construct attempted experimental disproof for them; second, to investigate integrative concepts such as diversity or stability. The next phase in the work in the north-east of Scotland was to use some functional relationships in constructing a theory for the control of growth that would explain how growth varied in this environment, and so might be the basis for prediction. The principal investigator, McLeod, changed from Plebian Induction to Aristocratic Induction where he used existing eco-physiological theories about the environmental control of tree growth, extending them to an analysis of his problems.

First, the scale of the problem had to be limited and a particular question from the empirical results was pursued further. Blyth and MacLeod (1981a,b) found that on some sites greater winter precipitation was a predictor of greater growth, which seemed to contradict one of the principal results across much of the area they investigated, that soil drainage was important. Jarvis, Mullins and MacLeod (1983) investigated the postulate:

Yearly increments of timber volume are regulated by soil moisture deficit.

for one forest stand over its years 21 to 31, at a site with a small rootable volume of soil, in the low-rainfall eastern region of the study area of Blyth and MacLeod (1981a,b), i.e., the empirical study was used to establish a probable domain in which the postulate would be valid.

This approach is inductive and confirmatory – using the suggestion of a limiting effect of dryness obtained from the empirical site survey. But it was important to try to quantify this effect, since the prevailing idea among practicing foresters at that time was that forest yield in the region was limited by too much moisture rather than too little. In this sense, the work was attempting to falsify a prevailing view, or at least the universality of application for that view.

It was clear from what was then known of water balance of forest stands that a simple natural concept, e.g., annual rainfall, or even rainfall in selected months, was unlikely to correlate with annual growth increment. By selecting soil moisture deficit, Jarvis *et al.* (1983) committed themselves to calculating the water balance of the site over a number of years and so estimating growth for those years. What they hoped to gain was a functional concept, *soil moisture deficit* ⇒ *stand timber volume growth rate.*

Calculation of soil moisture deficit was itself based on three functional concepts: transpiration (water loss through the trees), interception loss (the amount of water lost by evaporation from the canopy when it has been wetted by rain or snow), and evapotranspiration (the total amount of water that would be lost from the site when its surface is not wetted by precipitation, i.e., transpiration plus evaporation from soil).

The equation for transpiration used four physical constants, one hourly measurement available from a nearby weather station, and one function, canopy resistance that was extrapolated as an empirical equation from research into other forests. A system of equations for interception loss predicted the amount lost during precipitation under different types of meteorological condition (one measurement of daily rainfall and seven coefficients estimated at other forests). Evapotranspiration was calculated as though the canopy was always dry and then corrected for interception loss.

Measurements at the site showed that, once the available water in the root zone was depleted by 80 percent, trees failed to recover during the night the water that had been transpired during the day. This was postulated as the point at which trees came under water stress and that growth would be affected at this point. An index was calculated for each season as averages from three-day periods (value = 1 for conditions up to 80 percent of root zone water depleted (i.e., no projected effect, and a linear decrease from 1 to 0 for conditions between 80 percent and 100 percent depleted).

Wood volume growth of the trunk was estimated retrospectively after trees were felled. Annual ring width increments were measured along the bole of 12 sample trees, wood volume increment was calculated, and regression derived of volume increment on the ring width increments at the base of the tree. Basal ring widths were measured for all trees and stand volume increment estimated by applying the regression equation. In practice, only the upper canopy trees, i.e., those > 18 m tall, made appreciable increment and were used in the study. A similar pattern was found between the annual variation in growth and soil water deficit (Fig. 9.2).

This study illustrates difficulties in progressing from Plebian Induction to Aristocratic Induction. The interest of Jarvis *et al.* (1983) remained tree volume, an integrated measure of tree growth, and their study attempted to bridge a gap frequently found between partitioning reductions, e.g., research into only transpiration or only soil moisture, and that of growth. It illustrates common difficulties in ecological studies where attempts are made to synthesize research from partitioning reductions into a more integrated assessment:

(1)   Untested assumptions of generality. The task of constructing a comprehensive model, in this case of water balance, required *ad hoc* simplifications and adjustments as models were used that had been developed for

Fig. 9.2. A comparison of measured timber volume annual increment of upper canopy trees (Y) and calculated growing season water stress index (A). $Y = -5.8 + 27.8A$, $r^2 = 0.86$. (From Jarvis *et al.* 1983, with permission.)

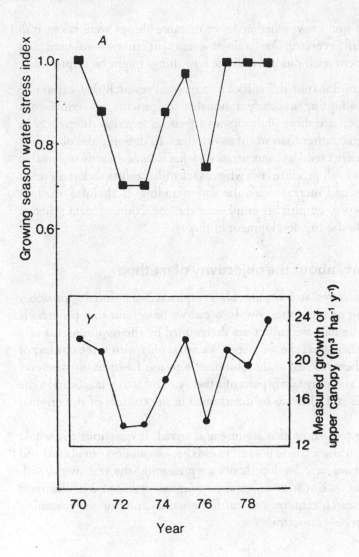

other sites. The simplifications are *ad hoc* in that their justification is argued for but the correctness of application at their site was not studied.

(2) Incomplete use of the causal explanation of component models. Even though the models that describe components of tree and stand water relations have a causal basis, it was necessary to use empirical functions where measurements of the actual processes were beyond the resources of the investigators, e.g., it was not possible to measure canopy resistance.

(3) Incomplete assessment of the component models. No judgement was made of whether the complete model is parsimonious or incomplete,

i.e., we do not know whether fewer or more things were taken into account than necessary. Only a single assessment criterion was used, and the assessment itself did not indicate how things might be improved.

These problems illustrate difficulties in ecological research that cannot be sidestepped by adopting a clearly rationalist, empiricist, or reductionist philosophy. In practice those philosophies are encountered as differences in attitude and degree rather than steadfast extremes. In defining the objectives of science, empiricists tend to concentrate on what is most directly observable or measurable and will produce laws whereas scientific realists seek to develop scientific theories and increase scientific understanding. In defining the best methods for science, empiricists emphasize the collection of data whereas rationalists emphasize the development of theory.

9.4 Uncertainty about the objectivity of method

To make progress in research requires both empirical and rational approaches – each depending on the other. But how can we be certain that progress is being made? If the data we collect are determined by the requirements of a theory we have, then what can we assess? We assess only within the context of that particular theory. Even if the objective is to use falsification wherever possible, we may still only falsify parts of a theory – and after a falsification the direction taken is still likely to be determined in the context of the original theory.

It is tempting to dismiss this argument as trivial. It questions the whole basis of scientific realism (there is a real world that we can find out about) and rationalism (that we can develop theories representing the real world with increasing accuracy). But it is an important argument because it illustrates that scientific research requires one further essential component governing the whole enterprise – criticism.

9.4.1 Criticism

By the mid- to late eighteenth century, there was confidence in science and logic to solve all problems. This philosophy dominated during the Enlightenment, the Age of Reason. At the same time, a debate had developed about the nature of our ability to perceive and reason that focused on precisely the weakness of Descartes' first rule and Bacon's assumption that he would be free from preconception. "What and how much can the understanding and reason know apart from all experience?" (Walsh 1975). Can the human intellect produce truths out of its own unaided resources, as opposed to when it cooperates with the senses? Churchman (1948) writes:

> "As science developed, it became clear that a failure to investigate the so-called self-evident propositions of reason or sense often led to unreliable and mean-

ingless results. The demand was met in part by criticism. Criticism is fundamental to science. It assumes that real progress can be made by developing the
dialogue between axioms and data."

This is the basis for the practical guidelines for developing a research proposal
(Chapters 2 through 4): that a dialogue between axioms and data can be
established, refined, and improved by critical analysis.

CRITICISM

Criticism asserts:

(1) For a correct beginning in research, axioms and the data on which they
depend must be known.
(2) The decision about what to research and how to conduct the research is
acknowledged to be subjective.
(3) The results of research must be exposed to criticism that is broad (not
limited to similar thinkers), unrestricted (not constrained by considerations that have nothing to do with the research), and fair (not motivated
by desire for personal attack).

There are real practical difficulties in elevating criticism to a principal
scientific philosophy. As individuals, we must do many things in a scientific
investigation that require the cooperation of others, and it is the people with
whom we cooperate or who have similar objectives who best understand the
work and so should be able to criticize it. Close colleagues may know how to
criticize details but may be less willing to criticize fundamentals if they
themselves accept those fundamentals. Furthermore, criticism is usually retrospective and does not, by itself, offer new solutions.

9.4.2 Relativism

RELATIVISM

The antithesis of criticism is relativism. The relativist argues that all truth is
relative: axioms are relative to the data used to support them, and data are
relative to the axioms and postulates that sparked the investigation. There is no
correct starting point for a scientific argument; the correctness of procedure
changes from individual to individual or from social group to social group.
Because no absolute criteria of "right and wrong" can be found either in
science or morals for the relativist, the choice of means to one's ends changes
widely: no point of reference can stabilize this constant change in plans of
action. If relativism is carried far enough, it leads to an anarchy that denies any
criteria in choosing means or solving problems.

Kuhn (1970) was a proponent of the idea that science is relativistic
(Chapter 11). He argues that scientists work in groups that have defined
techniques and address restricted problems. Major advances occur only at the

comparatively rare intervals when a revolution renders the techniques and problems of the group redundant. A new group of scientists then congregates around the new problems born of the revolution. According to Kuhn (1970), there is no continuity between the groups – so scientific advance is relativistic because new theory does not develop as a continuous logical sequence from previous theory. The counter-argument, advanced in Chapter 11, is that while scientists do work in groups this relativism is partial, not as complete as implied by Kuhn (1970), and that at some point it is countered by criticism.

Currently, a major charge against science is that it is relativist and so lacks validity as one of the principal activities in which humans should engage. The components of this argument are discussed in Section III.

9.4.3 Statistical experimentalism

But do we have to depend, ultimately, on criticism? Is no perfection of method possible that would guarantee progress? For some scientists, a specific definition of the experimental method occupies that exalted position.

Some consider Galileo (1564–1642) to be the originator of the experimental method (for a discussion, see Shapere 1974). Galileo wrote "in order not to proceed arbitrarily or at random, but with rigorous method, let us first seek to make sure by experiments repeated many times how much time is taken by a ball of iron, say, to fall to earth from a height of one hundred yards" (Galilei 1962). For scientists emphasizing experimentation it was this, rather than the empiricism of Bacon or the rationalism of Descartes, that was the major methodological advance at the turn of the sixteenth and early seventeenth centuries – and that opposed Aristotle and so caused what has been termed the scientific revolution. The experiment is a controlled, replicated manipulation and, in the last 70 years an extensive statistical methodology has developed for planning and analyzing it (Fisher 1966).

STATISTICAL EXPERIMENTALISM

Statistical experimentalism attempts to resolve the differences between empiricism, rationalism, criticism, and relativism by requiring the existence of some "ideal" against which scientific progress can be measured. There is an objective truth, we approach this through continued use of experiments but we cannot expect to reach it because experiments operate on samples or sub-sets of a complete system and produce answers within certain probabilities. (After Churchman 1948.)

The classic example is the replicated, randomized block experiment, analyzed by analysis of variance, that has proved so successful particularly in agriculture and the health sciences and usually is response-level experimentation (Chapter 6). The ideal of the statistical experimentalist is to remain completely methodical and free from the concerns of establishing a correct starting point for scientific argument, and so.avoid the problems of empiricism and rationalism. The ends of science are in some sense "fixed" (though subject to investigation), but the means of approaching those ends are not. This borrows some parts from other philosophies and may appear to resolve some of the differences between them. The assumption, taken from rationalism, that there is an objective truth, that we can find things out by experiment, is tempered by the acknowledgment that experimentation cannot give absolute answers, only probabilistic ones. In this, experimentalism borrows from statistical empiricism.

There are penalties to pay. Statistical experimentalism comes into play only after some decision about a starting point. There has to be some driving over-arching axiom, which is usually based on a previous finding. In this sense statistical experimentalism does work within a relativist framework. Although statistical experimentalism assumes an approach to the truth, the only truths that can be approached are those that can be contained within the controlled experiment or investigation. This is a weakness of statistical experimentalism as a philosophy: it presupposes what method is to be used without being able to investigate that method as part of its procedure. While the statistical experimentalists' philosophy resolves some important philosophical problems about whether there are correct starting points in data or theory, it can pursue only certain types of investigations and this makes it only partially useful to ecology and natural resource science. For example, if Hawkins were to show by experiment that certain behavioral processes of fish could produce isolation in breeding, there would still be the question of just how important behavioral processes are relative to other agents of isolation. These might include geographical or habitat separation and it might not be possible to investigate them by controlled, replicated experiment. Some of the most successful uses of statistical experimentalism have been in plant and animal breeding, where progeny have been assessed but why any variety is best is left unexamined.

Experimentation does require some reductionism to define an experimental unit. Experiments in ecology may be very useful in analyzing how a process operates. But experiments to tell you just how important that process is in an ecological system can be much more difficult. Ecologists have been urged to be experimentalists (Hairston 1989) – the implication is that the experimental approach is superior to others – but, as a philosophy of method, experimentalism in ecology needs further definition (Section II).

9.5 Discussion

It seems apparent that progress in ecology requires that different methods be applied to solve particular types of problem as and when they arise. No one philosophy and its associated methods are always superior.

The philosophy used in this book is that the solution to scientific problems requires analysis and synthesis of theory and data. There is no unequivocally superior starting point in the way assumed by empiricism or rationalism, but instead a starting condition in the relationship between axioms and data that must be established by conceptual and propositional analysis. Even where the research problem seems new, theories about other problems and/or even the most preliminary observations may influence the approach.

This means that axioms must be reexamined in the light of data; data must be reassessed as axioms change. This duality replaces the assumptions of empiricism and rationalism by acknowledging that scientific progress spirals between questions of theory and questions about data. There is no absolute reference point, only the continuous application of critical analysis, and, yes, this can be undermined by relativism. For scientists to do their job properly it is essential they understand how this undermining can happen (Section III). But, given that scientific investigation is a continuous process of analysis and synthesis, can this be placed in a methodological framework? Section II shows how this can be done for ecological research and particularly that integrative concepts can be developed that do not require either an assumption about the existence of emergent properties, or reduction of the ecological system of interest.

Given the different philosophies that exist in ecology, you can expect two kinds of criticism of your research. You will certainly receive criticism about the details of your investigation. For example, there may be disagreement about your classification of propositions as axioms or postulates, or about your construction of data statements. These criticisms are all about what can be known following the type of investigation you have chosen. But you may also face criticism about the whole philosophy of your investigation. Should you conduct an initial survey? Should you make experiments and if so of what type? It can be valuable to distinguish between these types of question in meeting criticism and deciding how to respond to it. Both types are valuable of course. But the response to a question about the philosophy of method cannot be met through adjusting the details of an investigation, and may be resolved only at a level broader than that of the individual project.

9.6 Further reading

A narrative introduction to philosophies of science is given by Chalmers (1982), describing origins of philosophical positions as well as recent developments. Losee (1993) presents a historical introduction to the philosophy of science that includes concise summaries.

Introduction to Section II:

Making a synthesis for scientific inference

In Section I some problems in making scientific inference are discussed. Chapter 7 illustrates that there is no single method of reasoning that can be followed invariable in research. Chapter 8 shows that statistical inference may be used in making scientific inference – but is not all that is required and may not always be possible. Chapter 9 gives examples of different approaches taken by scientists and shows that there is no reason for invariably preferring one to another.

Section II deals directly with how scientific inference can be made in ecology and how research using different methods of reasoning and techniques of analysis can be integrated.

SCIENTIFIC INFERENCE

A scientific inference is made for a specified scientific question using the following procedures and standards:

(1) A *synthesis* must be made of new results with existing theory.
(2) This synthesis must provide a *scientific explanation* of why something exists or occurs.
(3) The scientific explanation provided must be *coherent*, explaining both new and previously obtained information. The explanation may increase in coherence as new *scientific understanding* is obtained.

These component terms used in the definition of *scientific inference* are defined in this introduction and their application is illustrated in the following chapters.

Ecology provides some important challenges to making scientific inference. Recall four questions asked by beginning researchers about ecological research (Table 1.1):

22. How can I ask the right question and measure the right things? How can
     I sample a whole ecosystem? How can I avoid samples being influenced
     by unusual events?
23. How do we extrapolate from research on a limited system (laboratory or
     field plot) to give an understanding of a whole ecosystem?
24. How can I find a place (site) where a question can be answered? There
     may be difficulty matching a proposed theory with practical reality.
25. Which techniques are the best for measuring the response of a tree to
     environmental factors? For example, the development of a bioassay?

Questions 22, 23 and 24 raise issues about the nature of theories in ecology;
how they seek generality but must be built up from, and apply to, specific
circumstances. Question 23 asks how ecological knowledge can be constructed through synthesis. This problem of extrapolating from particular results
to more general theory is also at the root of answers to Questions 22 and 24.
Question 25 implies a similar difficulty. Trees are complex systems, i.e., we
do not completely understand how they function – or even how to measure
their response to the environment in a comprehensive way. The questioner
was considering developing a bioassay, e.g., placing seedlings at different
positions in a tree canopy and attempting to infer tree response and function
from seedling response. But can such information from seedlings be synthesized to explain tree response adequately?

The examples in Section I focus on questions about specific places or
instances. Whether piles of coarse woody debris on a cobble bar provide a
favorable habitat for small mammals. Whether climate and vegetation in
regions of Alaska had shown particular types of change during the late
Quaternary. What may cause an observed mid-summer depression in photosynthesis rate of mature white pine trees? Questions about generality of the
observations become more important as research progresses, and theories that
might explain them become more complex.

Such questions and answers might be as follows. Does coarse woody debris
provide habitat structure wherever it is found? A general answer requires a
synthesis of results about different types of woody debris in different habitats
to define general properties of habitat structure accounting for similarities
and differences between habitats. Can patterns of vegetation and climate
change in Alaska be explained by meteorological theories? Synthesis of
theories about vegetation and climate is needed. Climate changes for Alaska
proposed on the basis of ideas about vegetation change, must be coherent
with those for neighboring regions and based on other methods. Can a theory
for control of photosynthetic rate in mature pine trees be developed incorporating controls by both environmental and internal plant conditions? This
requires examining if mid-summer depressions of photosynthesis rate occur
in species of pine other than *P. strobus*, and in different environments, and

Fig. II.1. The process of synthesis requires a series of linked decisions: how data determine the fate of a postulate, what that decision may mean about the structure of the existing theory, and what that in turn means for the generality of the theory. Recall Fig. I.1 to contrast the process of analysis.

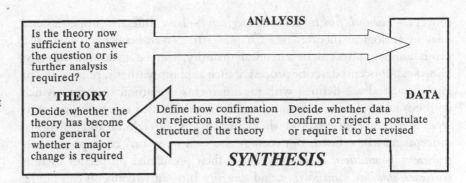

requires synthesis of the existing physical and chemical control theory with additional axioms.

We frequently start the process of synthesis by making a series of linked decisions about some new result (Fig. II.1). Do the data confirm or reject the postulate, or require it to be revised? This decision influences, and is influenced by, its consequences for the whole theory. When an unexpected result is obtained, particularly one requiring a marked change to present theory, there is a greater tendency either to repeat the investigation or to construct an additional data statement and hypothesis test for the postulate. Synthesis demands coherence encompassing the individual postulate, its significance for current theory, and any wider implications for generality of that theory.

Synthesis is a continuous process, not left until a large number of results have been obtained. Researchers constantly restructure theories in the light of new results – so why does this process need careful description and explanation? First, because ecology has particular difficulties with the structure of its theories and the concepts they use. Second, because making a correct scientific inference requires developing a coherent set of axioms and this is not always a simple task – a discovery may disrupt what has been a coherent theory and ecological theories may remain loosely constructed or incomplete for considerable periods.

SYNTHESIS

Combining constituent elements into a single or unified entity. Research introduces new concepts, changes the status of existing concepts, or defines new relationships between concepts.

The emphasis is on creating, developing, or extending a theory that will make a coherent explanation of the question and available data.

An important challenge for ecologists lies in the nature of the subject and defining how a theory can provide a coherent explanation. Three types of concept – *natural, functional* and *integrative* – have different properties and roles in ecological theories (see Chapter 10). *Natural* concepts, such as a plant, animal, physical or chemical quantity, can be measured directly. *Functional* concepts describe processes, such as photosynthesis, respiration, or predation, and are defined with measurements of natural concepts. Much synthesis takes place as functional concepts are extended to new instances or types of natural concepts, and when this happens both natural and functional concepts may be refined. But ecology also uses *integrative* concepts, such as *ecosystem, population,* or *community,* or their postulated properties such as *resilience, stability, equilibrium,* and *diversity.* Integrative concepts can not be measured directly but must be defined by sets of axioms using natural and functional concepts. General properties, defined by integrative concepts, must be built up through synthesis from many investigations. This process of building theory to define an integrative concept is called *upward inference.* Chapter 10 illustrates how increasing precision in concept definition advances ecological knowledge.

Questions about building general theories from research into particular instances are central to ecology, and have been since its inception as a science in the nineteenth century (McIntosh 1985). When you start working with an ecological theory you can not simply assume it applies generally – the *domain* of application of functional and integrative concepts must be considered (Chapter 10). A concept's domain is the limit of its application and domain specification is an integral part of an ecological theory using that concept.

The purpose of making a synthesis is to construct a coherent scientific explanation in the form of a theory and then make inference from the scientific explanation that the theory provides according to exactly how coherent you consider it to be. The more coherent, in terms of how much is explained and how well, then the stronger is the inference made. This means that scientific inference is not absolute (something is accepted or is not) but must be justified and developed in stages.

Scientific explanations have a particular form and constructing scientific inference requires a number of stages (Fig. II.2).

SCIENTIFIC EXPLANATION

A *scientific explanation* answers a why-type question constructed so that it:

> specifies the topic of concern,
> defines the contrasting set of alternatives in the question, and,
> defines the explanatory relevance required.

A scientific answer to a why-type question requires a synthesis that decides between contrasting alternatives and may include:

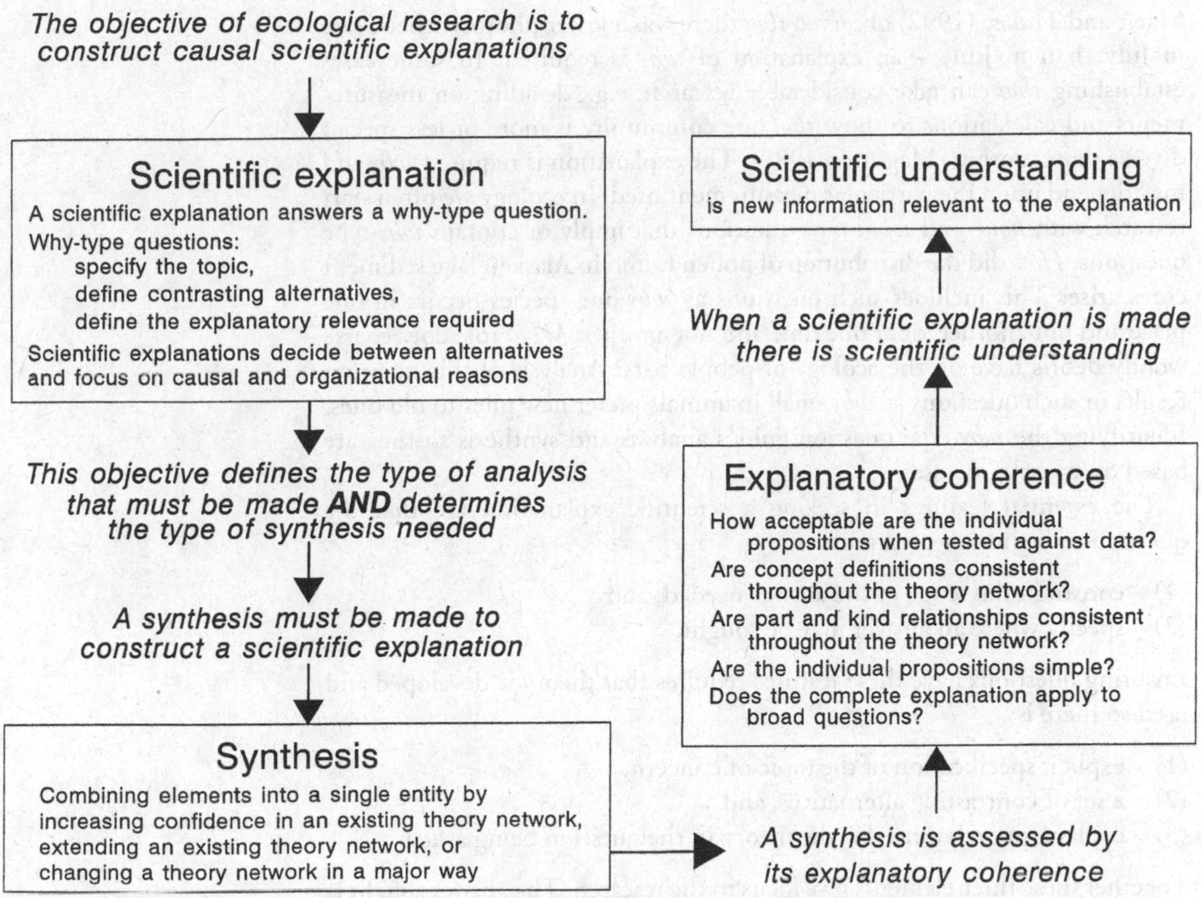

The objective of ecological research is to construct causal scientific explanations

Scientific explanation

A scientific explanation answers a why-type question.
Why-type questions:
    specify the topic,
    define contrasting alternatives,
    define the explanatory relevance required
Scientific explanations decide between alternatives and focus on causal and organizational reasons

Scientific understanding

is new information relevant to the explanation

When a scientific explanation is made there is scientific understanding

*This objective defines the type of analysis that must be made **AND** determines the type of synthesis needed*

Explanatory coherence

How acceptable are the individual
    propositions when tested against data?
Are concept definitions consistent
    throughout the theory network?
Are part and kind relationships consistent
    throughout the theory network?
Are the individual propositions simple?
Does the complete explanation apply to
    broad questions?

A synthesis must be made to construct a scientific explanation

Synthesis

Combining elements into a single entity by increasing confidence in an existing theory network, extending an existing theory network, or changing a theory network in a major way

A synthesis is assessed by its explanatory coherence

Fig. II.2. The constituent procedures used in making a coherent scientific inference.

developing increased confidence in an existing theory network,
extending an existing theory network, or,
a major change in a theory network. (After Lambert and Brittan 1987, and Thagard 1992.)

Scientific explanations are for particular why-type questions whereas a theory is a more general collection of axioms, postulates and underlying data. Scientific explanations use components of theories and developing a scientific explanation for a new why-type question may require development of the theory through new research.

WHY-TYPE QUESTION

Why-type questions ask about functioning of ecological systems on the basis of some observed phenomenon. Knowledge-that something happened is descriptive. Knowledge-why is explanatory.

Maeir and Teskey (1992) observed *that* there was a lower photosynthetic rate in July than in June – an explanation of *why* is required. In some cases establishing *that* can take considerable research, e.g., deciding on measurements and calculations to show *that* one community is more or less species diverse than another (Magurran 1988). The explanation is required *why* and may depend upon the particular measurement used. In ecology we often start research with *how-* and *what*-type questions that imply or contain *why*-type questions. *How* did the distribution of pollen found in Alaskan lake sediment cores arise? This includes such questions as *why* one species occurs in one place and not another, or at one time and not another. *What* role does coarse woody debris have on the ecology of pebble bars? Analysis of this question results in such questions as *why* small mammals prefer new piles to old ones. Identifying the *why*-type question guides analysis and synthesis so they are based on causal processes.

The essential features in seeking a scientific explanation are that the question to be investigated must:

(1)   constrain the scope of the answer needed, and
(2)   specify where an answer may be sought.

Ensuring questions have these features requires that theory is developed and used so there is:

(1)   explicit specification of the topic of concern,
(2)   a set of contrasting alternatives, and
(3)   explanatory relevance for the theory in the question being asked.

Together these three elements give focus to the research. The answer sought is explicitly defined as a development of theory sufficient to explain a contrast. This provides an essential restriction to the scope of the research and facilitates a defined scientific inference. The generality of a theory may build as successive questions are asked and answered, and this progress is defined by changes in the definition of concepts and increasingly precise specification of their domains. Chapter 10 shows why, and how, why-type questions are constructed to constrain the scope of answers needed and specify where an answer may be sought. An example is given of the progress of an investigation and how residual uncertainty and lack of precision in definition can be identified. Chapter 11 introduces some theories about how scientific progress is made and illustrates the process of scientific investigation in some fields of ecology developed over decades.

Using contrasts is essential. It is the way we can advance and establish knowledge – even though, subsequently, as a piece of knowledge becomes accepted and research proceeds to new questions, we may forget what the alternatives were. Contrasts are the foundation of component scientific procedures and techniques: experiments are manipulated contrasts, the strategy of multiple postulates emphasizes the need for many alternatives, and the foundation of statistical hypothesis testing is based on alternatives. Chapters 10 and 11 illustrate that our understanding of theories using integrative concepts, and the types of why-questions they can answer, is advanced through building investigations in contrasting circumstances. How those are chosen is a crucial task for the scientist.

Generally we prefer causal scientific explanations – although developing knowledge of causes may itself take considerable research.

CAUSAL SCIENTIFIC EXPLANATION

Causal scientific explanations have the structure and purpose of *scientific explanations* and

(1) Are based on causal and/or associated organizational reasons. (Defined in Chapter 7.)
(2) Are consistent. Under the same conditions the causal process will produce the same effect.
(3) Are general. To explain a kind of event causally is to provide some general explanatory information about events of that kind.
(4) When experiments are possible a designed manipulation or intervention of the causal process produces a predictable response.

In ecology our research is frequently into the organization of ecosystems, communities and populations but this is built upon axioms of causal processes. Similarly we may find explanations that unify apparently contradictory, or even just separate, pieces of theory. But these explanations also have underlying causal reasons. These types of explanation are discussed more in Chapter 15.

The objective is a coherent explanation, one that is consistent through all of its parts (Thagard 1992). Explanatory coherence requires specifying the full set of propositions against which the individual proposition(s) you have researched must be assessed.

EXPLANATORY COHERENCE

If a theory gives an effective scientific explanation of a why-question it is *coherent*. This requires:

(1) Acceptability of individual propositions, including that they have been tested with data statements.
(2) Concept definitions are consistent throughout the theory network.
(3) Part and kind relationships are consistent throughout the theory network.

(4) There are not *ad hoc* propositions that include or exclude special circumstances.

(5) Generally theories with fewer rather than more propositions are favored as more coherent explanations.

(6) The explanation applies to broad questions and circumstances.

While assessment of a postulate using a data statement is an empirical test (and statistical inference may be used) the coherence of a postulate is an assessment of its value in providing a scientific explanation and emphasizes its role within the theory. The *scientific inference* made depends upon these items of explanatory coherence and so scientific inference may be provisional and/or incomplete according to how many of the requirements of explanatory coherence are met, and how well they are met. It is possible to construct postulates that can pass an empirical test but have little or no explanatory value and so would not be given value in scientific inference.

Explaining with fewer rather than more propositions introduces the idea of simplification of explanations with axioms that have explanatory breadth. This can involve the type of explanation that seeks unification of previously disjunct knowledge. In ecology both functional and integrative concepts play important roles in defining unified explanations. Thagard (1992) suggests that if numerous propositions are necessary to explain some things then their coherence with each other is diminished. "There is nothing wrong in principle in having explanations that draw on many assumptions, but we should prefer theories that generate explanations using a unified core . . ." This is an interesting challenge that some ecologists feel is of central importance for the future of the subject (Chapter 16). Chapter 11 illustrates how two apparently opposing theories explaining the balance of plants, herbivores and carnivores in communities has been unified.

Chapter 12 is devoted to the construction and use of mathematical, statistical, and numerical simulation models. Much use is made of modeling in ecology – it is often seen as the way that synthesis can be achieved. This approach is discussed, particularly how the rules for constructing different model types can influence the synthesis made, explanatory coherence achieved, and so inference that can be made.

Scientific explanations yield scientific understanding (Fig. II.2) – and this too has a technical definition.

SCIENTIFIC UNDERSTANDING

To say a scientific explanation yields scientific understanding is to say it shows or exhibits some new piece of information relevant to what the explanation requires. (After Lambert and Brittan 1987.)

This distinguishes the concept of scientific understanding from the common language use of understanding. Scientific understanding is not a subjective increase in an individual person's knowledge of a scientific theory, nor is it simply the accumulation of scientific knowledge through the construction of laws based on measurements. You may know the theory about a subject yet still not know how it can be used to provide a scientific explanation for a particular observation or question – and until you do there is not a scientific understanding for the observation or question. Achieving scientific understanding involves the accumulation of new information, but not all new information provides scientific understanding; it has to be focused to answer a particular question. Salmon (1990) lists three ways in which scientific explanations can enhance understanding:

(1)   When knowledge is obtained of hidden processes, causal or otherwise, that produce phenomena that we seek to explain.
(2)   When our knowledge is so organized that we can comprehend what we know under a smaller number of assumptions.
(3)   When we supply missing bits of descriptive knowledge that answer why-type questions.

The objective of research is theories that provide coherent, causal scientific explanations. It is important to define a methodology to achieve this objective.

PROGRESSIVE SYNTHESIS

Progressive Synthesis has three principles:

  I.   Criticism. Standards must be applied to ensure just and effective criticism.
 II.   Definition. Precision is required in defining concepts, axioms and postulates, and data statements.
III.   Assessment. Explicit standards must be used to examine the relation between theory and data.

Progressive Synthesis has five component methods:

(1)   Analyze the question and seek to use contrastive techniques to focus research.
(2)   Expect to use different techniques of investigation as theories develop and new types of question are asked.
(3)   Refine both measurement and concept definitions.
(4)   Specify the new synthesis resulting from the research.
(5)   Define explanatory coherence of the synthesis to make scientific inference.

Notice that experimentation, surveys, or field investigations are not listed as component methods. These are procedures and techniques of component method (2). For example, an experiment is one way of obtaining a contrast (component method (1)) when enough is known to design a treatment and make appropriate measurements. At an early stage insufficient may be known to devise an experiment, and the appropriate contrast may be exploratory analysis of contrasting field situations even though the contrast may not be focused on just one difference. *Progressive Synthesis* is discussed in Chapter 15 but its principles and component methods are illustrated throughout this section.

10 Properties and domains of ecological concepts

Summary

There are three types of ecological concept. *Natural concepts* define and classify measurable or observable entities in the ecological world. These are frequently common objects, such as organisms, or features of the environment, such as rainfall. Natural concepts such as chemical compositions can also be used to imply some property.

Functional concepts define a measurable process, interaction, or structure, such as photosynthesis, migration, or eutrophication, that applies to natural concepts. New functional concepts arise to describe newly understood structures or interactions in natural concepts and research into functional concepts is constantly used to refine the definition of existing natural concepts and their classifications.

Integrative concepts describe organization or properties of ecological systems that can not be measured as absolute quantities and must be explained by a theory. For example, *ecosystems* are theoretical constructions, and ecosystem theory describes general organizational and functional properties of ecosystems But we can measure only particular instances of ecosystems, and not all instances are identical or share all of the supposed properties in the same way. Similarly, *resilience* of an ecological system is not measurable directly but is inferred from species persistence when changes occur. The interest in such concepts lies not in the measurement, which is indirect or comparative, but in the underlying causes of why one species may be persistent and another not.

Functional concepts must be used to explain integrative concepts and the interest lies in how constant explanations are in different circumstances, e.g., can greater resilience always be explained in the same way? The concept of a *domain* is used to specify when or where, or under what conditions, a functional or integrative concept applies and can be used to investigate the generality of functional and integrative concepts.

Scientific inference for integrative concepts is progressive, comprising the processes of constructing the requirement for a scientific explanation, making a synthesis, assessing the explanatory coherence of the synthesis, and so achieving scientific understanding. These processes are illustrated for a theory about diversity of lowland tropical rainforests.

10.1 Introduction

An objective in ecology is to define general theories providing answers to questions such as "Why are some ecological systems more diverse than others?" "Are some ecological systems more persistent than others – and what properties make them so?" An essential requirement for this is deciding what to call ecological systems because the names we choose themselves imply particular properties. For example, *ecosystem* and *community* are two extensively used terms, but definitions of both are still debated because their use implies particular types of relationship between organisms and their environment. Since the start of the subject, ecologists have used names for properties they thought may exist, and then tried to develop more precise definitions as research proceeded.

However, developing scientific explanations and inference about general properties has to be done without automatically assuming that there are general explanations to be had. The empirical basis for general questions is difficult to define. Can we use the same quantitative description of high and low diversity in different systems, e.g., forests and oceans? Is it reasonable to expect the same, or even similar, explanations for the causes of differences in diversity in all cases? Uncertainty about the definitions of terms we use and the scope of the questions we attempt to answer is a major difficulty for ecology because it means skepticism must be maintained about fundamentals of the subject.

Making inference about general properties of ecological systems, including what terms such as *ecosystem* and *community* mean, is called upward inference.

UPWARD INFERENCE

The process of developing inference for an over-arching theory from a set of specific investigations and where the theory contains concepts that do not have a direct equivalent concept by measurement.

Terms such as *ecosystem* and *community* are concepts defined by sets of axioms. We refer to *ecosystem theory* or *community ecology* as subjects with bodies of theory that may contain axioms of general properties and, crucially, axioms that define properties under particular circumstances. Currently a major problem in ecology is lack of consistency between scientists in what they mean when they use concepts that refer to general properties. This lack of agreement, discussed further in Chapter 16, usually reflects lack of precision in definition, and/or inadequately justified assumptions about generality between systems. Most importantly it reflects lack of a recognized methodology for making upward inference.

This chapter illustrates how upward inference can proceed in ecology through constructing scientific explanations (Fig. II.1). There are two requirements. The first is to classify concepts according to the type of information they represent. The second is to recognize limits in the application of concepts by specifying their domains. Because general information is being sought, which can not be confirmed or denied unequivocally by a single investigation, it is essential to follow a rigorous procedure in seeking scientific inference, and this is illustrated by an example.

10.2 Definition and purpose of ecological concepts

Ecology started with natural history observations of animals, plants, soil, water, rocks, rain, tree stumps, logs, and so on. In ecology we keep returning to observe these same things, but as research progresses we classify them in different ways according to properties that we discover about them. For example, a bird can be a member of a species, a food source, a predator, a unit of biomass, or a vector of a disease, etc.

NATURAL CONCEPTS

Natural concepts define and/or classify measurable or observable entities or events in the ecological world.

These definitions and classifications carry meaning for different theories. Development of the natural concept *coarse woody debris pile* is illustrated in Chapter 4. Attributes, such as *old* and *new* are added to it as concepts by imagination and concepts by measurement are developed. If the concept is confirmed, an additional classification of the natural world is made.

FUNCTIONAL CONCEPTS

Functional concepts define properties of natural concepts or express relationships between two or more natural concepts. Direct measurements of natural concepts can be made to define functional concepts.

Migration and *photosynthesis* are functional concepts. They have been measured and their properties defined under a variety of different circumstances. Although there may be much variation in the measurements, determining what causes that variation is itself an object of scientific research that continues to refine the definition of the functional concept.

At the *functional level*, theories are constructed because we have difficulty measuring what has to be explained about natural concepts. For example, determining whether small mammals used woody debris piles was primarily a question of measuring their occurrence in different places and at different times. If the entire habitat were transparent and each species of small mammal were a bright and different color, and extended observations could be made without disturbing the animals, then it seems likely that the postulates in Chapter 4 could be answered directly. Because this is not the case, concepts such as *use* are constructed for which a number of explanations may be possible. Similarly, alternative postulates about control of mid-summer depression of photosynthesis in white pine could possibly be resolved by comprehensive and simultaneous measurements of photosynthesis, growth, carbon metabolites, and water status carried out under experimental conditions. For these two research problems, if their measurement difficulties could be solved, and given sufficient time and resources, there seems a reasonable prospect of resolving the specific postulates entirely by measurement.

Many functional concepts can not be applied with absolute precision because their measurement remains imprecise – we can never measure the absolute rate of photosynthesis of a particular leaf or column of water in a lake because the measurement process influences the values obtained. Developments in measurement lead to refinements of functional concepts. For example, natural history observations of salmon led to the idea that return migration from the sea was to the fish's natal river. It was considered that the fish had not moved far into the sea and so the natal river was closest when migration occurred. The first requirement was to measure the range and consistency of movement (Scheer 1939) and salmon migration is now defined by a set of functional concepts describing not only migration patterns but how they vary, e.g., how both wild and hatchery fish may stray from their migration routes (Quinn 1993), and the underlying process, e.g., olfactory memory (Nevitt *et al.* 1994).

Natural concepts are used in defining functional concepts, but there is also a reciprocal influence where increasing research into functional concepts changes the definition and classification of natural concepts (Fig. 10.1), and particularly the definition of the *part* and *kind* relationships between them (Chapter 3). Piles of woody debris were classified by age and by location (Chapter 4). Those natural concepts of different types of pile provided scientific understanding of *use* of piles by different species of small mammal, so the classification of piles was confirmed as important. As photosynthesis was researched more, and variation between plant types was recognized and defined, then classifications of photosynthesis were made, i.e., C3, C4, and CAM (crassulacean acid metabolism, Ting 1985), that are used in ecological analysis of plant distribution and ecology (Black 1971, Ueno and Takedo 1992).

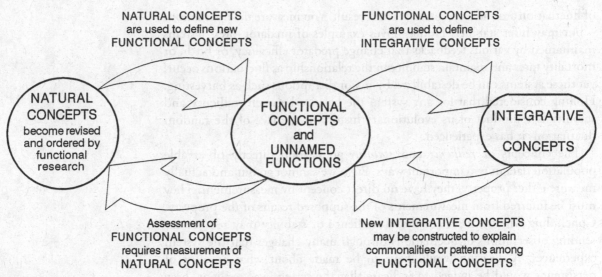

Fig. 10.1. Progress in research through developing concept definitions.

INTEGRATIVE CONCEPTS

Integrative concepts are theoretical constructions about the organization or properties of ecological systems. They can not be measured directly, and both their definition and detailed description must be synthesized from studies of a number of systems.

Many concepts describing the organization of ecosystems, populations, or communities are of this type. We use them to imply particular features of organization that are theoretical constructions based on generalizations from different studies. For example, Holling (1973) considered a range of possible outcomes for interacting species. His interest was in the situation where some constancy in numbers existed for each species, rather than an extinction, and he defined two different properties, *resilience* and *stability*.

Resilience determines the persistence of relationships within a system and reflects the ability of the system to absorb changes: resilience is the property of the system, and persistence is the result. You measure persistence – but may infer resilience. Holling (1973) gives examples of insect–plant systems such as the spruce budworm that feeds on balsam fir in the north-east USA and eastern Canada. The insect is generally endemic but becomes epidemic when the proportion of balsam fir in the forest increases: the number of balsam fir trees is then reduced owing to the insect damage. As the number of balsam fir decline, so do the number of budworms. However, both balsam fir and budworm do recover from the epidemic and subsequent crash – both are resilient.

Stability is the ability of a system to return to an equilibrium state after a temporary disturbance: stability is the property of the system, and the degree of fluctuation around a certain state is the result. You measure the fluctuations – but may infer stability. Holling gives examples of predator–prey systems maintained by various feedbacks that change predator efficiency, or birth, or mortality rates and maintain stability in the relationship as fluctuations occur; yet these systems can be destabilized by major disruptions such as harvesting. Holling considered that for any system the balance between resilience and stability is a product of its evolutionary history in the face of the random fluctuations it has experienced.

The concepts of *resilience* and *stability* go beyond directly observable population data in two important ways. First, we can not go out and actually measure either property, they have no direct concept by measurement. They must be inferred from measurements of the supposed results of the property. Concluding whether a system shows resilience or stability may require extending observations of fluctuations through many changes that the system experiences. A decision would have to be made about what amount of persistence would be judged to indicate that the system was resilient, how much fluctuation and/or what rate of return to a particular state would be judged to indicate stability. Second, these concepts are considered to apply to many types of population. While the specific causes of population increases or decreases that result in either resilience or stability may differ among populations, the theory that there will be such increases and decreases is common to them all. The decisions about what would be judged as resilience or stability depend upon investigation of many populations.

Integrative concepts such as resilience and stability have properties different from those of functional concepts such as photosynthesis or migration. We certainly have difficulty in measuring photosynthesis and understanding the precise causes of all the variation we can see – but we can at least measure photosynthesis well enough to study its variation. We may not be certain precisely how migration takes place, nor explain the variation we see within and among species – but some animals certainly do migrate repeatedly with at least approximately similar patterns. In contrast, we can not measure resilience or stability; we can only construct a theory that says populations will tend to be persistent or will be maintained within certain limits that themselves are difficult to predict. Nor can we measure an ecosystem or a community – we can define only what functional properties we think they have and attempt to measure those.

Integrative concepts are frequently proposed early in the development of an investigation without an extensive process of upward inference having taken place. In this way they provide a general heuristic, an indicator of what should be looked for, but they can remain largely conjectural for many decades. McIntosh (1985) describes questions about communities – whether they are open or closed, in equilibrium or non-equilibrium – that were discussed in the early decades of ecology and have been more recently rediscovered by theoretical ecologists, yet persist without resolution. Grimm and Wissel (1997) present an inventory and analysis of 163 definitions and 70 different *stability* concepts and suggest that the general term stability is so ambiguous as to be useless. They suggest it be replaced by:

Constancy:	staying essentially unchanged.
Resilience:	returning to the reference state (or dynamic) after a temporary disturbance.
Persistence:	persistence through time of an ecological system.

As research into integrative concepts progresses, increasing precision in the definition of meaning becomes essential and this requires defining the axioms of the theory and their domains.

An important problem is that common language names have frequently been proposed as integrative concepts and the concept may be teleological if the common language meaning is carried over in the ecological definition. For instance, the common language meaning of *community* is applied to humans and includes organization of individuals to achieve some common end often by particular means, e.g., sea-faring community. When applied to an assemblage of plants and animals in a particular place, this concept can hardly escape being teleological, i.e., that joint occurrence of species at a place, which may be called a community, can be taken to imply jointly held properties such as the ability to achieve resilience or stability. In the case of *community* the concept may also be anthropomorphic. We know some properties of human communities and we may assume parallel properties in assemblages of plants and animals, most particularly the interactions that ensure the survival of the community.

A further difficulty in establishing the scientific basis for integrative concepts is that we use only a few common language words to explain many types of thing. Perhaps the word most used in ecology with different meanings is *structure*. As a practical method for naming concepts about *structure* I recommend the use of modifiers, e.g., plant community spatial structure, root morphological structure, animal population demographic structure, animal community trophic structure. This is not being pedantic – it enables precision to be given to related but different definitions. Precision in defining concepts is particularly important where investigations proceed using different examples, and the ultimate problem is to define similarities and differences between the definitions of these concepts in the different examples and not just to assume the universal influence of an inadequately defined concept. Definition is also difficult because the details of practical investigation can be different between systems. It may take a long time to develop standards to assess resilience and stability that can be applied to different types of ecological system. Yet we require studies on different systems in order to define resilience and stability as integrative concepts.

Do we really need integrative concepts? Can we not be content with developing functional concepts and their theories? Goodness knows, that type of work is difficult enough! I want to illustrate why ecologists are led into constructing integrative concepts and the theories that define them.

We know, by identifying and counting them, that there are more species in some places than in others. For example, Currie and Paquin (1987) counted tree species in large scale quadrats (336 in total) over Canada and the USA where there is a wide range of environments from desert to warm moist forest. They sought correlations between tree species richness and environmental characters and concluded the strongest was with annual evapotranspiration. They fitted a non-linear model

$$TSR = 185.8/[1.0 + \exp(3.09 - 0.00432ARET)]$$

where TSR is tree species richness, ARET is annualized evapotraspiration in millimeters, and which accounted for 76 percent of the variance ($r^2 = 0.762$). However, this equation describes a trend from one extreme to another, in this case between 125 and 1025 mm annual evapotranspiration. It does not describe the wide range of species richness occurring within particular ranges of annual evapotranspiration. For example, between 100 and 200 mm annual evapotranspiration the range was 0 to 90 species; above 1000 mm the range was 40 to 178. Therefore, even between extremes of annual evapotranspiration the ranges of tree species richness overlap considerably. Furthermore, no ecological interpretation is given for the particular values of the constants in the equation – they are simply values producing the highest r^2 value. So this equation is not an explanation for "Why are there more species in some places than others?" but it does illustrate that there are differences between regions with different ARET values and differences within some regions.

In practice, valuable questions, those for which scientific explanations can be sought, arise when contrasts are made between situations that are generally similar but with a crucial difference (Method (4) of Progressive Synthesis). They may come from studying broad descriptors, such as those represented by the above equation, but do not depend on them. For example, tropical rainforests generally contain many tree species, but this is not always so. So we might ask:

> Question: Why do some tropical lowland rainforests have canopies composed of multiple species while some have canopies dominated by few species?

First, it is necessary to define what is meant by tropical rainforest because some are mountainous and some have seasonal rainfall, both conditions that may influence the numbers of species. Connell and Lowman (1989) restricted comparisons of species numbers to tropical rainforests at latitudes less than 18°, and elevations less than 1100 m, receiving at least 166 cm of rain per year and with very little seasonality. These criteria had been extensively researched by other scientists and were used by Connell and Lowman (1989) to define what should, and should not, be classified as *tropical lowland rainforest*. Notice that a natural history observation (rainforests in the tropics) is being classified as a natural concept (tropical lowland rainforest). Within this definition of the concept of *tropical lowland rainforest*, considerable differences do exist in the number of tree species that form the canopy – from more than 180 species over 10.5 cm diameter ha^{-1} (Whitmore 1984) to single species dominants (Connell and Lowman 1989).

In this situation much has been done to ensure that similar forests are compared – the major difference between them being tree species numbers. Unlike the research of Currie and Paquin (1987) there is no obvious single, or even few factor(s) that might be measured to explain these differences. A theory is required about the cause of the differences. One integrative concept to consider is *diversity* – but there are two problems with *diversity* that are characteristic of integrative concepts.

First, there are many ways of measuring and calculating and so defining it, and there is disagreement about which is best for particular purposes (Magurran 1988, Hawksworth 1995). Diversity is composed of two parts: species richness and the number of individuals there are of each species. Indices have been proposed that use these singly or in different combinations in an overall measure. These inform about different things and both may be useful depending upon the question being asked. So what is to be explained is not precisely defined in a way that can be resolved by a single measure.

Second, ecologists are not interested solely in the biometry of diversity, but that some property, or properties, of ecological systems cause them to be more or less diverse, i.e., why it occurs. For example, polluted or stressed environments show an increase in dominance and a decrease in species richness (Magurran 1988). In this sense *diversity* is about the comparative properties of high and low diversity ecological communities. For tropical lowland rainforests this may be definable when all, or at least many examples, are considered but not inferred from the study of just one.

So we need the concept of *diversity* as the name for a theory describing properties that we assume exist, but may not have defined, and determining why some communities have high species richness and some low, even when other things seem similar between them. Definitions and measurements of diversity based on numbers of species and individuals are comparative, not absolute, and at least initially we have no inkling from just these measurements of the very thing we need to know to answer the question, i.e., why greater diversity comes about and what is needed to maintain it. Furthermore, in order to discover whether there is an ecological theory to be described about diversity, we have to do research using natural and functional concepts that can be measured in particular forests.

10.3   The domain of functional and integrative concepts

In ecology, explanations can have two parts. One part answers questions about *why* something happens, usually in terms of direct causes, organization or structure, and is frequently represented by functional or integrative concepts. The second part specifies the domain, answering questions about *when*, *where*, or *under what conditions* a functional or integrative concept or proposition is important. Confusion and controversy have resulted when ecologists have tried to explain too much with a concept, i.e., have assumed, or claimed, too great a generality (for further examples, see Grubb 1992 and Chapter 11), or have not specified a domain. Much ecological research has the valuable objective of defining domains, and is frequently the most difficult part of research.

There is an important methodological reason for investigating domains. In ecology we seek to explain why a particular organizational structure exists or something happens. Frequently domains are set implicitly: (1) by the way an investigation is made – often we may justifiably choose a favorable situation to demonstrate a phenomenon in the first instance; and (2) by repeatedly considering a question within a limited, well-defined framework – this may be essential to avoid infinite regress, i.e., continually asking "why" to every offered explanation (Popper 1982). Unfortunately much of the confusion, and even contradictory definition, in the use of integrative concepts arises because the domains for particular uses have not been defined. That is: (1) having demonstrated the phenomenon does exist its universality is proclaimed without further investigations; (2) having posed, and answered, a question within a certain limited framework we may simply assume that the answer applies outside of that framework. Much ecological research concentrates on describing organizational structures or defining causes, but it is essential that domains be explicitly stated and not left as implicit. Only then can an explanation be complete.

DOMAIN OF A FUNCTIONAL CONCEPT OR PROPOSITION

The domain of a functional concept or proposition is defined as:

(A) the set of conditions that cause variation in a functional concept or proposition, or

(B) the conditions under which a functional process may influence an ecological system.

Example of (A)   In Chapter 6 four axioms were used to describe control of photosynthesis rate of foliage by the currently experienced physical environment. Three postulates were constructed suggesting causes why observed late

June and late July rates of photosynthesis were different, although the physical environments seemed to be identical. The postulates were not constructed to overthrow the four axioms of the physical effects on photosynthesis – but to specify conditions within which they may operate, and so affecting specific rates.

Maier and Teskey (1992) found that photosynthesis rate (P_{net}) was negatively affected by increasing atmospheric humidity deficit (AHD) *only* when pre-dawn xylem pressure potential (PXPP) was $< -1.0\,MPa$ (Fig. 6.10). Above that value of PXPP there appeared to be no relationship. So they had determined that the domain of influence of AHD on P_{net} was for conditions of PXPP $< -1.0\,MPa$. Finding this out leads to further questions such as "What conditions produce PXPP $< -1.0\,MPa$?" and "How frequent are conditions of PXPP $< -1.0\,MPa$?" and of course, "Why is the value of PXPP $< -1.0\,MPa$ critical?"

Observing a plant or animal in isolation, or in an artificial environment such as a glasshouse, growth chamber, aquarium, or laboratory, may never inform us of all the things it may do in its natural environment, or all of the things that may influence it. The domains of influence of different effects must be understood and experiments established with those domains in mind.

Example of (B)   In Chapter 5, it seems possible that mating both within, and between, species of cutthroat and rainbow trout could be influenced by a number of different processes and the research task was to determine their possible role. For example, we should ask both how the mate selection process may influence hybridization (defining the functional concept itself) **and** when that particular process may be important (defining its domain). This should be done for all processes that may influence hybridization.

DOMAIN OF AN INTEGRATIVE CONCEPT

The domain of an integrative concept has two components:

(A)   Specification of the extent of what the concept seeks to explain.
(B)   Specification of the set of functional concepts that an integrative concept uses.

This is the requirement to make a formal specification of the theory that defines the concept.

Example of (A)   Concepts such as *diversity, resilience* and *stability* have had general beginnings, but research requires precise definition of what is, and what is not, claimed by the integrative concept and if necessary qualifies the integrative concept by how it is named. For example, it should not simply be assumed that *diversity* of *lowland tropical rainforest* has the same causes as *diversity* of any other system, but the domain of each explanation should be defined. An important attribute of integrative concepts is their generality. But this must not merely be asserted but won through research in different situations – which defines the domain.

Example of (B)   In an explanation of a plant–herbivore system, such as balsam fir and spruce budworm, that is considered to exhibit resilience, there is a series of required functional concepts, particular about growth rates and dispersion systems that is necessary to explain how resilience actually occurs.

Two important conclusions from Section I were:

(1)   Multiple postulates should be considered, to avoid parenting a single postulate.
(2)   A postulate can be judged only in the context of its place in a theory – no independent scientific assessment can be made of a postulate.

Considering domains helps in both these cases.

1. Multiple postulates should be considered, to avoid parenting a single postulate

Where postulates are proposed about a process, then proposing a domain within which each operates can make a reasoned reconciliation between competing postulates more likely – if there is a reasoned reconciliation to be obtained. Furthermore, **not** to specify the domain of a process can imply that it has universal domain. It is probable that some processes in ecology can be influenced by multiple causes. The ecologist's task is not simply to assume one possible cause for a phenomenon, but to define how and when different causes occur.

Different examples of a community type have many similar properties, and yet variation does exist between them. What we can specify about different types or examples of the same kinds of things is their domains relative to particular functional concepts. For instance, throughout the world we can characterize many pieces of vegetation as saltmarsh, but we cannot predict exactly what species any individual saltmarsh may contain (Adam 1990).

2. A postulate can be judged only in the context of its place in a theory

The way in which we can judge a postulate in the context of its place in a theory is by specifying its domain. This means that ecological theories may contain descriptions of core processes, and a series of domains delimiting when, where, and how those processes operate. Not all the properties of a community may be understood completely by describing functional or integrative concepts. For example, although a saltmarsh habitat is characterized by periodic innundation of salt water, this does not specify the way in which plants and animals may survive it. We may find a range of adaptations to generally the same problem.

10.4 Example of use and development of ecological concepts and their domains

Development of ecological concepts (Fig. 10.1) takes place over many decades as they are defined, used, and refined or even discarded. Mycorrhizal fungi are an instructive example. They have been defined, and redefined, as natural concepts, while being used in developing descriptions of the functional concept of symbiotic mutalism. In its turn this functional concept has been used in constructing an explanation of diversity in tropical lowland rainforest.

10.4.1 Developing definitions of natural and functional concepts

The first illustration of a mycorrhiza was in 1840 by Hartig who published drawings of the fungal mantle and intercellular penetration by the fungus, the "Hartig net" (Trappe and Berch 1985), although he did not identify these structures as fungal. However, once the structures were identified as fungal, scientists disagreed about the functional relationship between fungus and plant. Some thought the fungus parasitized the plant. But in 1842 Vittadini observed that roots penetrated the sporocarps of *Elaphomyces* (a kind of truffle) and proliferated densely – and argued against parasitism. Vittadini observed that in other circumstances plant root accumulations occurred where nutrient concentrations were high, and so suggested that higher plants absorbed nutrients from the fungus – but he excluded the possibility of mutualism in which the fungus would absorb carbon compounds from the tree root (Trappe and Berch 1985).

Peffer first published the concept of mutualism (in 1877) based on analysis of orchid–fungus relations, but it was Frank in 1885 who first published a comprehensive morphological and anatomical analysis and defined the relationship (see Frank 1985). He had previously proposed the term *symbiotismus* for the fungus–alga relationship of crustose lichens (Trappe and Berch 1985), and he used this as an analogy for root and fungus. "The entire structure is neither root nor fungus alone but resembles the lichen thallus, a union of two different beings into a single morphological organ; it can be suitably designated as a fungus-root or mycorrhiza" (Frank 1985). In 1895 Frank demonstrated that pine seedling growth was suppressed in sterilized relative to unsterilized soil and that intermediate growth occurred if seedlings in sterilized soil became infected (Harley 1985), attributing these differences in growth to the presence or absence of mycorrhizae.

· So, by 1895, both the natural concept (mycorrhizal association) and the functional concept (symbiotic mutualism of the mycorrhizal association) had theoretical definition and empirical demonstration. While the first objective had been description and definition of the natural concept, that could not be achieved until the function was understood – at least in its main part. Much subsequent research has been needed to increase precision in the definition of both concepts – it has cycled round the left-hand side of Fig. 10.1.

There are two common types of tree mycorrhiza, each defined by a natural concept. Ectomycorrhizae have a fungal sheath round short roots, or "rootlets," and an internal Hartig net. What came to be called endomycorrhizae have distinctive structures within the root but no sheath, and were first observed in iris species by Nägeli in 1842 (Trappe and Berch 1985). Endomycorrhizae have two types of distinguishing fungal structure within the root: arbuscules, which are dendritic structures produced by the fungus within cells and surrounded by the cell's completely invaginated plasma membrane; and vesicles, which are spherical or oblong lipid-filled bodies within or between cells. These two features led to this type of endomycorrhiza being called a vesicular–arbuscular mycorrhiza (VAM).

Changing names of natural concepts, e.g., from endomycorrhizae to VAM, is a typical and valuable process if it identifies the natural concept with particular properties, structures or functions. In this example taxonomic definition is important. Recently it has been found that while all VAM have arbuscules, only some have vesicles, and the suggestion has been made that the concept VAM should be replaced by arbuscular. This type of debate may seem pedantic if you are busy using a currently accepted working definition of a natural concept. But it can represent an important change. In commenting on VAM taxonomy, Schenck (1985), while playing to a gallery of fungal physiologists and ecologists, jokingly declaimed "Taxonomists are a despicable group. They are obsessed with changing things and naming something new." But he answered his own taunt by noting that characterization and definition are essential for mycorrhizal research to progress.

The substantial hyphal sheath made the presence or absence of ectomycorrhizae easy to determine and to manipulate experimentally. Some ectomycorrhizal fungi can be grown in pure culture isolated from a host plant and such cultures can be used as specific sources for experimental formulation of mycorrhizae. From 1900 there was substantial research into the functional process of symbiotic mutualism of these mycorrhizae (Hacskaylo 1985, Harley 1985).

Experimental analysis of VAM, and so development of its associated functional concepts, proved more difficult. It was known by 1900 that they were the most widespread type, occurring in all but the most fertile soils, but evidence of their effects was difficult to obtain. Initially the only inoculum that could be used was infected soil or infected roots growing in soil, and quite often attempted inoculations of plants from these sources were not successful. In retrospect, an important problem was that explanations were being sought that would apply to all VAM (Mosse 1985). There had been inadequate classification into defined natural concepts representing the variation that occurs within VAM, and so apparently different results from functional investigations were obtained for what were thought to be similar things. It was not until the 1960s that it became clear that several morphologically different species of fungus formed VAM, that the host may be either obligate (always requiring the fungal association) or facultative (sometimes having the fungal association), and that only a relatively few VAM fungi infect the majority of plant species. Furthermore, among facultative hosts, colonization by the fungi did not always guarantee an increased growth response. It now seems that in facultative species a positive growth response by a plant to infection requires that the plant should be growing in a soil of low fertility (Janos 1980a,b, Allen 1991).

So it has proved difficult to define both the natural concept of VAM (what they are and when they occur), as well as the domain of symbiotic mutualism (under what conditions VAM fungi infect and what effect they have). The taxonomy of VAM fungi has been revised even recently. In practice, research into both the natural and functional concepts has had to proceed in a cyclical way (Fig. 10.1) and both technological and conceptual problems have taken longer to solve than for ectomycorrhizae.

10.4.2   Using functional concepts to define an integrative concept

A theory about tree species diversity in the canopy of tropical lowland forests is being developed (Janos 1985, 1996, Connell and Lowman 1989). It depends upon VAM, ecotomycorrhizae, and differences in their respective functional concepts of symbiotic mutualism and the extent that individual tree species form obligate and specific association with VAM or ectomycorrhizal fungi. Most importantly, this theory does not attempt to explain high diversity as a general phenomenon but seeks to explain the reasons for contrasts in diversity in a specified domain. While the majority of lowland tropical rainforests do have many tree species in the upper canopy (they are both species rich and there is not dominance by one, or even a few, species), some have one or just a few species and such dominance persists despite the presence of large species numbers in a neighboring region.

The mycorrhizal theory describes highly diverse forests as having VAM trees that are obligately mycorrhizal but do not form species-specific associations so that below-ground competition is minimal and success of individuals is a lottery and not dependent upon specialization, e.g., specific fungus associates. Low diversity forests are described as having ectomycorrhizal fungi that are species specific, and it is this specialization that results in dominance by one or a few species.

An alternative theory of high diversity in stable communities is that competition increases niche differentiation and species evolve to be specialists for those niches (Hutchinson 1959, Brown 1981). This was first suggested for animal communities but applying it to tree-species richness in tropical forests produces the reasoning that, because they have existed for a long time, the evolutionary process has resulted in competitively coevolved sets of niche-differentiated species each having a means of exploiting limiting resources that is sufficiently different to permit its continued persistence. The argument runs that this has continued for so long that there has been finer and finer partitioning of resources with greater and greater speciation. This might be considered as an alternative theory, in the sense of Chamberlin (1965), and has to be borne in mind, although it is not analyzed in detail here in parallel with the mycorrhizal theory. An important disadvantage is that it does not explain the observed contrast of high and low diversity within tropical lowland rainforest and it would have to be developed to account for that, i.e., explaining what prevented multiple-niche differentiation in some areas.

The first demonstration of mycorrhizae in the tropics was by Janse (1896), who documented VAM in a number of species and experimented into both the mechanism of infection and symbiotic mutualism. Richards (1952) suggested that mycorrhizae may be important in tropical forests, but not until recently has the importance of VAM in tropical forests been described in some detail, and a theory advanced based on field and experimental evidence that VAM are crucial for high diversity of canopy tree species. I wish to describe this from the perspective of a scientist, David P. Janos (1980a,b, 1983, 1985, 1987, 1988, 1992, 1996), working with functional concepts to define a theory for an integrative concept. This is not intended as a comprehensive review of diversity in tropical forests – there are many features to explain and multiple theories. Additional references are given in 10.6.

Before outlining the mycorrhizal theory, the background provided by four results of experiments and investigations needs to be described because these are used in defining axioms.

(1)   In a tropical forest soil with low nutrient availability VAM enable the survival and growth of tree seedlings.

(2)   Experimentally applied nutrient can increase total biomass, but decrease the total number of individuals of tree species that survive, and favors facultative over obligate mycorrhizal species.

> {For both (1) and (2).} Janos (1985) experimented with microplots of soil, 1 m × 2 m × 1 m deep completely contained by concrete sides and a fibre glass screen bottom, and used lowland tropical forest soil sterilized with methyl bromide (with 2 percent chloropicrin). Seedlings of nine

tree species were planted into four conditions: sterilized soil (without mycorrhizae (− MYC), without added fertilizer (− NPK), soils sterilized then inoculated with VAM (chopped roots) (+ MYC − NPK), soils sterilized then inoculated with addition of NPK fertilizer (+ MYC + NPK), and soils sterilized and fertilized (− MYC + NPK). While the mean total above-ground dry matter per plot was greatest for fertilized treatments, the greatest species survival was on the + MYC − NPK plots. In particular, the number of surviving individuals on the + MYC − NPK treatment of four tree species strongly responsive to mycorrhizae was twice that on any of the other treatments.

(3)   Rate of mycorrhizal infection determines growth of obligate mycorrhizal seedlings and infection is most rapid where the continued presence of infected vegetation maintains a source of mycorrhizal fungi in the soil.

Under field conditions the growth rate of mycorrhizal tree seedlings is greater than that of tree seedlings lacking mycorrhizae. Seedlings can be raised in pots without mycorrhizae and transplanted into field conditions, and the time taken for such plants to achieve the growth rate of similarly transplanted mycorrhizal plants can be measured. Janos (1988, 1992) used this as an indicator of site inoculum potential. The time taken to achieve similar growth rates was greater on forest clearcut sites occupied by non-mycorrhizal plant species than sites from which the non-mycorrhizal species were absent.

(4)   Soil fumigation with methyl bromide, which kills mycorrhizal fungi, results in lasting reduction in tree species diversity.

Twenty years after a one-time *in situ* soil sterilization of 10 m × 20 m field plots that had been cleared of trees, tree species diversity of the reestablished stand was lower than in neighboring plots that had been cleared but not sterilized (Janos 1996). Notice that this is another type of contrast established by experiment. Of course, organisms other than mycorrhizae are also killed by the procedure and that must remain as a caution in interpretation.

The existence of VAM and the positive effect they can have on tree seedling growth have not been in doubt. What these four results have been used for is describing how infection may occur under field conditions, the effects it may have on species composition of the community, and what is required to maintain sources of inoculum. These are additional aspects of the functioning of VAM needed to explain high and low diversity of canopy tree species for some tropical lowland rainforest, i.e., the research is attempting a cycle round the right-hand side of Fig. 10.1.

Results (1) and (2) suggest that at low nutrient levels VAM are important for growth, and under those conditions competition between trees may somehow be checked or reduced, though a functional explanation for that process is not apparent from this work. (One might speculate that above-ground competition is more limited in low as opposed to high nutrient conditions.) Results (3) and (4) suggest that a continuity of inoculum is necessary for VAM infection. These functional concepts about growth and infection are based on just a few investigations. Some scientists working with other species in other situations show apparently conflicting results for (3), others show some agreement (for a review, see Janos 1996). The uncertainties are not about the details of results (1) through (4) – but their significance in an explanation for high canopy tree species diversity in tropical lowland rainforests, and about how generally such results are likely to be found. That is, has scientific information been successfully developed for one cycle round the functional-integrative loop (right-hand side of Fig. 10.1), and is it likely that further cycles can be made?

The complete theory for explaining contrasting high and low diversity depends upon construction of functional concepts about symbiotic mutual-ism of the mycorrhizal association. How, and when, it occurs and the precise balance of benefit and cost to the trees in different conditions and the effect on their competitive status. Development of this theory to explain differences in diversity depends just as much upon the fate of postulates about natural and functional concepts as those that seem directly concerned with competi-tion and its effects on natural selection. Just as the refinement of natural concepts about mycorrhizae proceeded in tandem with the development of the functional concept of symbiotic mutualism, so development of under-standing about the integrative concept of diversity requires further under-standing of symbiotic mutualism and the development of additional functional concepts, e.g., about the infection process and what is required to maintain a source of inoculum.

The principal axioms and postulates are now laid out under three headings to reflect their distribution between different types of knowledge. A strength of the theory as a heuristic, i.e., as an investigative tool, is that it seeks to explain the causes that underlie a contrast – why some tropical lowland forests are more diverse in canopy tree species than others. Those which are more diverse are proposed to have VAM tree species; those that are less diverse are proposed to have ectomycorrhizal tree species.

Axioms and postulates about functional concepts and their domains

		Domain
Axiom 1.1	When soil nutrient status is low mycorrhizal association is a benefit to plants in acquiring nutrients	Functional
Axiom 1.2	A mycorrhizal association is a carbon cost to plants	Functional

| Postulate 1.1 | The carbon cost of VAM mycorrhizae is less than the carbon cost of ectomycorrhizae | *Functional* |

This is not a precisely worded postulate, since the concept of nutrient uptake per unit carbon cost may have to be introduced. But specifying what this entails may itself involve considerable research. An implication of Postulate 1.1 is that, unless there are special benefits in nutrient uptake for ectomycorrhizae, then species with VAM should have a competitive advantage over ectomycorrhizal species.

| Axiom 1.3 | Most tropical lowland rainforest tree species are obligately VA mycorrhizal but some are obligately ectomycorrhizal. | *Domain* |

This specifies the domain of a functional concept (type A specification). It establishes information about the relative frequency of the two types of natural and functional concepts that are mycorrhizal associations. While it is considered to be sufficiently true to be the basis of further research (it is an axiom), like many domain axioms it is imprecise (there is not a comprehensive species list of tree–fungus associations).

| Axiom 1.4 | There is no specificity between VAM fungi and tree hosts | *Functional* |
| Axiom 1.5 | Some ecotomycorrhizal fungal species associate with many host tree species but many have restricted host ranges | *Functional and Domain* |

As with Axiom 1.3, this is imprecise about the domain it specifies.

Axiom 1.6	Where nutrients are sufficiently available a facultatively mycorrhizal plant will grow more rapidly without mycorrhizae than with them	*Domain* *Functional*
Axiom 1.7	(a) In facultatively mycorrhizal plants, establishment of the mycorrhizal association is controlled by phosphate content of the host tissue	*Functional*
	(b) In facultative species, high phosphate levels tend to reduce mycorrhizal fungus infection	*Functional and Domain*

Axioms 1.6 and 1.7 establish limits on the effectiveness of mycorrhizal associations (set their domain of occurrence and effectiveness) to soils low in phosphorus (P).

Postulate 1.2   Continued infection by mycorrhizal fungi   *Functional*
                       of newly developing tree seedlings is
                       more likely where there are actively
                       growing mycorrhizal associations than
                       where they are lacking

A prediction from Axiom 1.4 and Postulate 1.2 is that VAM tree species may
provide inoculum for each others' seedlings, i.e., that multi-species VAM tree
communities would be effective in sustaining inoculum. In contrast the
ectomycorrhizal fungal species, with more specific tree hosts, are more likely
to be continued in stands with few tree species.

Axioms and a postulate that establish domains for VAM and ectomycorrhizae

Axiom 2.1   The majority of tropical clays have   *Domain*
                       substantial anion exchange capacity and
                       can bind P irreversibly and the
                       vegetation on them tends to be P limited
                       (Vitousek 1984)
Axiom 2.2   A minority of tropical lowland soils have   *Domain*
                       little available nitrogen (N) (as well as
                       low P) and the vegetation on them is N
                       limited as well as P limited (Vitousek
                       1984)
Postulate 2.1   Soils with both low P and low N present   *Domain*
                       a habitat requiring ectomycorrhizae
                       which tend to greater specificity of
                       association than do VAM

These two axioms and the postulate together are a type B specification about
the domain of functional concepts of mycorrhizae and symbiotic mutualism.
A prediction from Axioms 1.4 and 1.5 and Postulate 2.1 is that the more
specific ectomycorrhizae are more likely to occur where soils are both P and N
limited. However, developing measurements and a hypothesis for this postulate may have to include assessment of P and N availability.

Postulates using integrative concepts

Postulate 3.1   On soils low in both P and N, then   *Domain*

> *Integrative concepts of niche*
> *specialization and niche limitation*
> (a) Trees are ectomycorrhizal with
> specific fungal species associates.
> (b) Natural selection has resulted in high
> specialization by few species of tree.

This postulate specifies the domain of the low diversity condition. It is a type A specification for an integrative concept.

Postulate 3.2 now describes a process of coexistence (through a lottery process) and synergy (through mutual maintenance of a biological environment) rather than niche differentiation.

Postulate 3.2   On soils low in P but not low in N   Domain
(a) The controlling influence on success   Functional
or failure of a tree seedling is whether it
establishes VAM

> *Integrative concept of coexistence*
> (b) The principal source of inoculum in forested habitat is growing mycelium, whether any seedling becomes infected depends upon chance encounter, the chance will be higher as the number of existing trees infected with VAM is higher.
> (c) There is little niche differentiation between VAM trees with regard to competition for P because all are likely to be infected by the same limited suite of VAM fungi.

> *Integrative concept of synergy*
> (d) Maintenance of VAM fungi is required for continued existence of VAM tree species.
> (e) Because VAM fungi are not host specific then all host species sustain the source of inoculum that may infect each others' seedlings.
> (f) Natural selection has been to maintain VAM fungi, resulting in a high degree of convergence among obligate VAM rainforest tree species with respect to their ability to compete for P.

Postulate 3.2 is critically dependent upon the level of available P found in tropical lowland rainforest (the domain) and its effect on growth and continued propagation (functional concepts). With a high level of available P, mycorrhizal associations may not be maintained on facultatively mycorrhizal tree species. This could result in facultatively mycorrhizal species having a competitive advantage over obligately mycorrhizal species, and this in turn could result in reduced species diversity.

10.4.3 Making inference about an integrative concept

Postulates 3.1 and 3.2 present an interesting theory to explain diversity of tropical lowland rainforests – which took many years to develop. But can we infer that it is the best explanation? If not, what else must be done? Recall from Fig. II.2 constituent procedures in making scientific inference:

> Constructing the requirements for a scientific explanation.
> Making a synthesis.
> Assessing the explanatory coherence of the synthesis.
> Achieving scientific understanding.

In practice, scientific inference for integrative concepts is a progressive process. We can rarely say there is a crucial experiment or investigation (Chapter 11), though sometimes investigators like to claim that when writing a grant proposal for their next piece of work! I want to examine what can be said about each of the procedures (Fig. 10.2) in making a scientific inference for the question:

> *Why do some tropical lowland rainforests have canopies composed of multiple species while some have canopies dominated by just one or a few species?*

Constructing the requirements for a scientific explanation

We can say that a why-type question has been constructed for the following reasons:

> The topic of concern is the cause of high and low species diversity in tropical rainforests.
> The contrasting set of alternatives is the range of different levels of diversity that are found and in particular the fact that some forests have many canopy tree species while some have just one or a few.
> The explanatory relevance is defined by the postulates that high diversity is the result of lottery competition below ground and synergy in maintaining sufficient VAM to ensure tree seedling infection; low species diversity is the result of niche specialization in a restricted edaphic environment.

The most important thing about the construction of this why-type question is that it is restricted, and particularly restricted to a contrast. The research to answer this question may not apply to all questions about *diversity* – but it may actually answer some about *tropical lowland rainforest diversity*.

Making a synthesis

An initial synthesis has been constructed that brings together axioms and postulates about natural and functional concepts to define integrative con-

Fig. 10.2. The procedures for making a scientific inference and their status in assessing the mycorrhizal explanation for diversity in lowland tropical rainforests.

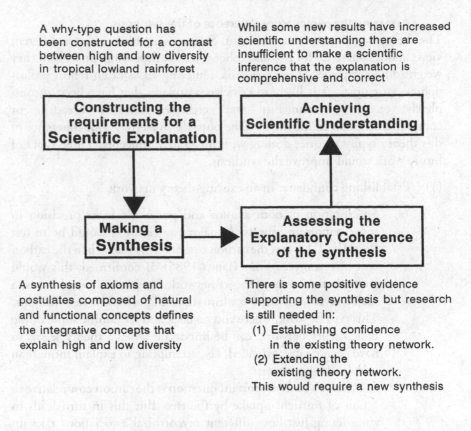

A why-type question has been constructed for a contrast between high and low diversity in tropical lowland rainforest

While some new results have increased scientific understanding there are insufficient to make a scientific inference that the explanation is comprehensive and correct

Constructing the requirements for a Scientific Explanation

Achieving Scientific Understanding

Making a Synthesis

Assessing the Explanatory Coherence of the synthesis

A synthesis of axioms and postulates composed of natural and functional concepts defines the integrative concepts that explain high and low diversity

There is some positive evidence supporting the synthesis but research is still needed in:
(1) Establishing confidence in the existing theory network.
(2) Extending the existing theory network.
This would require a new synthesis

cepts that explain high and low diversity. Its basic structure is:

High diversity is the result of coexistence between tree species rather than niche differentiation and competitive exclusion.
Coexistence is the result of:

(1)   lack of specificity in association between individual VAM fungal species and their host tree species,
(2)   synergy between individual tree species through continuous production of VAM fungi available to colonize tree seedlings of other species.

Low diversity is the result of specialization in the ectomycorrhizal fungus–tree host relationship enabling survival in low P and low N soils.

So the integrative concepts of diversity, coexistence, synergy, and specialization can be described in terms of functional and natural concepts that can be investigated in particular experiments or field investigations. Because the question about *diversity* has been restricted to a contrast, then it is possible to construct a synthesis.

Assessing the explanatory coherence of the synthesis

The theory is not completely coherent. Scientists are likely to have different views about coherence for this and other theories, depending upon how they weight different pieces of the synthesis. Differing views often reflect philosophies: an empiricist is likely to criticize postulates that have been incompletely tested; a rationalist may criticize lack of breadth, or comprehensiveness, in the theory. The possible concerns about coherence of this theory typify theories dealing with integrative concepts. Three types of future work would improve the synthesis.

(1) Establishing confidence in an existing theory network.

   (a) Specification of both axioms and postulates lacks precision in some important aspects. A major contribution would be to test Postulate 1.1 – that the carbon cost of VAM is less than the carbon cost of ectomycorrhizae (Janos 1985). If confirmed, this would strengthen the present theory network. If rejected, it would cause a major revision. Partial confirmation might lead to new insights. This type of work, where you go back and reexamine some of the basic theory network, can be important when a theory seems to have become overextended, i.e., attempting to explain more than it is justified in doing.

   Perhaps a more important question is the carbon cost relative to a unit of nutrient uptake by the tree. But this in turn leads to considering just how different mycorrhizal associations take up nutrients in different soil types and whether soil types are important in determining their functioning. Research is required to develop understanding of the functioning of these mycorrhizal associations.

   (b) Specification of the domains of some functional concepts lacks precision. Asking questions about a domain can be important, particularly where a universal domain is asserted or implied. For example:

   Axiom 1.4:   There is no specificity between VAM fungi and tree hosts.

claims universal domain for lack of specificity in VAM but

   Axiom 1.5:   Some ecotomycorrhizal fungal species associate with many host tree species but many have restricted host ranges.

is uncertain with regard to the claim of specificity for ectomycorrhizae. The claim for VAM is based upon limited investigations (Janos 1980a); although there are no grounds to reject it, it has not been exhaustive tested. One part of an assessment of this theory for any particular tropical lowland rainforest with many canopy species would be to determine (i) how many VAM species exist, and (ii) will each VAM species form mycorrhizae with all tree species. A complete set of multiple experimental crosses, even using tree seedlings, is likely to be technically impossible – but some approach should be made, perhaps by first classifying the tree species into ecological types, e.g., according to their seed type, or root morphology, and then examining representatives from each type.

(2)   Extending the existing theory network.

Extension can include adding a new concept or making a postulate that can be investigated. Many of the subpostulates listed under Postulate 3.2 are in this category, e.g., research into the mechanisms of VAM infection to establish the conditions under which seedlings do become infected, and whether this depends upon a source that has to be maintained. For example, are living hyphae much more effective in infecting seedlings than resting spores in these soils? This type of work makes a theory explain more effectively. It extends, or fills in, part of the theory network.

There exist unresolved exceptions, implying *ad hoc* propositions, to what the theory attempts to explain that may involve extending the theory network and/or possibly a major revision (see 3). Asian diptocarp forests are the most species rich of tropical lowland forests (Whitmore 1984). In contrast to the explanation suggested by Postulates 3.1 and 3.2 the dominant trees, the diptocarp species themselves, are ectomycorrhizal, although the subdominant species are VAM (Smits 1992). Although research might continue to investigate Neotropical and/or African tropical lowland rainforests for some time and obtain greater precision in the theory, eventually it must be reconciled with this apparent discrepancy in Asian tropical lowland rainforests. This type of discrepancy is unlikely to dampen interest in the mycorrhizal explanation for diversity because it may still explain differences in diversity in some conditions (domains), and diptocarp forests may represent a different domain. A further unresolved difficulty may be that in some situations, forests with one or few tree species as dominants are spreading naturally and replacing multispecies ones (Connell and Lowman 1989).

(3)   A major change in theory network.

This tends to happen less than the other two categories, since it requires work to have been done to establish the theory before it can be changed. Two types of major revision can be envisaged for this theory. One would involve a resolution of the differences between high and low diversity through information completely absent from the present theory – say, some additional factor that explains the difference in diversity in a different way, and explains the mycorrhizal difference as a consequence of that.

Another type of major revision may occur where theories are joined. The requirement for this is that the theory is incomplete in its explanation. As stated here, Postulates 3.1 and 3.2 imply that, where there is low soil P, lottery competition occurs. Increased limitation in the edaphic environment, low N and low P, results in greater specialization and so lower diversity, but the theory does not explain why this should be so; a VAM system may permit many species to coexist but it does not say why they actually do. The causal reasons are not sufficiently specified. Implying that ectomycorrhizal fungi are specialists, and tend to have species-specific tree associates, does not explain why the number of specialists should be limited.

Specific axioms or postulates about above-ground competition and how many different species may thrive nevertheless may come from other theories. Field (1988) gives a theory of the relation between N content of foliage and photosynthesis capacity for different foliage geometries. Similarly, although below-ground lottery competition allows for the possibility of diversity, because it is likely to reduce or restrict limiting specialization, it does not say why there should be high diversity. Janzen (1970) gives a theory on why clumps of the same individuals of a species may be restricted through herbivory and/or parasitism, which could add explanatory power about causes of high diversity. In contrast, Hubbell and Foster (1986) propose a theory for above-ground that also suggests reduction rather than intense species competition. They suggest that gaps in the canopy are important for regeneration but that uncertainty in where and when they would occur led to coevolution of guilds of generalists. Species may have accumulated in tropical rainforest because generalists may not be resistant to invasions and generalists may be difficult to eliminate once established.

At the moment it is not clear how general the VAM theory is. Its axioms are based on research with limited numbers of species on limited sites. The specific results obtained may not apply to other species and other sites. The concept of diversity, like many integrative concepts, is comparative (this forest is more diverse than that one), and for the theory to provide a comprehensive explanation for tropical lowland rainforests it must attempt to achieve generality, i.e., not merely be applicable to the forests in one region but also among regions. The four results listed prior to defining the axioms and postulates were obtained in the Neotropics. Repeating those investigations in other regions would entail using different species and would start to test generality, at least in what needs to be explained, and it may also reveal features of the theory not so far understood.

Scientific understanding

There are a number of pieces of information, particularly experimental results, supporting this explanation and so defining the scientific understanding. Mostly, scientific understanding increases incrementally. Each additional confirmation of an axiom, a contrary result that produces a revision, or acceptance or rejection of a postulate increases scientific understanding.

What scientific inference can be made about this theory? As already mentioned, at best, it only provides a partial explanation. Should we consider Postulates 3.1 and 3.2 so problematical that they should be completely discarded? No. The problems are typical of many ecological theories, although unfortunately (but not in this case) scientists sometimes do not focus on detailing those problems but concentrate on emphasizing what the theory would explain, if it were true. This theory should not be discarded because it has two valuable features.

First, it offers a heuristic, a way that things can be found out in future. The results obtained so far provide a pattern for future investigations where similar methods to examine parallel situations or test predictions can be applied. A valuable feature of this theory, as a heuristic, is that it contains axioms and postulates about contrasting situations – it does not seek to explain diversity as an absolute quality but rather attempts to explain, in part, why some types of forest are diverse and others not. Future research may be able to exploit the existence of these contrasts.

Second, it is possible to assess the postulates and seek confirmatory evidence for axioms in order to extend their domains and so work towards a more comprehensive assessment of the whole theory. Most important of all, the concept of diversity, as it applies to canopy tree species of tropical lowland rainforests, has been reduced to researchable components and the framework established for constructing inference to the best explanation, even if that point has not yet been reached for this aspect of diversity. The importance for ecological research for developing theories in this way is discussed further in Chapter 16.

10.5   Discussion

More confidence will be placed in the theory explaining contrasts between tropical lowland rainforests in canopy species diversity if further research is made to satisfy the three components of explanatory coherence: the acceptability of individual propositions; the simplicity of propositions and particularly that they do not define *ad hoc* rules; and the breadth of the explanation is increased – that it is demonstrated to apply to more and wider circumstances. There are no rules for deciding which of these three components should be worked on in a particular situation, except that some balance must be maintained in pursuing all three. This can require a long-term effort over decades, involving a number of scientists. The result of such work is likely to change the theory, including the details of the questions asked, rather than outright acceptance or rejection. This process is discussed in Chapter 11.

There are implications of this methodology for ideas about how ecological knowledge is gained. Some ecologists have suggested that knowledge should be built by researching a series of partitioning reductions (Chapter 9). Others have considered that this is not sufficient and that a holistic approach should be taken where the emphasis is on the properties of complete systems – the whole is greater than the sum of the parts.

The theory for low and high diversity of canopy tree species in tropical lowland rainforest uses integrative concepts of *niche specialization, niche limitation* (Postulate 3.1), *coexistence*, and *synergy* (Postulate 3.2). These are defined by functional concepts and their domains, along with conjectures about how these have influenced natural selection, that specify dependencies between trees and mycorrhizal fungi, and these are developed in the context of particular environments (domains). The approach taken is not one of reduction; research into natural and functional concepts has a clearly defined importance for integrative concepts. Neither does the approach invoke the concept of emergence, particularly that natural selection has operated on a system as a unit, in order to construct an explanation, nor does it consider a particular hierarchical organization.

We do not have to accept either derivational reduction and connectability or emergent properties (Chapter 9) as appropriate theories of knowledge (Maull 1977), or as the only possibilities that must dictate our method of analysis and synthesis. Levins and Lewontin (1980) suggest that holism and reductionism share a common fault, that they see "true causes" as arising at only one hierarchical level, although respectively different ones. They suggest that relationships should be looked for at all levels, that whole and part *do not completely determine each other*. The whole is contingent upon the reciprocal interaction with its parts, and with the greater whole of which it is a part. Note the similarity to the approach advocated by Shrader-Frechette (1986) (Chapter 9) that neither reductionism nor holism provides a self-sufficient philosophy that can yield adequate explanations.

What makes ecology distinct is the type of question asked, the type of synthesis they demand, and particularly the domains that determine functional and integrative concepts. The conceptual system outlined in this chapter is designed so that both precision and uncertainty of whole system properties can be described and researched. Explaining an integrative concept such as diversity in this way, does not mean that ecology can be reduced to biology, physics, and chemistry.

Natural selection works at the level of breeding populations, not whole systems such as forest stands, but the environment driving natural selection includes relationships to other organisms. Different functional concepts and their domains, used in the same explanation, may be defined at different scales rather than conform with a proposed hierarchy of relationships. Loehle (1988) points out difficulties in classification encountered when attempts are made to construct hierarchical classifications. A common hierarchical representation is:

cell – organ – organism – population – community – ecosystem – biome

but contains a mixture of types of organization. The first three refer to natural concepts, have obvious measurements, and have generally well-defined relationships between them. Population is a classification based on a number of functional concepts and may be used in different ways in different circumstances, i.e., there is no universal definition of what comprises a population in the same way that there is about what comprises an organism (though even for organisms there are exceptions). Similarly, ecosystem is a class where a particular theory of function is assumed, and it is not clear what an ecosystem is made up from, e.g., organisms, populations or communities, or mixtures of those and other things.

The technique of analysis and synthesis presented in this chapter – using natural, functional, and integrative concepts and specifying their domains provides a heuristic for the approach advocated by Levins and Lewontin (1980). They suggest, for example, that the community "is an intermediate entity between the local species, population and biogeographic region, [it is] the locus of species interactions." Whatever type of ecological system is being considered, the axioms that specify functional and integrative concepts define how processes work, while those specifying their domains define when and how the processes come into play or what may control their operation when they become important in particular circumstances. Crucially, domains do not have to be merely at larger physical scales than the functional or integrative concepts to which they apply. What determines the type of organization to be considered is the explanation required, and no attempt is made to require, *a priori*, a particular theory of hierarchical organization, or indeed that any such does, or does not, exist. Maull (1977) illustrates that important developments in science involve interactions, not between theories such as the theory of biology and the theory of physics, but between branches of science or fields from which a particular theory is constructed. Explanations must be constructed for particular problems and the scientific inference made from them does not belong to any particular subject.

10.6 Further reading

My definition of natural concepts is very close to that given by Dupré (1993) for natural kinds. Kiester (1980) discusses the role of natural kinds in ecological science relative to other science. Among philosophers of science natural kinds are things, like chemical elements or compounds like water, that follow laws (for a discussion, see Carr 1987). Species are frequently considered to be natural kinds but, as with many other things in biology and ecology, e.g., mycorrhizae and tropical lowland rainforests, our research purpose is to define properties and their variation, and it seems uncertain that there are laws, as opposed to theory (discussed in Chapter 16). Some people consider natural kind terms as rigorous designators of fixed properties (this is essentialism; for a discussion, see Carr 1987); because of this I have not used the term natural kind.

The frameworks we can use to build knowledge of integrative concepts are a subject of continued discussion. Similarities and differences in definition and use of holism and reductionism are illustrated in a series of articles in volume 58 of *Oikos* written by ecologists working on different problems. O'Neil *et al.* (1986) makes an extensive review of hierarchy theory and suggests a dual hierarchical structure for ecosystems of community – population – organism and biosphere – ecosystem – functional component.

In a Special Feature of *Ecology*, Karieva (1994) reports that, between 1981 and 1990, > 60 percent of the papers published in *Ecology* dealt with at most two species. The papers in the Special Issue discuss what higher-order interactions involving multiple species may be and how they can be detected. The requirements of environmental research have led to the development of such integrative concepts as *forest health* and *biological integrity* (Frey 1975). A difficulty with such concepts is providing measurements for them without making comprehensive descriptions of ecosystem function. For this reason scientists seek for indicators, e.g., Kerans and Karr (1994) discuss the presence of different species and this depends upon assumed knowledge of what controls these species and assemblages.

Pickett *et al.* (1994) discuss the importance of domains and their definition in relation to whole theories rather than just concepts (see also Chapter 11).

For additional readings on mycorrhizae and their role in tropical forest ecosystems see Baylis (1975), Malloch *et al.* (1980), Höberg (1986), Newbery *et al.* (1988), and Alexander (1989). Orians *et al.* (1996) provide a discussion of different aspects of diversity in tropical forest ecosystems. Huston (1994) reviews theories of how high species richness occurs and processes regulating diversity at various spatial and temporal scales.

11 Strategies of scientific research in ecology

Summary

Philosophers of science have investigated the progress of scientific research, how new discoveries are made, and have described the methods that scientists use. Philosophers' theories reflect different viewpoints about the relative importance of social processes and discussion and argument among scientists. The stage of development of a scientific field may also affect how progress takes place. Three theories are described and applied to examples of ecological research.

Kuhn (1970) suggests that scientific advance occurs through recurring revolutionary periods, when there are major changes in objectives and methods, separated by periods of normal research. Normal research follows a paradigm – when a distinct social group uses agreed methods and procedures to tackle agreed questions of importance. When the revolution comes, objectives, methods, and people involved in the research all change. The development of ecosystem research is analyzed with this theory.

Lakatos (1970) characterizes scientific research as a debate between data and rival theories, with theoretical advance continuous rather than revolutionary. Falsification is not a simple process but occurs through continuous debate and analysis. Scientists maintain a positive heuristic, a long-term research policy that anticipates possible refutations and develops a hard core of knowledge irrefutable by the methods of the research program. This is applied to the development of top-down and bottom-up theories about balance between plants, herbivory, and predation.

Initiation of a new field of research, or effecting a junction between existing fields, requires the investigation of domains (Shapere 1977). A domain includes related observations, an important problem about those observations, and the readiness of science to deal with both observations and problem. The first stage in theory development is refinement of the domain through investigating precision of measurements and extending the domain through finding what else may show similar patterns. The second stage is developing a precise theory. Analysis of domains is applied to the resolution of competing top-down and bottom-up theories.

These examples illustrate that different patterns of investigation do occur – although none of the theories exactly fits the chosen ecological examples. Two conclusions are drawn: (1) that debate and plurality of approaches are more likely to advance scientific inference than research under a rigorous paradigm, and (2) that different strategies of research are appropriate for establishing the detail of a theory as opposed to working out its domain of application.

11.1 Introduction

Making scientific inference about integrative concepts is a gradual process. Theories defining them need to be refined, and domain knowledge developed, through research in different situations. This can take considerable time, certainly involving the work of more than one scientist. So the question arises, "how do researchers, together, use and develop ecological theories?". Individual scientists may have different philosophies influencing what they select for study, how they work, and what conclusions are drawn. How can the contributions of different types of study be assessed? And how can a scientist decide what type of study is most appropriate at any time?

Answers to these questions define the level, and the standard, of criticism in a scientific field. Indeed, fields can be recognized by the logic and research techniques used, and pattern of critical development engaged in by contributing scientists, as well as the problem being tackled. In some fields social processes dominate both the types of question asked and the standard of explanatory coherence considered to be acceptable. This chapter concentrates on the methodological definition and assessment for theories, while Chapters 13 and 14 discusses processes of social interaction between scientists in more detail.

Three theories about research strategy developed by philosophers of science are described in this chapter. Their relevance to ecological research is discussed through examples:

> Progress through recurring revolutions (Kuhn 1970) applied to the development of ecosystem studies.
> Progress through competing theories in a research program (Lakatos 1970) applied to the development of the theories of "top-down" and "bottom-up" control of community structure.
> The emergence and development of theories from a domain of interest (Shapere 1977) applied to a possible resolution between "top-down" and "bottom-up" theories.

These theories of research strategy do not explain all the features or developments in the ecological examples to which I have applied them. But they do illustrate how a research strategy can determine progress and why there are

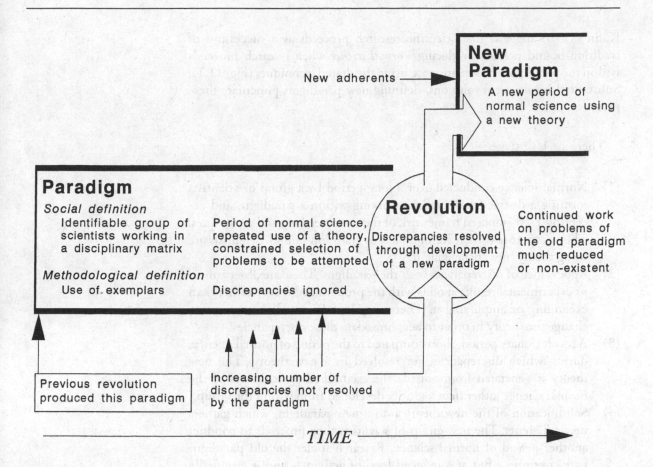

Fig. 11.1. Kuhn's theory. The progress of science through successive periods conducted within a paradigm punctuated by revolutions that resolve the discrepancies and form new paradigms.

social, as well as logical and technical, problems in making scientific explanations – particularly about integrative concepts.

11.2 Does ecological science advance through recurring revolutions?

Kuhn's (1970) theory of scientific progress emphasizes group functioning. There are two important concepts: *paradigm* and *revolution*.

SCIENTIFIC PARADIGM

A *scientific paradigm* has theoretical, social and technical components. When first introduced, the theoretical component provides a new view of some problem. This is adhered to by a group of scientists with a common attitude, ideal, and educational experience, using shared scientific techniques and exemplars to solve problems defined as important by the group.

Kuhn (1970) suggests that scientific research proceeds as a succession of tradition-bound periods producing *normal science* when research proceeds within the paradigm and scientists are using the shared techniques (Fig. 11.1). Substantial breaks, or revolutions, defining new paradigms punctuate these periods.

There are four stages:

(1) Normal science conducted over a long period by a group of scientists forming a distinctive discipline, working within a paradigm, and accepting a self-imposed framework of theory, objectives, and techniques. This acceptance is what Kuhn means by tradition-bound. Scientists are using theory, not challenging it.

(2) Appearance of discrepancies with the paradigm. These are observations or experimental results at odds with the propounded theory rather than expanding or amplifying it. There may be no immediate attempt to change the theory in order to accommodate these discrepancies.

(3) A revolutionary period, short compared to the period of normal science, during which discrepancies are resolved in a new theory. The new theory is generated from outside the established practitioners of the normal science rather than logically developed from within the group.

(4) Solidification of the new theory into a new paradigm, which gathers new adherents. The new group of scientists then proceeds to conduct another period of normal science. Research under the old paradigm may continue – but at a reduced level of activity – and it eventually ceases.

Kuhn's theory has had a major impact upon philosophers of science, social scientists and upon scientists themselves. For most scientists the description of researchers working and communicating in groups with shared objectives, ideals, techniques, and procedures rings true. We form professional societies or establish working groups that act as a focus for discussion. What was startling, when Kuhn introduced his theory, was the view that major scientific advance, the revolution, did *not* follow from careful, piece by piece, investigation but was the resolution of a mounting series of previously ignored discrepancies within the paradigm.

Ecosystem is one of the most important integrative concepts in ecology. After being introduced by Tansley (1935), the *ecosystem* concept became a major focus for research. In writing its history, Golley (1993) described ecosystem research as a paradigm and I want to examine whether ecosystem research fits Kuhn's description of a paradigm and the significance of that for developing coherent scientific explanations.

11.2.1 The ecosystem revolution

SCIENTIFIC REVOLUTION

Scientific revolutions have three characteristics:

(1) Previous awareness of anomalies within a paradigm.
(2) Gradual, and simultaneous, recognition of observational and conceptual problems.
(3) Development of a new paradigm, often with resistance from scientists of the prior paradigm.

Kuhn suggests a psychological basis for the advance of science through revolutions: "novelty emerges only with difficulty". Resistance guarantees change, so that when change does occur it is fundamental. People seek regularities, even when there are none, making them cling dogmatically to their expectations and constraining their ideas within a given paradigm.

Tansley (1935) proposes the *ecosystem* concept as counter to the *complex organism* concept (Clements 1905). The analog Clements had used was that vegetation "is an organic unit exhibiting activities that result in development, structure and reproduction" and that changes are progressive, periodic and rhythmic. "According to this point of view, the formation is a complex organism, which possesses functions and structure, and passes through a cycle of development similar to that of the plant" (Clements 1905). Vegetation scientists were interested in succession and progress towards climax vegetation; when the climax was reached, the community had stabilized its environment. Tansley's reaction was more to Phillips' (1934, 1935) support of Clements' ideas than to Clements himself. Phillips (1934) suggests "succession is due to biotic reactions only, and is always progressive . . . succession being developmental in nature, the process must and can be progressive only". It was the *complex organism* of vegetation that made this development take place.

Tansley rejects the idea of succession as inevitably towards a single climatic climax for each climatic region. He suggests that soil, physiography, and particular events, such as grazing and repeated fire, all influence vegetation so the "vegetation appears to be in equilibrium with all the effective factors present, including of course the climatic factors . . ." This led to Tansley describing vegetation as a system and particularly an *ecosystem*. For Tansley (1935):

"The whole complex of organisms in an ecological unit may be called the *biome*."

And

"The fundamental concept appropriate to the biome considered together with all the effective inorganic factors of its environment is the *ecosystem*, which is a particular category among the physical systems that make up the universe. In an ecosystem the organisms and the inorganic factors alike are *components* which are in relatively stable dynamic equilibrium. Succession and development are instances of the universal processes tending towards the creation of such equilibrated systems."

The anomaly observed by Tansley was that the climatic climax did not fit European vegetation. He does not reject the idea that vegetation modified its environment and considers that the analogy with the organism was useful if not pushed too far and he prefers the term *quasi-organism*. Tansley's ecological research expanded on vegetation classification (Tansley 1911) to description of *The British Isles and their Vegetation* (Tansley 1939) in which ecosystem is an integrating concept. *The British Isles and their Vegetation* contains many descriptions and diagrammatic representations of interacting climate, soil, animal, and other influences controlling vegetation types. He did not research detailed application of the concept to one particular vegetation type; that was not his research objective. He continued to use concepts of vegetation science such as association, sere, and climax to delineate particular types or conditions. Tansley (1935) is critical of attempts to widen the definition of *climatic climax* to accommodate observed variation and yet keep the organismal theory of vegetation intact, i.e., to shore up the existing paradigm. "I plead for empirical method and terminology in all work on vegetation and avoidance of generalized interpretation of what *must* happen because 'vegetation is an organism'" (Tansley 1935).

Golley (1993) describes development of the ecosystem concept by Tansley in social terms as an attempt to give ecology a more rigorous basis, particularly by presenting a physical theory founded on equilibrium concepts, i.e., that vegetation would come into equilibrium with all factors influencing it. Golley suggests that, prior to this, ecological concepts were inadequate and that the community function aspects of ecology had struggled for existence with plant and animal biology. However, Tansley was a long-established member of practicing vegetation scientists so this was not an externally fomented revolution.

According to Golley (1993), Lindeman (1942) was the first to implement Tansley's concept explicitly in "a quantitative effort to define the system and describe and understand its quantitative behavior". It was this work and the use made of it by Hutchinson and others (Hagen 1992, Golley 1993) that established a paradigm. Much of Lindeman's study of a small lake, Cedar Bog Lake, Minnesota, was similar to work by others, who had described lake metabolism and energy flow and suggested that the biota could be characterized as a network of organisms with groups of organisms linked by feeding. However, Lindeman went further than previous studies in linking the living and non-living parts of the system. He determined the food consumed by species using observation, experiment, and the literature, and crucially, organized species into food groups. His investigations focused on dominant species, and extrapolated from them to rare species on the assumption that they behaved similarly. He used published data of other scientists to convert biomass to energy and calculated rates of energy entering and leaving primary producers, primary consumers, and secondary consumers, and formulated these transfers as simple differential equations. Productivity was corrected to account for losses from metabolism, predation, and decomposition. Lindeman found the efficiency of energy transfer increased from producers to tertiary consumers and that the more remote the level was from the sources of energy the less dependent it was on one source.

Lindeman's was not a straightforward application of Tansley's concept of an ecosystem, nor could it have been because Tansley's definition did not specify a framework for how systems were organized, just that both organism–organism and organism–environment relations must be considered. Lindeman developed the *ecosystem* concept in a particular way. He gave its name to an extended version of the detailed analysis he and other lake ecologists had been doing over a period of years. Because he made a detailed study and attempted to measure, rather than just describe, the components of a particular ecosystem, Lindeman had to make important assumptions (axioms), e.g., exactly which organisms should be included in each trophic level. What he provided was a *model*, both conceptual and mathematical, of how a lake functions (Fig. 11.2) in terms of energetic and chemical interactions between trophic levels and the environment.

Initially Lindeman's paper was rejected by *Ecology* (Hagen 1992, Golley 1993) and his work considered as too theoretical, without sufficient empirical studies as a basis – he presented a generalized scheme, although his own detailed research was based on only one lake. His mathematical approach was derided by field biologists (Hagen 1992). The resistance, some of it vehement, is to be expected from Kuhn's theory. The major concerns were that: (1) a theory of ecosystem function could not be developed from the intensive study of just one lake, and (2) transfer between trophic levels could not be described mathematically. Similar criticisms of the energy flow and nutrient cycling aspects of ecosystem theory were still present 30 years later. The challenge that the objectors felt was as much about method, objectives, and the ideals of science in ecology, particularly about how general theories could be built from particular studies, as about the details of Lindeman's description of the ecology of a lake.

11.2.2 The progress of normal science

In Kuhn's theory " 'normal science' means research firmly based upon one or more past scientific achievements" (Kuhn 1970) and takes place within the paradigm. A *scientific revolution* produces some achievements that are

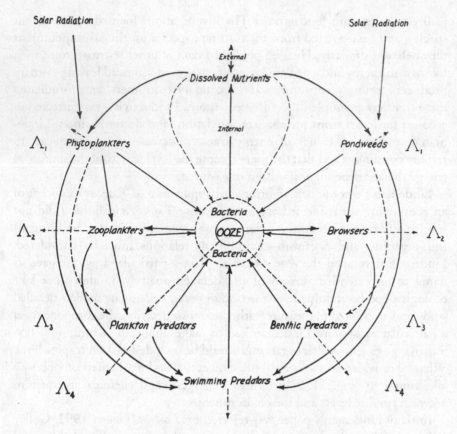

Fig. 11.2. Lindeman's (1942) diagrammatic model of ''generalized lacustrine food-cycle relationships''. The terms Λ indicate energy contents of the food cycle levels. Each level receives energy from the previous level and dissipates it through respiration, predation, and decomposition of dead material so that the change in energy over time is,

$$\frac{d\Lambda_n}{dt} = \lambda_n + \lambda'_n$$

Where λ_n is positive and represents the contribution of energy from Λ_{n-1}, the previous level, and λ' is negative and represents energy dissipation from Λ_n. Lindeman calculated values for terms he called λ_0 the rate of incident solar radiation, λ_1 rate of photosynthetic production, λ_2 rate of herbivorous consumption, λ_3 rate of secondary consumption, λ_4 rate of tertiary consumption. (From Lindeman 1942, with permission.)

sufficiently new and innovative to attract a group of adherents away from other scientific activity. At the same time this original achievement opens up many problems for a new group of practitioners. So the revolution produces a significant, but not complete, advance. Kuhn places particular emphasis on the instruction of the techniques of a paradigm to students through the use of exemplars.

EXEMPLAR

"The concrete problem-solutions that students encounter from the start of their scientific education whether in laboratories, on examinations, or at the ends of chapters in science texts" (Kuhn 1970). To this can be added the technical problems, and their solutions, that students examine in the scientific literature and laboratory during their research training.

Lindeman's work provided an *exemplar* for ecosystem science – how the concept should be applied to a particular ecosystem. Two essential over-arching axioms were:

(1) Plants and animals can be grouped into trophic groups ordered as trophic levels.
(2) Energy from the environment flowing through these levels can be calculated.

It was this focusing of the definition of *ecosystem* onto *trophic levels* and *energy flux* between them that provided an exemplar for similar studies. In each study of a new ecosystem the challenge for individual scientists was to develop functional concepts that defined trophic levels and develop measurements and methods of calculating energy flux and/or nutrient dynamics. By adopting this exemplar as the foundation of its methodology, ecosystem science became rationalist: there *is* an acceptable theory of trophic organization and it *can* be applied to an ecosystem.

The postulates of ecosystem science were at the level of the whole theory and of the type:

> There are differences between ecosystems in the ratios of energy transfer between trophic levels.

While the intent was to make comparisons between biomes, these were to be made *assuming* the axioms of trophic levels. Initially at least, there were no postulates about what might control differences between ecosystems.

Providing an exemplar was an important development. Scientists basing research clearly on Lindeman's exemplar could analyze ecosytems in a particular way. In the 1950s and 1960s: "Most [ecosystem] studies essentially repeated Lindeman's study of Cedar Bog Lake, in that they aggregated the diverse fauna and flora into a small set of trophic groups, determined flows of energy and materials between groups, and calculated ratios of input to output" (Golley 1993). In this sense, ecosystem research became what Kuhn calls "tradition bound". Kuhn describes why he considers working in a paradigm is successful "the paradigm forces scientists to investigate some part of nature in a detail and depth that would otherwise be unimaginable". Generally the logic of investigation within a paradigm is inductive, seeking to extend the application of a defined theory. But of course it is restricted in the questions asked and types of explanation that can be provided.

Lindeman's exemplar gave a narrower focus to ecosystem studies than did Tansley's. Tansley sought answers to varied questions about the past and future development of different vegetation types. His descriptive explanations were certainly not restricted to trophic levels and energy transfer. This seems not to have been recognized, or to have been of no concern, when Lindeman's exemplar was used in conjunction with Tansley's definition. In particular, historical features that may have influenced vegetation development, used by Tansley in his application of the ecosystem concept, were not included, nor were the types of organizational as opposed to energetic relationships that, for instance, grazers or fire may have (Tansley 1939). Perhaps these components were not felt to be needed because much early ecosystem research was conducted in other aquatic systems (e.g., a coral reef, Odum and Odum 1955; springs, Odum 1957, Teal 1957).

In addition to concentration of research effort provided by the exemplar, two other events gave ecosystem research important characteristics of a paradigm: a textbook written by Odum (1953), and the coalescence of a distinct group of scientists working within the ecosystem paradigm made possible by research funding programs.

Golley (1993) describes Odum's textbook as a marked change from previous ones that mentioned the ecosystem approach but still focused on population biology. A textbook is an important indicator of a paradigm (Kuhn 1970). Golley considers that after this text appeared "The subject was accessible and made sense to the educated layperson, yet its language and concepts reflected some of the most advanced trends of the 1950s", e.g., the languages of engineering and economics and the subjects of cybernetics and information.

The social component of Kuhn's definition of paradigm is particularly important. Social groupings of scientists share professional communication through journals and special meetings of professional societies. There is a common purpose driven by ideals, and group members will be unanimous in many of their professional judgements. Whatever their differences they have the common ideal that one particular type of investigation will provide advances unattainable in any other way. Common ideals give the impetus for pursuing particular research going beyond reasoning from the empirical content of existing science. Group members will definitely not study some things, or adopt some techniques, so excluding some scientists. Individuals may seek and approve a particular bias or emphasis to remain part of the group.

In the USA there was large-scale funding for ecosystem research through the US Atomic Energy Commission (AEC) in the 1950s and then the National Science Foundation administered International Biological Programme (IBP) in the 1960s. These funding programs were important and it was US ecosystem science, rather than that in other countries, that came closest to being a paradigm in the sense described by Kuhn. In Britain, for example, although there was research on productivity and its role in community development (Ovington 1961, Pearsall 1964) and on total soil processes (Macfadyen 1963), these were not efforts of large teams.

The AEC visualized direct applications of ecosystem theory to the applied problem of transport and fate of radioactive materials produced by weapons testing and the response of biological systems to ionizing radiation (Golley 1993). It provided relatively large funding to research groups that became permanent organizations at some US national laboratories and nuclear fuel and energy production sites. "The relation between ecosystem research and the military activity of the United States was never obvious, and ecologists seemed oblivious to the connection" (Golley 1993).

During the early 1960s the IBP was organized as an international, ecologically oriented research effort. The ecosystem concept was intended to unify the program. The major objective was comparison, between biomes of primary production, trophic structure, energy flow, limiting factors, and interactions regulating and controlling ecosystem structure and function. Systems analysis techniques were to be used as a technique for synthesis. Countries produced different programs. In the USA biome projects were established in grasslands, tundra, desert, coniferous forest, and deciduous forest, each receiving considerable funding and enabling the establishment of teams of substantial numbers of ecologists.

Golley (1993) analyzes the development of the grassland biome, the largest one in the USA. Hagen (1992) considers that this biome remained constant to the ideal of constructing a total ecosystem simulation model of productivity, energy relations, and nutrient cycling. The objectives were to define ecosystem components, measure the quantities of energy flow and nutrient cycling through the components, and to put these components into a model. This was the same basic logic as the exemplar provided by Lindeman, only now the systems being studied were considered sufficiently complex to require teams of specialists to make measurements. This was Big Science (Hagen 1992).

Golley describes the ELM model of the grassland biome that used trophic levels as its key feature as "structurally conservative". The real problem was that this structure was assumed from a particular theory and applied, in this case to grasslands, without sufficient exploratory analysis. It was a rationalist approach with the ideal – there **is** an acceptable theory of trophic organization and it **can** be applied to an ecosystem. It was perhaps fostered partly by the ideals of the biome leader, George Van Dyne, who had previously worked at an AEC laboratory. A central goal also helped to define and control the development of what was, for ecology, an expensive program that had to be conducted within a limited period of time and a defined budget. Each scientist investigated some component of the ecosystem, but they all had to fit their work to the central theory by providing information for the model. The two components of the paradigm, a social structure requiring a common ideal and the use of the exemplar, constrained the research.

11.2.3   Did a revolution terminate the paradigm?

Half way through the Grassland Biome program there was a social and academic revolution. Van Dyne was supplanted as leader and the modeling effort switched to developing different models for different types of grassland (tallgrass prairie, mixed prairie, etc.), and to developing models of component processes (abiotic, producer, root turnover, etc.), all partitioning reductions of the complete system. Three related difficulties could no longer be suppressed.

First, it had been assumed that a single grassland ecosystem could be defined for the biome. Golley (1993) refers to this as a spatial problem, i.e., that there was substantial heterogeneity in ecosystem function. It was also a methodological failure of attempting to apply a particular model without conducting exploratory investigations into the problems of *domain* (Chapter 10) – that is, of attempting to use a supposedly general theory without first asking exactly how relevant it is.

Second, the number of senior scientists involved, many with their own ecological specialization, made it difficult to sustain a uniform categorization of the grassland ecosystem into trophic levels. Methodologically the attempt to use a previously defined *trophic-level* concept was not accepted by these scientists. Furthermore, making a computer simulation model of the ecosystem required that the calculation of energy flows through the system should balance. This in turn focused on accurate determination of components, which inevitably led to questioning of the concept by measurement for trophic level, i.e., asking for increased measurement accuracy of what was really an inadequately defined integrating concept for that ecosystem.

Third, there was a conceptual problem with the systems theory of the time and how it could be applied in ecological research. Systems theory was introduced from engineering as a technique for synthesis and with the hope that the type of model-based explanation that Lindeman had developed could be extended to more complex systems. Watt (1966) describes the application.

> "In short we are describing a system in which everything affects everything else, and the complexity of the system of interlocking cause–effect pathways confronts us with a superficially baffling problem in scientific analysis. It is precisely this interlocking feature which is the most characteristic identifying aspect of a system. Indeed, for ecologists, a suitable operational definition of a system is 'An interlocking complex of processes characterized by many reciprocal cause–effect pathways'. Note that a principal attribute of a system is that we can only understand it by viewing it as a whole."

Watt (1966) describes two standards of the approach. First, "that extremely complex processes can most easily be dissected into a large number of very simple unit components rather than a small number of relatively complex units". Despite the general objective of understanding whole systems this is actually an assertion of reductionism, even though relationships between all the pieces will be considered. Second,

> "that complex historical processes in which all variables change with time (evolve) can be dealt with most straightforwardly in terms of recurrence

formulae that express the state of the system at time $t + 1$ as a function of the system at time t. Thus we try to understand the process, not in terms of its entire history but rather in terms of the cause–effect relationships that operate through a typical time interval."

This is an arbitrarily restrictive approach – at least as far as what is allowed in terms of causal influences, i.e., only those that can be counted in the short term, and, by using the word "straightforwardly" it is asserted, not justified, that this will be effective.

Technically the development of larger, faster digital computers permitted larger models to be run that dissected the system into many small components and simulated longer periods of time. Modeling in the AEC studies used analog computers (Olson 1964), where each component had to be wired in and so it was difficult make complex models. Modeling in the IBP used digital computers and software with a much greater capacity to express many components. This was used to achieve a greater reduction though not necessarily an effective synthesis.

The assertion that "everything affects everything else" made an impossible task for the modelers – when really identifying relative importance of different effects (functional concepts) under different conditions (domains) may have been a more appropriate strategy. Assumption of the dominance of short-term effects, and selection of units such as "grassland" without looking at its variation, were counter to Tansley's description and use of *ecosystem*. With these assumptions the attempt was made to use natural and functional concepts without developing and assessing integrative concepts such as *trophic level* and *ecosystem*. Wiegert and Owen (1971) describe the problem for model builders:

> "There is more than a mere semantic difficulty with the terminology employed in the Lindeman model. Producer, consumer and decomposer are ambiguous words that do not clarify the significantly different roles played by organisms in the dynamics of ecosystems."

In practice, *ecosystem* started to be used in a different way, to represent any set of interacting organisms and processes, e.g., *below-ground ecosystems* (Marshall 1977). This was a different type of use from that applied by Tansley and reflected the requirement for exploratory research into components, although conducted under the umbrella of biome-type investigations.

Kuhn's theory of scientific progress describes mounting discrepancies within a paradigm resolved through a new revolution (Fig. 11.1). There were certainly mounting discrepancies in the ecosystem paradigm, but there was not a revolution in the complete sense that Kuhn (1970) describes for the following reasons:

(1) Socially the paradigm did not break down. Funding for ecosystem

studies continued through the Ecosystems Program, separate from that for population biology and other aspects of ecological research. The Long Term Ecological Research (LTER) program continued funding research at a number of the biome sites and has added others.

(2) Ecosystem research did not decline but changed its explanatory objectives. The task became exploratory to define ecosystem components, rather than concentrate upon achieving a complete energetic and trophic synthesis for single ecosystems. For example, in concluding a synthesis volume about below-ground ecosystems, Marshall (1977) comments about modeling "As an organizing tool it is useful in ecosystem studies, but little is achieved without a basic understanding of the biology and ecology of the constituent organisms. That basic understanding has some way to go in the belowground ecosystem . . . .". Golley (1993) expresses the concern that, by 1993, many people used the *ecosystem* concept but few were doing whole-system ecosystem research.

(3) Whole-system studies were successful in some instances. One in particular was described by both Golley (1993) and Hagen (1992) as a continuing success: the work at Hubbard Brook of Bormann and Likens (1979). There, rather than asking the general question "How does it work?" and trying to describe that in terms of energy or biomass, researchers asked specific questions. For example, "What are the effects of deforestation on nutrient loss from the catchment?" and "What are the consequences of acidic deposition on nutrient dynamics?" These in turn were then translated into why-type questions by using contrasts. Detailed comparison of nutrient input and output of harvested and not harvested areas (synchronic contrast) were made. Examination of a time series during a change in acidity of the incoming precipitation (diachronic contrast) attempted to answer why acidity affected ecosystems. Moreover, because the Hubbard Brook ecosystem could be effectively delineated as a discrete watershed for the particular investigations, the question "What is an ecosystem?" did not force itself upon the researchers in the way it had done in the grassland biome. Most importantly, while the studies used part of ecosystem theory, they did not use *trophic level*, and so avoided the difficult problem of defining and measuring its components. So, although Golley (1993) claims Hubbard Brook studies as ecosystem research, they did not follow that paradigm in theory, methods, or social grouping. (a) They focused on nutrients and water, not energy, as Lindeman's exemplar had. (b) They used contrasts as the foundations of their methods. (c) The Hubbard Brook scientists were not part of the social group funded by the AEC and IBP. In fact, they explicitly kept apart from the IBP, and Bormann is reported (Golley 1993) as describing intense debate among the

Hubbard Brook scientists about future research rather than accepting an exemplar.

11.2.4 How useful is Kuhn's theory for understanding research strategy?

It is hard to overestimate the impact of Kuhn's theory on scientists, social scientists, and philosophers of science. Some scientists claim their discoveries are revolutions or establish new paradigms, and use Kuhn's language – although not necessarily with his meaning. Social scientists have developed substantial critiques of the whole scientific enterprise that are based upon the relativism Kuhn's description implies (Pinch 1990). However, studies of social organization in science show there to be considerable flexibility in both the structure of interacting groups and how tightly knit they are (Crane 1972).

Among the detailed critiques of Kuhn's thesis developed by philosophers of science, those of Shapere (1971) and Putnam (1981) helped to refine the ideas. Shapere developed two lines of criticism – the definition of *paradigm* and the *relativism* that Kuhn's theory implies – suggesting that is not how science really works, that scientists do not limit their ideas and methods nor restrict themselves into tight groups. Taken to its logical extreme and Shapere makes the point that this is what Kuhn does, there is no logical way that judgements can be made between paradigms. Kuhn's emphasis is on the sociological aspects that drive change.

Putnam (1981) also criticizes what he terms Kuhn's extreme relativism, noting that:

> "The whole purpose of relativism, its very defining characteristic is, however, to deny the existence of any intelligible notion of objective 'fit'. Thus the relativist cannot understand talk about truth in terms of objective justification conditions."

The development of ecosystem theory in the USA illustrates interplay between social factors and the objectives and methods of research. During the 1950s and 1960s ecosystem research funded by the AEC and US IBP took on the appearance of the Kuhnian paradigm. AEC funding, continuously given to a restricted group, encouraged that group to grow. What is particularly interesting, and most important methodologically, is the strongly rationalist stance this group took (they had a well-defined theory based on research using Lindeman's exemplar) and how they developed a relativist position (they formed a distinct group communicating among themselves rather than with outsiders). As Golley (1993), a participant at the time, notes

> "the concepts were not presented in the conventional form. Rather, they were derived from authority figures who frequently were the professors or the key

investigators. The ideas were often presented in authoritative language and, most important, as principles in the textbooks used to train the next generation."

In this sense Kuhn's description of a restricted social group conducting normal science within a paradigm and based on an exemplar with only limited development is correct for US ecosystem science as it developed through to the late 1960s.

The collapse of the exemplar, based as it was on a small lake, was due to major difficulties in defining trophic levels, establishing the importance of particular organism–organism or environment–organism interactions, and most important, defining exactly what should be considered as a functioning ecosystem. The synthesis fell apart when it was asked to provide scientific understanding for more complex questions. It was not a coherent scientific explanation from which scientific inference could be made.

Attempts to save the exemplar by revising the energy flow pathways, in particular ones that described more complex interconnections between trophic groups (Wiegert and Owen 1971), did not really tackle the problems of definition and measurement of such groups. The data could not be forced into the restricted theory and concepts needed to be developed.

The failure of the exemplar can not really be called a second scientific revolution. Certainly scientists from outside of the paradigm and with a strong natural history background suggested that there could be more to ecological control systems than energy and nutrients (Southern 1970). But scientists working *within* the paradigm also recognized the failure and the reasons for it. The most important problem, and one that Kuhn's theory helps to interpret, is the connection between adopting a strong rationalist approach and the ability to attract funding. This is discussed further in Section III.

Rapid change certainly occurs in science. But, for ecology, Kuhn's emphasis on changes being stimulated by unresolved anomalies that are solved from outside the field certainly does not fit all cases. Rapid change can also occur through the introduction of new equipment and techniques. For example, the advent of radioactive tracers provided new insights. Radiotelemetry, having undergone advances as microelectronics has advanced, is now becoming routine, and is having a major influence on animal ecology.

11.2.5 Scientific inference and the ecosystem paradigm

Some things about scientific inference can be learned from Kuhn's theory. When science is conducted within a paradigm the types of question that can be asked are restricted because only certain types of answer are considered to be suitable. The progressive development of new syntheses is inhibited – and as a consequence so too is scientific inference.

In the development of ecosystem science, it was considered important to take an approach that should allow conclusions about whole systems. Initially the approach was successful in producing functional classes. Macfadyen (1964) describes differences in the amounts and particularly ratios of standing amount of energy to productivity and roles of decomposers and herbivores in examples of aquatic, grassland, and forest ecosystems. The differences are large (Fig. 11.3). Standing energy of primary producer (stock) is greater in the grassland than the marine plankton: little material goes to decomposers from marine plankton, although a substantial amount goes from grassland. There are also differences in the herbivore systems: respiration is substantial for beef but negligible in zooplankton, while no beef goes to decomposers but substantial zooplankton does. These may seem obvious generalizations now but at that time this was an important synthesis. To make this synthesis Macfadyen (1964) used a number of scientists' research and the terrestrial systems he used are comparatively simple.

Subsequent research, particularly as a result of the worldwide IBP, revealed differences in, e.g., productivity and decomposition within biomes. This eventually led to contrastive analysis, but not until sufficient examples had accumulated within biomes that such questions could be asked as "Why is this piece of tundra (grassland/forest, etc.) more productive than that piece?" Heal and Grime (1991) suggest three types of approach:

(1) *Comparative analysis of ecosystems* requiring isolation of the components of different systems and comparing how systems may differ.
(2) *Cross-system comparison of components*, which may look at the presence, absence, or some measure of function of a component within a particular type of system but across different examples of it – say across some environmental gradient.
(3) *Cross-system comparison of ecosystems*, which first depends on a synthesis within each system. Heal and Grime (1991) suggest that a functional classification is required and, further, that there is a real prospect of developing this, e.g., for plants (Grime 1979).

The point for method is that once the cross-system comparison of ecosystems is faced with explaining differences within rather than between biomes, then there is need for development of the concepts used in constructing the initial investigations. Some may even need to be replaced.

Working within a close paradigm does not foster rigorous assessment of fundamental assumptions – and that can retard the development of theory. Claiming that a piece of research has established a new paradigm implies commitment to a theory or method rather than an emphasis on dispassionate

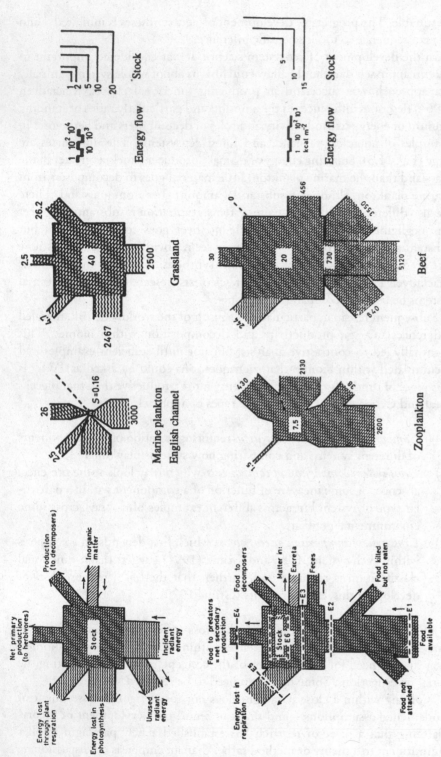

Fig. 11.3. Representative examples from Macfadyen's (1964) comparison of energy flow through plant (upper line) and herbivore (lower line) systems. On each line the left-hand diagram gives the general scheme, E indicates typical estimation points in herbivore systems. The units are 106 calories ha⁻¹. The flow scales are logarithmic and the stock (S) scales are proportional in area to the logarithm of calorific value. (Redrawn from Macfadyen 1964, with permission.)

assessment. Such claims are made more for social than for academic reasons (Section III).

11.3   The methodology of scientific research programs

Kuhn's (1970) theory was not accepted completely by philosophers of science. However, it intensified debate about how science does and should work by extending discussion to include social interactions among scientists as well as the particular logic of any research investigation. Lakatos (1970) draws sharp distinction between Popper and Kuhn's philosophies. For Popper, intellectual honesty in science requires specifying precise conditions under which you give up a position. *Belief* may be regrettably unavoidable, but *commitment* of the type attributable to those working within a paradigm is a disaster. Science should be perpetual revolution of bold postulates and austere refutations with *criticism* as the essential process. For Kuhn, it is only after a transition from criticism to commitment that progress and normal science begin and criticism of the dominant theory is a rare revolutionary process. There is a psychology, but not logic, of discovery.

Lakatos (1970) stresses the importance of this debate. He suggests that Kuhn was correct that science does not progress by *falsification*. The position that science grows by repeated overthrow of theories with the help of hard facts is *dogmatic* or *naive falsification* and rests on two false assumptions.

First, to say "this fact" falsifies "that theory" then a borderline must be established between theory and fact. Difficulty in doing that is discussed in Chapter 3, particularly that there is no such thing as *pure fact*. Scientists make many choices in constructing just one data statement for a postulate, and measurements have difficulties of effectiveness, accuracy, and precision (Chapter 6). All observations are made in the context of some question, however general that question may be. Establishing borderlines is a device within a theory and not absolute truth.

Second, propositions do not stand alone. They can only be derived from other propositions, and this context determines the type of data to be sought and frequently the methods used.

11.3.1   A strategy for continuous assessment

Lakatos proposed that tests of theory are not two-cornered contests between theory and investigation but at least three-cornered contests among rival theories and investigation. Furthermore some of the most interesting investigations result in confirmation rather than falsification. He proposed a scientific theory, T_1, is falsified if, and only if, another theory, T_2 has the following characteristics:

(1) T_2 predicts novel facts that are improbable for, even forbidden by, T_1.
(2) T_2 explains the previous success of T_1; that is, all the parts of T_1 that have not been refuted are contained within T_2.
(3)   Some of the information in T_2 exceeding that in T_1 has been corroborated.

The successive results of this method of investigation are a series of *problem-shifts*. For a series of theories $T_1, T_2, T_3, \ldots$ each has at least as much content as the unrefuted content of the previous one and differs from it by some additional proposition(s). A new theory is *theoretically progressive* if it predicts some novel, hitherto unexplained fact, and *empirically progressive* if that fact is corroborated. A theory in the series is falsified when superceded by a theory with higher corroborated content (it is more coherent). For example, succession theory for terrestrial vegetation no longer occupies the position it did before the advent of disturbance theory – but disturbance theory also permits the ideas of succession between disturbances. Lakatos distinguishes a *degenerating* problemshift in which a new theory offers only a content-decreasing *reinterpretation* of an observed contradiction, that the contradiction is resolved only in a semantic, unscientific way.

In Lakatos' scheme the idea of growth of theory and the requirement for empirical test are joined. The negative character of *naive falsificationism* is removed when science is viewed in this way. Certainly criticism and the process of refutation becomes more difficult because it must take account of alternative theories relative to some test, but it may be positive and constructive and not just provide a yes/no judgement. Successive theories are connected: they have continuity in their content. Lakatos says that falsification has historical character, replacing Kuhn's concept of a theory as static during periods of *normal science*.

11.3.2   The components of a scientific research program

Lakatos described this development as a *methodology of scientific research programs* (Fig. 11.4). Two important principles of the methodology are that scientists maintain a *hard core* of knowledge and use a *positive heuristic*.

HARD CORE KNOWLEDGE

Hard core knowledge is a collection of theoretical knowledge to be used and explored over time but not generally challenged.

An *over-arching axiom* might provide an important part of hard core knowledge. It is unlikely to be challenged as the result of a single investigation. The hard core also contains the knowledge used in specifying why some quantities are important. However, there is a subjective element about holding a hard

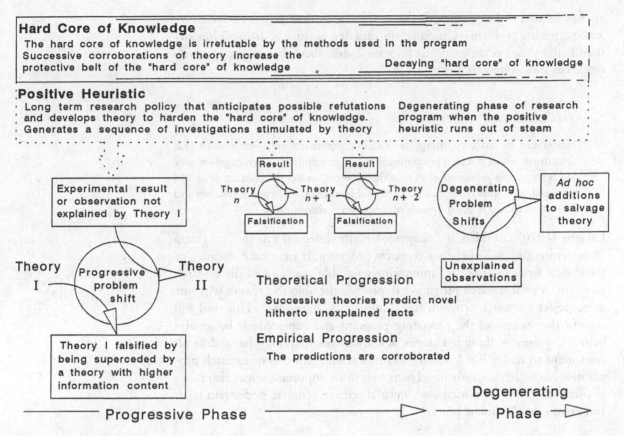

Fig. 11.4. Lakatos' theory. Science within a research program progresses by continuous replacement of one theory by another. Repeated investigations using the positive heuristic drive the process.

core: it is knowledge that a group of scientists favors because they investigate it and consider its associated questions important. There will be much discussion among scientists about how this hard core will be allowed to grow – particularly what will be allowed to join it – and this distinguishes it from the knowledge of a *paradigm*.

POSITIVE HEURISTIC

A heuristic is an art of discovery in logic, so a *positive heuristic* is an indicator of what to look for in a program of scientific research and how to find it. It may include a way of doing particular types of experiment, or constructing models of a particular type requiring particular types of data.

A science allowing development of the theory has heuristic power. This development illustrates, or at least gives encouragement, that some problems yet to be solved can at least be tackled, even though the precise content of the solution may not be predicted. Lakatos suggested that in most cases theories are not refuted – they decline due to a decrease in their power to solve new and interesting problems. Crucially, the idea of a positive heuristic is broader than Kuhn's idea of an exemplar because it describes what might be investigated as well as how to investigate it.

PROBLEMSHIFT

There can be no such things as crucial experiments or investigations that overthrow whole research programs. What an experiment or investigation may do is to cause a *problemshift*. A new theory incorporates the content of an old one and adds something new. Use of problemshift rather than *revolution* and *paradigm change* is based upon both criticism and continuity.

Lakatos (1970) considers it "dangerous methodological cruelty" to give a stern refutation to a fledgling research program. It may take decades of theoretical work to arrive at "interestingly testable" versions of the research program. A rival research program can provide the objective reason why one might reject a research program and eliminate its "hard core". This rival will explain the success of the preceding program and supercede it by greater heuristic power – though Lakatos acknowledges that may be a difficult assessment to make. For Lakatos: "Natural science consists of research programmes in which not only novel facts but, in an important sense, also novel auxiliary theories, are anticipated: natural science – unlike pedestrian trial-and-error – has 'heuristic power'".

11.3.3 Top-down and bottom-up forces in population and community ecology

In 1960 Hairston, Smith, and Slobodkin (HSS) proposed a theory about the interactions among community structure, population control, and competition. They argued from basic ecological theory and general observations, rather than from specific research results, that "the world is green" because herbivores are kept in check by their predators and so prevented from consuming all of the plants. The HSS theory was developed into the theory of "top-down" control. The alternative is "bottom-up" control, specifying that growth resource limitation controls plants, that plant resources limit herbivores, and herbivore resources limit predators.

Over the succeeding decades the top-down theory, and its competitors and successors, have motivated theoretical and empirical studies. This progress can usefully be compared to the methodology of scientific research programs proposed by Lakatos (1970) to examine how a theory can progress through its problemshifts, and developments in its hard core and positive heuristic, as well as how criticism develops both from within and outside of the program.

First specification and criticisms of the theory of top-down control

The HSS theory can be represented as a series of axioms and postulates making the first description of top-down control as an integrative concept.

Axiom 1:   All energy fixed in photosynthesis is dissipated through the biosphere.

The evidence discussed was that there is no worldwide accumulation of organic matter other than in local environments, e.g., peat bogs. From this HSS deduced Axiom 2.

Axiom 2a: All organisms, taken together, are limited by the energy fixed in the biosphere.

Axiom 2b: Decomposers, taken as a whole group, are food limited since they depend upon organic debris.

Deduction: Any population not resource limited must be limited to a lesser level than that set by resources.

Axiom 3a: Generally, terrestrial green plants are not depleted by herbivores or destroyed by meteorological catastrophes.

Axiom 3b: Occasionally, when herbivores are protected from their predators (HSS included parasites), then green plants are depleted.

Deduction: Generally herbivores are not limited by their food supply.

Postulate 1a: Predators (and/or parasites) control herbivore populations.

Postulate 1b: Herbivore populations are not controlled by the weather.

In addition to the evidence from population explosions of herbivores following release from their predators HSS considered that the runaway success of introduced species supported this postulate.

Postulate 2a: As a group, predators must be food limited and so density dependent in the control of their numbers.

Postulate 2b: Predators limit their own resources, the herbivores.

HSS acknowledge that territorial mechanisms may control densities within species but argued that this did not apply to the total predators of a system.

Postulate 3:   The community in which herbivores are held down in numbers, and in which producers are resource limited, will be the most persistent.

HSS argue that communities where producers were consistently reduced to be sparse would be vulnerable to invasions by other species.

> Postulate 4:   Interspecific competition for resources exists among producers, carnivores, and decomposers.

Murdoch (1966) criticizes this theory: (1) for its content, (2) the method of argument used in its construction, and (3) that it could not be tested due to both inadequate concept definitions and lack of possible tests.

(1)   *Content.* Murdoch suggests that not all green plant material might be edible and that there had been continuous evolution of plant defense mechanisms against herbivory. So for Murdoch, HSS Axiom 2a is incorrect and arguments based on energy flow through trophic levels are suspect. Murdoch also gives counter-examples to Postulate 1a. Organisms may be food limited without depleting their food supply, e.g., severe depletion may require high foraging efficiency and/or maintenance of herbivore populations up to the point of plant depletion. Not all herbivores are capable of depleting green matter, e.g., some herbivores are sapsuckers or fruit eaters, and Murdoch also notes that animals might be limited by requirements such as nesting sites and food quality rather than quantity.

(2)   *Method of argument.* Murdoch suggests inconsistencies in the HSS argument; in particular that *food limitation* is used in different ways with regard to herbivores and carnivores. HSS argue that herbivores are not food limited because green plants are not depleted. But it is also true that herbivores are seldom eaten out by their predators – so the same conclusion could be made about predators themselves, that they are not food limited (the opposite of Postulate 2a) and this would minimize the role of competition between them.

(3)   *Methodological issues.* Murdoch is critical of the lack of precision in the HSS arguments – particularly inadequate concept definition – and that precision must be improved if postulates are to be tested. Murdoch also criticizes the construction of a theory that could not be falsified – referencing Popper – and because of this the HSS theory should be rejected. While imprecise concept definition is a reasonable charge against HSS, though perhaps to be expected at the early development of a theory, to apply the methods of naive falsification is less appropriate, what Lakatos referred to as dangerous methodological cruelty.

These criticisms by Murdoch were important for "top-down" research. In future those scientists involved in the research could anticipate the types of refutation that might be attempted. Murdoch does not present detailed empirical evidence, so by Lakatos' definition no Theory II (T_2) will replace a Theory I (T_1) to cause a problemshift. Indeed Murdoch specifically does not seek to disprove the conclusions of HSS. The outline of his argument against the HSS proposal is typical of the way many scientists treat the type of metaphysics presented by HSS. In their brief reply to him (on the final page of Murdoch 1966) HSS offer no counter-arguments, although saying they could, and comment: "It is clear that observation and experimentation, rather than argument, will eventually resolve the question". At this stage Postulate 1a could best be considered as potentially an over-arching postulate for a research program into top-down control. Neither a hard core knowledge nor a positive heuristic for a research program had been developed. Research into bottom-up control developed into its own program, particularly into the effects of plant quality limiting herbivores. For almost 20 years researchers in each program developed understanding of systems that appeared to follow their own processes.

Emergence of an empirical method

Work followed providing hard core knowledge and a heuristic for the theory of top-down control. Paradoxically the first example that became extensively quoted by self-declared researchers into top-down processes itself quoted neither HSS nor Murdoch (1966). It was an investigation into whether high species diversity was due to intense competition or reduced competition through predation.

Paine (1966) studied species diversity of marine organisms in rocky inter-tidal zones. He found shores with more species also had proportionally more predators. At an experimental site in Washington State, USA, the main predator, a starfish (*Pisaster ochraceus*), consumed herbivores (two species of chiton, two limpets, one bivalve, three acorn barnacles, and a species of *Mitella*) constantly opening up spaces on the rocks that could be recolonized. The effect of this predation upon primary producers, which for mussels and barnacles is phytoplankton, was not measured. Competition was for space, and when starfish were removed experimentally for a period of years, barnacles (*Balnus glandula*) first colonized the bare rock but were then crowded out by mussels (*Mytilus californianus*) and a goose-necked barnacle (*Mitella polymerus*). Predator removal led to a decrease in number of species. Paine (1974) gives a comprehensive account of this process based on 10 years' observation following predator removal and defined the concept of *keystone species* (Paine 1969) for a species such as *Piaster*. He postulates (Paine 1966) that "Local species diversity is directly related to the efficiency with which predators prevent the monopolization of the major environmental requisites by one species". Prediction of diversity gives a more specific postulate (there is now something that can be measured) than Postulate 1a of HSS, which mentions that herbivores are controlled by predators – but does not say how it, nor what, could be measured. HSS use *control* as a holding concept.

Paine's experimental removal of predators was not replicated, but a control area (no removal) was contrasted side-by-side with the starfish removal area. The force of the experiment was in the contrast it achieved made clear by detailed analysis of the process of change within the treated area – this is a perturbation experiment (Chapter 6). The experiment was *repeated* on a different shore in the same area (Paine 1974) and in New Zealand at a seashore with different species but similar trophic structure (Paine 1971).

While Paine described a basic process, Dayton (1971, 1975) started to define its domain. He discusses environmental factors affecting predator influence, e.g., varying disruption of the sessile animals by log damage along a gradient of wave exposure. Logs washed down from rivers drift along the coast and may then smash into the rocks during storms, opening gaps in the mussel and barnacle cover. Paine (1966) suggests how predation processes may interact with other factors, particularly that "increased stability of annual production may lead to an increased capacity for systems to support higher-level carnivores".

Paine's results, that diversity could be determined by a major predator, were confirmed by a large experiment where artificial fresh-water ponds were manipulated for both nutrients and fish predators (Hall *et al.* 1970).

The causes of animal species diversity were a major focus of research during the 1960s and 1970s, with different theories and syntheses presented and discussed. Menge and Sutherland (1976) suggest that there was a different emphasis around "the seemingly contradictory roles of competition and predation in the determination of community structure". They proposed a resolution around the HSS theory, noting that "a realization of the fundamental lesson of their paper seems peculiarly absent from the controversy considered here". This is perhaps an unfair comment given that Murdoch (1966) had advanced substantial criticism of HSS, much of it telling because it introduced counter examples. Menge and Sutherland (1976) introduce direct experimental results that confirmed important aspects of HSS for particular situations but also produced the first problemshift in the research program.

First, Menge and Sutherland (1976) make some concept definitions and are critical of other researchers' investigations of diversity. For example, they define *species diversity* as the number of species in a community and specify that comparisons may be made within habitats (e.g., comparisons between different rocky intertidal communities) or between habitat types (e.g., between rocky intertidal and kelp communities). They specifically restrict *community* to collections of interacting organisms of all trophic positions in a given habitat,[1] suggesting that terms such as bird communities or plant

---

[1] Note the difference from the definition used in the discussion of lowland tropical rainforest diversity in Chapter 10. In practice a definition must meet a purpose. Problems really arise when results made with different definitions are placed together without the difference in definitions being taken into account.

communities are misleading and designating such associations as "guilds," assemblages of species using a particular type of resource. So the integrative concept, *community*, was defined specifically for use with *top-down* control. This process of definition, restriction of what should be considered, and explicit use of contrasts, established a why-type question. In practice, investigation into these rocky shore communities proceeded through increasing both the type and level of contrasts so that the basic exclusion experiment was repeated in different situations (type) and with numbers of alternatives (level). This, and the stepwise development of the research, would always keep the researchers in check because they had to explain to themselves, and each other, why "this is different from that".

Menge and Sutherland (1976) propose that competition between species regulated species numbers in a guild when they are near carrying capacity – usually true at higher trophic levels. Conversely predation regulates species numbers in guilds at lower trophic levels. They extend this to communities, postulating that in those with few trophic levels competition will be more important than predation and that as number of trophic levels increases predation will become relatively more important. As well as referencing the work of Paine and others in defining how predation can operate, they also reference literature describing how competition may operate in stable communities, where most species reach their carrying capacity and where, it was postulated, increased interspecific competition acts to reduce the array of habitats used by a species. This interspecific competition would result in selection for increased specialization, and species diversity would then be increased by successful invasion into such a community.

Menge and Sutherland (1976) use detailed information about species numbers and trophic levels from rocky intertidal communities to show how a resolution can be made between the competition and predation theories that explain high diversity. Nested contrasts are made between east and west coast USA, exposed and sheltered shores, and experimental exclusion of predators vs. no exclusion. They argue that greater temporal environmental variation on the East Coast produced a less stable community. So, defining the domain of the keystone species predator system in rocky intertidal zones extended the results described by Paine (1966, 1974) from removal experiments.

Does this match Lakatos' description of progress in a scientific research program? The HSS theory can be considered as T_1 (Fig. 11.5). The experimental and observational results that contained new information, i.e., about diversity rather than just control of trophic levels, as put forward by HSS, were explained by the Menge and Sutherland (1976) theory, i.e., equivalent to T_2. This is a new synthesis. Specifically, Postulate 1a of HSS could now be replaced by axioms stating conditions under which herbivore numbers and species diversity are controlled by predators – although they should be qualified by referencing the particular habitat, i.e., the rocky intertidal. In this sense T_2 does predict novel facts, though they were neither improbable nor

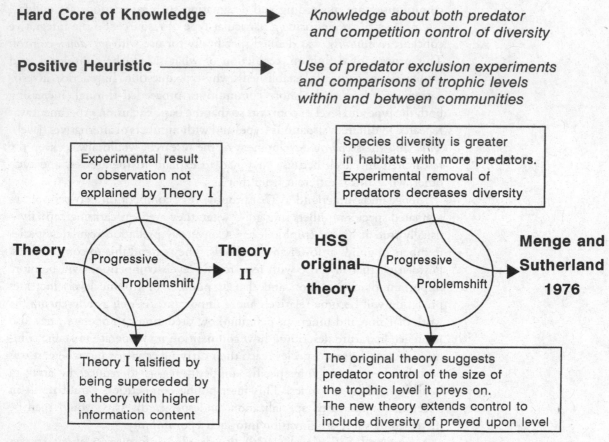

Hard Core of Knowledge ⟶ *Knowledge about both predator and competition control of diversity*

Positive Heuristic ⟶ *Use of predator exclusion experiments and comparisons of trophic levels within and between communities*

Experimental result or observation not explained by Theory I

Species diversity is greater in habitats with more predators. Experimental removal of predators decreases diversity

Theory I → Progressive Problemshift → **Theory II**

HSS original theory — Progressive Problemshift → **Menge and Sutherland 1976**

Theory I falsified by being superceded by a theory with higher information content

The original theory suggests predator control of the size of the trophic level it preys on. The new theory extends control to include diversity of preyed upon level

forbidden by HSS. Certainly T_2 seems to explain T_1, although it is based on data from just one habitat type. Some information in T_2 that exceeds that in T_1 has been corroborated, although really T_1 had no rigorous empirical base when first proposed. Lakatos' view of falsification as being more than naive falsification seems reasonable in this case. It is not simply that T_1 is incomplete. After the research of Menge and Sutherland, T_1 is an *inadequate* specification of top-down control, and in this sense false. From 1976 onwards scientists doing research into top-down control should not base their work on T_1 alone but should take account of T_2.

By *confirming* the HSS theory in a restricted but important way, Menge and Sutherland's work established a hard core of knowledge, i.e., that both competition and predation *may* operate to increase diversity. In future this result will not be challenged by the methods of the top-down research program. The new question is how the differences can be resolved through understanding of predator influence – into a coherent explanation of diversity. It is this acceptance of new driving questions that mark the real onset of a top-down research program. There is an acceptable hard core, established by

Fig. 11.5. An illustration of how research in ecology can cause a problemshift in a theory. The left-hand side of the figure illustrates the mechanism described by Lakatos (1970). The right-hand side shows this applied to the first use of experimental and observational results in the development of the theory of top-down control by predators.

Menge and Sutherland's synthesis, and a positive heuristic established by Paine's experiments that combined predator removal with detailed studies of changes in numbers and organization of prey species.

This work led to new questions. For example, Menge and Sutherland (1976) suggest that in structurally simple environments competition reduces diversity through competitive exclusion while predation first increases then decreases diversity because there are few refuges and hence overexploitation of a resource is more probable. In structurally complex environments competition may increase diversity through increased habitat specialization.

Fretwell (1977) contributes to the positive heuristic by postulating a way in which the theory may be generalized to different types of ecological system and offering some specific ways in which his predictions could be tested. He describes the system of predator → herbivore → primary producer as a three-link terrestrial system and attempted to generalize from one to four links. For example, he reasons that, in a two-link system of plants and grazers, the grazers, being at the top of the system, must be food limited – so the grazers limit the plants. This accounts for some systems where the world is not green, e.g., grasslands, which from Fretwell's perspective in Kansas were brown. Fretwell suggests that while three-link systems would be green, a four-link system, with a second carnivore layer at the top of the trophic levels, would have plants that were herbivore limited. He considers that the number of trophic levels would be determined by plant productivity and suggested that grazer- or predator-limited systems are short lived and mechanically unpalatable whereas food limited systems are long lived and may be poisonous or distasteful. However, because of the lack of specific investigations, Fretwell's (1977) contribution did not initiate a problemshift.

Although in their empirical studies top-down and bottom-up researchers were concerned with different ecological systems they were not academically isolated. White (1978) counters the idea of top-down control, suggesting "that animals live in a variably inadequate environment where many are born but few survive" leading to the theory of populations being "limited from below" rather than "controlled from above". He suggests that for many if not most animals, whether herbivore or carnivore, the single most important factor limiting abundance is relative shortage of nitrogenous food for the very young. Food quality is important, not the amount of energy flowing through the ecosystem. In stressing this, White (1978) argues for one of the counters to HSS advanced by Murdoch (1966) but giving it a more specific basis. If the world is green it is not because predators limit the grazers so that plants are not eaten, but because the ratio of nitrogen to fiber and carbohydrate in food plants is low – limiting herbivore growth and/or reproduction. He suggests predator control of herbivores to be the exception rather than the rule and reviews cases where food quality of herbivores controls various components of breeding success, and evidence that food supply controls carnivore numbers.

White suggests that the "green world" is not a paradox that must be explained by a theory of balance, i.e., dynamics of the predator → prey → plant system. Unpalatability of much plant material removes the problem of how the system is kept in balance. The usual shortage of adequate (i.e., palatable) food is sufficient, and White (1978) makes an explicit statement for bottom-up control: "the usual concept of control of populations from above might be better replaced by a concept of limitation of populations from below – from flows from one trophic level to the next". Without food shortage there can be no competition, but there can be shortage without competition. "Food may be in relatively short supply – so thinly spread through the environment that most individuals, although feeding continuously, and *whether or not other animals are seeking to use the same food*, do not get enough to survive" (White's italics).

It is interesting that at this stage both top-down and bottom-up research programs had good, if not extensively repeated, confirming evidence. Yet, though the ideas that were subsequently used in their attempted resolution were discussed by some scientists, strongly stated views defending one program and denying the other were produced and accepted by referees and journals.

Successive problemshifts in the research program

A feature of top-down research has been investigation of different types of community and these investigations started with different prevailing features of what was considered to control populations and communities. For example, Carpenter *et al.* (1985) start their exposition of cascading trophic interactions and its effect on lake productivity by mentioning the extensive work on nutrient supply as a regulator but comment, "However, nutrient supply cannot explain all the variation in the primary productivity of the world's lakes". They illustrate cascading trophic interactions in lakes as "a rise in piscivore biomass brings decreased planktivore biomass, increased herbivore biomass, and decreased phytoplankton biomass". Whilst stating that "Cascading trophic interactions and nutrient loading models are complementary, not contradictory," they also suggest that "The effects of food web structure are independent of those due to nutrient supply".

Data discussed by Carpenter *et al.* (1985) and described in greater detail by Carpenter and Kitchell (1988) are manipulations of lakes by removing and/or adding fish or planktivores and observing the effects on primary production and/or chlorophyll concentration. The results were complex, sometimes depending upon growth rates of residual populations after the removals, and movements of predators between on- and offshore. As animals grew they also changed in their abilities as predators and/or potential to be prey. In practice the use of nested contrasts was more difficult to achieve than in work on the rocky shore.

An important result of this work was to define and demonstrate trophic cascades in fresh-water lakes. Carpenter and Kitchell (1987) conclude that "substantial, sustained fish manipulations reconfigure food webs and modify primary production independent of nutrient loading". This is a second problemshift, particularly because direct investigation of primary production is incorporated into the study. The hard core is strengthened (more evidence for top-down control) and a positive heuristic confirmed (predator removal experiments provide answers to questions). However, Carpenter and Kitchell (1987) also mentioned that both top-down and bottom-up forces are important. Further, they commented that posing the question of control as a dichotomy is inappropriate and that a better question is "How much and when?", although their study did not attempt to answer that question.

Throughout this period of research, both top-down and bottom-up theories were improved (i.e., they became more coherent) for particular ecological systems so that scientific inference was made more secure. There certainly seem to have been problemshifts in top-down theory in the sense that new investigations incorporated something new into a developing theory. But the nature of ecological investigations is such that they are conducted on different systems, so successive empirical research does not replace that done earlier in quite the way implied by Lakatos. Explanations gain in coherence through parallel studies that confirm general applications of an integrating concept.

Crucial experiments and advancing or saving theories

Lakatos' (1970) emphasizes the growth of a research program through successive three-way disputes between two component theories within the research program and some data. Lakatos denies that there is such a thing as a crucial experiment in such work. His definition of falsification is not stringent, and uses comparative inadequacy between the theories rather than the total failure of one. Postulates can be falsified under some circumstances (Chapter 7 and 8), so why can we not do at least as well with theories?

CRUCIAL EXPERIMENT

A *crucial experiment* is established between two theories, T_1 and T_2, if T_1 predicts one thing will occur and T_2 predicts it will not and that this difference can be attributed to a specified difference between T_1 and T_2.

The real problem lies in the last part of the definition. Theories should contain clearly defined postulates. But they also contain many auxiliary postulates and axioms defining concepts and what might at first be considered incidental features and so not be included in a specification. Even a theory firmly based upon observations, e.g., Paine's (1966, 1974), contains auxiliary information. This can be of two types: that implying consistency within the theory and that implying boundaries to its application. Neither may be fully understood when the theory is first specified.

Attempts were made at other rocky intertidal sites to investigate application of the theory of top-down control. They failed to confirm it in a straightforward way. Paine (1966, 1974) had found that on more exposed shores high predation resulted in fewer sessile organisms. In contrast Menge (1992) found that on an exposed shore in Oregon, despite greater starfish density than at a sheltered part of the same shore, there were more rather than fewer sessile organisms. Experimental transplanting of mussels and direct observation of predation rates did show consumption at the exposed site was greater, but so too was growth of sessile organisms. Menge presents evidence and reasons that primary production was higher at the exposed site. So the difference with Paine's initial theory is resolved by adding an axiom that bottom-up forces may influence the relationship between predator (starfish) and prey (mussels and barnacles). There was auxiliary information in Paine's studies that was loosely, rather than precisely, specified. Precision in what should be specified may only become apparent after further investigations. So, rather than falsification of either one, two research programs were joined to define part of the domain within which this particular top-down process operates.

Quine (1951) recognizes that because auxiliary information is a part of theories it could provide a reason for not accepting a falsification. He writes: "Any statement can be held true, come what may, if we make drastic enough adjustments elsewhere in the system". Another related, although more sophisticated, description of the same problem was made by Duhem (1962) with respect to physics:

> "the physicist can never subject an isolated hypothesis to experimental test, but only a whole group of hypotheses; when the experiment is in disagreement with his predictions, what he learns is that at least one of the hypotheses constituting this group is unacceptable and ought to be modified; but the experiment does not designate which one should be changed."

And Duhem discusses how the physicist may choose to change either the theory or the implicit auxiliary information.

THE DUHEM–QUINE THESIS

The complex structure of theories, including their auxiliary information as well as specified axioms and postulates, leads to difficulties in testing theories because rejection of a single postulate may be declined in favor of appealing to alterations in linked propositions, or in previously unspecified auxiliary information.

This presents a difficulty in achieving explanatory coherence. Gillies (1993) notes that the Duhem–Quine thesis applies to theories with a high level of reasoned content. Theories using integrative concepts are in this category because they extend beyond a well-defined empirical basis. But Gilles also suggests that, when theories are tested, the group of axioms and auxiliary information under test is actually limited, and Quine's statement, though logically correct, is too extreme. Invoking failures in auxiliary assumptions can not save every postulate.

Gillies discussed the importance of this for the use of falsification. Theoretical statements can be distinguished in three ways: those that can be falsified, those that can be confirmed but not falsified, and those that can be neither confirmed nor falsified. The first two he classes as scientific, the last as metaphysical.

So confirmation **is** important. It may certainly be possible, and important, to attempt to falsify functional concepts used in the construction of integrative theory. But the type of decisions made in the three-way arguments within a research program, or where research programs collide, is between confirmation and failure to confirm in particular instances, but not complete falsification of a whole research program. When differences are found, typically they are resolved by further investigations that extend the theoretical content, eventually to provide possible resolutions of why and how different processes operate, i.e., to define domains.

By Gillies definition, the HSS theory as originally proposed (Hairston *et al.* 1960) is metaphysics because no explicit method for confirmation was detailed. Though Murdoch (1966) was right to criticize HSS, failure to specify a procedure for confirming would have been a more appropriate basis than that it could not be falsified.

Is the top-down research program degenerating?

Lakatos (1970) proposes that a research program degenerates when its positive heuristic runs out of steam, when unexplained observations are rescued by *ad hoc* additions to salvage the theory, and when the theory is reinterpreted to resolve observed contradictions. The hard core of the program decays and successive problemshifts degenerate as *ad hoc* explanations are added. This does not to seem to fit the recent developments in the top-down research program.

First, there are clear examples of dominant top-down forces. However, Strong (1992) argues that these are relatively unusual systems being restricted to "fairly low-diversity places where great influence can issue from one or a few species; the majority of examples of true trophic cascades have algae at the base and are aquatic". In more diverse systems, while top-down forces may exert themselves, their effect is buffered by defensive adaptations of plants to herbivory and by heterogeneity in the system that limits herbivore efficiency. Where cascades occur, the trophic system is ladder-like; where they do not, it is web-like, with omnivory and resource generalists that defy compartmental-ization into distinct trophic levels. Clearly the existence of top-down forces being exerted through trophic cascades marks a confirmation of the HSS theory – though not in the form they propose as a theory to explain *all* green places. But the hard core of knowledge when defined for these particular types of system seems likely to remain intact. Similarly the positive heuristic is valuable if there are similar types of system to investigate.

Second, the problem that exists for the top-down theory – that it does not explain everything – seems to be being resolved by defining its domain. This process, and the methods it may follow, is discussed later in the chapter. Menge's (1992) contribution to resolving apparent conflicts in results from the rocky intertidal has already been mentioned. Top-down theory is not degenerating, nor are there *ad hoc* additions, but its domain of application is being defined, e.g., it seems to apply in at least some aquatic systems with algae at their trophic base.

11.3.4 Criticisms of the methodology of scientific research programs

Lakatos' theory has been criticized by a number of philosophers of science both as a description of how research does take place and how it should take place. Berkson (1976) criticize Lakatos' exclusion of *naive falsifications* be-cause thay can actually be used and may contribute to scientific advance. He also asserts that the notion of "hard core" was not substantiated and every individual scientist has his own "hard core" and criticizes Lakatos for not providing a theory of actually how scientists should choose between theories; there is no explicit rationality of choice in Lakatos' *methodology of research programs.*

Musgrave (1976), a student of Lakatos' makes three criticisms. First, like Berkson, he criticizes the idea that scientists actually maintain a "hard core" of ideas that they do not challenge and indicates, by example, that the supposed "hard core" could be challenged and the theory revised. Musgrave suggests that the idea of a "hard core" of unchallengeable ideas is not one that scientists should follow. Second, the positive model followed by scientists is not as comprehensive or invincible as Lakatos suggested, and the requirement of a theory to cope directly with the results of actual investigations or experiments was strong. Third, Lakatos abandoned his attempt to provide a method by which research programs could be judged and resorted, in effect, to the position that "anything goes" – an anarchistic viewpoint. Musgrave quotes examples in which, by Lakatos' standards of judging research pro-grams, the superiority of one over another can be judged.

Some of this criticism is rather tough because what Lakatos did was to resolve some ideas that at the time appeared very different. He accepted science as a social process. But he proposed that there is not complete relativism between different time periods separated by revolutions. The continuity of theory, and the role it plays in continuous assessment, resolved the problem of the important role of falsification and how it could be balanced with theory development. On the basis of 32 years' development of top-down theory (1960–1992), the following points can be made.

Judging by the intensity with which the bottom-up theory was attacked, some scientists held top-down to be correct even though there was considerable evidence for the bottom-up theory when they made their attacks, though perhaps not about the systems they researched themselves. The attacks were reasoned, as far as they went. I take this as evidence that those scientists did actually hold something like a hard core of knowledge as a principle of their methodology and used it both to motivate and to rationalize their viewpoint. Further, given the confirmatory stance of much of the top-down research, even when attempted confirmation failed in some habitats, and no resolving explanation of the difference could be advanced, then much of that hard core knowledge was shared between numbers of individuals. Failure to confirm top-down control in a particular instance may simply not have been accepted by the group, who may have preferred to suggest existence of differences in the auxiliary information when instances of failure to confirm occur. The value of a hard core is clear, say, in the attempts to extend Paine's approach to other rocky intertidals. However, the importance of being able to criticize is equally clear in developing the resolution of apparent failures and differences into a specification of the domain for the top-down theory.

Certainly, as Musgrave (1976) mentions, the requirements to cope with external results are strong – but this took place in two stages. First, it was important to define the theory (top-down) and confirm its value in a number of instances (rocky intertidal, streams, lakes) for which the positive heuristic was essential. Obviously, this takes time. **Then** it was important to resolve its exceptions and in so doing gradually to specify the domain of the theory. In this respect, ecology may have different features as a science compared with physics and chemistry – which provided the examples for Lakatos. An essential part of ecology is understanding just how variation occurs in the structure of theory as it can be applied to different communities. This makes the Duhem–Quine thesis particularly important for ecology. It highlights the difficulties we have in falsifying and reinforces the importance of contrasting work in natural communities where the nature of the contrasts is close in all but a few things, and above all is well defined so that differences can be explained rather than left unresolved. If you have to use confirmation as a methodology, then it is essential to expose a theory to incremental tests.

Can there be a general method by which scientific research programs can be judged? Philosophers of science still dispute whether this is possible or even desirable (e.g., Worrall 1988, 1989, Laudan 1989). Of course there are some general methodological procedures, but the most standard and universal ones are about the details of method, e.g., statistical procedures. There are difficulties in applying such ideals of good procedure. For example, in the basic experimental treatment of removing predators the nature of rocky shores meant that while a control could be adjacent to a treatment, replication of treatment and control was not possible, so there was no estimate of measurement and sampling error. The requirement then is to study variation. For this, repetition on different sites was needed where reliable measurements could be made, e.g., sufficient rock of the same type within both treatment and control. Insisting on the replication of treatment and control as a prerequisite for "good science" would not be appropriate **in this case**. The essential feature of such experiments was not simply to observe a "result" as though it were a response-level experiment (Chapter 6), but to study the ecological processes following the predator removal so that interpretation had a substantial chain of reasoning. Scientific inference was incremental; statistical inference was not possible at the level of an experimental treatment.

A practical answer for ecology is that a research program is best judged on how coherent its explanations are, but for successive **particular** instances, not things in the abstract. Certainly, attempts should be made to falsify functional concepts where possible. But for more complex theories scientific inference progresses to maturity if explanations are enduring and confirmed by their value to other research programs. Even-well established research programs do not have completely independent theories, and scientists within the program define linkages to other research programs through concept definitions. Top-down theory used other types of research in its arguments, notably concepts from ecosystem science of energy flow and aspects of food web theory, and, to be accepted as a coherent explanation, it must not contravene those well-established definitions.

The top-down program has not degenerated – at least in terms described by Lakatos of development of an increasingly *ad hoc* theory where propositions are introduced that protect it by excluding particular conditions. The theory remains and is enhanced by the increasing understanding of its domain provided by the once competitive bottom-up program. Top-down control becomes part of a wider, more general, theory.

11.4 The investigation of domains

Both Kuhn and Lakatos give use and development of theory central place in their methodologies. In contrast, Shapere (1977) is concerned with how theory is developed from a set of information, which he refers to as a *domain* (Fig. 11.6). Four characteristics unify a subject matter and make it a *domain*:

(1)   Items such as observations or predictions from theories are related.

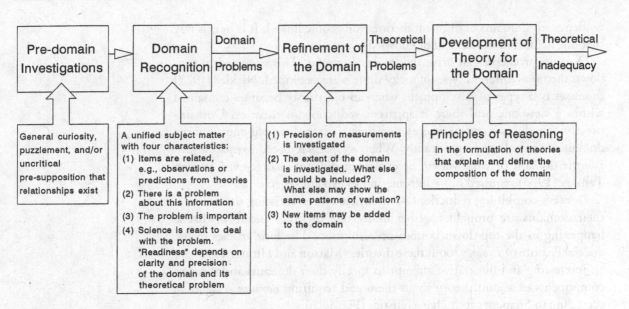

Fig. 11.6. Phases of development in a domain of scientific interest leading to specification of a theory as described by Shapere (1977).

(2) There is a problem about this information.
(3) The problem is important.
(4) Science is ready to deal with the problem.

Shapere (1977) identifies two types of problem: *domain problems* associated with the structure of the information, and *theoretical problems* associated with providing an account of what causes the observed patterns. Both problems are important – but without sufficient attention to *domain problems* then *theoretical problems* may not emerge with sufficient precision to be solved. This happens where a theory is developed without understanding the limits to its application, e.g., the situation when top-down and bottom-up contested for universal hegemony. Strong proponents of neither side appreciated the application limits of the theory that they advocated. Note that Shapere includes items other than simply observations as part of the domain – a domain is a body of information of various types that must be explained.

The *domain* of integrative or functional concepts (Chapter 10) defines limitations to their importance and application. In the development of top-down theory, description and definition of limitations grew, eventually becoming a body of information itself needing explanation. This is the problem that concerned Shapere. He suggests that different kinds of theory constitute science. Rather than trying to define what a theory is, or should have been, study of the "roots, roles and problems" of theories might resolve the problem of definition. A theory may be incomplete without being incorrect, incompleteness does not imply falseness, and continuing work on domain problems may explain inadequacies. In particular, Shapere notes a tendency for domains of different theories to become linked. It is in this way that our definition of the field of knowledge grows.

Domain problems can drive theory development. This occurred in top-down theory as different types of rocky shore were examined. Nickles (1977) discusses two types of development when an old theory becomes contained within a new one and there is apparent reduction in theories. Domain-preserving reductions occur where a new theory succeeds an old one – but the domain is not enlarged significantly. When *ecosystem* replaced *complex organism* the domain of what was to be explained did not change even though Tansley (1935) required more information to explain vegetation.

Domain-combining reductions occur where two preexisting theories and their domains are brought together under one theory – possibly what is happening in the top-down/bottom-up resolution. The four articles in the Special Feature of *Ecology* about these theories (Matson and Hunter 1992) are an interesting and illustrative attempt to specify their domains and develop components of a joint theory about them and requiring *domain recognition* according to Shapere's four characteristics (Fig. 11.6).

(1) Items are related   Both top-down and bottom-up forces may operate in a community – there is not a dichotomy between the two types of control (Hunter and Price 1992). Variation in top-down effects can best be understood "by superimposing their effects on a bottom-up view of trophic structure" (Hunter and Price 1992, see also Menge 1992). For Hunter and Price (1992), while removal of higher trophic levels certainly may influence lower levels, removal of primary producers leaves no system at all. Plant production should be considered in any trophic description; plant quality and quantity determine arrangement and patterns of spatial and temporal heterogeneity among herbivores.

Hunter and Price refer to this type of control as a cascade up the system proposing: (a) Systems can be dominated at any trophic level depending upon particular conditions, and variation in the level of domination is what has been found. (b) Feedback can occur throughout the trophic system, not just in one or the other direction. (c) Extensive herbivory occurs in natural systems and questions should not be changed from "Do natural enemies or does primary productivity regulate herbivore population dynamics?" to "Under what conditions of soil conditions, plant community structure, and abiotic variability do natural enemies dominate trophic interactions?"

(2) There is a problem about this information   While Power (1992) accepts Hunter and Price's general idea, and mentions other similar proposals (Sykes 1973, Oksanen *et al.* 1981), she poses a series of questions as to how the relative strengths of the two processes are to be assessed in a system. Among the factors regulating relative strengths are:

Efficiency with which consumers exploit their prey.

Whether predators compete among each other and/or are limited by resources other than food.

If there are substantial time lags between prey consumption and predator response.

Declines in edibility of food due to herbivory or predation.

Food availability for one consumer life history stage not influencing survival or fecundity of the subsequent stage.

Non-linear responses of primary productivity to grazing, e.g., a change in the stimulus to growth or palatability.

Effect of cover provided by plants on predator–prey interactions.

Power (1992) detailed four investigations required in determining relative importance of top-down/bottom-up forces are to be assessed in a community:

(a) The major sources of energy must be defined and quantified. Amount can control the type of system.

(b) Can trophic levels be distinguished or does omnivory obscure boundaries?

(c) Is the community stable and/or has recovered from any disturbance?

(d) The scale over which the system is to be measured must be defined.

(3) The problem is important This is clearly so, since it impacts on many applied and basic questions in ecology, e.g., causes of diversity in some community types.

(4) Science is ready to deal with the problem This is difficult to judge, other than that attempts are being made to specify questions that should be asked (Power 1992).

According to Shapere, the next stage should be refinement of the domain and then development of theory for it. Both Hunter and Price (1992) and Power (1992) present schematic diagrams showing how a resolution may occur. However, in both cases the attempts are presented primarily to explain results already obtained, which of course in itself is quite a challenge. Development of a theory for the domain structures of top-down and bottom-up control would require going through the processes of scientific inference (Fig. II.2).

There are similarities between Lakatos' idea of scientific progression and Nickles (1977) extension of Shapere's domain concept. In both accounts theories are not clearly rejected but grow and change, and both accounts express theory development as a logical process – at least that theories must be explanatory and account for observations in a sequential development and that new theories are intimately connected to old theories. The Nickles account is a more general and extensive view of from where new elements of a domain, or domains, may come. For Lakatos, new information is directed through use of the positive heuristic, whereas for Nickles information may come from any related source. These are not mutually exclusive ideas. Where we get information from depends upon the type of problem. We may use a positive heuristic to develop the details of a theory, define a theory's applicability by studying its domain, or develop completely new theories through making a synthesis of information from very different fields of inquiry (Darden and Maull 1977).

11.5 Discussion

Social influences play an important part in deciding what shall be studied and how research should proceed and inference made. Kuhn's description has charm in reminding us that scientists work in groups and gives drama in its emphasis on the importance of revolutions. However, the processes both of reaching agreement and maintaining disagreement are complex (Section III) and generally more continuous and parallel than Kuhn represents them. If a science is allowed to proceed as described by Kuhn then it has become too restricted a social process.

Lakatos' theory of the methodology of scientific research programs, while acknowledging social processes, describes developing scientific inference while some constancy of purpose is maintained. A group of scientists pursued work in top-down theory. The theory changed in important ways as their work progressed. They used a general heuristic, particularly removal experiments and its development to implantation experiments, and maintained their view of predator–prey relations as a hard core of knowledge. This separated them from scientists studying bottom-up processes, at least until the scientific resolution between the two theories became clear, and illustrated through specific investigations about particular places rather than general assertion.

The importance of such social groupings is high where confirmation is necessary. Usually there is no clear distinction between theory formulation and test, which is one reason why apparently contradictory theories such as bottom-up and top-down can coexist for considerable periods. Gillies' (1993) use of the Duhem–Quine thesis to illustrate that theories, rather than postulates, can only be confirmed is of major importance. For Lakatos, falsification of any theory involves its piece-by-piece change over a period of time, and techniques that are used to explore successive theories in the program, the positive heuristic, are themselves part of the overall research program. What we have to do is define where, or under what circumstances, a theory must be modified. What is its domain of operation? Once it has been established to have some explanatory power, and has some agreed confirming instances, then we must seek to understand the constraints on a theory rather than expect its complete downfall. Domain axioms should become incorporated into specification of a theory. So top-down theory has constraints, e.g., where there are food chains rather than food webs, and where herbivores have uniform access over plants.

11.6 Further reading

Giere (1988, his Chapter 2) provides an overview of theories about science and places those of Kuhn and Lakatos in a wider context than is given in this chapter. His own cognitive theory of science is illustrated by using examples that demonstrate evolutionary developments in scientific theories. Kuhn (1977) provides a background to his theory and the chapter "Second Thoughts on Paradigms" amplifies the paradigm concept. Margolis (1993) describes the cognitive processes involved in making changes in paradigms for individuals.

Research that studies the balance between top-down and bottom-up forces, rather than just one or the other of these processes, has become the general approach, e.g., for a grassland (Fraser 1998), for an insect herbivore system (Hunter *et al.* 1997), for the rocky intertidal (Menge *et al.* 1997).

12 Use of mathematical models for constructing explanations in ecology

Summary

Models are analogies representing important features of a system. By constructing and then studying a model we hope to explain how the system functions. However, mathematical models are made using particular rules and procedures of construction. These determine both the types of simplification made and how the model can be assessed. Rules and procedures for constructing three types of ecological model are reviewed and their use in ecological research is discussed.

Dynamic systems models have a defined mathematical form, usually differential or difference equations. They have a long history of describing idealized ecological interactions and speculating about general properties of ecological systems. Their use has been criticized because of difficulties in assessing how effectively they can represent particular ecological systems.

Statistical models have a mathematical structure specifically designed for fitting to data and calculating error. *Parsimony* is an important procedure in construction – using no more complicated a model than is needed to describe patterns in measured data. An example of a model of a *stochastic process* is described. Statistical models have been criticized because of difficulties in interpreting ecological significance of parameters and lack of uniqueness, i.e., that different models may fit a data set equally well. General explanations have to be built up by comparing models constructed in different instances.

Systems simulation models take advantage of increased computing power and flexibility in computer languages to make more comprehensive representations of ecological processes than either dynamic systems models or statistical models. However, the potential for complexity can lead to difficulty in deciding a model's bounds, what should, and should not be included, and how to assess the model.

We can not **depend** upon a model, of any type, to be a complete representation of a synthesis. *Synthesis* is a wider process. For each type of model the effectiveness of the modeling exercise (not just the model itself) can be judged by examining how the model increases explanatory coherence of the theory it represents.

12.1 Introduction

Construction and use of quantitative models is part of much ecological research. This chapter outlines three types: dynamic systems models, statistical models, and systems simulation models. Each has a different focus in the type of question they can be used to investigate and methods they use. Dynamic systems models are differential or difference equations that describe postulated relationships between components of an ecological system. They are used to explore details of these relationships and typically the model is assessed by its description of important characteristics of the system rather than whether it fits any particular data well. Statistical models are parsimonious, i.e., containing only sufficient variables necessary to explain variation in observed data. Their mathematical structure enables calculation of goodness of fit and, provided data is available, they can give understanding of variability between systems. Typically, the object of constructing a system simulation model is to synthesize information from theory and empirical investigations. These models tend not to have a unified mathematical structure, as do dynamic systems models and statistical models. More flexible rules are used in their construction, but their resulting complexity can make assessing them difficult.

Modeling is the source of much controversy. Mathematical models are used in speculation about possible ecological relationships and so generate what some consider to be ecological theory. Such speculation has been criticized either as not being testable or, if tested found to be wrong, yet the models continued to be used (e.g., C. A. S. Hall 1988). The extent to which models can be tested has been questioned, particularly whether a model can be validated, i.e., accepted as correct (e.g., Konikow and Bredehoeft 1992, Oreskes *et al.* 1994). This criticism has been leveled primarily at systems simulation models. In these models, as the number of parameters estimated during calibration increases, so also does the chance that parameter values may accommodate deficiencies in model structure.

Quantitative methods of model assessment are discussed for each type of model. These have strengths but also weaknesses. What is considered important is how models of different types can be used in constructing scientific explanations and ultimately model assessment should take this into account.

MODEL EXPLANATIONS

Model explanations are based on the principle of analogy. A model describes important features in a simplified representation of a system and can be used to

Fig. 12.1. A quantitative model is based upon an ecological theory and constructed using rules and procedures of the chosen modeling type. Output from the model is assessed against data, in varying ways, again depending upon the modeling type. Inference about the ecological theory, based on model assessment, must include understanding of how rules and procedures may influence model structure and/or the type of model assessment that can be made.

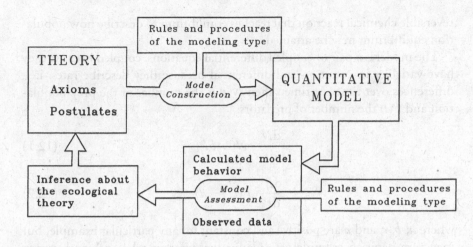

illustrate how interactions may take place to produce particular outcomes. By studying a model ecologists attempt to understand how actual systems behave under different conditions.

There is a common methodological problem for all model explanations.

It is essential not simply to assert the value of a model explanation.
The explanatory analogy itself must be defined, analyzed, and tested.

The construction of a model is not equivalent to the construction of ecological theory. Model construction follows particular rules and procedures of the modeling type (Fig. 12.1). These rules and procedures translate theory into a model, frequently involving a simplification, both in the relationship between components of the system and in the fact that the model may represent only part of the theory network. The translation may also involve complication when the modeling type requires particular types of mathematical equation. How these simplifications and complications can affect the explanatory coherence of different types of models is discussed in this chapter.

12.2 Dynamic systems models

12.2.1 Simple differential equation models

Calculus is used because it can model changing rates of ecological processes and how they affect each other. One of the first mathematical models developed in ecology attempts to describe the relationship between the numbers of prey and their predators and how this changes over time. This is the Lotka–Volterra model, named after the two mathematicians who, independently, first described it. Volterra (1926) wished to explain the oscillatory levels of fish catches in the Adriatic. Lotka (1925) used the analogy of a reversible chemical reaction that reaches equilibrium to describe how population equilibrium may be attained.

The model is a pair of coupled differential equations, coupled because they have variables in common and differential because they describe rates, i.e., differences over time. At time t then $N(t)$ is the number of the prey population and $P(t)$ the number of predators:

$$\frac{dN}{dt} = N(-bP) \qquad\qquad (12.1)$$

$$\frac{dP}{dt} = P(cN - d) \qquad\qquad (12.2)$$

where a, b, c, and d are positive and constant for any particular example, but may vary between examples to account for different relationships between predator and prey.

Predators consume, and depend upon, prey, so there are underlying causal relationships, i.e., there are matter and energy exchanges. Population numbers of prey and of predators determine their interactions, i.e., organizational relations. However, the model uses neither of these directly but defines the dynamics simply in terms of:

the more there is of N (prey) then the more P (predators) will increase, and
the more there is of P (predators) the more N (prey) will decrease.

So this model is a simplification both in how the relationships are represented and what is considered, i.e., other factors that may influence either prey or predators are not included.

The model defines rates of change, enabling predictions to be made about predator–prey cycles. It predicts oscillatory behavior (Fig. 12.2). As the number of prey increase so do the number of predators – but lagging behind those of their prey. An increase in predators reduces the prey, eventually to the point where numbers of predators themselves are reduced and so the cycle repeats itself. The model illustrates two features of populations that ecologists have felt, important: populations fluctuate in response to ecological interactions, but there is overall stability in that neither prey nor predator is driven to extinction.

DYNAMIC SYSTEMS MODELS

A *dynamic systems model* has a defined mathematical form, most usually differential equations, but not including logical switches or stochastic forcing.

The Lotka–Volterra predator–prey model is a dynamic systems model. There are no switches of the type "If prey decreases to a certain value then no more

Fig. 12.2. Schematic periodic solutions for prey, $u(\tau)$, and predator $v(\tau)$ for the Lotka–Volterra model. u and v are non-dimensional representations, $u(\tau) = \dfrac{cN(t)}{d}$ and $v(\tau) = \dfrac{bP(t)}{a}$, with $\tau = at$. Note the oscillating patterns, with predator lagging prey and represented as having reduced amplitude. (From Murray 1989, with permission.)

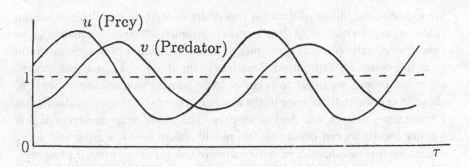

are consumed by predators". Such limits of this type are difficult to apply in this model type. The rules of calculus permit conditions to be expressed using indicator functions, and this might be used to simulate switching but would complicate the equations. Switches are more typically found in systems simulation models, described later in this chapter. Stochastic forcing is where a random variable affects components of the model. (*Stochastic* is defined in the next section of this chapter.) A stochastic forcing might be to make prey intrinsic growth rate, *a*, depend upon a random variable that is supposed to represent year-to-year variations in quality and/or quantity of food and affecting reproduction rate. Such a change would, in turn, require a different model structure. The Lotka–Volterra model is deterministic, i.e.; variation in any component is totally determined by the system dynamics as defined in the model and excludes stochastic effects.

DETERMINISTIC MATHEMATICAL MODEL

A model is *deterministic* when it is assumed that all possible behaviors are determined by the set of equations comprising the model.

The simplicity of the Lotka–Volterra model is its attraction, but this very simplicity has been criticized. Murray (1989) defines some model assumptions:

(1) In the absence of predation then prey grow unboundedly described by the aN term.

(2) Predation reduces the prey's per capita growth rate by a term proportional to the prey and predator populations, the $-bNP$ term.

(3) In the absence of prey the predator population has exponential decay, the $-dP$ term.

(4) The prey's contribution to the predator's growth rate is cNP, proportional to the available prey as well as the size of the predator population.

The NP terms represent the conversion of energy from one source to another; bNP is taken from the prey; cNP accrues to the predator. These assumptions are the result of the simplification procedure needed to apply the modeling rules, i.e., representation of the process as a coupled differential equation. The assumptions raise obvious questions. It is unlikely that a prey species would grow in a completely unbounded manner in the absence of a predator species, other limitations are likely to occur at some point. Predators are unlikely to show an exponential decrease in the absence of prey – they may all die within a finite time, or seek and find other prey. The model is so constructed that neither condition can occur, i.e., there will always be some prey and some predators, but towards the extreme values of few surviving prey or predators, the form of these terms produces the rate of return in the cycle.

Murray (1989) describes many attempts to apply the predator–prey Lotka–Volterra model. He discusses difficulties in the procedure in one application, and problems due to mathematical rules that make models of this type difficult to use. The application was the attempt to analyze interaction between snowshoe hare and lynx populations, as estimated from the number of pelts of the two species in the fur catch records of the Hudson's Bay Company from 1845 to 1934 (Leigh 1968). There were obvious cycles of these species in the pelt records. However, as Gilpin (1973) points out in a paper entitled "Do hares eat lynx?" in some cycles the maximum of the lynx population **led** rather than lagged the snowshoe hare population. That simple assessment is sufficient to conclude that those cycles are not described by predator–prey interactions of the Lotka–Volterra type. Finnerty (1979) reports that the snowshoe hare pelt data were from eastern Canada, near Hudson Bay, while the lynx data were from western Canada, so a causal relationship underlying these cycles based on consumption and availability of food might not be expected. An alternative explanation is that fur trappers themselves caused, or at least contributed to the cycles (Weinstein 1977). They may have hunted both animals until their populations were reduced; then, as catch per unit effort decreased, they changed target species to other than lynx and snowshoe hare, so allowing both populations to recover.

For Murray (1989): "The moral of the story is that it is not enough simply to produce a model which exhibits oscillations but rather to provide a proper explanation of the phenomenon which can stand up to ecological and biological scrutiny". In terms of explanatory coherence (Fig. II.2) this model aimed at a broad explanation by using simple propositions. But those are not acceptable when tested against data, and the concept definitions, as translated into limits to growth, seem unconnected to how predators or prey may grow. So this model is not coherent and we should not infer that it represents an effective explanation of snowshoe hare–lynx predator–prey system.

The rules of calculus permit substantial mathematical development of differential equation models. May (1981) examines what he considers to be a more realistic prey growth function for the Lotka–Volterra model,

$$\frac{\mathrm{d}N}{\mathrm{d}t} = rN\left(1 - \frac{N}{K}\right) - PF(N,P) \qquad (12.3)$$

where K is a carrying capacity for prey, i.e., without predation it does not have unrestricted growth. The first term on the right-hand side of equation 12.3 ensures a reduction in growth as total numbers increase, rather than unbounded prey growth, and May considers alternatives for the $F(N, P)$ function. May's interest was in what type of population dynamics might be produced as parameter values and the formulation of the right-hand expression changed. For some parameter values cycles did occur, but for others prey and predator reached equilibrium. Data for naturally occurring predator–prey systems showed only one, the snowshoe hare–lynx system, to have cycles (and we know that they may not actually be causally related as predator–prey) and the rest to have equilibrium points. The difficulty is that, although the rules of dynamic systems modeling, i.e., calculus, permit introduction of such terms, this is an *ad hoc* procedure. The basic model does not represent the process of predation. It describes only some supposed consequences of the dynamics – that neither predators nor prey become extinct and the supposed idea that there are predator–prey cycles. Manipulating the form of the equation and adding additional terms is an attempt to overcome deficiencies in the model using the rules of calculus rather than through a detailed scientific analysis of the problem. Notice also that, although the intention is to set an upper limit to prey, this is not done by opening the analysis to factors that might cause that limit but by setting a carrying capacity of animals. This preserves important simplifications in the model – no new units are introduced and the number of terms is kept to a minimum – but it does not improve the explanation the model can offer.

Simple models have continued to be used, even though they have not been successfully tested quantitatively, or, if assessed qualitatively they have been shown not to apply. This was C. A. S. Hall's (1988) major concern. He showed that three models, the Lotka–Voterra predator–prey model, the logistic growth equation, and the Ricker recruitment curves used to predict fish catch and population recruitment, were each suspect in their application to actual ecological situations. Nevertheless continued attempts were being made to use them in resource management. Hall's point was that in their simplicity these models are too wrong, or too uncertain in their foundation, to be useful in resource management. Unfortunately, the failure of some simple models has not led to change in the definition of concepts, the way they are represented in propositions, or development of a rigorous assessment procedure for field data. Rather, their mathematical complexity has been developed, i.e., the standpoint is taken that the model is not wrong, it has only to be made more complex, and the danger then lies in assuming that the first synthesis is sufficiently correct to be improved upon by modification.

This is equivalent to assuming that a general predator–prey theory has been discovered, without confirming instances.

Caswell (1988) argues that mathematical models have an important role. He lists suitable tasks for them as exploring the consequences of a theory, demonstrating the connection between apparently unrelated theories, examining different models of the same phenomenon, and designing empirical tests. These are essential functions of the first aspect of metaphysics in science (Chapter 9), i.e., the process of analysis of the internal character and self-consistency of a problem, to inquire what is real with argument. It should make why-type questions precise and indicate what type of explanation to investigate (Fig. II.2). Most scientists engage in metaphysics as they consider their future research, but mathematical modeling has formalized a particular type of speculation and so some models have become enshrined as theory, which is precisely what they are not. C. A. S. Hall's (1988) position is that continued use, and even development, of models which are found to be wrong is inappropriate – it is poor metaphysics.

Caswell (1988) gives analysis of the relationship between *complexity* and *stability* as an instance where mathematical modeling led to development of those concepts. Elton (1958) suggests that more complex systems were more stable but mathematical analysis by a number of investigators led to distinguishing two types of complexity, four types of stability, and three levels of resolution (Pimm 1984). Caswell (1988) insists that model development and analysis illustrate "if nothing else, . . . the situation must be more complicated than Elton thought it was". But mathematical analysis has not been content with just that claim. In a review of *stability*, Grimm and Wissel (1997) make their central thesis that:

> "the fundamental cause of the terminological confusion stems from the following conflict. Stability concepts derived from mathematics and physics are only suited to characterizing the dynamic behavior of simple dynamic systems, but ecological systems are not simple dynamic systems . . . By 'simple dynamic system', we mean a system in which, for instance, state variable, reference states, and possible disturbances are, because of the simplicity, unambiguously defined."

They illustrate that ecological systems are described using many variables and states and that disturbance can affect them in different ways. The projection of an ecological system onto a simple model makes simplifications that overrule interactions essential to considerations of stability (Grimm and Wissel 1997) and in this sense the metaphysics produced in using these models has been counter-productive.

12.2.2 Using dynamic systems models to predict the unexpected

Three levels of model assessment can be considered. The first is whether a model meets its design criteria by fitting the observations or data motivating its construction. This might satisfy the requirement that individual propositions are represented effectively in the model. Second is whether the model can predict a system response, within the same general framework used in its construction, where particular circumstances differ. The model might then be considered to have some generality. Third is whether a model predicts something which is new and unexpected but is subsequently found to be true. This would markedly increase the exploratory coherence attributable to the model. However, the extent to which a model may meet any of these levels of assessment is influenced by the rules and procedures used in its construction. Comprehensive assessment cannot depend solely upon these levels but requires detailed analysis of the explanatory coherence of the model.

There are technical difficulties in fitting differential equation models to data so that errors in the model, and in the measured data, can be assessed separately (but for such a treatment in the application of the Lotka–Volterra competition model to laboratory culture data, see Pascual and Karieva 1996). Sometimes such models are used to aim for the third level of assessment without satisfying the first two. Mathematical models, sometimes created in the physical sciences but with structures thought to be applicable in some area of ecology, are found to generate a new result that may help to explain some supposed feature of ecology. An attempt is made to see whether those features do exist and can be explained by the model. An example is catastrophe theory (Loehle 1989a), which found application in explaining the epidemiology of spruce budworm outbreaks in *Abies balsamea* forests (Ludwig *et al.* 1978).

Similarly, chaos theory has been suggested as having application in ecology, and research is in progress to see whether this is so. Chaos theory has a well-defined domain. It does not mean that a phenomenon is completely random, or jumbled, but that it follows a pattern with a particular type of variability. It is applied to difference equation representations of dynamic systems. For example, the logistic model represents changes in population numbers:

$$N_{t+1} = rN_t\left(1 - \frac{N_t}{K}\right) \qquad r > 0, K > 0 \qquad (12.4)$$

where r is an intrinsic growth rate and K a carrying capacity (Murray 1989). This equation can be rescaled by writing $u_t = N_t/K$ so that carrying capacity is 1.

$$u_{t+1} = ru_t(1 - u_t) \qquad r > 0 \qquad (12.5)$$

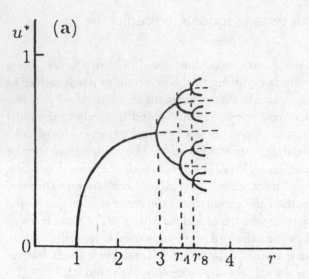

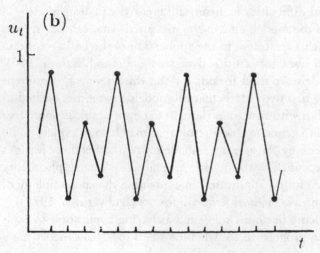

Fig. 12.3. (a) Stable solutions (schematic) for the logistic model as r passes through bifurcation values. The sequence of stable solutions have periods $2, 2^2, 2^3, \ldots$ (b) An example (schematic) of a four-cycle periodic solution where $r_4 < r < r_8$ where r_4 and r_8 are the bifurcation values for four-period and eight-period solutions respectively. (From Murray 1989, with permission.)

The starting value, u_0, is assumed $0 < u_0 < 1$. For values of $0 < r < 1$ the equilibrium value of u attained, called u^*, is 0, i.e., the population proceeds to extinction. But as r increases to < 1 then u^* increases (Fig. 12.3a). However, at $r = 3$, two possible stable solutions become possible, and, as r increases further then 4 stable solutions occur for r_4 and r_8. When r lies between r_4 and r_8 then u_t has a four-cycle periodic solution (Fig. 12.3b). As r increases further then multiple states become possible, what are called chaotic solutions, exist for $r > r_c$. Exactly at what values of r these occur depends upon the particular equations used. Figure 12.4 illustrates the equation $x_t + 1 = x_t + rx_t(1 - x_t)$ for $1.9 < r < 3$, which by suitable rescaling can be written in the form of equation 12.5 (Murray 1989).

Chaos is a well-developed mathematical theory, i.e., there are many instan-

Fig. 12.4. Long term asymptotic values for the discrete equation $x_{t+1} = x_t + rx_t(1 - x_t)$ (ordinate) for $1.9 < r < 3$ (abscissa). These are typical of discrete models that exhibit period doubling and eventually chaos as r is increased. (From Peitgen and Richter 1986, with permission.)

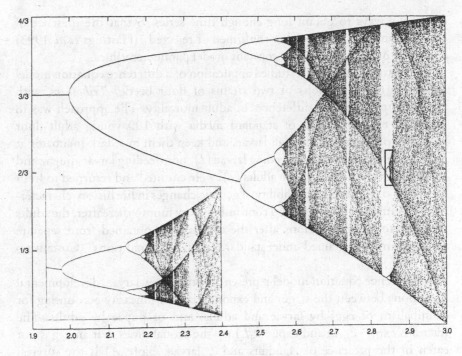

ces of how this applies to difference equation models. Kot *et al.* (1988) illustrate how this might apply to epidemics of measles and chickenpox in major cities. The types of equation to which chaos theory applies, and patterns of divergence that occur, are understood. It has been established that some models of ecological systems exhibit chaotic behavior, e.g., three-species food chain models (Klebanoff and Hastings 1994). However, it is important to define the domain of chaos theory. Hastings *et al.* (1993) describe a number of misunderstandings that have arisen in ecology, particularly that chaos is equivalent to observing the effect of random events on a system. This is not so. In all systems of equations exhibiting chaotic behavior recurring patterns are observed over time. Chaos theory describes a particular type of variation – not that there is complete randomness.

If ideas about chaos in ecology are to progress from speculation to development of ecological theory it is necessary to establish that models to which chaos theory applies are effective representations of ecological systems. Chaotic properties appear when population models are run over extended numbers of generations with the same parameter values, implying that the controlling features of population development are consistent. Alternatively, if the control of populations proceeds through different domains so that different parameter values, or even model equations, may apply, then chaos as a mathematical theory may not be relevant. There is a difficult empirical requirement to meet. "To study chaos in ecology, experimentalists need to determine ways to obtain long enough time series so that the presence of chaotic dynamics can be either confirmed or rejected" (Hastings *et al.* 1993) and these must be series with constant model parameter values.

Costantino *et al.* (1995) studied application of a difference equation model separately to populations of two strains of flour beetle, *Tribolium*, with experimentally applied differences in adult mortality. The approach was to inoculate replicated 20 g of standard media with 100 young adult flour beetles, 5 pupae and 250 small larvae, and keep them in a dark incubator at 31 °C. Every two weeks the feeding larvae (L), non-feeding larvae, pupae and callow adults (P), and mature adults (A), were counted and returned to fresh media. "To counter the possibility of genetic changes in life history characteristics, beginning at week 12 and continuing every month thereafter, the adults returned to the populations after the census were obtained from separate stock cultures maintained under standard laboratory conditions" (Costantino *et al.* 1995).

A difference equation model represents birth of new larvae, developmental transitions between the stages and exponential non-linearity accounting for cannibalism of eggs by larvae and adults, and of pupae by adults. The fractions $\exp(-c_{ea}A_t)$ and $\exp(-c_{ei}L_t)$ are the probabilities that an egg is not eaten in the presence of A_t adults and L_t larvae; $\exp(c_{pa}A_t)$ is the survival probability of pupae in the presence of A_t adults. The set of difference equations is:

$$L_{t+1} = bA_t \exp(-c_{ea}A_t - c_{ei}L_t) \tag{12.6}$$

$$P_{t+1} = L_t(1 - \mu_1) \tag{12.7}$$

$$A_{t+1} = P_t \exp(-c_{pa}A_t) + A_t(1 - \mu_a) \tag{12.8}$$

where μ_1 and μ_a are, respectively, larval and adult probabilities of dying from other than cannibalism.

Costantino *et al.* (1995) were unable to observe population changes over the long time periods suggested by Hastings *et al.* (1993). Instead, they transformed the above equations into stochastic equations, estimated parameters using the experiment data, and then examined stability properties of the models with the different parameter values estimated for different levels of imposed mortality. As μ_a increased from low to high values populations showed a transition from stable fixed point, to periodic cycles and then aperiodic oscillations. As Costantino *et al.* (1995) note, their conclusions about the relevance of chaos theory for the population biology of the *Tribolium* depend upon the adequacy of the model.

As with many new theories the proponents of chaos theory can be enthusiastic; "The challenges are many, yet the chance of success is high" (Hastings *et al.* 1993). Schaffer and Kot (1986) describe some of the far-reaching consequences if chaos theory does apply in ecology (rather than merely to some dynamic systems models that have been constructed). In particular the usefulness of equilibrium-based theories would be questioned, as would the operation and importance of competition as a driving force in ecological systems. Similarly, Karieva (1995), commenting on Costantino *et al.*'s (1995) results suggest that understanding underlying population dynamics is essential to understanding the consequences of environmental perturbation. The high mortality rates imposed on the flour beetles produced variation in population numbers, not simply suppression of the population, which might otherwise have been expected. Karieva considers this an important result and concludes "Costantino *et al.* show us how the marriage of nonlinear models and experiments can help accomplish the task".

If chaos is found in ecological systems then it will be an example of the third level of model assessment, where a model explains an unexpected result. However, the rules of modeling, particularly the requirement of representation as a closed system of difference equations, and the procedures, particularly the requirement for a long time series for assessment, both raise difficulties. Costantino *et al.*'s attempt to short cut the assessment process by solving for model parameters over limited time periods itself needs to be tested as a method. They introduced axioms, both in the model form, which they note, and, importantly, through the method of investigation, i.e., the mathematical model is examined using a synthetic construction. These introduced axioms and methods of investigation limit the scope of the why-type question to the restricted circumstances of the flour beetles manipulated in a particular way. Note particularly the requirement to maintain the genetic structure of the population. Working in a restricted domain it is often essential to establish the basic axioms of a theory – but it is equally important to recognize what those restrictions are, and not automatically transfer the result to a more complex domain.

12.2.3 Fitting dynamic systems models to ecological systems

Dynamic systems models can be made from actual systems. Carpenter and Kitchell (1987) constructed such a model of cascading trophic interactions in lakes. Starting with the problem that only about half of the measured variation in lake primary production can be explained by variation in nutrient loading they asked whether the residual variation could be explained by variation in predation, i.e., top-down effects cascading through trophic relationships. They synthesized their own, and others', previous research of the relationships between phytoplankton of different types, herbivores on the phytoplankton and predators on the herbivores (Fig. 12.5).

Their model is a series of coupled differential equations expressing what the rate of change of each compartment depends on. The mathematical formulations used depended upon the requirements of the relationship to be

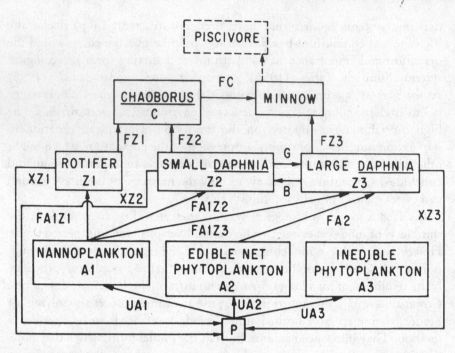

Fig. 12.5. Groups in the trophic system of a lake are depicted by compartments representing planktivorous fish (minnows), an invertebrate planktivore (*Chaoborus*), three size classes of herbivore (rotifers, juvenile *Daphnia*, and adult *Daphnia*), three sizes of phytoplankton, and phosphorus, the nutrient limiting primary production. The phytoplankton classes represent groups of algae that respond differently in grazing experiments. The basic unit of the model is grams of carbon. Interactions between compartments are shown by arrows between them and the abbreviations indicate functions representing the interactions. (From Carpenter and Kitchell 1987, with permission.)

expressed and were formulated from experiments and direct investigations of the relationships. For example, the instantaneous rate of change of *Chaoborus* (*C*) declined due to the rate of predation by minnows (*FC*),

$$\frac{dC}{dt} = FC \qquad (12.9)$$

Carpenter and Kitchell (1987) also consider that non-predatory mortality of *Chaoborus* might occur, but set that to zero for their analyses. Predation on small *Daphnia* by *Chaorborus*, represented in Fig. 12.5 by FZ2 was

$$FZ2 = a[1 - e^{-bZ2}]C \qquad (12.10)$$

which includes effects of the sizes of the *Chaoborus* and small *Daphnia* populations. Predation on rotifers was represented by a similar equation. The interactions are too complex to be solved as equations so the model was simulated on a computer proceeding with a time step of 0.04 days and using the Adams–Bashforth integration method. The complete model contains 16 coefficients, all estimated by Carpenter and Kitchell from their own and other scientists' experiments and field investigations.

Solely from the equations given here, only a small part of the complete model, one can appreciate that changes in predation by minnows on *Chaoborus* will affect the size of the small *Daphnia* population and this in turn

Fig. 12.6. Predictions of the model and observations in Peter Lake, 5 June 21 to August, 1984. Top, zooplankton biomass, mg dry mass m^{-3}; middle, chlorophyll a concentration in the epilimnion, mg m^{-3}; bottom, primary production in the mixed layer, mg C m^{-3} day^{-1}. Vertical bars denote 95 percent confidence intervals based on three mixed-layer samples. (From Carpenter and Kitchell 1987, with permission.)

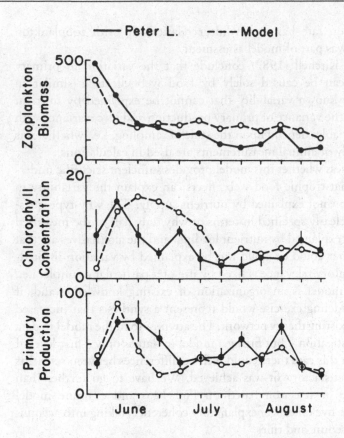

— Peter L.   - - - Model

may affect the amount of phytoplankton. The model represents a network of trophic feedback.

The first result presented by Carpenter and Kitchell (1987) was from a simulation when phosphorus levels and predation intensities of *Chaoborus* and *Daphnia* were set to those found in Peter Lake, Michigan, and time courses of zooplankton biomass, chlorophyll concentration and primary production were both measured and simulated (Fig. 12.6). Carpenter and Kitchell state: "We conclude that the range of dynamic behavior expressed by the model is a reasonable representation of the dynamics occurring in the mixed layers of lakes during summer stratification". Clearly their focus was on general patterns of variation in the three output variables. However, each variable shows consistent, not just random, deviation between simulated and measured. For example, zooplankton biomass was underestimated for the first three time periods and overestimated for the last nine, and for both chlorophyll concentration and primary production the simulated values were more often outside the 95 percent confidence intervals of measurements than within them. Data from the model outputs was also used to illustrate relationships that Carpenter and Kitchell thought reasonable, e.g., that primary production rate has a curvilinear relationship with zooplankton biomass and this was part of model assessment.

Carpenter and Kitchell (1987) conclude that the variation in primary production that can be caused solely by food web effects is similar in magnitude to the known variability that cannot be explained by nutrient supply. Also, that the variance of primary production and its covariance with other dynamic variables depend upon the scale of sampling, i.e., whether sets of weekly, monthly or annual measurements are used in calculations.

How can we assess whether this model provides sufficient scientific understanding to say that trophic food web effects can explain the variability in primary production not explained by nutrient supply? The why-type question (Fig. II.2) is clearly specified in terms of why only half of the measured variation could be explained by nutrient loading, and the alternatives are that the remainder either could or could not be explained by variation in predation. The explanatory relevance is fixed by use of a particular example, i.e., Peter Lake. The model is an organization of existing knowledge and, if successful, the modeling exercise would represent a synthesis that increased confidence in an existing theory network. The attempt to fit the model to data is a scientific investigation – but fitting a model to data is only a first level of assessment, and in this case there could well be differences between scientists as to whether a satisfactory fit was achieved. We have to go further than simple acceptance or rejection based on Fig. 12.6 and examine model performance in the five criteria of explanatory coherence, taking into account the modeling procedure and rules.

(1) Acceptability of individual propositions The model is a quantitative synthesis of previous empirical research and there is considerable evidence in support for the propositions. There are important axioms in the construction of the model about what needs to be included and how it should be represented, e.g., considering *Daphnia* in two size categories and considering phytoplankton in three size categories known to respond differently in grazing experiments.

Carpenter and Kitchell were satisfied with model–data fit, for them it provided confirmatory evidence that top-down effects could be responsible for variation in measured primary production rates. However, systematic deviation of simulated results from measurements suggests that the model structure, the form of some relationships, some parameter estimates, or combinations of all three, were only partly correct. No quantitative rules for accepting or rejecting the model were used – generally they are not defined in applications of dynamic systems modeling in ecology.

Taking the view that this model is sufficient for its purpose, i.e., that it illustrates that variation in piscivore predation could be sufficient to account for unexplained variation in primary production, highlights the question of what accuracy should be required in the model–data relationship. The deviation between model and data could have a number of contributing causes, including faults in model structure.

(2) Consistency of definitions between theory and model Mathematical modeling forces precision in concept definition that may extend beyond that specified by the theory. In a model, quantities have to be defined, as must relationships such as increase or decrease, even though the theory may not be explicit about those definitions. Carpenter and Kitchell estimate many relationships through additional data – but the form of those relationships was chosen as a best fit to each particular data set. This may not necessarily be effective even though it may be the only available procedure at some stages. Empirically based relationships could distort the theory being modeled.

(3) Consistency in part and kind relationships between theory and model Carpenter and Kitchell make an important distinction for their model between size classes of *Daphnia* and types of phytoplankton, i.e., these become different kinds. Failure to maintain such distinctions i.e., inappropriate aggregation of variables, often in an attempt to simplify a model, can lead to the model failing to explain the phenomenon.

(4) Simplicity of propositions in both theory and model The basic propositions in Carpenter and Kitchell's (1987) model are simple in describing already-researched relationships between trophic compartments and nutrient and primary production rate. But a concern, as with many models, is whether the detailed forms of the equations are accurate and can be interpreted in an ecological sense. For example, the exponential term in the equation to describe predation on small *Daphnia* by *Chaoborus* (and a similar term in the equation describing predation on rotifers) accounts for the effect of changes in the size of the *Daphnia* (and rotifer) population. Carpenter and Kitchell (1987) do not justify the exponential term explicitly but follow previous researchers in using it.

Although models are frequently considered to be simplified representations of theory, this example illustrates how complexity occurs because of the modeling process. In effect the rules of modeling make each equation an additional axiom. In a verbal description we say, "As the amount of small *Daphnia* decreases the absolute amount of predation on *Daphnia* decreases". Then, as we model the relationship, we are forced to make a quantitative description, adding an axiom of the type "This decrease can be represented by an equation of such and such particular form." These problems of quantification must be explored – regardless of the extent to which a model "fits" the data. Model fit may be the result of compensating deficiencies between different equations and apparent lack of fit may be influenced by compounding interaction between the forms of the equations – even though the model is correct in general structure.

(5) Breadth of the model explanation The model is constructed primarily from investigations into particular lakes. Claims that the model has general applicability were not made on the basis of a detailed analysis of model results for different lakes. Carpenter and Kitchell make the case for general application on a general argument about the types of lakes that were the basis of the research, and how the inclusion of other factors into the model may, or may not, alter its predictions. This is the appropriate way to use such models rather than assert that this one model represents all such systems.

The value of Carpenter and Kitchell's model is that it illustrates the dynamics of a complex system – even though we are uncertain of the details of the representation. With dynamic systems models, even those representing actual systems, as this one does, assessment is typically made on the ability to predict something that seems reasonable. There may be no explicit quantitative standards and in some cases, as with rejection of the Lotka–Volterra model for the snowshoe hare–lynx cycles, none may be needed. At this stage in the investigation it is a reasonable to say that predation is a possible explanation of variation in lake productivity not explained by differences in nutrition – but is not sufficiently strong to be called the only, or a certain explanation. (For further research, see Carpenter and Kitchell 1993a,b.)

12.3 Statistical models of dependence

Statistical analysis is based on models. For example linear regression analysis assumes the relationship between x and y can be modeled by the formula $y = a + bx$. The important feature is that a and b, and their variances, and the residual between fitted model and data, can be estimated. When fitting a linear regression equation we assume that the model holds and become interested in the parameters, their variances, and the residuals: assessment is an automatic consequence of model fitting. Scientists can investigate for better-fitting models, e.g., to see whether non-linear regression, or multiple regression, give better fits.

Sometimes a distinction is made between statistical models, for which parameters have no causal meaning, and process (or mechanistic) models for which they do. In fact, statisticians have developed models where the parameters can have biological or ecological meaning. One important example is time series analysis modeling. Its overall purpose is to analyze how ecological systems are *dependent* over time, e.g., conditions today (this year, etc. . . .) influence conditions or events tomorrow (next year, etc. . . .). In ecology many things, repeatedly measured over time, do show distinct patterns suggesting dependence. The size of a population in one year may have an important effect on its size next year and a sequence of yearly measurements may show this. Variation in successive tree rings is correlated – the effect of markedly good or bad weather on tree growth lasts for more than one year (Fritts 1976).

The statistical theory of *stochastic processes* is used to model situations where such dependencies occur. In discussions between ecologists I have heard the word *stochastic* used to describe variability, measurement error, and situations where there is only partial understanding. This confusion is unfortunate. The word stochastic has a distinct and closely defined meaning in statistics that can usefully be applied in ecology to describe our understanding of certain types of causal process.

STOCHASTIC PROCESS

A *stochastic process* is a continuous causal process in time, space, or both, responding to variation in an external influence, and producing a varying series of measured states or events.

STATISTICAL MODEL OF A STOCHASTIC PROCESS

A *statistical model of a stochastic process* represents dependence of successive and/or neighboring events in response to variation in an external influence on the process. These models are parsimonious, using the fewest number of parameters capable of explaining quantitative variation in some observed data.

12.3.1   Modeling dependence in time series as a stochastic process

A requirement for time series analysis is repeated measurements of the output of a system at equally spaced time intervals, e.g., daily measurements of the length of a growing conifer shoot during spring and early summer in a temperate latitude, called here shoot extension (Fig. 12.7). So there is a series of values:

$$x_1, x_2, x_3, \ldots, x_n.$$

and the questions are: "What makes x vary? Can we model how the variation is produced?"

Underlying shoot extension there are causal processes, i.e., photosynthesis, translocation, cell formation, cell expansion, and other aspects of growth that we expect are influenced by weather conditions. However, a detailed model structure is not specified in advance. Modeling starts with output from the system (daily shoot growth), and inputs (daily weather variables) postulated to influence the output, and the task is to calculate how the causal process may

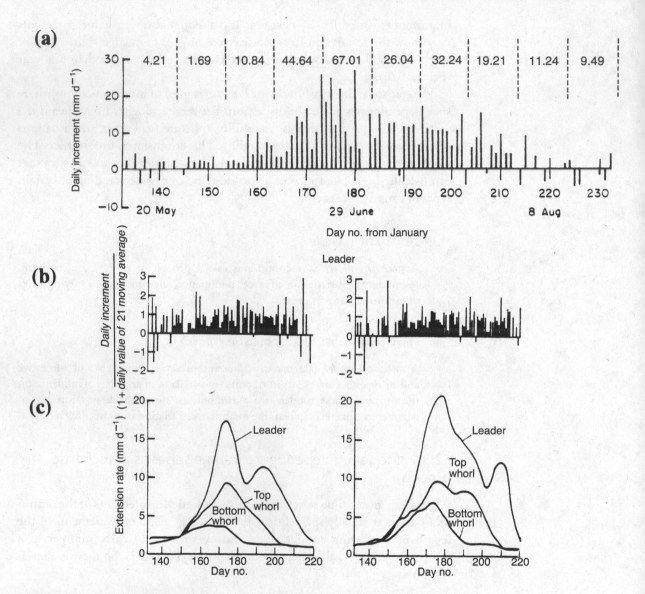

Fig. 12.7. (a) Measured daily increment from the leading shoot of a *Picea sitchensis* tree in a plantation in south-west Scotland. The variance for 10-day intervals, between the dashed lines, increases then decreases over the growing period; (b) detrended daily increments for the leading shoot of the two trees; and (c) trends in the extension rate of the leading shoots and shoots from the top and bottom live whorl of two trees, as defined by 21-point moving averages. Note general similarities between trees, in both detrended increments and trends. Quantitative models must be calculated for individual trees and interpreted to define a theory. (Redrawn from Ford *et al.* 1987a.)

respond to the inputs in producing the output (Fig. 12.8). The model is calculated as a transfer function, i.e., an equation that transfers the patterns in the weather variables to patterns in shoot extension. It does not express direct

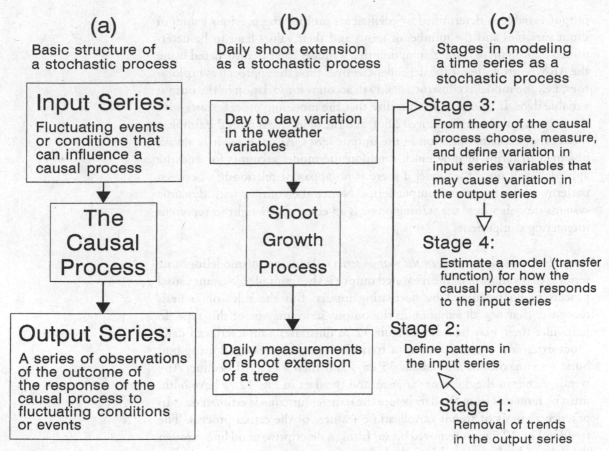

(a)
Basic structure of
a stochastic process

(b)
Daily shoot extension
as a stochastic process

(c)
Stages in modeling
a time series as a
stochastic process

Input Series:
Fluctuating events
or conditions that
can influence a
causal process

The
Causal
Process

Output Series:
A series of observations
of the outcome of
the response of the
causal process to
fluctuating conditions
or events

Day to day variation
in the weather
variables

Shoot
Growth
Process

Daily measurements
of shoot extension
of a tree

Stage 3:
From theory of the causal
process choose, measure,
and define variation in
input series variables that
may cause variation in
the output series

Stage 4:
Estimate a model (transfer
function) for how the
causal process responds
to the input series

Stage 2:
Define patterns in
the input series

Stage 1:
Removal of trends
in the output series

Fig. 12.8. (a) General
description of a stochastic
process that produces a time
series of observations. (b)
Measured daily shoot extension
of a tree depends upon daily
fluctuations in the weather and
the way that growth processes
respond to that variation. (c)
The sequence of modeling a
time series as a stochastic
process. Note that investigation
starts with the output series,
i.e., the observed performance
of the process of interest.

relationships between the component causal processes, but it does indicate
how they may function.

In practice the value of x at any time, t, may depend in part on what
happened in previous time intervals and in part on the new conditions that
are influencing performance of the causal process in the current interval. For
example, shoot extension today may depend upon whether the last few days
have been sunny and so whether sufficient photosynthate is available for
growth, as well as today's temperature and/or solar radiation.

Statisticians have developed a particular type of model structure, and rules
for fitting it to data, so that the dependence between inputs and outputs can
be defined quantitatively and the error in the model can be estimated. The
models most useful to ecologists and biologists are usually referred to as of the
autoregressive–moving average (ARMA) type.

Modeling such a time series as a stochastic process usually involves a stage
to ensure the data is in a suitable form for this type of analysis (Stage 1, Fig
12.8c) and three stages for model identification and fitting (Stages 2 through
4). Transfer functions have a particular form where by the current value of the

output variable is determined by coefficients multiplying previous values of input variables and the number of terms and their values have to be determined. While there is the assumption that a model can be constructed using the ARMA structure for the dependencies over time the approach is exploratory, i.e., the model is constructed only to account for variation in the output variable data. It is important to realize that the modeling process starts with the output, not the inputs (Fig. 12.8c), and the objective is to fit the simplest model that explains variation in the output series. An assessment is always obtained in terms of how much variation the model accounts for and the approach will not fit a model if there is no apparent relationship between patterns in the input and output series. Notice the contrast with dynamic systems models where the starting point is an equation thought to represent interacting components.

Stage 1: Removal of trends in the output series   An axiom of modeling time series is that variation in the measured output is the result of a constant causal process being perturbed by fluctuating inputs. But the rules of analysis recognize that not all variation in the output series may be of this type, in particular there may be trends. Figure 12.7a illustrates a time series of daily shoot extension of a tree that has a trend, an overall increase following bud burst to a maximum after about 35 days and then a gradual decline. Any trends, either in the mean or variance, and the data in Fig. 12.7a have both, must be removed from the data before the transfer function is estimated.   · In practice analysis of trends can illustrate features of the causal process. The trend in Fig. 12.7a was removed by: (a) fitting a descriptive trend line through the data and calculating the residuals (actual measurement minus trend line value at that time), which corrected for variation in the mean; and (b) dividing each of these residual values by (1 plus trend line value), which corrected for the variance (Fig. 12.7b). The two corrections have biological significance. The descriptive trend line defined the characteristics of periodicity in the shoot meristem, i.e., when it started to grow, its time of maximum activity, and when it stopped, and trend lines for shoots on different positions of the tree showed small differences in the time of initiation of growth and times of maximum extension, but large differences in the time of cessation, the shoots at the top of the tree growing for longer (Ford *et al.* 1987a). The trend in variance indicates greater sensitivity to fluctuation in the environment when growth rates are high.

Stage 2: Define patterns in the output series.   The next task is to see what types of dependency may exist in the output series. The first statistic to calculate on a detrended series is the *autocorrelation function*; the correlation between individuals and their first neighbors in the series, then between individuals and their second neighbors, third neighbors, etc. Notice it is the autocorrelation *function*, i.e., there are a number of terms, each one representing the correlations between the individuals and neighbors at successive distances apart, 1, 2, 3, etc., referred to as successive lags, 1, 2, 3, etc. The first term of the autocorrelation function is the correlation of a time series with itself, i.e. it has a value of 1 at lag 0. The sample autocorrelation function $r(j)$ where j is the order of lag;

$$r(j) = \frac{\sum_{r=j+1}^{T} (x_t - \bar{x})(x_{t-1} - \bar{x})}{\sum_{t-1}^{T} (x_t - \bar{x})^2} \qquad (12.11)$$

The use of this statistic is illustrated in Fig. 12.9, where it is applied to detrended daily mean temperature and large oscillations can be seen in the data that are clearly reflected in the autocorrelation function. Analysis of the autocorrelation function led to the series being represented by an equation with the form:

$$X_t = b_1 X_{t-1} + b_2 X_{t-2} + \varepsilon_t \qquad (12.12)$$

b_1 and b_2 were fitted as parameters, and ε_t represents a random noise series.

The autocorrelation function is effective in identifying patterns, even in series where there is considerable variation, as there is in the detrended shoot extension (Fig. 12.7b). Generally autocorrelation functions of replicate detrended shoot extension series were zero or negative at lag 1 and were significantly positive for lags between 2 and 5. It turned out that the effects of temperature and solar radiation on shoot extension were complex and, while themselves having distinctive patterns, together they produced variation in shoot extension that had less visually obvious, but still analytically detectable, pattern.

Stage 3: Choose and measure input series that may cause variation in the output series The choice of input variables reflects the scientists' view of the theory underlying the investigation and a quantitative assessment of that choice is made during data analysis. In this example the input variables were chosen when planning the investigation. The scientific literature emphasized the importance of temperature but not radiation, and the generally accepted theory was that shoot extension was not dependent on current photosynthesis but could use stored carbohydrate. This was under investigation test in this research – although other variables were also measured, and there was, in effect, a postulate that each may have an effect on shoot extension on this species in this environment.

Daily measurements of temperature (Fig. 12.9a) and solar radiation themselves had substantial but different patterns. The temperature data has clear

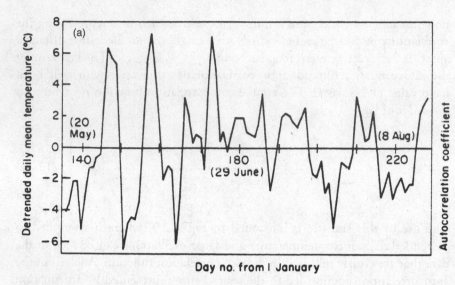

Day no. from I January

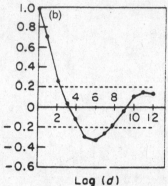

Lag (d)

Fig. 12.9. (a) Detrended daily mean temperature, and (b) its autocorrelation function with estimates of significance for departure from zero, $P = 0.05$ (-----) at the same plantation in south-west Scotland, and over the same period as given in Fig. 12.7. (Redrawn from Ford *et al.* 1987b.)

oscillations, and an autocorrelation function with positive terms at both lag 1 and lag 2 (Fig. 12.9b). (If it is warm/cold today, it will tend to be warm or cold tomorrow and the next day.) Solar radiation had positive terms only at lag 1. (If it is sunny/cloudy today, it will tend to be sunny/cloudy tomorrow.) The characteristic wave-like variation in temperature reflected the passage of warming cyclonic weather across south-west Scotland in early summer and was of longer cycles of positive correlation than for solar radiation.

Stage 4: Estimate a model (transfer function) for how the causal process responds to the input series   The cross-correlation function is used in analysis of inputs on an output, e.g., of temperature and solar radiation on shoot extension rate. This is similar to the autocorrelation function but is calculated from the detrended values of input series (e.g., temperature or solar radiation) and the detrended measured output series (shoot extension). At this stage there are technical rules to follow that are not described here (see Box and Jenkins 1970, Diggle 1990; computer programs are available, e.g., S-Plus, Statistical Sciences, Inc. 1993).

The model produced, called the transfer function, is an equation that when multiplied with the input series (in this case the daily series of temperature and radiation) produces an estimate of the detrended output series (the daily shoot extension rate); that is, it "transfers" the patterns in the input series to the output series. It describes the effects of the causal process linking the two. Where daily shoot extension rate is represented by y (mm), temperature by T (°C), and solar radiation by R (MJ m^{-2} d^{-1}) the transfer function model for shoot extension of topmost branches of one tree during the first 35 days of growth was:

$$y_t = 0.133 T_{t-1} - 0.042 T_{t-2} + 0.0107 R_{t-2} + 0.0150 R_{t-3} \qquad (12.13)$$

The parameters of this equation do have a precise ecological meaning: increases in temperature had a positive effect with a one-day lag, and smaller but negative effects at a two-day lag. Increase in daily solar radiation had a positive effect at lags of two and three days, but notice there is no effect for solar radiation at $t-1$. On average, across the range of shoots investigated, 80 percent of the variation in daily shoot extension could be accounted for with models of this type and there were no consistent differences between shoots in different positions on trees, or between trees.

12.3.2   Assessing a stochastic time series model as an explanation

The rules of time series modeling are clearly laid out for application as a statistical technique. It proceeds sequentially by modeling patterns in input and output data and calculating the transfer function between them, and assessing at each stage. The only models produced are those that fit data, at least in terms of removing patterns in the output data, and their effectiveness can be assessed in terms of variance reduction. But do such models provide a good explanation when assessed by the standards of explanatory coherence?

As with many statistical models the explanatory relevance (Fig. II.2) is clear: variation in a particular data set must be accounted for. The contrasting alternatives are whether or not it is necessary to include variation in temperature and solar radiation in the model, although these alternatives are not specified in detail before the modeling procedure is undertaken. The topic is specified by the data and the why-type question is implicit and of the form "Why do variations in daily shoot extension occur?" The model-based synthesis is in the form of a transfer function.

(1) Acceptability of individual propositions   A strength of time series modeling is the model selection and estimation process. Conjectures are made that effects can be detected by including particular measurements and/or measuring at a particular frequency with specified accuracy. However, the exact model produced, while being sufficient to explain a particular data set, may not be replicated precisely for other data sets. Time series analyses require an interpretive phase that must use what else is known about the system.

In the example, modeling was valuable in offering new insights into causal processes. It produced a new result in shoot extension studies: that solar radiation had a positive effect but lagged two days and more. Previous analyses, frequently using simple, or specifically one-day lagged, correlations, but not the complete auto- and cross-correlation functions, had found effects of temperature but not solar radiation, supporting the (possibly wrong) conclusions that shoot extension is not affected by recent photosynthesis. The negative effect of temperature with a two-day lag is interesting (if it was a warm/cold day two days ago, shoot extension will tend to be less/more). It may be that rapid growth, which can take place for one day, then limits subsequent growth and vice versa. Similar equations were obtained for other shoots, so replication was achieved.

However, the transfer function applied only to the first 35 days of growth when there was substantial variation in temperature and solar radiation. A transfer function could not be obtained for the period following the first 35 days. This probably does not mean that shoot growth is not influenced by temperature and solar radiation variation, but that these inputs did not vary sufficiently to have a measurable effect, temperature variation in particular was small, and/or that other processes became more important. For example, in this forest, as the summer proceeded, soil moisture could be reduced and affect growth (Deans 1979) – and increasing water deficits caused increasing tissue shrinkage during the summer (Milne *et al.* 1983), but these processes were not measured for this analysis.

This modeling approach can also be limited by measurement accuracy and precision. In this example the percentage of variation that could be explained by a time series model decreased for shoots in the lower part of the tree where shoots grew less in total amount and the relative measurement error was greater.

(2) Consistency of definitions between theory and model   Transfer functions define effects, but not how that effect is achieved. In this example the transfer function is written in units of the effects of two environmental variables directly upon daily shoot increment. The complete theory of environmental influences on shoot extension would consider the physiological components of growth, e.g., radiation on photosynthesis and plant water status and temperature on physiological rates. The concept of *effect* is not defined in the same way between model and the larger theory, and transfer functions are sometimes referred to as black box models.

(3) Consistency in part and kind relationships between theory and model   The transfer function construction can limit part and kind connections between different parts of a theory and must be interpreted in the light of underlying theory.

(4) Simplicity of propositions   The strength, and at the same time the limitation, of time series modeling is that because it is driven by a quantitative estimation process it is constrained by available data. The recommended principle of model construction (Box and Jenkins 1970) is parsimony, i.e., to fit the simplest model describing variation in the data. Only what is necessary to explain this variation is included in the transfer function. In this sense the transfer function is simple. There are two difficulties.

First, the procedure of model identification (how many terms, at what lags, and for what inputs) followed by model fitting (precise determination of transfer function parameters) does not always produce uniquely best models, i.e., transfer functions with different combinations of input variables may fit almost equally as well, although some models are certainly rejected. A choice may have to be made in exactly how parsimonious to be. In this example, models were constructed for 25 shoots from different trees and at different positions on them. For some, though not all, the variance accounted for could be increased, though not a great deal, by increasing the number of terms, e.g., to include three or four previous days of both input variables rather than merely two. Achieving a fit to the data is not an entirely objective process and an interpretation had to be made to produce a general scientific inference. In this case, by examining models fit to different shoots, Ford *et al.* (1987b) conclude that: (a) temperature affected shoot extension with a lag of one or two days, and solar radiation, with a lag of two to three days; (b) temperature had a larger effect than solar radiation; the approximate range in daily temperature during the period was 10 deg.C which would give a change in daily shoot increment of close to 1 mm whereas the range in daily solar radiation was approximately $10 \, \mathrm{MJ} \, \mathrm{m}^{-2}$, which would give a change of close to 0.2 mm.

Second, time series models require that the range of variation in the input series has perturbed the causal process sufficiently to vary the output series. That the same model did not fit for the second half of the shoot extension period (so the model fails the test of prediction) does not mean that the same causal processes were not operating but that they did not dominate the system in the same way. We would not reject the model because it failed the test of prediction, but rather set a limit to its domain of application.

(5) Breadth of the complete explanation   The use of time series modeling is based on specific data sets. The focus of assessment is on fitting but different data sets should be used to explore variation in model structure and/or parameter values. General conclusions must be made through synthesis of a sequence of modeling results with other information. The approach contrasts with that sometimes taken using dynamic systems models in ecology when the equations are proposed as general explanations and research must be made to test that. Time series modeling is an empirical approach, where improvements in measurements, and increasing the type of variables measured, can improve models, and successive investigations can extend the generality of conclusions. But this requires an underlying process-based theory, e.g., that shoot extension is controlled by plant physiological processes. This type of modeling is valuable for examining how component processes may be linked to provide explanations in different circumstances.

There are practical difficulties in using time series analysis, particularly that a considerable number of equal interval data points are needed. Thirty is a suggested minimum for calculating the autocorrelation function, but this permits only the rudimentary nature of the dependencies to be described in simple terms. Usually, and particularly if it is required to do *input → output* modeling, more data points are necessary.

Time series analysis can be valuable as an exploratory method for uncovering possible dependencies, even without going as far as calculation of a transfer function. An examination of pattern extraction from a number of different ecological series is given by Ord (1979). Francis and Hare (1994) use time series analysis of salmon catch records and ocean temperature measurements to propose a system of decadal shifts in ocean processes in the north-east Pacific Ocean.

12.4 Systems simulation models

SYSTEMS SIMULATION MODEL

> *Systems simulation models* are hybrids using the flexibility of computer programing languages to represent changes in states and conditions, as well as changes in rates, by different types of mathematical function, empirically based relationships, and incorporating stochastic forcing.

For systems simulation models speed and storage capacity of computers are used to examine how the details of model construction, and differences in fitted parameter values, influence model outputs.

Systems simulation models can represent a wide range of component processes extending beyond that of either dynamic systems models or statistical models. The model is defined by the computer program algorithm rather than a set of equations. Carpenter and Kitchell's (1987) dynamic systems model produced biased outputs relative to measured data and we might consider improving on that with a systems simulation model, perhaps by representing some relationships with non-calculus-based formulations and including factors in addition to trophic relationships. Ford *et al.*'s (1987b) time series transfer function model of shoot extension fit the first, but not the second, half of the measured growing period. Perhaps, by constructing a systems simulation model with direct representation of photosynthesis and the growth process and including effects of soil moisture and rainfall, it might be possible to explain fluctuations in daily shoot extension for the whole growing season. However, as the number of model components and associated parameters increases, so too do the problems of model assessment.

Through the example of WHORL (Sorrensen-Cothern *et al.* 1993), a model of competition between trees, the requirements for constructing a systems simulation model are defined, techniques used in model construction and assessment are illustrated, and the extent to which a systems simulation model can help in developing explanatory coherence for a theory is examined.

12.4.1 Objectives, theory, and model design

The overall purpose of WHORL was to develop theory for the process of aboveground intertree competition in single species, even-aged, stands. Generally, intertree competition models had been developed using statistical models (usually regression). They described how trees of different sizes, and distances apart, influenced each other's future growth. However, no generally accepted mathematical equation of this relationship was available (Ford and Sorrensen 1992), i.e., the models did not support a general theory for the dynamics of intertree competition. There were two important problems.

First, the representation of tree size simplified what may make a tree successful in competition with a neighbor. Although it was generally assumed that competition was for light between neighboring crowns, trunk diameter at breast height (dbh), or some transformation of it, was frequently used as a surrogate for crown size because of positive correlation between the two and dbh is routinely measured in forest research. There was no clear agreement as to which surrogate measure could be used most effectively, and a single measure of overall tree size alone may not describe to what extent a crown affects neighbors (Ford and Sorrensen 1992). Sorrensen-Cothern *et al.*'s (1993) objective was to dispense with surrogates and represent the process of competition between the foliage of tree crowns directly.

Second, models were for the most part fit to data from older stands with trees widely, and frequently evenly, spaced. Such data typically become available when you can easily walk through a commercial forest. The objective was to develop a model for heterogeneously spaced trees competing at the early stages of canopy development when the stand is a dense thicket and competition may actually be more intense than in later stages of stand development. The opportunity arose because permanent plots had been established in dense (80 000 trees ha^{-1}) stands of irregularly spaced, naturally regenerated *Abies amabilis*, Pacific silver fir, 3–5 m tall. Individual trees had been mapped, so spatial interactions could be investigated, and growth and death over a 10-year period had been recorded.

A body of theory existed about morphological plasticity (e.g., tree crowns and branches growing with different shapes as competition occurs, changes in height increment as trees become shaded) and physiological acclimation (e.g., foliage photosynthesis rates differing with increasing shading, increasing efficiency of light interception by more shaded foliage) of trees growing in closed stands (Sorrensen-Cothern *et al.* 1993). Investigations of factors controlling the amount and distribution of branches and foliage, and photosynthesis rates, had shown that morphological plasticity and physiological acclimation occurred in *Abies amabilis* at the site, and our objective was to use this information in the model to examine its possible effect on the competition process. The system simulation model was to investigate a more complex theory of above-ground competition than is assumed in regression-based models. The why-type questions (Fig. II.2) involved why and how the complex phenomena of mortality, dominance and plasticity determine the outcome of competition. Contrasting alternatives existed in the form of the model needed to fit the data, particularly whether it was necessary to simulate morphological plasticity and physiological acclimation. Answering this required modeling foliage at different parts of the tree separately. Theories that plants could be considered as modular were of current interest (e.g., Hardwick 1986). So a further question was: "Do branches and foliage act as independent components?". However, while competition might be represented as the interaction between modular components, in this case foliage-bearing branches, there was also the requirement to simulate some whole-tree properties, particularly height growth. The explanatory relevance was given focus by a particular data set. This has advantages when a new complex is being explored but obviously limits the generality of an explanation.

The technical difficulty in constructing systems simulation models, such as WHORL, brings us face to face with two important and related questions:

(1)   How much should be included in a systems simulation model?
(2)   How can we assess the effectiveness of such a model?

Choices made in what should be included, and how components should be represented, determine the model's *bounds*. In WHORL, competition is for light, not for nutrient or water, which would require simulation of soil processes. This is an example of how WHORL is bounded above, i.e., how much is represented in the model. In WHORL, light absorption by foliage is converted to values for tree height increment and branch increment, conversion rate depending upon foliage characteristics that differ according to foliage position in the canopy. Foliage lower in the canopy intercepts a greater proportion of light received (morphological plasticity) and may convert that to increment at a a higher rate (physiological acclimation). However, known details about the conversion process, e.g., photosynthesis rates, respiration, translocation, metabolism of conversion of translocated material at the growing branch and tree apices, are not represented. This is an example of how WHORL is bounded below, i.e., how much of the details of a component process are represented. Both interception rate and the conversion rate are aggregated parameters.

The way aggregation should be done is governed by the purpose of the model – but wherever the bounds are drawn becomes an axiom created by the modeling process. Bounding limits the number of parameters to be estimated, but it may result in a crucial process not being represented effectively. The process of bounding means that the model may be incompletely specified, particularly by aggregated parameters, and/or underdetermined, where choices are made about the processes to include and there may be

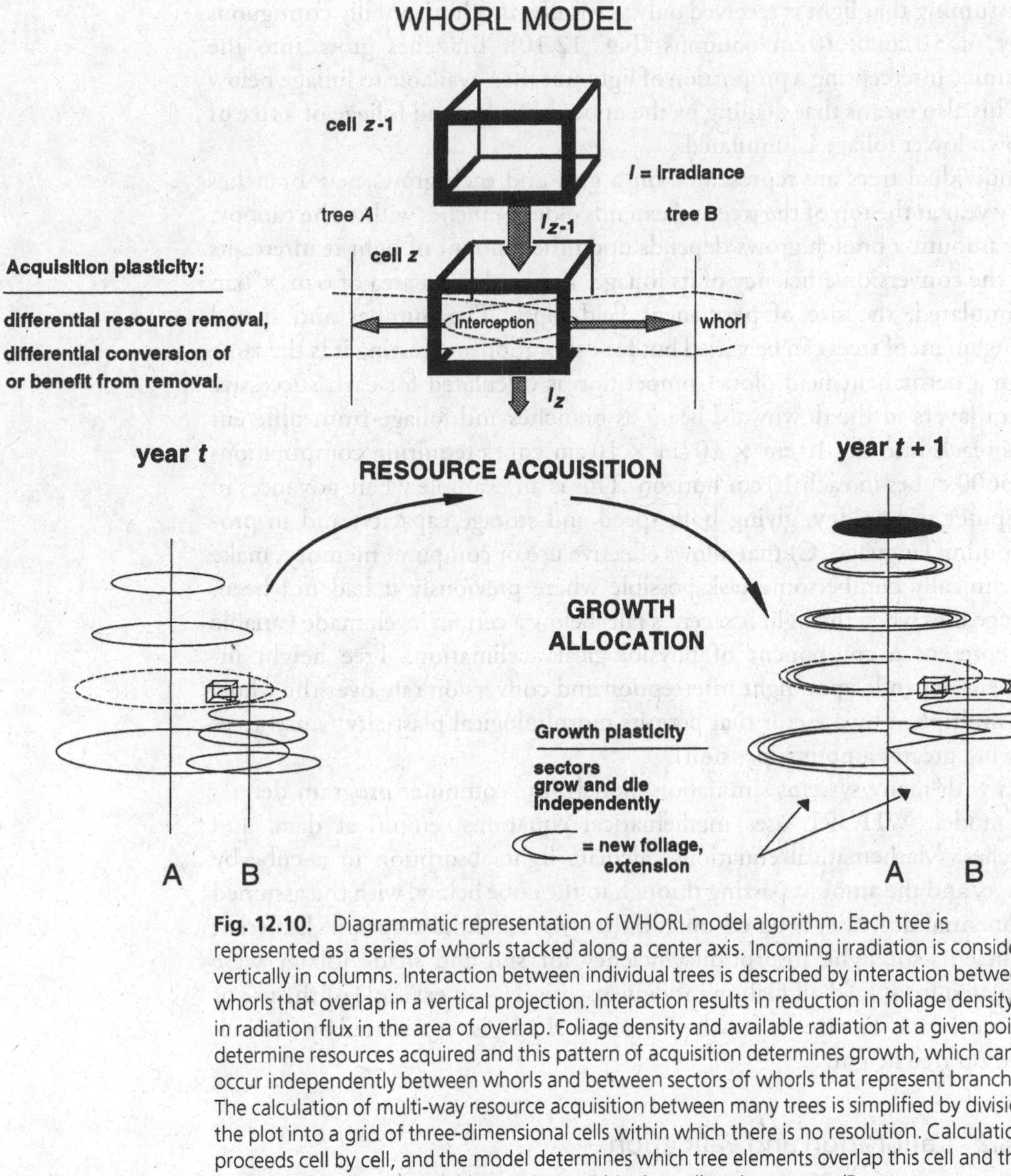

WHORL MODEL

Fig. 12.10. Diagrammatic representation of WHORL model algorithm. Each tree is represented as a series of whorls stacked along a center axis. Incoming irradiation is considered vertically in columns. Interaction between individual trees is described by interaction between whorls that overlap in a vertical projection. Interaction results in reduction in foliage density and in radiation flux in the area of overlap. Foliage density and available radiation at a given point determine resources acquired and this pattern of acquisition determines growth, which can occur independently between whorls and between sectors of whorls that represent branches. The calculation of multi-way resource acquisition between many trees is simplified by division of the plot into a grid of three-dimensional cells within which there is no resolution. Calculation proceeds cell by cell, and the model determines which tree elements overlap this cell and then assigns acquisition of available resources within that cell to those trees. (From Sorrensen-Cothern *et al.* 1993, with permission.)

reasonable alternatives but we can not choose between them on the basis of available data. Incomplete specification and underdetermination are problems that can remain unacknowledged problems in modeling.

The assumption in WHORL is that competition for light is determined by interactions between branches and the foliage they support. This is simulated by assuming that light is received only vertically, in a horizontally contiguous array of 10 cm × 10 cm columns (Fig. 12.10). Branches grow into the columns, intercepting a proportion of light not then available to foliage below it. This also means that shading by the upper branches and foliage of a tree of its own lower foliage is simulated.

Individual trees are represented on a grid and each grows new branches every year at the top of the tree and extends older branches within the canopy. The amount a branch grows depends upon the amount of light it intercepts and the conversion efficiency of its foliage. In practice, an area of 6 m × 6 m is simulated, the size of permanent field plots. The number and spatial arrangement of trees can be varied but for calibration and testing it is the same as on a permanent field plot. Competition is calculated for each successive 10 cm layers in the downward beam as branches and foliage from different trees reach into the 10 cm × 10 cm × 10 cm cubes requiring computations for 3600 cubes in each 10 cm horizon. This is an example where advances in computer technology, giving both speed and storage capacity, and in programming language (C) that allows effective use of computer memory, make a technically cumbersome task possible where previously it had not been. Foliage dies when the light it receives falls below a certain level, made variable to represent a component of physiological acclimation. Tree height increment depends upon light interception and conversion rate over the whole tree multiplied by a factor that permits morphological plasticity (small trees growing greater amounts per unit).

As with many systems simulation models, the computer program defines the model. WHORL uses mathematical equations, empirical data, and switches. Mathematical equations calculate light absorption in a cube by foliage, and the amount passing through to the cube below, with the assumed exponential decrease, is based upon a theory for light absorption. Different coefficients for light absorption efficiency for sun and shade foliage were calculated from field and laboratory investigations – an example of the use of empirical data. The tree type, whether sun or shade, is represented by a switch based on tree height.

12.4.2 Calibration and validation

As the number of processes included in a model increases, then so too do the number of variables. Some are estimated independently from the modeling exercise as coefficients (as Carpenter and Kitchell (1987) did for all the parameters in the lake trophic interactions model). But for other variables there may be no information from other research and they must be estimated by fitting the model to a data set in a calibration procedure. Systems simulation models vary in the numbers of coefficients and parameters. WHORL is intermediate in this respect, eight parameters were estimated by fitting the model to field data for the growth of the forest and dynamics of competition in the stand. These include parameters determining efficiency of the rate of conversion from light absorbed by foliage to tree height growth and branch growth.

MODEL CALIBRATION

"A test of a model with known input and output information that is used to adjust or estimate factors for which data are not available." (ASTM 1984.)

To start the calibration process, initial ranges for each parameter are estimated. In some cases this is comparatively easy because information exists in the literature, at least for other species. In other cases, e.g., the conversion of light intercepted into growth, it is more difficult, since there is no empirical research available for what is an aggregated quantity.

The simulation process then starts. With an initial value for each parameter selected, and initial conditions of tree numbers, heights and foliage amounts, the stand is grown within the computer for 32 years, the last age when the stands were measured. Different combinations of parameter values produce different resulting simulated stands. A calibration is achieved when a set of parameters produces a simulated stand with the same characteristics as the measured one.

This should not be a haphazard procedure. The technique of *optimization* is used for many models. The results of simulations from successive sets of parameter values are compared to the actual structure of the stand to calculate the best (optimal) set, i.e., with the smallest difference between data and model output (Fig. 12.11). Of course, just having calculated an optimal set of parameters does not guarantee that this is a scientifically reasonable set. It is important to examine the effect of variation of parameter values, within their acceptable ranges, on the output of the model. This technique, called *sensitivity analysis*, can take place both before, and after, calibration, depending upon the particular model.

MODEL SENSITIVITY

The degree to which the model result is affected by changes in a selected parameters.

In the early stages of constructing WHORL we discovered, through sensitivity analysis, that unless small trees were permitted to have higher rates of conversion of light absorbed into growth (physiological acclimation) and greater rates of height growth per unit conversion from light absorbed (morphological plasticity) than large trees the simulation did not produce anything close to the observed stand, in particular mortality was

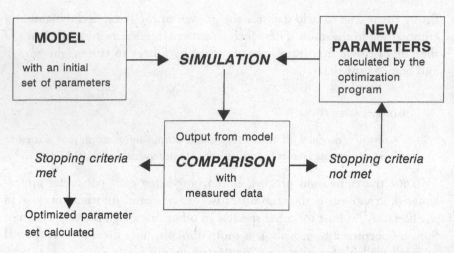

Fig. 12.11. Parameters in a model are *optimized* so that model output matches measured data. An initial set of model parameters is used first, and the difference between model output and measured data is calculated. A stopping criteria is provided to the optimization procedure which is the maximum difference allowed between simulated and measured data. If that has not been attained then the optimization procedure calculates a further combination of parameters likely to produce an improved fit. A further simulation is made with these and the comparison repeated. Different types of comparison, stopping rules, and methods for calculating the most likely new parameters can be used according to the particular type of model being investigated.

overestimated and the morphology of simulated tree crowns bore little relation to observed crowns.

There are important difficulties in calibration. Typically in systems simulation models many parameters (usually more than five) are estimated from a calibration data set. Parameter estimates may accommodate structural inadequacy in the model, i.e., they allow a model to fit the data even though some of its component functions are not correct. There may be no independent check possible for whether a parameter value is reasonable, other than that the model is deemed to have worked. The question also arises of which model output and corresponding data should be used in calibration, particularly when optimization procedures are employed that typically used a single variable. As assessment of the lake trophic dynamic model illustrated (Fig. 12.6), separate assessments against multiple parameters can be useful.

After calibration the next stage is validation.

VALIDATION

A test of the model with known input and output information that is used to assess that the calibration parameters are accurate without further change. Preferably, it should represent a condition different from that used for model calibration (after ASTM 1992).

Not all ecological systems simulation models go through this process, since a different data set may not be available, or even impossible to obtain. There has been considerable discussion about validation of ecological models (e.g., Rykiel 1996). The process of examining whether a model can simulate a second data set, i.e., predict the outputs within reasonable bounds given particular inputs, is reasonable, but there is dispute over what can then be said about the model. The term *validation* implies that the model can be considered true. However, if parameter estimates accommodate structural inadequacy that will not be identified through validation. Of course the confirmation of a validation is useful but falls short of being able to say that the model is true.

12.4.3 Assessment using multiple outputs

With these difficulties in mind, a six-stage process of analysis and testing (Fig. 12.12) was developed and applied to WHORL (Reynolds and Ford 1999) to see how effectively WHORL could simulate more than one output from the model at the same time – in this instance 10 were used. Parameter values might still compensate for structural inadequacy in the model, the process of accommodation – but it is less likely, and more likely to be noticed, when a number of assessment criteria are used. The 10 outputs were specifically selected to represent both the outcome of competition, e.g., by using outputs of mortality and height frequency distribution of surviving trees as assessment criteria, and the crown structure of trees, e.g., through outputs of crown ratio and crown apex angle as assessment criteria. These choices were made so that the model was required to simulate structure of an open grown tree crown effectively with the same parameters that were effective in simulating the results of competition. Ranges of acceptability were established for each output value, i.e., a simulation was not expected to get mortality (or any other output) precisely correct but within a certain range, e.g., for mortality this was set at ± 5 percent. This recognized that there was likely to be some variability in measurements of the outputs, although we were not able to determine precisely what it was. Ranges for outputs used in assessment are given by Reynolds and Ford (1999).

A large number of different combinations of parameter values were tried. WHORL could achieve all 10 outputs within their acceptable ranges, but never all of them with one set of parameters. This highlights the difficulty. If any single model output had been chosen, e.g., tree mortality, then a successful parameterization would have been found – but only at the hidden cost of accommodation, distorting the model so that it did not simulate other aspects of crown growth and the competition process. In practice, groups of parameterizations were able to achieve different combinations of outputs within their specified ranges. Sixty-five parameterizations in Group 1 (Fig. 12.13) achieved the first 8 outputs but not the last 2; 18 parameterizations achieved the first, fourth through ninth, but not the second, third and tenth, etc.

The set of groups in Fig. 12.13 is called the Pareto Optimal Set, i.e., those parameterizations that simulate at least one group of assessment criteria better than others. The next stage (Stage 3, Fig. 12.12) is to see whether the parameter spaces had really been adequately searched and then to assess for

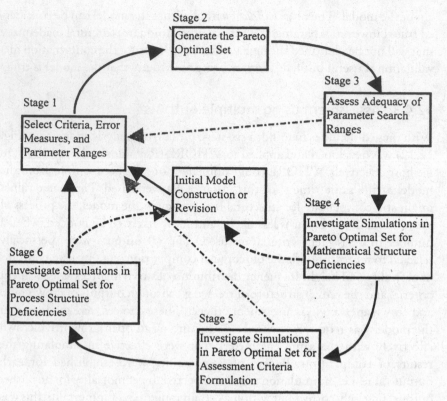

Fig. 12.12. A six-stage process of analysis and testing for ecological process models using the Pareto Optimality criteria. (From Reynolds and Ford 1999, with permission.)

deficiencies that explain why no group of parameterizations achieves all 10 outputs correctly. These deficiencies occur in:

(1)   mathematical structure (Stage 4), i.e., whether the actual mathematical representations were reasonable,

(2)   the particular model assessment criteria, e.g., impossible criteria may have inadvertently been set (Stage 5), and,

(3)   the models own process structure (Stage 6), i.e., how it represented the competition process.

Problems were found at each of these stages (Reynolds and Ford 1999) leading to construction of a new model, WHORL2. The most important change was in how morphological plasticity and physiological acclimation were represented. This was changed to be entirely dependent upon the position of the foliage in the canopy rather than whether the tree it was attached to was large or small in height. WHORL2 was then used to produce a new Pareto Optimal Set. Although still no single parameterization achieved all 10 outputs, the proportion achieving 8 increased. It was recognized that the two outputs measuring the rate of large tree height growth and the r^2 estimate of its regression on large tree size were influenced by factors other

Fig. 12.13. Pareto Set for the WHORL model of competition in a dense, naturally regenerated stand of *Abies amabilis*. Sixty-five parameterizations successfully simulated the first 8 assessment criteria and constitute Group 1. Eighteen parameterizations successfully simulated the first and the fourth through eight assessment criteria. The eight groups achieve unique combinations of assessment criteria, e.g., Group 2 is more successful than Group 1 because it achieves criterion 9, but less successful because it does not achieve criteria 2 and 3. Other parameterizations achieved only some criteria, but fewer than members of at least one of these groups and they are Pareto dominated and rejected. The two criteria *Growth rate slope* are regressions between tree height and its growth rate over the last two years of simulated growth, the two criteria *Slope R^2* are the r^2 values of the growth rate slopes and that are calculated separately for suppressed and dominant trees. (From Reynolds and Ford 1999, with permission.)

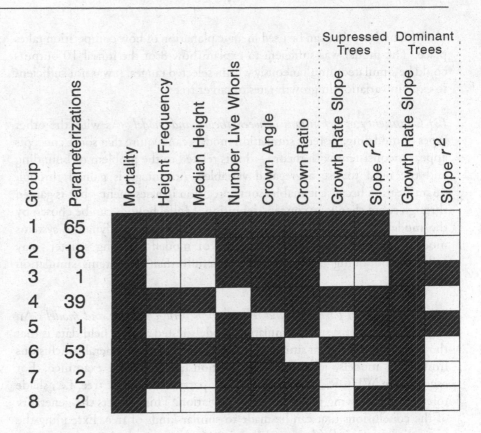

than competition for light and so were not effective assessment criteria.

Does WHORL2 provide a good explanation when assessed by the standards of explanatory coherence? Recall that the purpose of constructing the model was to make a new synthesis for the above-ground competition theory applicable to a dense young stand of *Abies amabilis*, a shade-tolerant tree.

(1) Acceptability of individual propositions Assessment using multiple criteria and constructing the Pareto Optimal Set allowed examination of acceptability of individual propositions and how appropriate they were within the model. Developing the model led to a series of changes in model structure, i.e., the theory of competition was altered and incorporated into WHORL2 and WHORL2 was simpler than WHORL. Errors in some technical assumptions, i.e., the axioms of the modeling process, were also corrected. Of course this process is limited by the bounding decisions. Some propositions are simplifications of the process, e.g., the details of the tree growth process. The important issue is whether propositions are necessary and sufficient to explain the phenomenon. The modeling exercise (not just the model itself) answered that it was necessary to include plasticity and acclimation and enabled an interpretation of theories about independence of branches in whole tree

crowns and how that can be used in an explanation of how competition takes place. The model was sufficient to explain how 8 of the initial 10 outputs could be simulated simultaneously within selected ranges. It was not sufficient to explain variation in growth rates of larger trees.

(2) Consistency of definitions between theory and model   As with the other types of modeling, systems simulation models can require that some concepts appear inconsistent with theory – this is related to the problem of bounding and the need to use aggregated variables. Advantage is gained through quantification because it enables some precision in assessment – but is gained through some change in concept definition. Model bounds can be chosen by the modeler and this should be an explicit decision. In dynamic systems models and statistical models the rules of modeling, being stricter, may influence bounding decisions more forcefully than in systems simulation models.

(3) Consistency in part and kind relationships between theory and model   An obvious feature of systems simulation models tested against field data is that they apply to a particular situation. In attempting to draw general conclusions from such models, part and kind relationships must be examined. For example, in WHORL, *Abies amabilis* is a particular kind of tree, i.e., shade tolerant and showing plasticity and acclimation. This restricts the generality of the conclusions that can be made to similar kinds of tree. Extending the generality requires further modeling exercises to see whether other species and/or conditions behave in the same way.

(4) Simplicity of propositions   Systems simulation modeling does not place a high priority on simplicity. Parsimony is not a prized characteristic of these models. The objective is to represent a theory within predetermined bounds rather than construct the simplest model that will fit a particular data set. Unnecessary complication can be kept in check by ensuring that multiple assessment criteria are selected to test the functioning of important model components.

(5) Breadth of the complete explanation   There are two limitations to the breadth of explanation that systems simulation models provide. First, calibrating the model with data from a particular site or situation limits the interpretation that can be made. In this example the model would have to be fit initially to other plots within the forest, to see whether model structure was adequate and parameters values remained within their specified bounds, and then to other types of forest where we would expect to change some elements of model structure. Second, the bounding process focuses the model on particular phenomena; that is, onto a particular part of the theory network.

Differences in the detail of bounding conditions may lead to similar, but different, models and results. In this example judgement of whether the model represents a theory that is the best explanation of the single species competition process requires additional examples.

12.5 Discussion

There are two values in constructing quantitative models. First, they can provide a defined representation of a theory, particularly one describing an intricate system, enabling complex dynamics to be studied in a way not possible if theory representation is confined to written-word definitions. Second, when models define relationships between measurable concepts then calibration and assessment with data may contribute to the objectivity of the theory. However, there are difficulties with both these values.

Models as defined representations of a theory
We can not depend upon a model, of any type, to be a complete representation of a synthesis. Synthesis is a wider process. We must ask has the model:

increased confidence in an existing theory network,
extended an existing theory network, or
changed a theory network in a major way.

Carpenter and Kitchell's (1987) model increased confidence in the role of top-down processes to explain variation in trophic structure. Ford *et al.*'s (1987b) model extended the theory of environmental control of shoot growth to include solar radiation as well as temperature. WHORL, and its development to WHORL2, by illustrating the importance of modular representation of foliage and quantifying the role of morphological plasticity and physiological acclimation, confirmed additional axioms of competition theory, at least for closely grown *A. amabilis*. In each case the effectiveness of the modeling exercise (not just the model itself) can be judged by examining how the model increases the explanatory coherence of the theory it represents.

But models are not equivalent to theories and it is unfortunate that sometimes modelers view their work in this way. Due to rules and procedures of the modeling type, models contain axioms additional to, and/or changed from, the theory on which they are based. Where models are to be fit to data some generality may be sacrificed to achieve precision for a particular case. When using models to construct a scientific explanation we take into account how the model was constructed and assessed. This is particularly important in developing explanations using integrative concepts. For example, results from WHORL, and its development to WHORL2, defines *above-ground competition* using specific descriptions of functional concepts, particularly results from modularity of tree growth, morphological plasticity and physiological acclimation. But full definition of *above-ground competition* requires comparative analysis between what is found in *A. amabilis* and in other species and under different circumstances. This more complete definition of a theory of *above-ground competition* can not be made solely on the basis of WHORL2 – although a definition should not exclude its results.

The contribution of models to the objectivity of a theory

Model assessment against data provides a test of a model's effectiveness. Three levels of assessment can be made for complete models: *fitting, predicting*, and *revealing different results*.

Fitting is not a strong assessment criteria – yet it can be difficult to achieve and when it is achieved there has to be understanding of how that was done. Parameter values may be sought ensuring model output is close to observed data – but where there are multiple parameters in a model and assessment uses a few, or sometimes only one, model output, then multiple parameters may accommodate deficiencies in model structure. This can be appreciated both from the time series example, where differences in the transfer function exist for different shoots. Clearly, if model structure is completely inappropriate the fitting procedure may not produce an accommodation – but beyond this a successful "fit" for a few model outputs may provide insufficient information about the model structure to judge whether it is necessary and sufficient as a representation of the process.

Prediction is often considered as more valuable than fitting and is widely used in both statistical modeling and systems simulation as validation. For example, in statistical modeling it is frequent practice to divide available data into two, fit the model to one part, and then see whether the model produced can predict the second part. Blyth and McLeod (1981a,b) actually collected a second data set to test their models (Chapter 9). When a model fails to fit the new data, it does not predict. This was a failure of the shoot extension model – it did not fit the second half of the growing season. But success in prediction may indicate only that a model has accommodated the data while failure to predict may not indicate that the model is wrong, in the sense that it should be completely rejected, but that it may be incomplete relative to the new data. The second set of data has no greater, or lesser, weight than the first. When prediction is used in a confirmatory way it is stronger than fitting as an assessment process because of what it tells you about your modeling techniques, i.e., the effectiveness of the accommodation process in fitting parameters to a choice of outputs, but it may tell you no more about the effectiveness of the model in representing the theory than fitting did.

Revealing results of a different kind is the strongest assessment, i.e., that the model predicts something not expected and when searched for through more research, is found to occur. This criterion is sought in assessing dynamic systems models for which detailed fitting and prediction are difficult and rarely used. For example, chaotic behavior of particular classes of dynamic systems models is a result of a different kind from what might be expected. But incorporating such results into general ecological theories is likely only if fitting and/or predicting have already been attained.

These three levels of assessment are rarely achieved all at the same time and there can be difficulty in achieving fitting alone. Model assessment, like assessment of postulates, is a progressive procedure and requires a comprehensive approach of going into the details of its explanatory coherence. We can never say that a model has been validated in the sense of it being a true representation. We can not claim this for theories (Chapter 11) and models have additional difficulties owing to the rules of model type used and procedures of construction.

12.6 Further reading

Haefner's (1996) textbook on modeling biological systems gives examples of a range of model types as well as a critical analysis of the modeling process. For further development of the model described by Carpenter and Kitchell (1987), see Carpenter and Kitchell (1993c). Carpenter *et al.* (1994) fitted time series models to predator–prey data and concluded that experimental manipulations, or other substantial perturbations may be essential for definition of non-linearities in interactions.

Hilborn and Mangel (1997) provide an illustration for ecologists of how some developments in statistics and modeling can be applied, particularly the role of modeling in examining multiple hypotheses. Rykiel (1996) reviews issues of model evaluation.

Introduction to Section III:

Working in the research community

Recall nine questions asked by beginning researchers about working in the research community:

26. Can I get funded?
27. Funding!
28. I am interested in funding and individual recognition.
29. How do I choose a committee that will give me a broadly based opinion?
30. How can I get my committee to stop suggesting things to do?
31. What if my committee steers me incorrectly?
32. How can I analyze, integrate, and present to the outside world?
33. How can I choose a research topic of value? Or if you have chosen a research topic how can you persuade someone to fund it?
34. How can I make research relevant to current topics of theoretical and practical interest?

Obtaining funding, achieving recognition, selecting and working with a committee, finding a topic of value and relevance, and seeing work through to publication involve social processes. Funding decisions are frequently made by panels of scientists and, while a research proposal should have scientific merit, the panel must be persuaded that what it offers is better than comparative proposals. Achieving recognition requires more than doing effective research – the research must be acclaimed by others. Working with a supervisory committee requires actively seeking criticism and support for a research proposal, and for the research itself. Writing a paper for publication requires that you present an argument that persuades referees of its correctness and value, and that it will be understood by the scientific community reading the particular journal. Ensuring relevance for your work requires not only that it be good science but that it be recognized as contributing to the solution of some problem that others, whether researchers, resource managers, civil servants, industrialists, or general public interests, have defined as important.

Social influences must not be thought of as something surrounding science – they influence the details of how work is conducted. Most importantly they impact criticism. Just and effective criticism is essential for science to progress. Criticism must be just, in the sense of being neither a spoiling argument seeking to discredit individuals, nor biased by weight of social prejudice against different viewpoints or scientists working in particular organizations. It must be effective in being sufficiently detailed to point to inaccuracy, inadequacy, or incompleteness and so to the possibility of improvement. But just and effective criticism is difficult to achieve. Chapter 13 describes how investigators of science, whether sociologists, anthropologists, or scientists themselves, have repeatedly found social processes reducing the effectiveness of criticism brought to bear on research.

Section III moves beyond recounting these social problems to define how effective criticism can be achieved, though not guaranteed, and what this implies for the definition and development of objective knowledge. Four stages of *scientific criticism* are identified and defined. *Direct analysis* is applied to individual investigations by colleagues involved with, or close to, the work. *Testing through repetition* requires additional work using the results of an investigation, either directly or indirectly is stronger when not conducted by the investigator who made the initial discovery or claim. *Refining through extended use*, involves use and critical analysis by scientists in related but not identical fields, so that criticism is made from a larger scientific social circle. These first three stages are described and analyzed in Chapter 13.

The fourth stage, *standpoint criticism*, is markedly different from the first three. It involves criticism and development through work where objectives of investigators, as well as their methods, may be very different. For ecological research this stage is frequently influenced by relationships between science and society's needs, involving resource and environmental managers, conservationists, and, in a wider context, government and its funding agencies. Scientists, collectively through their own societies, and as individuals, assert their own standpoints in arguing what the objectives of research, resource management and government policy should be. Particularly at issue is what should be studied, who the critics should be, how decisions about science objectives should be decided, and how science should find its place in decision-making. These questions have a major influence on the growth of objective knowledge and are analyzed in Chapter 14. They are not remote from the working scientist and have direct bearing on research objectives, how research is criticized, and whether the results are applied.

13 Scientific research as a social process

Summary

Scientific research involves a social process, and social interactions have an important influence on what we classify as objective knowledge. Within the scientific community there are norms about how scientists should behave in claiming priority for discoveries, giving credit to others through the citation process, and maintaining emotional neutrality during research. Counternorms can be recognized for each norm, and social processes influence the balance between norms and counternorms and the relative degrees of skepticism and dogmatism maintained within groups of scientists.

Social processes are particularly important in production of scientific papers. Scientists simplify accounts of their work as they write and respond to criticism and use particular forms of rhetoric to construct arguments in the most favorable way. These rhetorical conventions are standardized within different branches of science. Peer review pays particular attention to the types of knowledge claims made by authors and how these are expressed within the rhetorical conventions of the subject.

Social studies of science have challenged the historical view of science as intrinsically objective – with individuals proceeding with humility and skepticism for their own ideas and results. They have shown that by determining who is involved, and what attitudes they have, social systems can limit not only the amount, but also the types of criticism permitted. Among scientists in related fields of research the degree of objectivity increases and the nature of knowledge changes, as scrutiny and debate pass through three stages of criticism. (1) *Direct analysis*: groups of scientists cooperate, discuss, debate directly, and peers review each other's work. (2) Testing through *direct repetition*: scientists within the same field repeat research, either directly or indirectly, and confirm or change knowledge. (3) *Refining through extended use*: as theories developed in one field are used in another concepts are refined and theory domains are specified in greater detail.

Table 13.1. *Bacon's four classes of idol that impede scientific progress and their present day interpretation. (After Losee 1993)*

Bacon's description	Present-day interpretation
"Idols of the Cave . . . everyone . . . has a cave or den of his own, which refracts and discolours the light of nature owing to . . . his education and conversation with others, or to the reading of books, and the authority of those he esteems and admires . . ."	Upbringing and education result in personal bias in what a scientist chooses to read and study, and limit individual attitudes
"Idols of the Tribe have their foundations in human nature itself and in the tribe or race of men . . . human understanding is like a false mirror, which, receiving rays irregularly, distorts and discolours the nature of things by mingling its own nature with it."	People associate and form groups. They adopt similar attitudes and approaches that reinforce social cohesion of the group
Idols of the Market Place. "For it is by discourse that men associate . . . But words plainly force and overrule the understanding, and throw all into confusion, and lead men away into numberless empty controversies and idle fancies."	Undefined differences in definitions of concepts and theories: use of common language terms for complex and partially understood scientific ideas
"Idols of the Theatre . . . all the received systems are but so many stage-plays, representing worlds of their own creation after an unreal and scenic fashion."	The received dogmas and methods of various philosophies construct caricatures of the real world

13.1 Introduction

Within science is a social process. Scientists debate with each other, both directly during work in the laboratory or field, at seminars and conferences, and indirectly through sequences of publications. What becomes accepted in science is the result of this debate. This social process determines scientific progress, and concern that it can result in bias and confusion is as old as scientific research itself. In the early seventeenth century, Bacon considered that science was confounded by four classes of Idol (Table 13.1) and urged his empirical method as a way to combat them. We still have similar "Idols" today, although the philosophy and methods of Baconian empiricism are no longer considered adequate to overcome them.

Sociologists, anthropologists, and scientists themselves have investigated the social structures and processes in science and shown that social interaction determines what we consider to be *objective knowledge*. Scientists may feel uncomfortable to be studied as subjects – or be concerned that the cognitive merit of their work may be challenged by studies of science as a social process – but they can not afford to ignore the results of such studies. Recall that *objective knowledge* is defined as knowledge that has been researched then scrutinized, and debated among, scientists. Each is a social process that can make science successful or unsuccessful and there is no guaranteed objectivity. Science can progress only through criticism. Social studies of science provide a more comprehensive definition of what needs to be criticized and the obstacles likely to be encountered.

In this chapter I discuss theories about social interaction between scientists and how it influences research and the creation and use of scientific literature. A method for assessing the progress of scientific knowledge through three stages is documented: analyzing through peer review, testing through repetition, and refining through extended use.

13.2   Social influences and social structures

Just as scientists debate theories, argue over conflicting results and discuss the way to make progress, so too do sociologists and others who study science. Two views of the nature and importance of social influences have already been described (Chapter 11). Kuhn (1970) emphasized science as a social process, with *normal science* being undertaken by a cohesive group of scientists researching similar problems within a *paradigm* and using the same *exemplars*, until terminated by a revolution from outside the group. This was countered by Lakatos' (1970) description of scientific research programs, within which some ideas form a hard core belief, and a general methodology, the positive heuristic, is used, but where progress occurs through continuous analysis, debate, and successive *problemshifts*.

Both Kuhn and Lakatos were philosophers of science and came to their descriptions through retrospective analyses of particular discoveries. Investigations by sociologists have examined the social process of science directly during current investigations, have described social aspects in greater detail, and have attempted to define how they influence scientific achievement. An important impetus to their work has been adoption of a scientific approach to the study of science itself (Bloor 1976).

13.2.1   The balance between norms and counternorms in scientists' behavior

The purpose of science is to find things out. Merton (1957a) illustrates that socially a scientist is continually reminded of this requirement – that the primary behavior valued by the institution of science is originality. Creativity enables a person to design new investigations or define new theory and find out about the unknown. Recognition and esteem accrue to those exercising their creativity and making original contributions. Merton (1957a) states: "Interest in recognition therefore need not be, though it can readily become, simply a desire for self-aggrandizement or an expression of egotism. It is, rather, the motivational counterpart on the psychological plane to the emphasis on originality on the institutional plane."

But what should the rewards be, and how should they be given? Merton suggests the primary award for originality in science is that other scientists acknowledge priority. The token of this award is citation of the publication in which discovery is reported. Once a scientist has made a contribution he or she has no more exclusive rights to the ideas – they become public. The only "property" right is that of the discovery.

Science, like other social institutions, incorporates potentially incompatible values, or norms, particularly "originality, which leads scientists to want their priority to be recognized, and the value of humility, which leads them to insist on how little they have been able to accomplish" (Merton 1957a). These opposing values call for opposing kinds of behavior and reconciling them is no easy matter. Merton suggests that for scientists who have internalized both values this creates an inner conflict by generating a distinct ambivalence to claiming priority. He describe the case of Charles Darwin, who, after working for more than 20 years on the theory of evolution through natural selection, but not publishing, was stunned to receive a manuscript by Russel Wallace outlining the theory. Darwin wrote to his friend Lyell (Darwin 1925), "So all my originality, whatever it may amount to, will be smashed . . .". Darwin wondered about publishing a short version of his text but wrote, "I cannot persuade myself that I can do so honourably" and wished to accede priority. Friends intervened and arranged a joint public presentation by Wallace and Darwin. The reason we now emphasize Darwin's contribution rather than Wallace's is partly because his was the more substantial and comprehensively reasoned text, partly because Wallace recognized this and did not push for priority, and partly because humans seem to need to attribute historical events to "great" personages. We have elevated Darwin in this way. Nevertheless, the priority, in the strict sense of who was first, rather than who was comprehensive, really should have belonged to Wallace.

Priority as a driving force in science has three weaknesses:

(1)   the urge for priority can lead to rushed, poor, or even fraudulent science.
(2)   Most supposed priority is, at best, technical, since multiple and independent discoveries may be separated by incidental amounts of time

relative to the complete cycle of science.

(3)   Priority, as a reward system, can be amplified by other social factors, particularly continuing fame based upon previous, but not necessarily current, work.

Each deserves additional comment.

The urge for priority leading to poor science   The search for recognition can lead, at the extreme, to outright fraud. This is rare, since results are likely to be checked by other scientists, either explicitly in repeating the work, or implicitly by attempting further research on the supposed result. So discovery of fraud is more likely the more apparently valuable the fraud would be if it were true. More frequent is what were termed by Babbage (1830) trimming and cooking. The trimmer clips off "little bits here and there from observations which differ most in excess from the mean and [sticks] . . . them on those which are too small . . .". The cook makes many observations and selects only those that agree with preconceived ideas. This may be as simple as not reporting failures. Even scientists who have discovered something through *bona fide* research have been known to fabricate additional support (Kubie 1953).

Multiple and independent discovery   This is a well-documented and repeating phenomenon (Merton 1961). The instance of Darwin and Wallace has already been cited. Another notable and considerably researched case is the invention of calculus by Newton and Leibnitz. Newton's struggle to persuade others that he had priority led him to "outright deceit" (see Merton 1957a, who documented published research on the argument). For any discovery a second independently researched instance is an extremely important confirmation that practicing scientists may require before incorporating a finding into their own work. The elevation of priority to be a primary force personalizes the progress of research, which actually profits from independent repetition.

Parallel discovery occurs because the scientific knowledge of a time contains sets of results and observations that numbers of individual scientists are trying to explain and to which they are adding incremental components. For example, many ideas about species, and theories about their origins, were advanced in the 50 years before Darwin and Wallace published their works. Even a major new synthesis, such as the theory of evolution through natural selection, becomes increasingly inevitable once a set of observations and approaches have been defined, i.e., a domain has been described and work is in progress. At this stage the problem is well enough defined by a group of scientists, not just the individual making the discovery, so that a solution can be recognized by the group when it is made. Isolated discoveries can go unrecognized for many years, e.g., Mendel's discovery of principles of heredity.

Multiple independent discoveries are reassuring for science as a whole – even if disappointingly undermining for those who set fame as a motive for their work. Clearly, definition of the theory of evolution through natural selection did not depend upon Darwin. Wallace had worked in isolation from Darwin and his work was equally likely to have had the same general influence. Description of a race to discover something, e.g., Watson's (1968) description of the race to define the replication process, even though it may emphasize a winner for dramatic effect, illustrates the closeness of such races between scientists. A small amount of isolation between competitors can lead to parallel discoveries. Merton (1961) reviews both the social theory of discovery, i.e., science is a continuing process of cumulative growth with discoveries tending to come in their due time, and the heroic theory, where discovery is the work of genius and without that one person there would have been no discovery. He suggests an intermediate position. Analysis of multiple independent discoveries showed that some scientists were involved in more of them than others, so that there are some people who tend to make many discoveries, but these are, nevertheless, made in the context of a continuing incremental advance.

To be effective as a driving force priority of discovery must be rewarded to motivate sufficient scientists to compete and ensure that solutions are obtained. However, rewards do not follow priority in a straightforward way. Merton (1968) defines the "Matthew Effect" by reference to the Book of Matthew (13:12) in the New Testament of the Bible:

> "For whosoever hath, to him shall be given, and he shall have more abundance: but whosoever hath not, from him shall be taken away even that which he hath."

This can operate in science where equally productive scientists working in small institutions receive less recognition than those in large and famous ones. Merton (1968) comments on interviews with Nobel prize winners, suggesting that recognition is skewed in favor of established scientists in cases of collaboration, and of independent multiple discoveries. He suggests that, because association of an idea with an established scientist may heighten its visibility, the effect could be advantageous to science as a whole. The Matthew Effect operates in grant awarding. Taubes (1986) describes a case of deliberate publicity seeking by a scientist. The scientist employed a publicity agent and this led to accrual of resources so that only he was in a position to conduct critical research in a contentious field where others not able to gain resources opposed him. In some instances, granting authorities deliberately establish programs to overcome aspects of the Matthew Effect, e.g., grants specifically for young scientists, or specifically for untried ideas.

Merton (1957a) summarizes the tension that can occur as the result of the social norm of originality and its token of priority as follows:

> "When the institution operates effectively, the augmenting of knowledge and the augmenting of personal fame go hand in hand; the institutional goal and the personal reward are tied together. But these institutional values have the defects of their qualities. The institution can get partly out of control, as the emphasis upon originality and its recognition is stepped up. The more thoroughly scientists ascribe an unlimited value to originality, the more they are in this sense dedicated to the advancement of knowledge, the greater is their involvement in the successful outcome of inquiry and their emotional vulnerability to failure.
>
> Against this cultural and social background, one can begin to glimpse the sources, other than idiosyncratic ones, of the misbehavior of individual scientists. The culture of science is, in this measure, pathogenic. It can lead scientists to develop an extreme concern with recognition, which is in turn the validation by peers of the worth of their work. Contentiousness, self-assertive claims, secretiveness lest one be forestalled, reporting only the data that support an hypothesis, false charges of plagiarism, even the occasional theft of ideas and in rare cases, the fabrication of data, – all these have appeared in the history of science. They can be thought of as deviant behavior in response to a discrepancy between the enormous emphasis in the culture of science upon original discovery and the actual difficulty many scientists experience in making an original discovery. In this situation of stress, all manner of adaptive behaviors are called into play, some of these being far beyond the mores of science."

Scientists recognize these tensions and, to varying degrees depending on the individual, consider that progress depends upon them (Mitroff 1974a). For each value, or norm, that scientists hold as essential for development of objective knowledge, there also exists a counternorm of subjective behavior which, at the very least, must be recognized (Table 13.2) and which many scientists interviewed by Mitroff considered as essential. So, while recognizing that science must be a rational activity (Norm 1) the origin of much science lies in subjectivity (Counternorm 1). If we are to weigh evidence then, we require emotional neutrality (Norm 2) – but to conduct an arduous investigation we also require emotional commitment (Counternorm 2). We require that the assessment of a scientific idea should be neutral to whoever makes the claim (Norm 3), but still may give greater warrant to individuals whose work we know to have been effective previously (Counternorm 3). Open exchange of ideas and data is essential if science is to work effectively (Norm 4), but we may have to work in a solitary and secluded way during the development of a new idea or method (Counternorm 4). We expect that scientific achievement is its own reward (Norm 5) – but scientists award each other status in academic societies and congratulate each other with prizes and medals, and for some people a desire for these pushes science forward (Counternorm 5).

Table 13.2. *A list of the norms and corresponding counternorms of social behavior held by scientists. (After Mitroff 1974a)*

	Norms	Counternorms
(1)	Faith in the *moral virtue of rationality* (Barber 1952)	Faith in the *moral virtue of rationality and non-rationality* (Tart 1972)
(2)	*Emotional neutrality* as an instrumental condition for the achievement of rationality (Barber 1952)	*Emotional commitment* as an instrumental condition for the achievement of rationality (Merton 1963, Mitroff 1974b)
(3)	*Universalism*: "The acceptance or rejection of claims entering the lists of science is not to depend on the personal or social attributes of their protagonist; his race, nationality, religion, class and personal qualities are as such irrelevant. Objectivity precludes particularism . . . The imperative of universalism is rooted deep in the impersonal character of science." (Merton 1957b)	*Particularism*: "The acceptance or rejection of claims entering the list of science is to a large extent a function of who makes the claim" (Boguslaw 1968). Social and psychological characteristics of the scientists are important factors influencing how their work will be judged. The work of certain scientists will be given priority over that of others (Mitroff 1974b). The imperative of particularism is rooted deep in the personal character of science (Merton 1963, Polanyi 1958)
(4)	*Communism*: "Property rights are reduced to the absolute minimum of credit for the priority of discovery" (Barber 1952). "Secrecy is the antithesis of this norm; full and open communication [of scientific results] its enactment" (Merton 1957a)	*Solitariness*: Property rights are expanded to include protective control over the disposition of one's discoveries; secrecy thus becomes a necessary moral act (Mitroff 1974b)
(5)	*Disinterestedness*: Scientists are expected to achieve satisfaction, prestige, and their self-interest in work through serving the whole scientific community (Barber 1952)	*Involvement*: Scientists are expected by their close colleagues to achieve satisfaction, prestige, and their self-interest in work through serving their special communities of interest, e.g., their invisible college (Boguslaw 1968, Mitroff 1974b)
(6)	*Organized skepticism*: "The scientist is obliged . . . to make public his criticisms of the work of others when he believes it to be in error . . . . no scientist's contribution to knowledge can be accepted without careful scrutiny, and that the scientist must doubt his own findings as well as those of others" (Storer 1966)	*Organized dogmatism*: "Each scientist should make certain that previous work by others on which he bases his work is sufficiently identified so that others can be held responsible for inadequacies while any possible credit accrues to oneself" (Boguslaw 1968). Scientists must believe in their own findings with utter conviction while doubting those of others (Mitroff 1974b)

We must be critical of all scientific work (Norm 6), yet believe wholeheartedly in our own work (Counternorm 6).

Through extensive interviews with 43 practicing scientists working in a NASA-funded program on lunar geology, Mitroff (1974a) demonstrates that different scientists hold these norms and counternorms to different degrees. The most exasperating, to the majority, were three scientists most firmly committed to their ideas, i.e., tending to hold and practice more counternorms than norms, and were judged by their peers to be both the most creative and the most resistant to change. Mitroff (1974a) makes the point that each norm is restrained by the social dynamic of interacting scientists. Both norm and counternorm of each pair do not operate equally in every situation, norms dominant in one situation can be subsidiary in another.

13.2.2 Cooperation and competition between individual scientists

Scientists cooperate in groups of different sizes for different purposes. Hull's (1988) contention was that small research groups of three to five scientists occur periodically as the locus of innovation and initial evaluation and are important in generating rapid change. Members publish jointly and share techniques and problems. *Groups* increase the speed of conceptual change, regardless of the content of the program, and have at least short-term effects on the content of science. *Group* allegiances are local but local change must occur before the wider spread to the *deme*. Scientists may join groups, even though they do not agree entirely with all issues. Hull described the *deme* as a larger group using each other's ideas more frequently than those outside and with considerable mutual citation – but that the social interaction within *demes* is not as tight as within research *groups*.

Hull (1988) studied scientists working in systematics in zoology from the mid-1970s. He was actively engaged in that research, but also worked as a philosopher, conducting interviews with scientists about their research objectives and methods, and making a study of editorial policy and refereeing practice of the journal *Systematic Zoology*. He proposes:

(1) Social relations of cooperation and competition determine scientific progress. Both are essential, and science works as well as it does because of the interplay and balance between them. Cooperation is important because conceptual advance, as well as practical and organizational aspects of much work, is helped if individuals work together in *groups*. Within the group there can be acceptance of ideas while they still are being examined. Competition ensures the essential criticism and checking necessary to establish results. For Hull, innate curiosity in scientists is a given, and the whole process of science is driven forward by

individuals' need for priority of discovery, resulting in social recognition. Objective knowledge results, not despite bias and commitment, but because of the stimulus it gives to other scientists. Also, because of competition, altercations in science are inevitable. So, for Hull it is the counternorms (Table 13.2) that are, and should be, dominant.

(2) Scientific theories can not be tested, in any final sense, because many versions of the same theory exist, with each scientist having particular definitions of one or more concepts. Success is gauged in terms of use of concepts and their particular definitions by other scientists and is the result of competition. For Hull (1988), change in the definition of concepts is the very essence of scientific advance. He gives the examples of Darwin's idea that *species* change, where previously they had been considered as things that were constant, and Darwin's use of *evolution* to describe this process of species change, where previously that concept had been used in embryology with a very different meaning. Hull (1988) suggests: "the tide of terminological change can no more be arrested by protests of those of us who are more terminologically conservative than can the flow of molten lava from an erupting volcano be stayed by prayer". And "[t]he rigidity with which scientists insist on their preferred terminology is matched only by the semantic plasticity of that terminology" – scientists insist on their terms but are loath to define them too closely, leaving themselves room for maneuver. For Hull science is, and should be, elitist, not egalitarian. He suggests that few scientists publish much in their lives, and most of the advance comes from a few people – those who have been successful in the competition.

Hull suggests error is inevitable in science. Scientists can not tell the whole truth and nothing but the truth because, at the forefront of scientific debate, no one can be precisely sure what the truth is. Science does not depend upon scientists presenting totally unbiased results but on other scientists, with different biases, checking them. Scientific theories are not easily confronted with data – individual scientists take all sorts of decisions necessitating compromise, in measurements, sampling, and general methods. Research under such conditions gives scientists ample opportunities to try out different combinations of limiting assumptions, and they do not remain completely consistent in their methods and measurements in successive investigations. Furthermore, scientists are not precise in defining what they mean. Much conceptual change in science does not occur in the open but is hidden. A defined concept may be taken and used but in a way not envisaged when the original definition was made, and with the definition certainly not encompassing the new use, e.g., ecosystem (Chapter 11). Sometimes scientists are aware of what they are doing; often they are not.

Hull (1988) suggests that, because of the importance of checking, and the promise that it may happen, it is not necessary that an experiment *has* been replicated, but only that it *could* be replicated, and so the potential for testing exists. Because scientists look for support, testing is reserved only for those things that challenge support. He considers: "As unseemly as factionalism in science may be, it does serve a positive function. It enlists baser human motives for higher causes".

There are important counter-arguments against factionalism. As Hull (1988) himself acknowledges scientists can frustrate the scientific process, if, as contestants in a priority race, they abandon avenues of research that would have presented greater long-term payoff and go for quick results. Some scientists go beyond competition of ideas and practice competitive exclusion. They use any influence they can muster to exclude non-adherents to their views from publishing in particular journals, or receiving research grants (see Hull's (1988) own account). This holds science up. It limits what is researched, and, by unnecessarily fragmenting the scientific world, it restricts the scope of critical comment.

Rather than taking Hull's (1988) viewpoint that unbridled competition is necessary for scientific progress some observers and researchers have taken the intensity of competition, and the excesses that are part of it, as discrediting science. Mahoney (1977) catalogs many illogicalities and examples of bad scientific practice, suggesting that the idealized view of science (Table 13.3) hinders scientific progress. Scientists exist in groups and argue between themselves without there being sufficient empirical scientific foundation to resolve the issues. Mahoney's concern is for a direct counter to this relativism. The only effective counter is genuine criticism grounded in detailed analysis of the science rather than the more typically observed justification that is the actual outcome of competition. Intense competition that is decided socially by one scientist's being "right" may not lead to a resolution which gives forward impetus to new research.

Hull (1988) suggests that of Mahoney's observations are "incredibly naive". He appeals to implicit tradition:

> "If generations of scientists behave in certain ways, then possibly there is
> something to be said for this behavior. The inference is even stronger if there is
> a correlation between the prevalence of a behavior and the success of scientists
> who exhibit it. Scientists are far from infallible in the methods they employ in
> their investigations but they must be doing *something* right."

Vindictive bickering between the researchers that he studied influenced Hull's views. In contrast, Crane (1972) emphasizes cooperation and its importance for the diffusion of knowledge. She analyzes social interactions by direct questioning and study of citation networks and defines research areas where members are linked by both direct communication and mutual

Table 13.3. *Mahoney's (1977) comparison of views of science*

Topic	Idealized view of science	Revised view
Nature of knowledge	True (provable) assertion	Warranted (but fallible) conjecture
Ultimate authority for science	Logic (reason) and sense experience (data)	There is no ultimate authority
The nature of data	Certain, immutable facts	Factual relativism
Role of theory	Secondary to data	Prerequisite for data
Nature of progress	Accumulation of facts	Reconstruction of "facts"
Way to discovery	Logical (by induction)	Psychological (by induction)
Theory evaluation	Instant rational assessment of one theory at a time (via reason and data)	Simultaneous competition among theories (via data, rhetoric, conceptual elegance, politics, etc.
Science practice	Monolithic enactment of a logical script, totally explicable by philosophy	Poorly understood complex of conjecture, tacit knowledge, and personal belief. It is only explicable if psychological and social influences are acknowledged

citation, the latter defining more connections. Scientific productivity varies between individuals within the research areas and citation connections are greater between more productive than less productive scientists who also remain in the research area for longer periods of time. The more productive scientists are important nodes in networks of communication and form an *invisible college*. Crane (1972) suggests that lack of such an invisible college might inhibit the development of a research area. When a high proportion of researchers work individually they find it difficult to evaluate either their own research or the importance of their areas in relation to other research areas.

Crane (1972) emphasizes the importance of networks in facilitating the diffusion of ideas. She likens diffusion to a fashion-like process in which influence is transmitted through steadily expanding networks. The exchange of ideas between members of different research areas is important in generating new lines of inquiry. One important function of interdisciplinary research is to facilitate this, not just through transference of techniques but also in theory. There must be a balance:

"Some degree of closure is necessary in order to permit scientific knowledge to become cumulative and grow, while their ability to assimilate knowledge from other research areas prevents the activities of scientific communities from becoming completely subjective and dogmatic." (Crane 1972.)

Hull's argument conflates the necessary criticism and comparison of ideas with personal career-oriented competition, which can be pushed to the point of competitive exclusion. Social forces and personal investment then actually minimize ideas, counter to interests of scientific progress – though scientists involved would not admit to such. Career-oriented competition, if successful, can be positively disrupting to science if it leads to accrual of necessary resources to few scientists. This may destroy or prevent development of the invisible college and the network of interactions. Unless a variety of approaches, consistent with empirical results, is pursued, published, and maintained within a field, then there is no longer an available *comparison* of active ideas. Hull's assertion that science, motivated largely by competition, has been successful begs two questions. Has science progressed as rapidly and broadly as it could have: are there enduring unsolved problems? Has science been focused effectively: are there unresearched topics of importance to science and society? It is interesting that over the period Hull studied zoological systematics the cladistic approach gained hegemony – but this is not taken by all scientists to be advantageous to the subject.

13.2.3 Fraud and misconduct in science

Studies of fraud in science illustrate the detrimental effects of intense career-oriented competition. Stewart and Feder (1987) studied an unusual and unfortunate case where a biomedical scientist had fabricated results, and had done so even as an undergraduate (Culliton 1987). They examined his publications for the three years during which he had admitted to fraud, to see whether his colleagues, referees and editors of journals could have identified the errors and discrepancies before publication. The scientist had apparently been particularly productive. He published 18 full-length papers and numerous abstracts for conferences and book chapters over a three-year period. The analysis conducted by Stewart and Feder (1987) illustrates a consequence of what Merton (1957a) terms the pathogenic culture of science – an extreme concern with recognition, not just by the fraudulent scientist but by those around him.

Stewart and Feder found many of the papers to be "flawed in ways that could have been recognized by those concerned with their publication". In the 18 individual research papers as many as 39 errors or discrepancies were found, with a mean of 12. Stewart and Feder are particularly critical of co-authors who should have read the papers sufficiently thoroughly to uncover errors and discrepancies. Thirteen of the 18 papers had what Stewart and Feder term "honorary" authors – scientists who had no direct involvement in the research but were senior scientists who provided grant support or who were in overall charge of the laboratory. The 13 papers with at least one honorary author had significantly more errors than the 5 without.

Stewart and Feder consider it particularly striking that 7 of the 18 papers contained statements and data that impeded accurate reconstruction of the way in which experiments were carried out. They also itemize instances of republication of data. In their conclusions they are critical of emphasizing numbers of publications and the role of laboratory heads in encouraging this because it undermines quality and rigor. They cite, as an example, a memorandum from a laboratory head (not associated with the case of fraudulent research) tying employment of technicians to production of papers on a three-monthly basis. The memorandum stated: "There is no demand that these be literary masterpieces in first line journals; journeyman works for publication in second, third or fourth line archival publications will be quite satisfactory".

Stewart and Feder's (1987) paper is unusual in that both authors were practicing scientists in the National Institutes for Health and took a number of years from their normal research activity to make their analysis (Culliton 1987). They had extreme difficulties in publishing this work. Some journal editors rejected their paper on grounds of scientific merit; others declared that, while the paper was important, potential litigation hampered them (Culliton 1987). The head of one laboratory in which the fraudulent scientist had worked hired an attorney. Stewart and Feder's paper was eventually published in *Nature* with the unusual editorial comment: "Some editorial changes have been made in this manuscript without the consent of the authors. A reply from Braunwald follows on page . . .". Braunwald (1987), a senior professor at one of the medical schools involved and a co-author of the fraudulent work, while agreeing with some concerns expressed by Stewart and Feder, challenges some of their data and conclusions. In particular he suggests that to catch some errors co-authors would have to have recalculated data for themselves. He denies that co-authors were honorary and stated that he had been involved with the research. Braunwald concluded, in particular, "it must be understood by all that to be a scientist is a privilege and that society invests a special trust in all scientists".

The problem of inadequate detailed involvement of senior authors was also recognized as enabling fraudulent research in another case (Marshall 1986). That fraud was discovered through the diligence of a scientist who, in reviewing the work when it was submitted in support of a promotion application, noted discrepancies leading him to further detailed analysis. In that instance the university medical school required co-authors of the fraudulent researcher to defend all their papers. In this case there was concern to follow the norms of science rather than to protect action based on counter-norms.

These two cases involved clearly fraudulent work in biomedical science. The individuals were culpable. But the dangers of the laboratory system were also exposed. The desire for publications, and/or the research grants that may be dependent upon them, can lead to a breakdown of critical discussion and dominance of counternorms over norms. These outright frauds were caught but work that might be considered only slipshod or unreliable might not be caught, resulting in unreliable papers being submitted and, unless the refereeing system is excellent, they might be accepted. It is interesting, and in retrospect essential for revealing the system that enabled published fraud to occur, that some scientists spent considerable time investigating these cases. Clearly this was not an easy task, either socially or intellectually. It required both diligence in execution and fortitude in withstanding self-serving criticism.

In ecology the laboratory system, where a senior academic is responsible for obtaining grant and contract funding for a considerable number of scientists, is not developed to the same extent as in medicine. But large programs are funded and develop a requirement for continuous grant and contract funding and the pressure to publish and extol the value of those publications is increased. In ecology there can also be problems of academic responsibility in large-scale research. Multi-author publications are produced and praised as illustrating a willingness to work together on the part of the individuals. This may be so, and may certainly be advantageous in some respects, but multi-authored papers have the potential danger of diffusing responsibility for a piece of work.

The essential lesson to learn from the analysis of fraudulent science is that co-authors must be able to justify all aspects of a scientific paper, not to defend from egregious fraud, which is generally quite a small risk, but to ensure appropriate detailed criticism of the research itself.

The National Science Foundation (NSF) defines misconduct in science as:

> "fabrication, falsification, plagiarism, or other serious deviation from accepted practices in proposing, carrying out, or reporting results from activities funded by NSF; or
> retaliation of any kind against a person who reported or provided information about suspected or alleged misconduct and who has not acted in bad faith." (Buzzelli 1993a.)

The Office of Inspector General in the NSF receives around 50 allegations of misconduct a year (Buzelli 1993b) mostly associated with NSF funded research. Of these 50 percent are charges of plagiarism and 10 percent fabrication and falsification. Other allegations concern such things as breaches of confidentiality of peer review and tampering with experiments. Some 19 percent are miscellaneous and not misconduct in science.

Resolution of cases takes various forms. Allegations to the NSF may be investigated by the institution in question, or by the NSF, and with varying outcomes including barring a scientist from holding federal grants for a period of time. Outcomes can not simply be categorized as guilty or not guilty. Certainly some are upheld and others are not, but comments on particular cases, described in successive NSF semiannual reports to Congress (e.g., National Science Foundation 1994, 1997a) reveal personal difficulties between collaborators, and between supervisors and graduate students. There is considerable difficulty in defining plagiarism. Buzzelli (1993c) concludes, "the scientific community should review its standards and practices concerning plagiarism". While the extreme of copying another person's original text unquoted is generally agreed as plagiarism, other instances may be the result of squabbles and misunderstandings.

The number of cases of misconduct reported to the NSF is small but the full extent of the problem is difficult to estimate (National Science Foundation 1990), since much information is based on surveys of scientists and the methods used have not been standardized.

13.2.4 The role of gender in scientific debate and discovery

The social structure of science influences the balance between cooperation and competition and the type of criticism applied to research. This has an important influence on the development of objective knowledge. There may also be substantial differences between individuals in how they give and respond to criticism, the most studied and commented on being those based on gender. Hull (1988) acknowledges that "there may be a sense in which the social organization of science is male based".

Historically there have been substantially fewer women making careers in science than men, and those who did had lesser academic qualifications, entered jobs with less status and pay, and were promoted less rapidly for similar qualifications (Zuckerman 1991). The situation is changing with more women earning doctoral degrees and proceeding to employment in science, although proportions vary considerably between academic subjects. Zuckerman (1991) notes that women are becoming assistant professors at about the rate to be expected among new degree holders but that "Gender parity in hiring new Ph.D.s in academia or industry has not, the evidence suggests, been around for long".

In a presidential address to the American Association for the Advancement of Science, Widnall (1988) analyzes the group behavior of scientists. Her concern was why fewer women than men completed graduate school, particularly through to the Ph.D. One survey she cites showed that, of those choosing science, a larger proportion of women than men completed their BS degree and that women entered graduate school in proportion to women BS graduates. However, attrition and stopping at the MS level created a major drop.

Widnall (1988) comments on two surveys of male and female graduate student progress from Massachusetts Institute for Technology and Stanford University. She saw little difference in Graduate Record Examination scores (a USA nationwide examination required for admittance to graduate study) and Grade Point Average (a US system of assessing undergraduate performance). She suggests that, to be successful, "the graduate student must run both an academic and a financial support gauntlet" and that the financial support "gauntlet" is less specific than the academic "gauntlet". To be successful, "[t]he task for the student is to find a spot in a functioning research group, work on a topic central to the interests of the group with sufficient financial resources to carry out the research, and work with a faculty adviser who will both supervise the research and guide the educational and future career development of the student". The Research Assistantship (RA) is the favored method of support for this to be realized, and Widnall suggests that a smaller percentage of female than male graduate students in all fields of science are supported as RAs. She also suggests there is greater loss of self-esteem among female than male students as they progress through graduate school. She comments:

> "In view of the importance of the hidden agenda [i.e., the financial and support 'gauntlet'] that uses structured professional experiences to elicit independence in the student, some significant fraction of the women students is less equipped to seek out, to engage, and to profit from these experiences . . . Many women students report discomfort at the combative style of communication within their research groups . . . For many women this style of interaction is unacceptable, either as giver or receiver."

These differences may be important, not only for their impact on women, but for the scientific process. Criticism is essential in science, but it is not achieved if excessive competition encourages scientific rhetoric rather than effective debate, particularly where critics attempt to "prove" a point rather than investigate a problem. This fosters self-confirmatory bias. On the other hand, where it can, criticism must go to the heart of a problem and this certainly can be sharp and uncomfortable to receive.

More than 50 studies, in a range of subjects, showed women publishing only 50 to 60 percent as many papers as men of the same ages do (Zuckerman 1991). Gender differential in publication rate grows as scientists become older. Some postulates explaining this difference have been rejected following investigations: particularly that women publish fewer co-authored papers, and that women's scientific careers are disrupted by marriage and childbirth. Analyses showed married women scientists actually publish more than single women scientists. It is not clear whether women receive small but cumulative disadvantages and lack of encouragement that reduce their rate of publication (Cole and Singer 1991). Or, as Loehle (1989b) suggests, women publish less than men do because generally they are less combative and more reflective in their work. So papers are not published as repeated or multiple tokens of scientific rhetoric in a competitive game but simply as reports of work done. For both men and women a few individuals produce large numbers of papers, and Keller (1991) considers that if those women prefer to lead small rather than large teams that may explain differences in publication numbers.

Harding (1991) suggests that while formal restraints to women's participation in science have been lifted, structural obstacles remain, particularly those associated with differences in male and female socialization processes, and the same argument can be extended to racial differences. The identification and pursuit of science as a competitive game has itself determined the type of science carried out and results obtained. Science is not a detached, value-neutral, enterprise, but is intimately connected with social and political goals through the pursuit of success by individuals. "Modern science has been constructed by and within power relations in society, not apart from them" (Harding 1991). How should women participate in science? Should they accept the current system and work to strengthen its norms rather than counternorms? Or should there be a movement to change its procedures, structure and objectives?

A weakness of Hull's argument for competition as the essential and correctly dominant force in science is that it leads to an elite that determines not only the methods of, but also the agenda for, scientific investigation. Creating objective knowledge requires that research be scrutinized and debated. For this to be effective then different types of research must be conducted – made from different standpoints. Criticism must not just be about the details – were there sufficient replicates, were measurements comprehensive? It must also be about fundamentals and for this different types of research must be conducted. An example of research being conducted from a different standpoint is McClintock's discovery of jumping genes (Keller 1984). Diversification of the people involved in science leads to more types of problem being tackled and this in itself is a necessary component of criticism of any dominant group, from wherever it arises.

13.3 Creation and use of scientific literature

Through the critical process of peer review, scientific papers should become accurate accounts of effective research. However, while the scientific literature is essential, it cannot be perfect in quality. The difficulties of research itself, of writing an effective account, variable editorial and refereeing standards, and the techniques authors use to accommodate to social requirements, all combine to make some scientific literature misleading and some, at worst, wrong. To make best use of the scientific literature it is essential to appreciate how the publication process works as a social process.

Scientific publications should be evaluated in three stages. First, colleagues with knowledge and understanding of the field should review draft manuscripts – both specialists and generalists can make useful comments – and substantial improvements take place at this stage. Second, when a paper is submitted for publication, peer reviewers and journal editors evaluate it. Reports produced within an organization or published in journals without a recognized peer review system are often referred to as the "gray literature". Third, once a paper appears, it can be challenged, used, or ignored by other scientists and subsequent reference to it, in literature citations, influences how a paper is viewed. Each stage is a social process influenced by the norms and counternorms of science.

13.3.1 Constructing a scientific paper

Writing a scientific paper is technical writing – but it is nevertheless an art (Perry 1991). Rules of grammar and standards of English usage (Strunk and White 1979, Fowler 1996) and style of presentation (Anon. 1993) must be followed, but the principal task is presentation of ideas in a format and language making them clear to the reader. Most scientific ideas are complex and not easily reduced to a linear description.

Technical writing is hard work because it is a creative act. The art is to take a collection of ideas and data and transform them into something greater than the simple sum of the parts, something with form and meaning. We must "think on paper," simultaneously refining ideas and developing the form of the written text at all levels, from choice of words through to the overall structure. For most of us this requires substantial, and repeated, rewriting as we examine the relationship between the questions we wish to answer, the data and observations we have, and the interpretations and qualifications we must put on that information. Scientific writing must follow journal conventions and respond to criticism from co-workers and peers as well as referees. Detailed research shows that paper writing requires considerable debate and negotiation among authors and co-workers, resulting in simplification in ideas and data represented, and use of rhetoric intended to persuade readers of the importance of the work.

Knorr-Cetina (1981) made a detailed study of a laboratory research process and of writing a scientific paper. The foundation for the paper was a series of graphs and tables made as the work progressed. Certainly these were reported as the central feature of the paper but both the reasons for doing the work, and the detailed methodological choices made during it, were not reported. The Introduction was constructed retrospectively to provide what Knorr-Cetina calls a "plausible script" but omitted the real practical reasons why the work was undertaken. It constructed relevance for the work. Methods were described without discussing the multiplicity of choices made and much tacit knowledge was ignored, e.g., things of important practical significance about measurement procedures. This hindered readers from perceiving and analyzing the operations of the laboratory and prevented them from repeating the work.

Two processes are at work. (1) With few exceptions, papers must be short. A recurring response of referees, and particularly editors, is that papers are too long. "Busy people do not have time to read all of this" is the type of comment received by authors about their papers. Some journals invite referees to tick off a box indicating a required percentage reduction in length. In practice the constraint on length may be as much motivated by journal page limits and the desire of editors to include more papers in a volume than by lack of succinctness on the part of authors. (2) Journals, and the scientists reading them, require scientific papers to follow particular styles, both in layout and language usage, and these reduce research reporting to a form of shorthand.

The upshot is that scientists simplify their findings and papers become persuasive argument rather than dispassionate reporting. Star (1983), working and observing the scientific process in a laboratory, describes simplification as the outcome of a complex social process. The actual work of building a chain of inference is not recorded in the paper, or, if it appears in early drafts tends to be removed by the final version. '[I]ll-structured problems are thus transformed into well-structured problems by the expedient of ignoring progressive complications or qualifications. In some cases they are simply declared to be well structured and worked on as if they were" (Star 1983). Simplifications are made in describing the interrelationship with other problems and particular types of result may be selected, or emphasized. In Star's (1983) view: "A *fact* emerges which is simultaneously stripped of its complexities and isolated from its relationship to a larger work/historical context". In addition to pressures of the publication process, Star (1983) describes other pressures leading to simplifications. These include situations where scientists with different specialties work together and arrive at compromises in descriptions of each others' work, and "conclusion" pressures where scientists are expected to give results, e.g., to satisfy funding agencies, before, by their own standards as scientists, sufficient information has been gathered. Beware: simplification is not synonymous with clarification.

Connor and Simberloff (1979) complain that if research in ecology is to proceed by "conjecture and refutation," then the factual basis for the refutation must be made available in publications. They note that in *Ecology* editorial policy changed with increasing restrictions in publications of tabular data that might be included. Connor and Simberloff (1979) recount additional problems and financial costs of insisting on publication of more than the usually tolerated amount of tabular data. More recently it has become possible to lodge data in publicly accessible data bases.

Scientists are interested, even determined, that their work should become a component of the regularly cited literature. Schuster and Yeo (1986) suggest that at the level of technical debate, "[scientific] arguments are pieces of practical reasoning rather than formal reasoning, more akin in their public, printed manifestations to legal briefs than to solid chains of strictly valid inference". They say that scientific argument is essentially persuasive argument and should be considered *rhetorical*. This view of scientific arguments as rhetoric, and aimed at persuasion rather than analysis and further test, implies an intent actually to avoid criticism. For example, the very reduced form of Methods sections was termed by Knorr-Cetina (1981) the "rhetoric of objectivity" because the choices, and how they were resolved, were excluded – yet they were essential in determining the results. Medawar (1963) criticizes the normal style then used in experimental biology as producing a fraudulent representation of the thought processes accompanying, or giving rise to, the work. His concern was that the format of Introduction, Methods, Results, Discussion was inappropriate, particularly where the Results section was expected to contain factual information with no interpretation.

In some ecology journals an emphasis is placed on stating what editors or referees term "hypotheses". This too is a stylistic device leading to a particular form of rhetoric. The question represented as a "hypothesis" may not be the question with which the research started but has been constructed, at least in part, during the writing process and after the research is complete. Stylistic devices, whether of the inductive type criticized by Medawar (1963), or of the "hypothesis" format, are an integral part of scientific writing but are only social conventions. They are not essential in the strict sense of reporting the work. For much useful work it may be difficult to construct papers using a particular style. This can hinder new approaches and is one reason why new journals are formed. Conventions vary between subjects, change over time within subjects, but they have the common purpose of enabling scientists to develop particular forms of persuasive argument, or rhetoric (Bazerman 1988).

13.3.2 Peer review

The basis of peer review is that the journal editor sends submitted manuscripts to scientists knowledgeable in the topic for comment on both the content of the work and clarity of presentation and writing. *Ecology* currently sends guidelines to reviewers and questions for them to answer about the paper (Box 13.1). These are clearly designed to encourage referees to be fair, not to use the cloak of anonymity to be destructive, and to make specific and detailed comments that will assist the author(s) in improving the paper.

Journals vary in how peer reviewers are selected. In some cases journals publishing a wide range of subjects have an editorial board comprising

Box 13.1. Ecology, Ecological Monographs, and Ecological Applications

Publications of the Ecological Society of America
GUIDELINES FOR REVIEWERS

CONFIDENTIALITY—The enclosed manuscript is a privileged communication. Please do not show it to anyone or discuss it, except to solicit assistance with a technical point. Your review and your recommendation should also be considered confidential.

TIME—In fairness to the author(s), please return your review within 3 weeks. If it seems likely that you will be unable to meet this deadline, pleasure return the manuscript immediately or call the Managing Editor today.

CONFLICTS OF INTEREST—If your previous or present connection with the author(s) or an author's insitution might be construed as creating a conflict of interest, but no actual conflict exists, please discuss this issue in the cover letter that accompanies your review. If you feel you might have any difficulty writing an objective review, please return the paper immediately, unreviewed.

COMMENTS FOR THE AUTHOR(S)—What is the major contribution of the paper? What are its major strengths and weaknesses, and its suitability for publication? Please include both general and specific comments bearing on these questions, and **emphasize your most significant points**.

General Comments:

1. Importance and interest to this journal's readers
2. Scientific soundness
3. Originality
4. Degree to which conclusions are supported by the data
5. Organization and clarity
6. Cohesiveness of argument
7. Length relative to the number of new ideas and information
8. Conciseness and writing style

Specific Comments:

Support your general comments with specific evidence. You may write directly on the manuscript, but please summarize your handwritten remarks in "Comments for the Author(s)." Comment on any of the following matters that significantly affected your judgment of the paper:

1. <u>Presentation</u>—Does the paper tell a cohesive story? Is a tightly reasoned argument evident throughout the paper? Where does the paper wander from this argument? Do the title, abstract, key words, introduction, and conclusions accurately and consistently reflect the major point(s) of the paper? Is the writing concise, easy to follow, interesting?

2. <u>Length</u>—What portions of the paper should be expanded? condensed? combined? deleted? (Please **don't** advise an overall shortening by X%. Be specific!)

3. <u>Methods</u>—Are they appropriate? current? described clearly enough so that the work could be repeated by someone else?

4. <u>Data presentation</u>—When results are stated in the text of the paper, can you easily verify them by examining tables and figures? Are any of the results counterintuitive? Are all tables and figures clearly labeled? well planned? too complex? necessary?

5. <u>Statistical design and analyses</u>—Are they appropriate and correct? Can the reader readily discern which measurements or observations are independent of which other measurements or observations? Are replicates correctly identified? Are significance statements justified?

6. <u>Errors</u>—Point out any errors in technique, fact, calculation, interpretation, or style. (For style we follow the *CBE Style Manual*, Fifth Edition, and the ASTM Standard E380-92, "Standard Practice for Use of the International System of Units.")

7. <u>Citations</u>—Are all (and only) pertinent references cited? Are they provided for all assertions of fact not supported by the data in this paper?

8. <u>Overlap</u>—Does this paper report data or conclusions already published or in press? If so, please provide details.

FAIRNESS AND OBJECTIVITY—If the research reported in this paper is flawed, criticize the science, not the scientist. Harsh words in a review will cause the reader to doubt your objectivity; as a result, your criticisms will be rejected, even if they are correct!

Comments directed to the author should convince the author that:

1. you have read the entire paper carefully;
2. your criticisms are objective and correct, are not merely differences of opinion, and are intended to help the author improve his or her paper; and
3. you are qualified to provide an expert opinion about the research reported in this paper.

If you fail to win the author's respect and appreciation, your efforts will have been wasted.

ANONYMITY—You may sign your review if you wish. If you choose to remain anonymous, avoid comments to the authors that might serve as clues to your identity, and do not use paper that bears the watermark of your institution.

specialists in different areas who assist the editor in selecting reviewers. Usually two are chosen and normally reviewers' names remain confidential to the editor, although sometimes a reviewer will sign a review, perhaps if he or she feels that knowledge of their identity will help the authors to interpret and respond to comments made. If peer reviews differ substantially then an editor may seek additional review. In cases where a paper is contentious, or specifically critical of other work in an important way, or very new in what it says, then multiple reviewers may be used. The editor, or sometimes the specialist member of the editorial board, will collate peer review comments and make a decision on whether the paper will be published. Even for those papers accepted for publication the editor will usually make acceptance conditional upon authors responding positively to reviewer's comments. Once a paper has been accepted and the final manuscript submitted, then the editor or a technical editor will ensure consistency within the paper. For example, references within the text must match those listed as literature cited. Figures and tables are numbered consistently and their captions explain what is being shown. Editors also check for journal style, e.g., that the literature cited follows the format required by the journal.

Ideally peer review should ensure high quality publications. However, it certainly is not a perfect system. In selecting published papers for students to review and discuss in class, I have found two things. First, while the majority of papers in peer-reviewed journals are of an acceptable standard, it is not difficult to find papers with considerable discrepancies, errors and lack of clarity, at least sufficient to undermine one's view that the paper provides information on which one can build further research. Second, if I give a paper to students with the straightforward instruction "Please read that paper and we will discuss it" they will generally not find errors or inconsistencies that subsequently, when pointed out, are obvious to them. We do seem to read for content, to find the story line, and skills of effective criticism take time to develop. Part of the reason may be that we trust the review process, and part may be that criticism requires first understanding what the author(s) are attempting to say, and this itself can take considerable time and effort.

The peer review process can substantially alter manuscripts. Myers (1985) recounts progress into publication of separate articles by two experienced biologists advancing what they considered radical ideas. Both papers were initially submitted to journals seeking groundbreaking articles of general interest, i.e., *Nature* and *Science*. Both papers were rejected, even after review and resubmission and the papers subsequently appeared, much altered, in more specialist journals. Myers describes reviewer's comments on successive versions of the manuscripts as a negotiation on the status of the knowledge claims – continuing the process started by co-workers described by Knorr-Cetina (1981). Myers emphasizes that knowledge claims are socially constructed, particularly through the influence reviewers had on the form of the article, e.g., the proportion of commentary on theory relative to presentation of data and the placement of different aspects of the work within the papers. He suggests that the review and writing process have an important consensus-building function that maintains the homogeneity of the scientific literature.

Studies of journal peer review provide overwhelming evidence for exercising caution when reading and interpreting papers in peer-reviewed journals. Cicchetti (1991) collated and summarized studies of peer review for journals from behavioral sciences, medicine and physics, where comparisons were made between referees refereeing the same paper both for overall judgement on acceptance or rejection and scores given to individual questions, using categories similar to those in Box 13.1. Cicchetti's paper was followed by comments from 34 journal editors and people involved in the publication process, about the studies themselves, and about implications for the publication process. The dominant feature was a low agreement between referees reviewing the same paper. Considering only overall assessments, whether the paper should be published or not, then for generalist journals there was more agreement that a manuscript should be rejected; for subdiscipline journals reviewers and editors agreed more about acceptance than rejection. Quantitative assessment of detailed comments used intraclass correlation coefficients, and 0.3 was a typical value for a category of question within one journal and does not indicate substantial agreement between referees. There was less agreement for journals publishing more general articles than for those specializing in subdisciplinary fields of study.

A wide range of opinions was expressed by the commenting editors, reflecting variation in how referees are, or should be, used in the editorial process, and how referees should be selected, e.g., referees likely to have similar or dissimilar perspectives or specialist comments to make about a paper. Cicchetti (1991) stresses that proper selection of reviewers, so that a good match is made in technical interest between submitters and reviewers, should increase reliability, i.e., agreement between reviewers. However, this should not be sacrificed in favor of ensuring that the review is correct, and for this reviewers should, between them, cover different aspects of the article. In comments following Cicchetti's paper, many suggestions were made for improving peer review: using multiple reviewers, author anonymity (i.e., blind reviewing), rewarding referee contributions, and training reviewers. Some considered blind reviewing to be unworkable because reviewers are likely to identify authors from self-citation or other features, e.g., use of unique locations in the research. However, McNutt *et al.* (1990) found blinding (the authors were not identified to the reviewer) worked in 76 percent of cases in a medical journal and significantly improved the review quality. The case for maintaining confidentiality of referees is that it allows scientists in the same research program to offer criticisms while maintaining a good working relationship (Hull 1988).

In addition to its technical difficulties, peer review has been severely criticized because of social abuses. For example, Van Vallen and Pitelka (1974) assert in a Commentary in *Ecology* that intellectual censorship existed, particularly that mathematical ecology was pursuing a policy of competitive exclusion, having recently become part of the ecological establishment. Previously it had been a new branch of the subject and its adherents had had difficulty in having their papers published. Van Vallen and Pitelka cite a number of abuses: rejection of a paper giving an empirical refutation to a major theory of mathematical ecology, selectivity in reference citations in review papers, and withholding data from publication because of lack of agreement with predictions of mathematical models. They interpreted their assertions in a broad social framework. Some problems in publication arise because of subjectivity in approach to scientific issues, some because referees, either consciously or unconsciously, exert selectivity as part of a group activity. Their question is "Does this person belong?" And work is not judged solely on its merits.

Peters and Ceci (1982) conducted a dramatic experiment of resubmitting previously published papers to psychology journals for review. The papers were originally written at prestigious institutions and were resubmitted to the same journals that had previously published them, but at the second submission fictitious names and addresses were used. Of the 38 editors and referees involved, only 3 detected the resubmission, which is an interesting commentary in itself. The rejection rate of these previously accepted papers was 80 percent, the grounds for rejection frequently being described as "methodological flaws".

Peters and Ceci (1982) discuss the possibility that rejection was the result of a bias against the fictitious and therefore unknown authors and institution. Alternatively there could be a positive bias for acceptance of the original paper with the prestige of the original institutions acting to blind referees against the methodological flaws that were subsequently reported. They reviewed articles with similar evidence of defaults in the peer review system and their article is followed by an extensive series of written comments from editors of journals from a range of subjects. Horrobin (1982) recounts the complaints made by women to the Modern Language Association, indicating that there had been few articles by women in the Association's journals in comparison to the number of women who were members of the Association. When a blind reviewing process was subsequently introduced the number of papers by female authors rose dramatically.

Mahoney (1977) studied confirmatory bias among referees of the *Journal of Applied Behavior Analysis*. A topic of current interest in the field was selected and a paper was artificially concocted in three different versions reporting positive, negative, or mixed results. In the case of mixed results, the paper was supplied with either a positive or negative Discussion section. The papers were sent to different referees, each receiving only one paper. They were asked to rate the paper for different criteria.

Referee ratings for Topic Relevance did not differ between versions cast with different results. In contrast, the Methodology sections, with identical experimental procedures, were rated better if accompanied by positive rather than negative results. There was a similar bias with regard to Data Presentation and overall Scientific Contribution. Manuscripts suffered very different prospective fates of recommendation for acceptance or rejection, depending on the direction of their data. Negative results earned a significantly lower evaluation. Manuscripts with mixed results were consistently rejected without any apparent influence of the interpretations made. One dramatic and unplanned aspect of the investigation was the accidental inclusion of a typographical error leading to contradictions between the methods and results. Of the referees who read manuscripts with positive results, only 25 percent noted the error, compared with 71 percent of the reviewers who read manuscripts with negative results ($P < 0.05$).

13.3.3 Problems of quantity and quality

Without publishing, one's ideas and work will neither be criticized nor contribute to the scientific debate. Nevertheless, despite this essential requirement, there are frequent and authoritative complaints that too much is published and relatively little is ever used. For example, over half of the papers appearing in journals covered in *Science Citation Index* were not quoted in the five years after their publication (Hamilton 1991). Values for subdisciplines ranged from only 9.2 percent uncited in atomic, molecular, and chemical physics, 78.0 percent for applied chemistry, to 86.9 percent for engineering.

There seem two opinions as to why the situation is unsatisfactory. Newman (1986), a member of the Editorial Board of the *Journal of Ecology*, suggests that there is a strong tendency for editors to select papers sound in evidence but only moderate in originality – original papers conceivably have less strong evidence and may be heavily criticized and rejected during peer review. In reply, Moss (1986), also writing with editorial experience, suggests originality is not necessarily suppressed by editors or referees but by a "changed system of values". Career requirements for publications has encouraged the JPU, the Just Publishable Unit, a paper with enough reliable data that some editor must accept it. Authors do this to maximize number of publications achieved for a unit of research effort. Such papers are in line with what Moss terms "current fashions" so as not to undermine "the intellectual security of the most pedantic referee that the author can imagine". McGrath and Altman (1966) describe the consequences of tying career advancement to numbers of publications. It can lead to an emphasis on methodological rigor rather than substance and theory, particularly where the same technique or procedure can be applied repeatedly to only slightly different questions or circumstances. However, development of theory based on empirical results is time consuming and not suitable for publication in numerous segments, which can lead to the weakening and/or fragmentation of a research network (Crane 1972).

Both Moss and Newman detect aspects of the subjectivity described by Mahoney (1977). However, Moss (1986) adds to that the needs of the publication industry, describing this dynamic: "Almost everything can be published somewhere in the plethora of journals and symposium volumes which houses the mutualism of scientist-insecurity and publisher-profiteering".

Attempts have been made to counter this. The NSF issued a notice (National Science Foundation 1989) in which they state:

> "Evaluation of scientific productivity must emphasize quality of published work rather than quantity. To ensure this emphasis, NSF will now limit the number of publications considered in reviewing a grant application."

They specified that only five publications most relevant to the research proposed and up to five other significant research publications would be used in merit review.

Marshall (1986) reported that after analysis of fraudulent research the investigating committee in the medical school involved recommended for the school's own purposes of promotion review:

(1) that peer review should focus on the quality, not quantity, of a researcher's work,

(2) that each department should develop a means to identify the type and degree of participation of every faculty author in each published work,

(3) that co-authorship should "reflect scientific involvement and imply responsibility for the work reported," including a responsibility to defend co-authored papers if called upon, and,

(4) that the medical school should develop clearer guidelines for supervising

trainees and "realistic" standards of productivity. (After Marshall 1986.)

For the working scientist advances in abstracting, particularly electronic access, have reduced some of the difficulty owing to the quantity published, but problems of quality remain. These can be countered only by developing a rigorously critical attitude as you read.

13.3.4 Literature citation and its analysis

The *Science Citation Index* (SCI) produces a computer data base of all papers cited in articles in the journals it reviews. For example, a paper by J. Doe (1990) may have been cited three times in 1995 and a reference to each of those citations can be found, under Doe, J., and the particular paper in the SCI for 1995. If you identify a particularly important paper for your research, this is a valuable way of being able to trace subsequent uses of it. However, the value depends on how much relevant citation there has been. One can generally use titles of the papers that have made the citation to select whether you should read them. In some cases you can identify people working on the subject, perhaps in a slightly different way from you, and of whom you were not aware.

The SCI has been extensively used for making analysis of influences, e.g., which scientists are most influential, and networks, e.g., identifying groups of interacting scientists. Tagliacozzo (1967) identifies two types of paper: those with a short citation life and those that are sustained in the literature. Presumably, the ideals of Moss (1986), Newman (1986), and the NSF (National Science Foundation 1989) would be to increase the sustaining literature. The half-life of citation for a paper varies markedly between research areas: that of the experimental sciences being shorter than others. Citation practices differ markedly between countries. For example, the multi-national nature of science in the Netherlands, Denmark, and Switzerland is apparent in both publication and citation patterns (Inhaber and Alvo 1978).

The SCI has been used in assessing productivity of scientists for performance reviews and promotion (e.g., Cronin and Overfelt 1994). Two relationships, Lotka's Law and the Ortega Hypothesis have been investigated using publication and citation records. Lotka's Law (Seglen 1992) states that the number of papers produced by individual scientists is skewed, with few scientists producing larger numbers of papers. Where x is the number of papers produced and y is the number of scientists producing that number of papers, then $x^n \cdot y = c$. The values of n and c must be computed for individual data sets. A relationship of this form can be fitted to data from a wide range of fields, although values of n and c both vary (Pao 1985, 1986).

Despite the generally skewed distribution of scientific productivity, the

Ortega Hypothesis states that the research of more productive scientists does depend upon that of the less productive. Cole and Cole (1972) challenge this on the basis of a citation analysis illustrating that more productive scientists predominantly cite other productive scientists and they conclude that the majority of scientists could be dispensed with without impeding scientific progress. MacRoberts and MacRoberts (1987) challenge this conclusion by questioning the assumptions of citation analysis (see also the commentary papers following their article).

First, there is no evidence that bibliographies are complete and unbiased lists of influences of scientific work. Indeed, detailed analysis of citation specifically shows them not to be. Only 30 percent of influences may actually be referenced (MacRoberts and MacRoberts 1986a), and reference may be made to secondary rather than primary sources (MacRoberts and MacRoberts 1986b). Some work is used but rarely or never cited, some is cited mainly, or only, through secondary sources, while some is cited every time it is used (MacRoberts and MacRoberts 1986b).

Second, persuasion (Gilbert 1977), rather than a desire to give credit where credit is due, is the major motivation for citing (Brooks 1986). Cronin (1984) reviewed work suggesting that citation is selective to serve scientific and personal goals. Detailed analyses of scientific papers illustrate window dressing (citing papers with which you wish to associate your work rather than which you need as a component of your argument), padding (citing more references than you need to ensure that you miss no scientist you think important), and sprinkling references as an afterthought to enhance the respectability of a paper. Hull (1988) made an extensive study of the editorial records of the journal *Systematic Zoology*. The vast majority of citations in journals are positive, giving support to a theory. For Hull the "primary purpose of positive citations is not so much to give credit where credit is due but to gain support for one's own work". Hull considers that negative citations, where work is cited because it is thought to be incorrect, are mainly to establish social boundaries between groups. MacRoberts and MacRoberts (1987) suggest that many scientists contribute to scientific advance but receive little or no credit for their contributions.

Analyses have been made of both the type and context in which citations are made. Classification schemes for citations have been developed (for a review, see Liu 1993) distinguishing between purposes and contexts of citation. "Citation, ultimately, is a personal practice. It is difficult to believe that a wholly satisfactory theory of citing or citation behavior will soon be devized" (Liu 1993).

13.4 Developing and using explicit standards of criticism to construct objective knowledge

Criticism is a technical process conducted in a social framework. Effective criticism requires detailed knowledge of the scientific arguments and methods of investigation. It focuses on research quality and relevance to a developing theory or study of a domain. Just criticism is unprejudiced about scientists conducting the work and can acknowledge that different types of study contribute to scientific advance. We need effective criticism to strengthen the network of objective knowledge and just criticism to ensure its rapid and unbiased growth.

SCIENTIFIC CRITICISM

The degree of objectivity that we can accord to knowledge, changes as scrutiny and debate pass through four stages:

(1) *direct analysis,*
(2) *testing through direct repetition,*
(3) *refining through extended use,* and
(4) *standpoint criticism.*

The first three stages are made in the context of justification and focus on how things are researched. The fourth stage, standpoint criticism, analyzes the context of discovery, what things are researched, and what approaches are taken.

Standpoint criticism is defined and discussed in Chapter 14. It involves criticism where objectives of investigators, as well as their methods, may be very different and frequently offers the most radical criticism.

Direct analysis, has the most explicit procedures for scientists and is applied to individual investigations particularly during planning and conduct of the research, and preparation and peer review of manuscripts. It has strong technical components. Do experiments have adequate controls? Are procedures for statistical inference based on adequate exploratory analysis of assumptions? Do definitions conform to usual usage and if not are the exceptions justified?

Testing through direct repetition requires additional work, perhaps in a different habitat or with a different organism, or doing new research that uses the previously obtained result. The criticism or support that such work can give to a new idea is stronger when it is not conducted by the investigator who made the initial discovery or claim, or someone with a close academic relationship. But this second stage is still made from within the same invisible college of scientists. Hull (1988) suggests that repetition is not essential but only that research is conducted that can be repeated. This is insufficient for ecology. Theories must be defined not only by central axioms but also by specification of the domain within which those axioms operate. Because we can not assume universal domain for our theories, it is essential to test through direct repetition.

Refining through extended use involves use and critical analysis by scientists in a related but not identical field, so that criticism is made from a larger scientific social circle. This stage often involves the use and so refinement of concepts and defining the domain of a theory, rather than analysis through a particular series of investigations. It places results and theory in a more general context.

It is valuable to develop explicit standards for making a critical analysis, particularly for reading manuscripts or published papers:

(1) To counteract your own biases. Particular interest in the work can focus your attention on the support it gives to your own ideas. It is rare that the full information in a paper, particularly the limitations to the work, is appreciated at the first reading.

(2) You can not depend upon peer review to have revealed all inconsistencies, illogicalities or mistakes in conducting the research.

(3) You can not take for granted what authors say about the significance of their work – what is claimed may not actually have been realized.

Mitroff (1974a) suggests that there are two types of scientific problem: well-defined and structured, e.g., where agreed sampling procedures and methods of analysis have been established; and poorly defined, e.g., where a domain is under investigation and both theory and methods are uncertain. Mitroff comments that "Whereas the conventional norms of science are dominant for well-structured problems, the counternorms appear dominant for ill-structured problems". We can expect criticism to follow a regular path for well-structured problems but to be more difficult for poorly defined problems.

It is rare that the logic of the investigation is laid out in a complete and sequential manner that starts with assumptions (axioms) and moves through to conclusions. This is to be expected because, for example, some assumptions about the measurement techniques used, or the experimental or investigation design may be mentioned only when those aspects of the study are discussed. Yet they can be crucial to the study. An author may use a particular stylistic device, e.g., starting the paper with an attractive question, and look forward to what improvement could be achieved if that question is answered. With this type of approach the assumptions on which a question is based may not be dealt with comprehensively or in order of importance. It can take some time to untangle the logic of the scientists' argument from a paper that follows a stylistic device, e.g., the inductive format or the hypothesis format. Following is a 10-point checklist to assist critical analysis.

(1) *Establish the overarching axiom of the paper.* This is important in order to appreciate the set of assumptions that may be used without them being explicitly mentioned.

(2) *List the axioms on which the research is based that are stated in the paper.* Many authors list their assumptions as they make their literature review. It is valuable to turn those assumptions into axioms to point up the logical structure of the work. Of course what can be important in doing this process is your own knowledge of the literature. But even if you are reading a subject with an incomplete knowledge of the literature it can still be important to check the axioms and their internal consistency and whether they support the postulates.

(3) *Define any axioms missing from the paper.* You may only come to appreciate missing axioms after you have read the paper completely. Authors, referees and subject editors may all implicitly consider the same propositions to be axioms that need not be stated. That does not automatically mean they should be taken as axioms.

(4) *List the postulate(s).*

(5) *What are the major concepts and can you define them from reading the research?* This is extremely important. Different researchers may use concepts in different ways, even though they may agree on a written definition if requested to make one. Lack of precision in definition and use is a major difficulty in ecology (Chapter 16).

(6) *The effectiveness of concepts by measurement.* Are you satisfied that the measurements were effective? (a) Does what was measured match what should have been measured? (b) Were the measurements competent?

(7) Inferential procedure. If statistical analysis was conducted are you satisfied that it was appropriate?

(8) *What is the result of the paper?* Is it confirmation, negation, revision or improved definition of a concept, revision of a theory, a report of a new phenomenon, or development of a new classification? (This may not be a complete list.)

(9) *Can you identify the factors that made this study possible?* For example, new ideas, a new measurement, access to a site previously not investigated?

(10) *What would you consider to be the next piece of work that the investigators should carry out?* This can be particularly important. A study that ends by pointing forward to a new piece of work is acknowledging the scientific process.

13.5 Discussion

Study of scientific research as a social process contains elements of agreement and disagreement among its practitioners – as does any science (Shapin

1995). But two results are established with sufficient generality that scientists should consider them seriously.

First, there is no guarantee of objectivity in science by following particular research procedures or methods. Philosophers of science, as described in earlier chapters, have also established this, but studies of the social processes of science illustrate a further dimension. What we call *objective knowledge* is a negotiated product between the producers of pieces of work and those involved in reviewing the work through to publication and then its use. This does not imply that our results and conclusions are valueless. However, they are achieved in the context of justification. Individual skepticism – even about your own completed and published work – is most important for enhancing the degree of objectivity. Yet it is the quality most challenged by social processes.

Second, scientists establish social structures that can be exclusive, conforming to the group, and on occasions predatory in their competitive attempts to achieve dominance. In this respect scientists are no different from other people – and that is precisely the point. Science deserves no special elevation as a human activity and scientists are not immune from the forces making individuals engage, consciously or unconsciously, in group activities, establishment of hierarchies, exclusion of different types of person, and other social phenomena. Unfortunately, these group processes, with their social objectives, frequently inhibit effective criticism of research progress, or may restrict alternative lines of research and so restrict the development of knowledge as objective.

MacRoberts and MacRoberts (1996)[1] contrast what they term the traditional view of the scientific process with a review of the theory that knowledge is socially constructed. In the traditional view:

> "Science consists of a unique set of institutional arrangements, behavioral norms, and methods that enable it to exist and function effectively. Ultimate answers are Nature's; man is only a mediator and a passive observer. Scientists combine the methods of science with objectivity, disinterestedness, humility, universality, and skepticism, and stand back and let Nature tell her tale unimpeded; they are trained to perform the task of objectively observing Nature 'as she is'. Scientific disputes are settled on the basis of evidence and rational discussion. The content of science is independent of personal and social forces and is, therefore, outside the realm of sociology and psychology."

MacRoberts and MacRoberts' synthesis of the alternative view contains the following points. The story book scientist, the objective, disinterested, humble, universal, and skeptical seeker of truth simply does not exist. Science is "closely enmeshed with prevailing cultural history and beliefs" and "is

[1] In the following quotation MacRoberts and MacRoberts (1996) own footnotes to references have not been included.

constantly being rewritten, sanitized and fictionalized to make it appear autonomous and rational. Scientific knowledge is socially negotiated, not given by nature" (MacRoberts and MacRoberts 1989). Furthermore this construction is a collective process not the product of a few heroes – even though scientists themselves may choose to elevate individuals, usually dead scientists, as untouchable icons under whose protective glory they can shelter. Most importantly: "Scientists' beliefs explain natural reality, natural reality does not explain scientists beliefs" (MacRoberts and MacRoberts 1989). Scientific disputes are decided not by "one set of results matching nature and thus being right' and the others side's data not matching nature and thus being wrong' but instead is a complicated business involving consensual judgements and interpretations emerging from argument and negotiation". Harding (1991) suggests that the ideal of a value-neutral science, as maintained by the traditional view, offers hope to scientists, and the whole scientific enterprise – that it can produce claims that will be "regarded as objectively valid without their having to examine critically their own historical commitments, from which – intentionally or not – they actively construct their scientific research".

Science progresses through both measurement and the development of theory. In what are sometimes called the hard sciences, theories are usually constructed so that those questions about them are resolved through measurements that most, if not all, scientists recognize as conclusive. Social processes are important in determining which research problems are considered important, which group will receive resources, and precisely how the problem is constructed. In ecology we have additional problems. Many theories are about integrative concepts. Questions about them can not be resolved by single investigations but research must progress at a number of sites, or with different organisms, and resolution of similarities and differences requires complex negotiation about both core axioms and their domains. Consequently, for our subject, just and effective criticism is central to its development.

13.6 Further reading

Brannigan (1981) develops ideas about multiple parallel discoveries and discusses general aspects of the social basis of discovery. Primack and O'Leary (1989) document differences in publications between men and women in a selected groups of ecologists. Roy (1979) gives a succinct criticism of peer review and its effects on science. Sinderman (1982) provides a scientist's view of various social activities and interpersonal interactions that are part of the conduct of science and gives advice on how to succeed in them. The volume edited by Miller and Hersen (1992) outlines general issues of research fraud, details some specific instances, discusses contributing causes and the editors conclude by making proposals.

14 Values and standpoints and their influence on research

Summary

Our value systems help us to maintain consistency in our lives and lead us to adopting identifiable standpoints on particular issues. People with different value systems are likely to adopt different standpoints. The fourth and most fundamental stage of criticism applied to a piece of research is standpoint criticism – how it looks to people with different value systems. This is particularly important in defining the most basic assumptions made.

From their differing standpoints, scientists, resource managers, and policy-makers can each limit the type of research that may be attempted and the management or policy to be implemented. Some scientists value scientific research for its own sake: some politicians value science in terms of its tangible benefit for solving identified problems in society. These contrasting values lead to different standpoints in the way that science should be organized and funded and particularly in the relative importance of peer review in the assessment of research programs and individual proposals.

Decisions about environmental policy and management of natural resources are made from particular standpoints. How scientific research can influence these decisions most effectively remains uncertain. Examples are given where scientific analysis was central to a particular resource management policy, and where it was excluded. Major changes became inevitable in both cases. In the first, society's demands changed the objectives of management – despite scientist's concerns. In the second, environmental catastrophes occurred resulting in policy changes that some scientists had been urging.

The relationship between science and society is discussed from the perspective of ecology and how this relationship may influence what science is done and how it should be analyzed critically.

14.1 Introduction

The three stages of criticism discussed in Chapter 13 use techniques practiced routinely by scientists. However, as Harding (1991) comments: "The methods of science . . . are restricted to procedures for testing of already formulated hypotheses. Untouched by these careful methods are those values and interests entrenched in the very statement of what problem is to be researched and in the concepts favored in the hypothesis that are to be tested."

We all maintain values – often very strongly. Usually our values are related, in a value system, determining or justifying particular standpoints. One person may value and praise personal economic self-sufficiency, the traditional skills and hardiness of foresters, and close-knit rural communities. Another may value and praise parsimony in human use of natural resources, preserving the integrity of existing ecosystems, and coordinated local, national, and international action in favor of environmental conservation. These two sets of values, although not mutually exclusive, would probably be associated with different standpoints. On logging policy for old-growth forests the former is likely to urge continued harvesting and the latter to urge preservation of remaining old-growth forest. Individual value systems are frequently, though not exclusively, identifiable with a person's self-interest, they may be held with varying fidelity, and they are rarely entirely self-consistent. Groups holding similar value systems may express them as an ideology and represent and advance them through institutions such as political parties, special interest citizens groups, or scientific societies.

Value systems determine the standpoint from which particular scientific investigations are conceived. For example, those wishing to eradicate a source of pollution may suggest a research program different from those prepared to consider mitigation but unwilling to halt the activity causing the pollution. Standpoint criticism is the fourth stage of criticism in development of objective knowledge. It is concerned with what things are researched and what approaches are taken. This is particularly important in ecological and environmental research where science is used not only to answer agreed questions between scientists but to influence peoples' values and bring new adherents to a political, economic, resource management, or conservation standpoint. Differences in value systems are also apparent with respect to science itself. Some politicians with influence or control over national funding agencies think government expenditures should be explicitly for the immediate national good. Further, that in government-funded research accountability should be taken to the level of individual research programs. Some scientists think that science should be funded for its own sake – and that that is best for the national good.

STANDPOINT

A *standpoint* is a subjectively held viewpoint about an issue justified by an individual person or group's value system.

The first section in this chapter, 14.1 Standpoints in science, management, and policy, illustrates that ecologists and resource managers have standpoints – with members of each group viewing scientific research and environmental and resource management in particular ways. These standpoints are not automatically right or even best for society. The second section, 14.2 Reviewing and funding scientific research, illustrates scientists' standpoints about science. These are often forcefully expressed as support for the peer review system in deciding research directions as well as in reviewing grant and contract proposals. Peer review has been strongly criticized as beneficial neither for scientific progress nor for deciding research for policy questions.

The third section, 14.3 Science, scientists, and society, describes how societies' requirements from science are changing, and with that change come explicit demands that scientists change both how they behave and what they do – that they should modify their standpoints. However, the nature of science is complex and there is ambivalence in societies' demands between wishing to steer science yet, at the same time, expecting scientists to lead because they have technical knowledge. This ambivalence sets a dilemma for scientists in how they should conduct themselves, both in their work and in relation to society. Most important is how scientists define their own values, determine their standpoints on particular issues, and construct effective standpoint criticism.

14.2   Standpoints in science, management, and policy

Scientists, and policy and management professionals, influence what science is done. The types of influence and the way they are exercised have to be carefully understood. Ecologists should examine whether their research is really aimed at a required task, or rather at investigating their own theories, or supporting environmental use or conservation policies that they value. Managers should examine whether they are constraining research or the implementation of its conclusions, to be within the bounds of a current management approach or to suit a particular management style rather than analyzing the scientific basis of policy or management.

14.2.1   Scientists' standpoints

Scientists' values can become apparent in the conduct of national research programs. The National Acid Precipitation Assessment Program (NAPAP) was a decade-long program in the USA through the 1980s, costing an estimated $600m. Its objective was to assess the quantity and source of acid precipitation and possible effects on a range of agricultural, ecological, human health, amenity, and property values. The program was controversial. US government agencies, including the Department of Energy, the Environmental Protection Agency, and the Department of Agriculture, had some allocated research funds channeled through NAPAP. Not infrequently these agencies have different standpoints on environmental and ecological issues, but in NAPAP there was an attempt to develop a unified, comprehensive, scientifically based assessment of acid precipitation effects.

The program was intended to emphasize assessment of effects rather than basic research. The distinction between scientific research for discovery and for assessment was difficult for some scientists to appreciate. NAPAP was at the interface of science and public policy – its role was to provide information for policy-makers and policy analysts. For Russell (1992) a lesson to be learned from NAPAP was that policy-makers, policy analysts, and scientists must "understand, respect, and police adherence to their special roles". These are:

> "that policy-makers define what is important to inform the ultimate decision, and then make it after getting the information they require. Policy analysts define and explicate the options for the policy-makers, bringing to bear the relevant scientific information along with other considerations of relevance to the decision. Scientists provide estimates of causes and effects under alternative conditions based on their research, fully disclosing all the uncertainties and areas of ignorance." (Russell 1992.)

In a program such as NAPAP it is essential for assessment questions to be clearly defined and articulated. This requires careful negotiation among parties with different goals and agendas because the questions set can imply anticipated policy outcomes. "Setting the questions is essentially a policy process. It follows that the proper role of science is to advise on what is practicably achievable, and with what expected level of certainty. *It is not to seek to influence, based upon inherent scientific merit, what the questions should be*" (Russell 1992, emphasis added). The NAPAP Oversight Review Board (ORB) made a major criticism.

> "It is our judgement that the assessment function to which NAPAP's scientific and technical findings were to contribute was not as fully successful as the research process. The assessment function should have been the central focus of the NAPAP endeavor from the first, but it was not adequately addressed, supported, or carried through at key points during the program. As a consequence, the assessment portion of NAPAP was incomplete in its coverage of elements essential to a comprehensive weighting of alternatives. The appearance is that the NAPAP scientific efforts were guided to an excessive degree to resolve interesting scientific questions rather than by the potential to improve policy decisions, given the limited resources available to the program over its life." (ORB 1991.)

A lesson from NAPAP is that it is difficult to persuade scientists to stick to defined policy issues.

As described in Chapter 13, much science is conducted with personal success in mind. This has a major influence on topics that scientists will select, and how they approach them. Many scientists will not give high priority to the repetition of investigations – even if the repetition is proposed for different species or environments. Repetition may not gain acclaim from other scientists yet the increased certainty that may come through repetition was often what was needed for NAPAP's assessment task, rather than the excitement of developing new components of theory.

Scientists' standpoints were reinforced through administrative procedures. Much of NAPAP's budget was from funds that agencies were required to reallocate from their own budgets, so frequently the issue became one of asking particular scientists in those agencies to change what they were doing and work on acid precipitation assessment issues. Neither the scientists nor the agency were always interested in doing that. Scientists wished to continue their career development in a chosen field – agencies did not wish to have their previously conceived research programs disrupted. Compromises were made in setting the objectives of research. Where contracts were placed outside of agencies they were judged primarily by peer review and what appealed to scientists, rather than what may have been needed for assessment, was likely to be favored.

Some scientists thought NAPAP an unnecessary expenditure. They had already decided from previous results what the effects of acid precipitation were and what remedial action should be taken. Public opinion became sufficiently decided that policy on sulfur emissions could be changed before the final NAPAP assessment (NAPAP 1991). However, most of the results from NAPAP were available and understood by that time – and these influenced public opinion. But for some people the evidence presented up to and including the NAPAP assessment was insufficient. Russell (1992) notes:

> "Political success is measured in finding broad consensus that a course of action is reasonable, even though contending parties may still prefer somewhat different outcomes. For this to happen in our society it seems essential that major stakeholders should be confident that their positions have been heard and seriously considered. They also need to be comfortable that the decision was arrived at after a thorough, honest evaluation of alternatives, and that, taken as a whole, the decision serves a national purpose, and is fair."

This process was pursued in NAPAP. Through a series of drafts the content of the final assessment document (NAPAP 1991) was extensively debated between NAPAP staff, scientists in the contributing agencies, and representatives of major groups who would be influenced by policy introduced as a result of NAPAP. The debate between the document writers and some agencies continued right up until the day before publication. To dismiss the arguments of those who are not convinced, even if you feel their concerns are motivated by economic or political self-interests rather than scientific reasoning, is to place your standpoint, and the values motivating it, as more important than theirs.

Cowling (1992), who along with Russell, served on the NAPAP Oversight Review Board, comments that science and technology alone could not provide the wisdom to make wise choices.

> "The responsibility of scientists and policy analysts in a democracy is to understand and clearly communicate the scientific facts and uncertainties and to describe expected outcomes objectively. Deciding what to do involves questions of society's values where scientists, as scientists, and policy analysts, as such, have no special authority. For this reason the processes of risk assessment are very different from the processes of risk management or scientific reviewing."

Scientists in ecology can have very firm ideas on what they think should happen to management of a natural resource or in an environmental policy question. They may use scientific arguments to justify their opinion – but how a management or environmental policy is determined is, and should be, more than a scientific matter. In the case of NAPAP the debate about both the amount of acid deposition and its effects, and the policy change to be made, had to go on for a longer time than many scientists thought necessary and involved additional research, particularly repetition.

In some cases individuals or groups of scientists become involved with practical problems of a particular ecological system. In one such case, Livingston (1991) describes the ecological analysis of the Apalachicola River estuary in north-west Florida and its effects upon resource management policy. In 1972 the economy of Franklin County depended on the Apalachicola estuary fishery, particularly oysters and shrimps. An alliance was made between university researchers and the fishing community that funded an interdisciplinary research program designed to provide the basis for sustained management of the resource. Early results showed Apalachicola Bay to be extremely productive owing to a combination of geomorphological characteristics, salinity distribution, and nutrient relationships. The bay is shallow with salt water entry restricted by barrier islands. Productivity is phosphorus limited, and wind-driven water turbulence in shallow water leads to mixing of bottom sediments and phytoplankton production in the euphotic zone.

Scientific results were important for both fisheries management and land use policy. A proposal to construct a series of river dams was dropped, partly because they would have preferentially retained phosphorus that could have restricted estuary productivity. Pesticide use was regulated despite vigorous opposition from state agencies responsible for coastal spray programs. Research linking river wetlands with the estuary was publicized through radio and television shows, newspaper stories, educational tapes, and input to secondary school curricula. Franklin County adopted a regional plan that included the following. (1) Purchase of environmentally critical lands in the drainage system. (2) Designation of the Apalachicola system as an Area of Critical Concern, a legal definition of guidelines for development within the region. (3) Creation of cooperative research efforts to determine potential impact of activities such as ongoing forestry management programs. (4) Provision for aid to local governments in the development of comprehensive land use plans. Livingston noted that while the floodplain forests and river fisheries were of intrinsic interest it was the economically important commercial fisheries of the estuary that justified the public purchase of environmentally critical lands.

Land purchases were increased to include areas in the immediate vicinity of the estuary as well as river margin lands. These included purchases on the barrier islands to complete a ring of publicly owned lands around the most environmentally sensitive areas of the estuary. The objective was maintenance of natural drainage delivering fresh water from fringing wetlands, preservation of important habitats, particularly seagrass that provides temporally important carbon inputs to the estuary ecosystem, and maintenance of populations of oysters vulnerable to effects from urbanization and agricultural activities. Livingston summarizes this whole activity as unprecedented, the most ambitious management effort in the USA, and that bold decisions were taken by local, state, and federal administrators to protect a natural resource before it was destroyed.

This activity was opposed. Starting in 1982 a civil suit, claiming $60m, was filed against Franklin County Commissioners and their advisors by a land developer of the principal barrier island of the estuary. At issue was the right of developers to bring high-density development to the barrier island. After four years a federal judge dismissed the charges with the opinion that the plaintiff had used the suit "to harass and intimidate the defendants". Although an order was made for the plaintiff to reimburse the defendants costs, this was never paid. Livingston (1991) continues his account.

> "Over those years of costly litigation, the original group responsible for the management plans was broken up through death, sickness, and harassment by those who wanted to develop the last unpopulated coast of Florida. A new group of Franklin County officials, with the support of powerful state officials and developers, moved to consolidate political power in the hands of those who were behind local and regional residential and commercial development."

Livingston considers that by 1991 deliberate efforts of local politicians, sympathetic state bureaucrats, and local and regional news media, to reverse the planning process of the 1970s and early 1980s dominated the situation. Land purchases by developers were made on the barrier islands on condition that adequate sewage treatment was provided. However, that requirement had not been enforced. The important oyster fishery declined owing to overfishing and natural disasters, particularly a hurricane and a following drought. The loss of that industry removed an important and influential political force. Livingston (1991) notes:

> "Over the past decade, the relative importance and effectiveness of the research community in the development and implementation of planning processes in the Apalachicola region have deteriorated considerably. There has been a gradual shift from the use of objective scientific data to political and bureaucratic manipulations as the primary influences in the management process. As part of this trend, the research arm of the National Apalachicola Esturine Reserve (formerly Sanctuary) has been largely eliminated by local, state, and federal officials . . . The deliberate elimination of significant environmental research in the Apalachicola system has set the stage for the initiation of major municipal development. In this way, such development can proceed without the impediment of scientific impact analysis."

The manipulation of scientific research programs and dissemination of their results is a common feature of environmental issues. Harassment clearly occurred in the case and was judged to be illegal. But how should decisions be reached? The change in Franklin County officials was part of the democratic system. Scientists may claim objectivity in their science – but use of science in an extensive educational program to sustain a particular resource management policy can certainly be interpreted as political – "the chief contribution of the scientific program was to provide an objective basis for constructive political action" (Livingston 1991). It can equally be argued that provision of housing on the barrier island is constructive – and giving it priority over estuary conservation illustrates a different value system. Livingston suggested that a failure in preserving the habitat of the Apalachicola estuary was that too great an emphasis of the research was support of the oyster industry rather than emphasis on the intrinsic value of the estuary as an ecological system. This itself is an environmental standpoint – that ecosystems should be conserved for their intrinsic worth – and to pursue that requires political support.

Ecologists do have values and take standpoints on particular issues. While these vary in detail between individuals, important elements are reflected in the comments of Russell (1992) and Cowling (1992) and well illustrated by Livingston (1991). The necessary scientific basis for a policy or management decision may require greater confirmation than scientists may usually be prepared to obtain. What scientists may call objective knowledge may not seem so to managers or policy-makers. This is important standpoint criticism. Some of it may seem unjust to scientists, but there are important points to emphasize. Where a science-based change in policy or management is to be made, more confirmation may be required than some scientists may think necessary. In practice the amount of confirmation needed has to be judged in the policy forum, not by scientists, even though gathering the information falls within the realm of scientific investigation.

As with other scientists, ecologists can show a firm conviction about the theories they propose and continue with this conviction into its application in policy or management. But policy changes are not the domain of scientists alone.

> "The proper role for both scientists and policy analysts is to provide advice and counsel – each in their areas of special competence – to those who are charged by our society to make policy decisions. It is not a proper role for either a scientist or policy analyst, as such, to seek to make (or even have special influence on) society's decisions." (Cowling 1992.)

However, the institutional process for utilizing scientific knowledge is not well defined.

14.2.2 Managerial standpoints

Natural resources are frequently administered by national agencies whose work involves setting policies for management, explaining them to the public, and implementing them consistently. There are many examples where scientific research has contributed to the solution of management problems. However, large agencies can find difficulty in responding to research results that suggest a change in policy or management procedures. The reason may be partly due to the nature of bureaucracy: once a policy is set a change can require substantial retraining and possibly financial investment. It may be partly due to differences in interpretation of scientific results: the number and types of investigation necessary to convince an organization that a change is needed, and can be implemented successfully, may be greater than, and different from, what scientists typically produce. Implementation of scientific results can also be influenced by the very nature of bureaucracy itself. Once a policy is in place, and has been publicized, change may be opposed by the agency lest it appear embarrassed by having to make a change. There can be a feeling within the agency that the public, other agencies, or related professional groups, will consider that if a policy has to be changed it must have been wrong in the first place, that mistakes were made, and no doubt could be again.

Schiff (1962) describes the dynamics of instituting a policy change within the United States Forest Service (USFS) for controlled burning of forest in the south of the country. In the early twentieth century, Forest Service employees were avid enthusiasts for forest conservation rather than the previous norm of exploitation. As part of this new conservation ethic a policy of fire suppression was adopted. This was reinforced by the occurrence of some massive forest fires in the first decade of the century just after the formation of the USFS. However, some foresters outside the Forest Service suggested that some species, notably longleaf pine, *Pinus palustris*, were dependent upon fire, both for regeneration – removal of organic material exposes mineral soil as a seed bed, and early growth – and for grass suppression. Fire favors longleaf pine saplings, which are able to recover from grass fires. Although research outside of the Forest Service, notably by H. H. Chapman (1926, 1932), illustrated the importance of fire for the species, and that it was a natural part of the ecological system in maintaining southern pine forests, the Forest Service did not recommend a policy of controlled burning.

From the turn of the twentieth century through to the 1930s the Forest Service was preoccupied with establishing fire protection systems in the south of the country. State forest authorities were encouraged through the Clark–McNary Act providing federal grants-in-aid for fire protection. Schiff detailed examples of how the results of experiments and investigations into the positive effects of prescribed burning, conducted by both Forest Service and state forestry employees, were either suppressed or quietly forgotten. It was more important for the organization to push forward with protection measures, and particularly to stop uncontrolled forest burning by the public, than to examine how controlled burning might be used. Managers were concerned that any statement in favor of controlled burning might be interpreted as an excuse for indiscriminate burning by those indifferent, or opposed, to protection. Schiff notes this implies that the Forest Service itself would be unable to explain and implement a more sophisticated policy. Schiff describes the way in which, during the 1930s, articles from government scientists that recommended controlled burning were subjected to excessive review periods and suggests:

> "Commitment to absolute exclusion now made any retraction of publicized shibboleths difficult. The Service had been willing to permit research on controlled burning; disseminating the results was quite another matter. Belief persisted that promulgation of information would lower the prestige of federal and state organizations. At the same time, it was thought such material would be misinterpreted as sanctioning indiscriminate woods burning." (Schiff 1962.)

The USFS opposed landowners experimenting with prescribed burning, threatening to withhold their grants-in-aid. The ecological effects of fire suppression were accumulation of undergrowth, an increase in hardwood species in the forest, and reduction in the amount of longleaf pine. The accumulation of undergrowth made the forests vulnerable to major conflagration and ultimately led to a policy change.

> "Catastrophe precipitates action as nothing else can. A protracted drought

during the 1942–43 winter fire season, combined with a shortage of trained fire-control men and heavy fuel accumulations, resulted in the most appalling damage experienced in a decade. Stunned by disastrous blazes, administrators turned to prescribed burning as a solution . . . Of course, prescribed burning's protective aspects, not its silvicultural features, had greater salience for administrators." (Schiff 1962.)

This change in Forest Service policy in 1943 was still opposed by some state organizations. The new policy was not specifically to encourage regeneration and early growth of longleaf pine, but to reduce accumulation of wood that could cause extensive, hot burning, and massively destructive fires. By 1949 the Forest Service recognized the value of prescribed burning, both for regeneration and protection, and the average area burned annually by wild fires dropped. However, some states' forest policy still did not publicize prescribed burning until after disastrous fires in the 1954–55 winter fire season, when markedly greater burn occurred on private lands that had not previously been prescribed burned.

Considering the whole episode Schiff (1962) suggests:

"Ignoring early caveats, the Service tragically slipped into a rut from which escape proved difficult and embarrassing. Thus had evangelism subverted a scientific program, impaired professionalism, violated cannons of bureaucratic responsibility, undermined the democratic faith, and threatened the piney woods with ultimate extinction."

This may seem a harshly worded statement of what is a difficult problem – how a bureaucracy can develop a working relationship with science that is likely to, and perhaps should, challenge its fundamental procedures and policies. However, it can be extremely frustrating for scientists, working to solve a serious problem in resource management, to have their work rejected or even disparaged by a management agency. It seems essential that some researchers outside of the agency should be able to work consistently on important issues in the way in which H. H. Chapman was able to – but of course this depends upon obtaining adequate funds for research.

The use of scientific information made by resource managers can vary markedly. Management styles have changed in US federal forest and marine fisheries resource agencies (Miller and Gale 1986). Decisions are science based but the objectives of management have changed, or are under pressure to change, as environmental issues are recognized by the general public to be more important. Because of this, the range of people interested in management decisions made by the agencies has increased. Miller and Gale (1986) were particularly concerned with the line manager who has administrative, supervisory, and policy-making functions at the regional and local levels. They state that line managers necessarily operate within and beyond the limits of science. "The difficulty, of course, is that the limits are poorly marked." They quote Henning (1970):

> "complications emerge when resource managers, by upward mobility through the administrative hierarchy, reach levels where policy is made and as a result can determine what the public interest is in resource decisions. On the one hand, their decisions are no longer technological, but rather value judgements. On the other hand, they can take part in brokerage politics with other interest groups while they still claim respect and detachment because of their profession."

Miller and Gale (1986) delineate four professional styles prominent at various times in US federal forestry and fisheries agencies, each with a different approach to the role of science in the implementation of policy.

The *forstmeister* has a professional style characterized by an intuitive understanding of natural resources. The participants in the management system are controlled as a family and the style implies commitment to the resource; the means (resource management techniques) are not separated from the ends (resource management goals). The *specialist* has a style rooted in professional education. The management system is controlled as a family and the style implies commitment to the applied sciences, disciplinary approach to policy decisions; the means justify the ends. The *politician* has a negotiating style most attuned to extra-scientific problem solving involving conflicting constituencies. The management system is controlled by the minimization of conflict and the style implies commitment to the constituencies, negotiating approach to policy decisions; the ends justify the means. The *executive* has a non-resource-specific style of facilitating cooperation whatever the bureaucracy. The management system is controlled as an organization and the style implies commitment to the bureaucracy, synthesizing approach to policy decisions; there is coordination of means and ends.

Miller and Gale (1986) suggest that the *forstmeister* style of resource management was preeminent in the USFS for the first half of this century as a viable management style. There was clear federal title to forest lands, forest units were comparatively isolated, and there were few resource conflicts. In the 1950s, as the demand for wood products increased, the dominant style became that of the *specialist* committed to professional timber management, and specialists in engineering and botany were more numerous than those in soil science or landscape architecture. Miller and Gale suggest that with the increasing importance of environmental issues, particularly from the start of the 1970s, then the *politician* style became more in evidence.

At its outset, marine fisheries were based on a farming concept. There was not a time when the resource had to be managed simply because of its existence as territory in the way that forest management was required. Consequently, the *specialist* was the dominant management style, with the objective of protecting the brood stock and predicting yields. In the USA the 1976 Magnuson Fishery Conservation and Management Act extended national jurisdiction over fisheries to 200 nautical miles (about 370 km) and established eight quasi-federal Regional Fisheries Management Councils. Miller and Gale note that this act legitimized marine fisheries management on the basis of "optimum" rather than "maximum" sustained yield, and suggest growing acceptance that fisheries policy is social policy. They put forward the idea that the Regional Councils act as political forums and that politicians are appointed to senior fisheries management positions. They suggest that specialization has run its course in fisheries management.

Miller and Gale also comment that, although there are some managers with an *executive* style within fisheries management, there are few such in forest management. Generally, the executive style is not apparent in either agency. They quote Barnard (1968) that "executive work is not that *of* the organization, but the specialized work of *maintaining* the organization in operation".

The current requirement for scientific knowledge in environmental and resource management is much broader than previously needed. As the issues have extended, so too has the requirement for scientific analysis and synthesis. However, lack of clear and effective organizational structures and practices for decision-making that incorporate scientific analysis accentuates argument conducted from particular standpoints. Scientists can be drawn into this argument or exploit it for their own purposes. This is not surprising, since much political argument is about how decisions will be made, and not just what decision will be taken. It is important for scientists to analyze their standpoints. It is important for government management and policy-setting agencies to consider the most effective way science can be involved.

14.3   Reviewing and funding scientific research

Most money for scientific research comes from government sources, and typically scientists make funding decisions. Along with technical competence, the norms of scientific behavior of emotional neutrality, disinterestedness, and organizational skepticism are values leading scientists to the standpoint that peer review is the most appropriate assessment procedure. However, just as there are doubts about the effectiveness of peer review in judging potential publications (Chapter 13) there are doubts about its effectiveness in funding decisions. Some of these doubts have a similar basis – that counternorms can disrupt the effectiveness of peer review and that its decisions are socially rather than academically determined. There are additional causes for concern. First, that the process of peer review, when combined with shortage of funds and intense competition, leads to conservatism in awards and is not in the best interest of science. Second, that scientists' peer review is not the appropriate way to assess research designed for policy or management problems because it is made from an exclusive standpoint.

14.3.1   Research proposals and their peer review

There is considerable variety in the structure, required content, and maximum permitted length of research proposals, depending upon the funding agency and program being applied to. Some agencies such as the National Science Foundation (NSF) in the USA, or the Natural Environmental Research Council in the UK, have the express purpose of funding academic research, use the competitive proposal mechanism, and have standardized the format of the research proposal and how it is processed. The NSF has funding programs in ecology with regular, twice yearly, application deadlines for receipt of proposals. It also has special programs of limited duration, often in response to particular issues, e.g., global change, sometimes run jointly with other agencies. It has programs for Long Term Ecological Research, instrumentation and graduate thesis improvement. There are restrictions on who can apply for research funding – typically faculties from universities and scientists working for non-profit institutions, although industry is involved in some programs. Copies of the essential text of successful proposals can be obtained from the NSF, although this is only free for a limited number of proposals.

Proposals to the NSF follow a particular format (Table 14.1). Most US universities have an administrative system that checks essential points of the proposal before it can be sent, particularly the form of the budget request, although not usually the content of the proposal itself. Typically, within the agency, there is a manager for each funding program whose responsibility is to find peer reviewers for each proposal. Within the NSF many although not all programs also use panels of scientists drawn from the scientific community at large. The panel assesses the proposals for each round of applications and this provides a group discussion peer evaluation. Each panel member is expected to read a number of proposals and discuss their merits at a panel meeting. Typically between three and five panel members read each proposal and an individual panel member may read up to 15 proposals. Panel meetings may last up to three days and membership on a panel may be for three years. Proposals are ranked according to consensus of merit, with exceptions noted by the program manager and possibly taken into account when final decisions are made.

Decisions on funding are made by the program manager, usually in consultation with other NSF staff, and using both the panel evaluation, if there has been one, and the written peer review comments. To avoid conflicts of interest, the NSF has strict rules of procedure. Proposers are required to provide lists of collaborators who will then not be used as reviewers. Re-

Table 14.1. *Required contents of a proposal to the National Science Foundation. Each of the listed sections is specified as a form in National Science Foundation (1995a,c) which are issued together as single booklet. NSF has a (currently) experimental electronic proposal system (http://www.fastlane.nsf.gov)*

Information about the Principal Investigator/Project Directors	Specification of gender, citizenship, ethnicity, and whether or not disabled, for each Principal Investigator. This is requested but not required information
Cover Sheet for the Proposal to the National Science Foundation	Specification of addresses and telephone numbers of the organization and principal investigators submitting the proposal. This sheet has space for NSF administrative comments. This must be signed to certify statements in the proposal are true
Project Summary	This is a one-page statement of objectives, methods to be employed, and the significance of the proposed activity to the advancement of knowledge or education
Table of Contents	List the page numbers and sequences of the proposal in a specified format
Project Description	A clear statement of the work to be undertaken including: objectives, expected significance, relation to longer-term goals of the Principal Investigator's project, relation to present state of knowledge and work in progress elsewhere, description of methods and procedures, plans for data and specimens. May not exceed 15 pages including visual materials
References Cited	A specified format is recommended
Biographical Sketch	Biographical sketches of each Principal Investigator, limited to two pages, and may include a list of 5 references relevant to the work and 5 auxiliary references
Proposal Budget	A detailed budget must be given and a budget justification should be made
Current and Pending Support	A list of the grants and contracts the Principal Investigators are receiving and have applied for
Facilities, Equipment and Other Resources	A description of the facilities and equipment available to the investigators that are not part of the proposal and are essential to the proposed work

viewers are requested to return proposals if they feel they have a conflict of interest. During panel discussions panel members are required to leave during discussions of a proposal from the same university or organization or if the applicant has any connection with the panel member. Applicants are sent a summary of the panel and program manager's comments along with the individual reviews and their ratings.

The numbers of regular proposals funded/considered by the Ecology Program of the NSF were: Fall 1993 23/87 (26.4%); Spring 1994 28/104 (26.9%); Fall 1994 22/100 (22.0%); Spring 1995 27/116 (23.3%) (T. Frost, personal communication). The rate of acceptance of dissertation improvement grants over the same two years was 29 percent with around 60 applications each year. For NSF proposals individual reviewers and panel members are asked to rate proposals on a scale: excellent, very good, good, fair or poor. Unless a proposal is in the "excellent" or top of the "very good" categories its chances of being funded are low and in panel discussions a proposal needs to find one or two members who really champion it.

One view is that an acceptance rate of around 25 percent is reasonable. Peer review and panels raise important questions that must be responded to before a proposal is funded and it may take several resubmissions before a proposal is funded. Lawton (1994) gives an alternative view: "The problem is relatively simple. Because money is tight, official (i.e. government-supported) grants committees and panels have to find reasons to reject a high proportion of requests for funding". Under this intense competition proposals that seem risky are likely to be rejected. Innovative research, and maybe interdisciplinary research, which are difficult for individuals to assess, may disappear (Lawton 1994). Some, though not all, reviewers place a strong emphasis on preliminary results in judging whether the proposal will be successful and in such a highly competitive situation one or two low scores can cause a proposal to fail. Comments such as: 'This is a trust me proposal and I am a show me reviewer' and "Once some of the papers under review are accepted, a stronger case could be made for funding" are not uncommon. The result of this emphasis on preliminary or even final results is that reviewers and panel members are more likely to make positive judgements if a proposal is for increments to an established and accepted program of research.

Lawton suggests that the response by some scientists to this situation is analogous to the tortoise. Cover yourself with armor and move very slowly, i.e., propose nothing that may alarm reviewers by its novelty and be very explicit about precisely what you will do. Be more detailed in the proposal than you really can be in the research itself, but at least pretend that you have mapped out your time in detail (Lawton 1994). One ecologist, with an unnerving ability to write proposals that are funded, described his strategy as writing proposals at "the cutting edge of conventional wisdom"! That is, appear to be excited by what the mainstream of ecologists thinks is important in your particular research area and propose only what they will understand.

Does it matter that the system may not explicitly foster innovation? Will science not progress just as effectively – and perhaps with more caution? Three concerns have been expressed: the system is dishonest, experienced researchers with proven records tend not to be funded, and interdisciplinary research is not funded.

Dishonesty may occur because with an emphasis on preliminary results the work being applied for may be already substantially complete. The grant will be used for at least some new ideas not even mentioned in the proposal – a process sufficiently common to have its own name, "bootlegging". The office of the Inspector General of the NSF considered that a scientist who submitted a proposal representing research already completed as work to be done under an NSF award committed misconduct (National Science Foundation 1994). In that case no new work at all was proposed – but there can be less clear instances. Peer reviewers can pay particular attention to preliminary results and the boundary between preliminary results and completed work may itself not be clear-cut. Van Vallen (1976) states "Exciting work by honest people does not get funded". He suggests that research into ideas not mentioned in a proposal is inevitable. New ideas are most likely to change during progress of any research – that scientists do not research exclusively what they propose and nor should we expect them to. Goodstein (1995) further suggests that the high ethical standards demanded for a peer review system to work effectively are eroded by the intense competition for grants. "Most scientists do hold themselves to high standards of integrity, but as time goes on, more and more referees have their ethical standards eroded by the unfair reviews they receive . . ." He considers that peer review was suited to the period of expansion in funding but not to present conditions.

Lawton (1994) observes that many though not all of those that he considers to be the best ecologists failed to get funding from government agencies. Detailed analysis does show a weak relationship between science citation rates, proposal review ratings, and proposal funding decisions. Cole *et al.* (1978) considered 1200 proposals to the NSF from 10 different fields. There was only weak correlation between proposal review ratings and number of journal citations made to the proposer's previous work. For the ecology program, proposer's citation rates explained only 1 percent of the variance in proposal ratings. Recent publication record explained none of the variance. Individuals who had published none, one, or two publications in the previous 10 years (lowest quintile of publication rate) actually received more good or excellent ratings on their proposals than individuals who had published 19 or more publications (highest quintile). "Note that in ecology those scientists who have published the most papers are less apt to get favorable ratings on their proposals than those who have published the least papers . . ." Cole *et al.* (1978). That previous record of publication should have such a low influence may seem surprising, since reviewers are requested to take previous research into account. Ecology had the largest coefficient of variation among reviewer ratings of the 10 programs examined by Cole *et al.* (1978).

Abrams (1991) substantiated the weak relationship between past achievement, as judged by citations, and funded proposals. He argues that individuals who have produced good science would continue to do so. He found a high positive correlation between Scientific Citation Index citations in different periods for members of an NSF panel in ecology ($r^2 = 0.778$) and for scientists cited in an ecology textbook, i.e., citation rate is consistent over time. Abrams (1991) acknowledges difficulties with science citations as an indication of achievement. He suggests that even so: "The fact that reviews of research proposals have a low correlation with past scientific performance, and that past scientific performance is highly correlated with future performance suggests that proposals may not be good predictors of future performance". He puts forward two possible reasons why research proposals may be poor predictors of future scientific accomplishments. First, that the ability to produce a highly rated proposal has little correlation with the ability to carry out and publish high-quality research. Second, that the standards used in proposal review are not those used in assessing publications. Cole *et al.* (1978) comment that some productive scientists may simply write poor proposals. Abrams considers there is no *a priori* reason to believe that ability to make research questions both easily comprehensible and exciting to a reviewer, or panel member, reading under severe time constraints is correlated with ability to produce a high-quality scientific paper based on original research.

Van Vallen (1976), Abrams (1991) and Lawton (1994) all suggest that performance-based awards would lead to shorter proposals and more innovative work. "Real advances in any subject are as likely to happen by accident as by careful planning; creativity is a chancy, stochastic game, helped by sharp eyes, deep knowledge and hard work" (Lawton 1994). The origins of scientific discovery certainly are difficult to characterize. Retrospective attempts to assess the effectiveness of government support of science have also uncovered difficulties in developing predictors (e.g., Kostoff 1994).

Interdisciplinary research also seems to require special consideration. It is difficult to define – some people consider that all ecology is interdisciplinary because botany, zoology, and the physical and chemical sciences may all be involved. From the perspective of classifying research, a more focused definition is that interdisciplinary research uses approaches and techniques from different fields to make analysis and answer questions in another field and these proposals may not fall clearly into one funding program. This type of work is particularly difficult to assess at the proposal stage because there will be few people who can span the disciplines involved.

Difficulties with judging the merits of proposals by peer review are apparent to funding agencies, as well as to some scientists, although the perceived difficulties can be different. Government funding of science has to serve the national interest and there are substantial difficulties in determining how this is achieved. Unfortunately scientists do not always seem to be effective at considering the national interest other than that the best peer-reviewed science is good. Between 1981 and 1997 peer reviewers of NSF proposals were asked to use four criteria in their assessment: (1) research performance competence, (2) intrinsic merit of the research, (3) utility or relevance of the research, and (4) effect of the research on the infrastructure of science and engineering. Problems were encountered, particularly that reviewers overemphasized points (1) and (2). In a survey, fewer than half of reviewers questioned said that they usually commented on all four criteria and as many as 20 percent said they ignored the criteria all together (National Science Board 1997). As Lawton (1994) notes, competence and productivity in research (criteria (1)) is not a deciding criterion. Typically it is mentioned in reviews and at panels, it is seen as a requirement, but the substantive judgement is made on criterion (2) as reviewers determine it from the proposal. NSF program officers reported more difficulty in getting response from peer reviewers about criteria (3) and (4) so that the mission of NSF (Box 14.1) was not being met as well as it should be. A Task Force proposed new criteria for research proposal assessment (Box 14.2) that were adopted in 1997. The Task Forced commented:

> "NSF is increasingly asked to connect its investments to societies' value, while preserving the ability of the merit review system to select excellence within a portfolio that is rich and diverse. Having two criteria, one for intellectual quality and the other for societies impact, should serve to reveal the situations where proposals have high quality but minimal potential impact (and vice-versa). Quality will continue to be the threshold criterion, but will come to be seen as not sufficient by itself for making an award." (National Science Foundation 1997b; http://www.nsf.gov/nsb/merit2.htm)

The bias toward criteria (2) is an example of scientists exerting their own standpoints, as scientists, to the exclusion of others, and the Task Force is attempting to redress the balance. A new form has been developed for proposal reviewers that asks for separate comments on the two criteria, and separate ratings between "excellent" and "poor". An important question is whether peer reviewers, panels, and program officers will take notice of the new criteria, or whether a more radical change has to be made in the required format and balance of proposals themselves. In panel discussions, people focus on what is common to them all. If proposals are mainly a written text of work to be done then that is likely to be the focus of a review.

In discussions, scientists will usually admit that there can be problems or abuse of peer review but tend to maintain that it is the best available system. This is a standpoint not a truth. Peer review has not always been the way in which federal funds were distributed and became popular only after World War II (Guston 1996). Between the two World Wars private foundations provided a major proportion of research funding in the USA and peer evaluation of proposals was not usual. It was specifically rejected by the

Box 14.1. The three long-range goals of the NSF and the set of core strategies employed to reach those goals in the Foundations Strategic Plan (National Science Foundation 1995b, summarized in Appendix C of National Science Foundation 1997b)

The National Science Foundation has three long-range goals:

> *Enable the U.S. to uphold a position of world leadership in all aspects of science, mathematics, and engineering.* This grows from the conviction that a position of world leadership in science, mathematics, and engineering provide the Nation with the broadest range of options in determining the course of our economic future and our national security.
>
> *Promote the discovery, integration, dissemination, and employment of new knowledge in service to society.* This goal emphasizes the connection between world leadership in science and engineering on the one hand and contributions in the national interest on the other.
>
> *Achieve excellence in U.S. science, mathematics, engineering, and technology education at all levels.* This goal is worthy in its own right, and also recognizes that the first two goals can be met only by providing educational excellence. It requires attention to needs at every level of schooling and access to science, mathematics, engineering, and technology educational opportunities for every member of society.

To move toward the achievement of these goals, NSF employs a set of core strategies. These strategies reaffirm the Foundation's traditions, especially its reliance on merit review of investigator-initiated proposals, yet at the same time point to new directions for the Foundation.

> *Develop intellectual capital.* Selecting the best ideas in research and education and the most capable people to carry them out is at the heart of NSF's programmatic activities and the merit review system with which we implement those programs. Opening opportunities for all Americans to participate fully in an increasingly technological society is an essential part of NSF's mission.
>
> *Strengthen the physical infrastructure.* NSF's programs support investments in new windows on the universe, through facilities planning and modernization, instrument acquisition, design and development, and shared-use research platforms.
>
> *Integrate research and education.* NSF aims to infuse education with the joy of discovery and to bring awareness of the needs of the learning process to research, creating a rich environment for both.
>
> *Promote partnerships.* For NSF, success requires collaboration with many different partners, including universities, industry, elementary and secondary schools, other Federal Agencies, state and local governments, and other institutions. We also carry our partnership across national boundaries."

Box 14.2. Guidelines issued for proposal review by the National Science Foundation, July 1997 (National Science Foundation 1997b). Reviewers are asked to rate a proposal on each of these criteria. The notice also asks for a statement of potential conflict of interests, defines obligations of reviewers to protect the confidentiality of proposal contents

Instructions for proposal review

Please provide detailed comments on the quality of this proposal with respect to each of the two NSF Merit Review Criteria below, noting specifically the proposal's strengths and weakness. As guidance, a list of potential considerations that you might employ in your evaluation follow each criterion. These are suggestions and not all will apply to any given proposal. Please comment on only those that are relevant to this proposal and for which you feel qualified to make a judgement.

CRITERION 1. WHAT IS THE INTELLECTUAL MERIT OF THE PROPOSED ACTIVITY?

Potential considerations: How important is the proposed activity to advancing knowledge and understanding within its own field or across different fields? How well qualified is the proposer (individual or team) to conduct the project? (If appropriate, please comment on the quality of prior work.) To what extent does the proposed activity suggest and explore creative and original concepts? How well conceived and organized is the proposed activity? Is there sufficient access to the necessary resources?

CRITERION 2. WHAT ARE THE BROADER ASPECTS OF THE PROPOSED ACTIVITY?

Potential considerations: How well does the activity advance discovery and understanding while promoting teaching, training, and learning? How well does the proposed activity broaden the participation of underrepresented groups (e.g., gender, ethnicity, disability, geographic, etc.)? To what extent will it enhance the infrastructure for research and education, such as facilities, instrumentation, networks, and partnerships? Will the results be disseminated broadly to enhance scientific and technological understanding? What may be the proposed benefits of the proposed activity to society?

Please provide an overall rating and summary statement which includes comments on the relative importance of the two criteria in assigning your rating. Please note that the criteria need not be weighted equally.

Rockerfeller Foundation on the grounds that scientists would only push their own research. Of course program managers within foundations came to have great influence and many sustained excellent programs of groundbreaking research (Kohler 1991).

14.3.2 Scientific research with policy implications

Natural resources policy is the concern of wider and wider groups. Consequently the scientific analyses upon which management decisions may be based are the subject of intense scrutiny from many different standpoints. Different groups used different criteria in their examination of any work. Clark and Majone (1985) outline two important problems about applied research. (1) In practice, scientific inquiry can not necessarily discover most of the things that a policy-maker might like to know and much of what is discovered is incomplete or remains uncertain. (2) Scientists themselves are likely to disagree, particularly where a topic is of current research interest. The scientific community may operate a *laissez-faire* approach that permits emergence of new theory when it is able to fight its way through the sociological processes of the scientific community. However, this may not assist in deciding on the best course for a practical problem given incomplete information.

Clark and Majone (1985) were particularly concerned to develop a wider and more effective set of institutional mechanisms for the judgement of scientific work for policy purposes. In their view the role of anonymous peer review of science for policy purposes is more limited than in science itself. Useful peer communities are rare in the analysis of science for policy purposes and peer review may actually be more part of the problem in achieving a review for policy purposes than part of a solution. A priority should be exploration of new institutional mechanisms for the critical appraisal of science in a policy context. A problem is that research for a policy issue may follow academic fashion using one particular type of scientific approach.

Clark and Majone (1985) stress the need for appropriate criticism. In their view, criticism either from the purely scientific or purely political standpoints is inadequate. Partial perspectives slight the integrative and synthetic considerations so essential to useful inquiry on practical problems. They describe three types of critical attitude necessary in the assessment of science for policy purposes:

> *Rational criticism*, focusing on the development of models for use in policy.
> *Practical criticism*, focusing on the evaluation of policy per se.
> *Ethical criticism*, dealing with the proper role of scientific knowledge in society.

Clark and Majone (1985) suggest that there are two important questions in the analysis of science for policy use: "Criticism by whom?" and "Criticism of what?"

A major problem is when the critical roles played by individuals or groups become mixed – one can frequently see all three of the critical attitudes being adopted at one and the same time. This is to the disadvantage of the total process. Individuals or groups with clear opposition or support for the introduction of a policy, i.e., a *practical* objection to proposed or current policy, will choose to engage in unjustified discrediting or excessive support of a scientific analysis. This was symptomatic of the debate about the research conducted into acid rain and its effects on lakes and forests. Opponents of proposed legislation to restrict sulfur emissions criticized the science in such a way that could have stultified the development of theories that could be tested. For example, a demand that only tree death or decline be considered as an indicator of pollution influence, or an insistence on applying particular forms of statistical inference to the measurement of tree growth, can prohibit consideration of such things as changing soil chemistry as an indicator of influence (Loucks 1992). Clark and Majone (1985) consider that the most common case of inadequate criticism occurs when an inquiry designed for one role receives its sole criticism from those involved in making judgements in another. They cite the example where academic ecologists may be asked to review environmental impact statements mandated for federal projects.

Clark and Majone (1985) describe different modes of criticism that can be adopted: output, input, and process modes. The *output* mode is the most common, focusing on the products of an inquiry, i.e., scientific results and conclusions. This is the commonsense view of criticism. Attempts are made to answer such questions as "Can the hypotheses be rejected?" The intuitive appeal of this mode of criticism is often misleading. It can not be used effectively where policy goals are ambiguous or not defined. It may be compromised where policy goals may be multiple or contested. One of the most important objectives of scientific analysis in a policy context may be to provide enlightenment. Construction of apparent hypotheses can certainly obscure this. The output mode of criticism can also fail where the results of a scientific inquiry can not be known within a meaningful time frame for the policy decision, a common situation. The output mode of inquiry assumes that the means of inquiry are neutral, but really appraisal of scientific inquiry for policy purposes also requires that the means of inquiry be critiqued.

The *input* mode of inquiry focuses on the data, methods, people engaged in the work, and the actual activity of the inquiry. Who is doing the analysis? What are their histories? Also, the maturity of the sciences involved should be considered as well as the practical aspects of time and money that has been available. Clark and Majone suggest that this form of critical appraisal could be particularly important during the progress of a project.

The *process* mode of inquiry concentrates on the procedures used, the provisions for quality control, and questions that govern participation in, and conduct of, the inquiry. Clark and Majone (1985) note that one of the most important functions of a scientific inquiry could be the provision of procedures for resolving partisan arguments. Has "due process" been exercised in the conduct of the science? This mode provides information that the other two miss, but is a costly type of assessment because it requires intimate knowledge of the process and direct and extensive observation.

Clark and Majone present a classification of critical criteria likely to be applied by different types of people, scientist, peer group, program manager, policy-maker, and public interest group, for the three different modes of criticism, output, input and process. From this table, they present four *metacriteria* that they considered cut across the modes and roles. These are *adequacy, value, effectiveness*, and *legitimacy*. These four criteria together provide guidelines for assessment of scientific investigations in policy relevant circumstances.

Criteria of adequacy   Are the operating procedures adequate within which scientists will do their work? Critical disputes and debates must be channeled to well-defined categories of scientific argument. Clark and Majone note, "Most scientists engaged in policy-relevant work have experienced the frustration of . . . uncontrolled debate where rhetorical style counts for more than technical accuracy. Worse, many have found that their efforts to establish 'ground rules' (i.e., criteria of adequacy) before the debate commences are viewed with suspicions or downright hostility".

Criteria of value   Is the choice of study governed by its value to the management or policy issues? A variety of biases may be encountered where scientists attempt to bend the problem to their pet theories or where program managers devise programs of investigation that simply fit into the disciplinary orientation of the scientific staff they have. Clark and Majone discuss three value criteria: internal, external, and personal. Internal value criteria are concerned with how well the scientific investigation is likely to be carried out, and the answer to this question comes from other scientists within the scientific community. External criteria of value come from people outside of the field of inquiry who may use the results. This may include other scientists through to the general public. Will this research help to solve the problem? Clark and Majone argue that scientists who will do the work must be involved in developing criteria of value, otherwise sociological processes are likely to usurp other aspects of the value-setting procedure.

Criteria of effectiveness   Does scientific inquiry actually help to resolve practical problems? Clark and Majone focus on the need to insure those potential policy changes or development can be effective. If they are not, then associated science will not be effective in the resolution of the problem. The criteria of effectiveness are most likely to be applied at the level of the research program, rather than at an individual set of experiments or investigations. What is the contribution of scientific inquiry to the policy agenda as compared with its contribution to the policies themselves?

Criteria of legitimacy Is scientific research a legitimate requirement for this policy-making issue? It is incumbent on scientists and policy-makers to establish the legitimacy of any scientific investigation to a practical problem. In some cases this may appear to be straightforward but this can not be assumed to be so. An important consequence is that if legitimacy can be defined and established then science need not forfeit its requirement to uncertainty on various points. Clark and Majone describe a change in the attitudes of the general public towards science from one of acceptance to one of questioning. It is to the benefit of scientific argument within policy discussions to outline where science can and can not be of value.

14.4 Science, scientists, and society

There are difficulties in defining research programs, conducting the research, and utilizing results in applied ecological research. Standpoints of scientists, managers, and policy-makers together determine these difficulties. What should the relationship be between science and government?

Byerly and Pielke (1995) reviewed some historical aspects of science policy, particularly in the USA (Radford Byerly was chief-of-staff to the United States House of Representatives Science Committee). An important initiative was the social contract with scientists suggested by Vannevar Bush, as Director of the Office of Scientific Research and Development, when he proposed the setting up of the NSF after World War II. Central to this were the assumptions that scientific progress is essential to national welfare and that science provides a reservoir of knowledge that can be applied to national needs. "[S]cientific progress on a broad front results from the free play of intellects, working on subjects of their own choice, in the manner dictated by their curiosity" (United States Office of Scientific Research and Development 1945). Bush argues that because scientists can best judge science, the direction of research should be their responsibility. Good science alone justifies support from society. Byerly and Pielke suggest that this emphasized pure science over applied science and devalued the connection between science and its environment. Nevertheless the attitude expressed by Bush determined government policy and most importantly set a cultural environment for science. Similar attitudes were prevalent in other countries at that time.

Current needs of society, and attitudes within it, have changed the environment under which science operates. This is explicit in the NSF's goals (Box 14.1). Science went through a period of exponential growth in the 1950s and 1960s but more recently it has had to adjust to stationary, or even decreasing budgets. The precise role that scientists should have in policy and management decision-making, and how they should exercise it, is unresolved. Cowling (1992) and Russell (1992) clearly felt that scientists had had an influence in NAPAP that gave the government requirement of assessment less than its due importance. On the other hand, in developing a management policy for the Apalachicola estuary, Livingston (1991) felt that scientists should provide political leadership and use science in persuasive argument.

Byerly and Pielke (1995) argue that the social contract, as advanced by Bush, needs to be renegotiated. To be sustainable, science must meet two related external conditions: (1) democratic accountability, including accountability to goals of society; and (2) sustained political support. Under democratic accountability science should be consciously guided by goals of society – denial of their importance encourages elitist isolation. The basic question of science policy can be phrased as how non-scientists make scientists do what we all, as citizens, have decided on. However, the central difficulty in this is the asymmetry of information between those who would govern science (the principals) and the researchers (the agents) (Guston 1996). While there is asymmetry in all principal–agent relationships, scientists claim uniqueness for science due to its extraordinary technical and theoretical complexity, the unpredictability of advance, difficulties in discerning the consequences of discovery, delivery of products through intermediaries (often as general benefit to society) rather than directly to the government.

While society in general should be the principal, in practice it is legislators and the agencies that act as principals. Much science policy considers the most effective way the principal–agent relationship can be organized. Sometimes the emphasis is on expenditure, other times on results. Guston (1996) notes two difficulties. First, the problem of "adverse selection". Because of its own lack of expertise the government, as principal, has difficulty in selecting which scientists, as agents, most completely share its goals. Even the appointment of government scientists does not solve this problem entirely. Second the problem of "moral hazard". Delegation by the principal not only provides incentive to perform the required task but "to cheat, shirk, or otherwise act unacceptably" or simply be ineffective. Chubin (1996) discusses these problems in a broader context including the style and effectiveness of communication between scientists and the community and the need for major change in the way that traditional institutions work.

There are important elements of principal–agent relationship in the working of the National Science Foundation. Its present goal is certainly not to support science for its own sake, or with the general hope that benefits will accrue sometime in the future. Yet beyond the general goals and strategies in the Foundation's Strategic Plan (Box 14.1) there are obvious difficulties in deciding what programs to initiate and, within each program, which proposals to fund. In the formulation of goals the question arises how much centralization is effective. In the USA the National Science and Technology Council attempts to ensure priority setting in all research and development. Government agencies are encouraged to conduct strategic planning. The Government Performance and Results Act of 1993 requires tying expenditures to goals in a specific manner.

Government goal setting and monitoring is a complex process. Science requires the involvement of more and different types of people, and is increasingly conducted across organizational and national boundaries (Hicks and Katz 1996). We can no longer categorize research simply as "basic" or "applied", nor can we identify universities or government institutes as doing distinct and rather separate types of research. Unforeseen and important linkages occur between different fields of research and can rapidly create whole new approaches to important problems. Science policy-makers have to respond to these changes, but the heterogeneity of science, i.e., its complexity and dependence on intricate knowledge, and detailed communication and cooperation, makes it difficult for them to impose their goals.

Rip and van der Meulen (1996) describe two systemic aspects of national research systems. "Steering" is the extent to which the scientific system is sensitive to attempts by the government to implement its objectives. "Aggregation" is the organization of processes of agenda building within the research system itself to guide the national science effort. These two processes exist in different ways and to different degrees between countries (Table 14.2).

Rip and van der Meulen (1996) describe the components of national research systems, from policy-makers through to scientists, as a mutually interdependent system. Incentive structures, which are fundamental to steering, have been created by modern national states, intentionally or as side effects of other measures, e.g., support measures for higher education. National research councils or foundations mediate between science and the state and can be "both a government bureaucracy and a parliament of the scientific community" (Rip and van der Meulen 1996). Aggregation in building research agendas for policy questions is a complex process. It can be bottom-up, but may also be induced by other parties and supported by a socially distributed infrastructure of institutions and procedures. The number and type of organizations has increased, e.g., research councils, advisory bodies, programming bodies, and standing review panels. As well as being involved in determining agendas such bodies may mediate between the state and performance of research, though with only delegated authority. They frequently experience dual allegiance, to the state and to relevant scientific communities.

Rip and van der Meulen (1996) compared seven industrialized nations for the relative amounts of steering and aggregation in their national research

Table 14.2. *Criteria used to make synthesis scores given for two qualities of national research systems, steering and aggregation. (After Rip and van der Meulen 1996)*

Score	Steering	Aggregation
5	The state is emphatic about its own aims, and does not depend for them on the scientific community. Sanctions are built in. Intermediary bodies, e.g., research councils, are held responsible for outputs related to the state's aims, and tend to reproduce a similarly strict principal–agent structure in their relations with research performers	There are system-wide institutions with explicit aggregation tasks, which are recognized as such and which function well. There are specific networks and other linkages connecting different kinds of researcher (universities, research institutes, and industry) with actual agenda-building activities
4	The state has discretion to decide its aims, but checks potential support. Sanctions are possible, as well as incentives. Contracts, which are strictly kept, are an important instrument	Institutions throughout the system have aggregation/agenda-building as an important or sole task. Overall coordination and agenda-building might happen as a result of communication and cooperation, but it is not institutionalized
3	The discretion of the state to decide on its aims is limited to missions. Calls for tender are used, but so are network interactions. Contracts with research performers may be made but there are no strict performance checks	Some organizations with a key role in the research system, often at the intermediary level, include aggregation/agenda building among their tasks, or have this as a side effect
2	While the state has a final say, especially when missions are concerned, it listens to researchers and stimulates bottom-up agenda-building processes. Dialogue is more important than contracts. Intermediary bodies have their own responsibility, within overall limits	Aggregation occurs within segments of the research community, but not across them and merely as a side effect
1	The state prefers to support the bottom-up agenda, but checks whether agenda-building has been done properly. It is active in identifying opportunities for, and implementing, strategic turns in the research system. There is some dialogue and a great deal of *ad hoc* intervention	There is no institutionalized aggregation. Aggregation is limited to traditional scientists' channels (societies) and researchers position themselves individually and compete in this way

systems (Table 14.2). A two-dimensional map for the seven countries (Fig. 14.1) was produced. Rip and van der Meulen (1996) also produced vignettes for each country, three of which are reproduced in Table 14.3. These illustrate, at a practical level, what the differences between steering and aggregation mean in terms of institutions and the way science operates. Rip and van der Meulen (1996) comment that highly institutionalized aggregation, while productive for normal purposes, might create closed-shop situations, e.g., "old-boy" networks in funding. The UK is an example of a research system with high steering and low aggregation. This was the result of a strongly specified and pursued standpoint imposed by the Cabinet, creating strong competition for resources and privatizing government laboratories where possible. Rip and van der Meulen argue that, because of the heterogeneity and continual changes in science, a research system is desirable in which aggregation is favored over steering and that it is possible to evolve this structure, which includes the development of diversified funding of research.

In response to their own critique of science in the USA as being too isolated from the needs of society, Byerly and Pielke (1995) suggest a national debate, accompanied by empirical evidence, i.e., a process of aggregation in Rip and van der Meulen's terms. This should answer two questions: (1) in what ways does science contribute to the national welfare? (2) How can science be marshaled to assist in addressing society's specific problems? Byerly and Pielke say that the debate should not be limited to scientists but that the academies and professional societies should lead it and Congress, universities, laboratories, industry, non-governmental organizations and individuals should participate. It is interesting that, in response to their own analysis of science as being too isolated, Byerly and Pielke do not suggest greater steering of science by government.

Both Byerly and Pielke (1995) and Rip and van der Meulen (1996) propose that scientists take a major role in science policy. Yet the problems articulated by Cowling (1992) and Russell (1992), that scientists did not place sufficient emphasis on government's requirements, can not be ignored. Nor can the very specific proposal for science review made by Clark and Majone (1985). Kantrowitz (1975) states: "It is, for example, unthinkable in a democratic society that scientists would actually be endowed with the authority to assume full moral responsibility for the social impact of science". He sees the problem not only that scientists represent only a small fraction of society but that, as experts, they have characteristics that make them unsuitable for making decisions. Expert's judgements are necessary but not final and Kantrowitz (1975) quotes Laski's (1931) description of problems with experts.

"For special knowledge and the highly trained mind produce their own limitation which, in the realm of statesmanship, are of decisive importance.

Table 14.3. *Characteristics of the three national research systems with different styles of organization. (After Rip and van der Meulen 1996)*

Netherlands	USA	UK
High aggregation/low steering	Intermediate aggregation and steering	Low aggregation/high steering
The state has few aims of its own. "Making knowledge relevant for wealth creation" or "assessing academic research on its societes relevance" remain slogans, with no real attempts to force these views on the research system. In so far as steering occurs, it is in modulating and shifting ongoing aggregation processes, and creating incentives for what is considered to be strategically interesting. Steering is seen as doing what is good for the research system, often in their terms, with occasional attempts to break through the conservatism. Socially distributed agenda-building is part of Dutch culture and is taken up by a number of organizations in the research system. Research council activities rely heavily on input from researchers. The Sectorial Advisory Councils, tripartite and with agenda-building as their main remit, have been very successful, and all scientists continue to contribute freely to their work, which is an indicator of productive institutionalization of the aggregation process	A dual situation exists in which the conduct of basic research is measured in terms of the model of private universities, jealous of their autonomy (even if they depend on the state and other sponsors for research support), and mission-oriented research steered by federal government departments. There is no overall science policy-making in spite of recent attempts to link science with national/societal objectives. National laboratories have to become more active, now that they cano not rely any more on support from their federal ministry or agency. The state plays an increasingly active role in stimulating and orienting research. Within several institutions aggregation occurs, for example the National Institutes of Health, e.g., with a strategic plan deriving from interaction with biomedical communities. There is some aggregation within networks of departments, but in North American political culture, the government departments preserve their discretionary authority.	From the early 1970s, starting with Rothschild's customer–contractor principle, the tendency has been to reinforce the privilege of the state, which sets its own aims, for its own reasons, and is increasingly strict in implementing them. A mission-statements exercise, and rearrangement of research councils, directed from the Office of Science and Technology, and with strong industrial/business presence at the top is an example. A research assessment exercise for universities, and top-down budgetary decisions, have created a situation in which universities, and departments within universities, struggle to get high scores to safeguard their funding. They are prepared to spend to buy professors who will hopefully enhance their score. Privatization of public research institutions, and market testing of services, is part of the ideology. Strong protests, arguing that research is to some extent a public good and should be organized and funded as such, influenced some measures but not the ideology. Aggregation was effectively destroyed, but setting up a foresight exercise created some bottom-up agenda-building.

Fig. 14.1. Seven industrialized nations compared in relation to the steering and aggregation in their national research systems. (From Rip and van der Meulen 1996.)

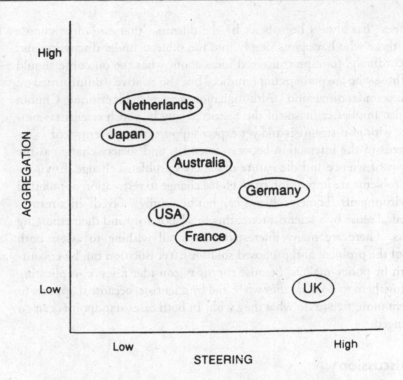

Expertise, it may be argued, sacrifices the insight of common sense to intensity of experience. It breeds an inability to accept new views from the very depth of its preoccupation with its own conclusions. It too often fails to see round its subject. It sees the results out of perspective by making them the centre of relevance to which all other results must be related. Too often, also, it lacks humility; and this breeds in its possessors a failure in proportion which makes them fail to see the obvious which is before their very noses. It has, also, a certain caste-spirit about it, so that experts tend to neglect all evidence which does not come from those who belong to their own ranks. Above all, perhaps, and this most urgently where human problems are concerned, the expert fails to see that every judgement he makes not purely factual in nature brings with it a scheme of values which has no special validity about it. He tends to confuse the importance of his facts with the importance of what he proposes to do about them. . . . The expert tends, that is to say, to make his subject the measure of life, instead of making life the measure of his subject. The result, only too often, is an inability to discriminate, a confusion of learning with wisdom. . . . There is no illusion quite so fatal to good government as that of the man who makes his expert insight the measure of social need."

Kantrowitz (1975) suggests that the moral responsibility that many scientists feel very deeply can easily affect their judgement as to the state of scientific fact, especially when the pertinent scientific facts are not thoroughly clear, as is often the case in science. The selection of governmental scientific committees "has always been beset by the dilemma that one must choose between those who have gone deeply into the subjects under discussion and who, accordingly, have preconceived ideas about what the outcome should be, and those who are perhaps unprejudiced but also relatively uninformed on the subjects under discussion". Although more genteel in his critique, Chubin (1996) also implies criticism of the restricted way in which scientists communicate with non-scientists and yet expect support from government.

Problems in the interaction between scientists and society change as the complexity of science and the nature of society's problems change. Environmental problems are important, from global change to restoration of polluted local environments. These problems can not be simply "solved" in a restricted technical sense by a scientist retreating to a laboratory and then emerging victorious. There are many interested parties, all wishing to assess both analysis of the problem and proposed solutions. Yet isolation can be encouraged both by policy-makers, because they may consider it less complicating and freeing them to do what they wish, and by scientists, because it appears to leave them more free to do what they wish. In both cases standpoints can go unchallenged.

14.5 Discussion

Along with other groups in society scientists have standpoints. Part of the value system of some scientists is that science is beneficial, in and of itself, leading to the standpoint that the science to be done should be judged by scientists through peer review. In the past this has been the practical foundation of the scientific enterprise and institutions were designed to use and support peer review. Increasingly the correct role for peer review in funding decisions is debated. Some scientists are concerned about its effectiveness for encouraging innovation. Governments express concerns for the accountability of scientific research as an investment for society. Science policy analysts are concerned with the most effective way that science can interact with society.

A standpoint, adapted in various forms by some ecologists, is that ecologists can, and should, represent what is good for the maintenance of ecosystems. That humans depend upon ecosystems, and so what is good for ecosystems is good for humans. Like other standpoints this can not be assumed as an incontrovertible truth but must be criticized and debated. If this standpoint is strongly held it may lead to representing elements of ecological knowledge with more certainty than is warranted. It can be interpreted as a demand for influence in society in the role of expert.

A different standpoint is that society depends in part upon science for analysis and suggestions of solutions to important problems and that this requires just and effective criticism of scientific knowledge both from scientists and from society. Science would benefit from the criticism but there are major difficulties in achieving it. The most important is that of rising above the role of protagonist, or expert, both by being absolutely clear where the necessary subjectivity of management and policy-making must start, and enabling non-scientists to understand uncertainties in research results. Time and again this comes down to how we, as scientists or managers, approach our work and the institutional structure within which science operates. If the results of scientific research are to be effectively used in management and policy-making, then the nature of decision-making, and how science is used in it, must be considered.

14.6 Further reading

Issues discussed in this chapter, and Chapter 13, are the subject of ethical considerations. Shrader-Frechette (1994) describes the content and rationale for codes of ethics for scientific research. Erwin *et al.* (1994) provide an anthology of articles dealing with ethical issues in scientific research and Stern and Elliott (1997) provide a guidebook for developing a graduate level course in research ethics.

M. S. Hall (1988) presents valuable information, checklists, timetables, and a model for proposal development while Redfield and Crowder (1989) provide advice, much of it social, on what they term "grantsmanship" for ecological research.

Bella (1996) presents analysis, and further references, to the way that organizations seek confirming instances and minimize criticism. He debates organizations making discussions about environmental problems, and universities. Barber (1961) documents cultural and social sources of resistence to scientific discovery.

Introduction to Section IV:

Defining a methodology for ecological research

In this section, components of research described in previous sections are synthesized as a methodology, Progressive Synthesis (Chapter 15) and some published criticisms of ecological research are reviewed (Chapter 16). Criticisms of ecological research have included concerns over what should be studied, along with concerns of how research should proceed. These criticisms are made from different standpoints that are based on different ideals. Chapter 15 starts by defining the standpoint of Progressive Synthesis on which practical suggestions about methods for ecological research are based.

15 The methodology of Progressive Synthesis

Summary

The motivation for ecological research is to make scientific inference based on the coherence of a scientific explanation for a why-type question. Research may be required in first describing phenomena, and scientific inference is unlikely to be definitive, but the aim of making explanations and so advancing ecological theory drives research.

Three types of explanation are outlined: causal, organizational, and unifying. In ecology we base our fundamental working assumptions on causal processes. We frequently seek to understand the organization of ecosystems, communities and populations. We intend that our integrative concepts will provide unified explanations.

Progressive Synthesis is a methodology for scientific investigation in ecology that acknowledges the requirement to make upward inference about integrative ecology. It defines how different techniques available for investigating ecological problems, e.g., surveys, experiments, modeling, can be used in constructing objective knowledge. Progressive Synthesis has a philosophy, three principles, and five components to its method.

The philosophy is pragmatic realism. Science aims to provide the best explanatory account of natural phenomena, and acceptance of a scientific theory involves the belief that it belongs to such an account. This steers a course between two views: belief that scientific theories really are true representations of the world, and an opposing belief that all scientific theories should, and can, do is account for the information gathered so far.

To produce the best explanatory account, three principles must guide investigations. (1) Criticism of increasing breadth must be applied to objectives, methods and results. (2) Precision in definition is required to develop the coherence in breadth and detail of an explanation. (3) Postulates must be assessed with explicit standards so that their merit and the degree of objectivity of the theory of which they are part can be criticized. The five components of method acknowledge that science is a continuing process, that we seek better and better explanations of individual phenomena and how explanations of different phenomena are related. Two particularly important elements of strategy are emphasized: using contrasts to develop explanations rather than trying to explain something in isolation, and continuously refining and defining concepts and measurements made for them.

15.1 Introduction

The examples in Section II illustrate that ecological theories develop through repeated attempts to find explanations for why-type questions. Ecological research rarely (if ever) produces uniformly applicable laws but it does produce theories that grow and develop in their definition of functional and integrative concepts and that can be used in constructing scientific explanations. Development of the theory of top-down control illustrates this. Paine's (1966, 1974) results were clear-cut. If precisely the same result been found wherever predator removal experiments been repeated then a law would have been established. This would have been a covering law (Salmon 1990), one that could be used to describe the effects of all predators on their prey populations. But precisely the same results were not found. The particular conditions operating where predator removal experiments were done, including the balance of species present, influenced the results obtained. A theory was developed that provided a coherent explanation of both the original and successive experiments and investigations. The introduction to Section II outlined a procedure for making scientific inference in ecology. (1) A *synthesis* must be made of new results with existing theory. (2) The synthesis produces an *explanation* for a why-type question. (3) The explanation must be *coherent*, explaining both new and previously obtained information.

This chapter defines a methodology for the procedure and the analytical stage preceding it. Progressive Synthesis is a methodology comprising five components of method used in constructing scientific reasoning in ecology and three principles defining how they should be applied (Fig. 15.1). The five component methods detail how ecological theories can be analyzed, how different techniques of investigation are used, the types of progress that can be expected, what is involved in making a new synthesis, and how to define explanatory coherence of ecological theories.

Two principles must be continuously applied to the five components of method: precision is required in defining concepts, axioms, postulates and data statements, and theories, and explicit standards must be used to examine the relation between theory and data. These principles in turn can be effective only when combined with the principle of continuous application of just and effective criticism.

15.2 The standpoint of Progressive Synthesis

The standpoint of Progressive Synthesis is that scientific inference is made by constructing explanations to why-type questions. Two groups of assumptions

Fig. 15.1. Philosophy, principles and components of method of Progressive Synthesis.

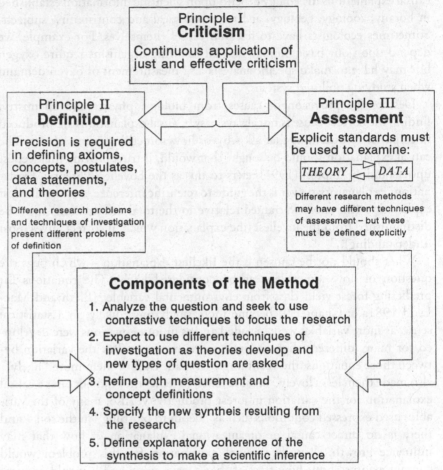

Philosophy:

Science aims to provide the best possible explanatory account of natural phenomena; and acceptance of a scientific theory involves the belief that it belongs to such an account

Principle I
Criticism
Continuous application of just and effective criticism

Principle II
Definition
Precision is required in defining axioms, concepts, postulates, data statements, and theories

Different research problems and techniques of investigation present different problems of definition

Principle III
Assessment
Explicit standards must be used to examine:

$THEORY \rightleftharpoons DATA$

Different research methods may have different techniques of assessment – but these must be defined explicitly

Components of the Method

1. Analyze the question and seek to use contrastive techniques to focus the research
2. Expect to use different techniques of investigation as theories develop and new types of question are asked
3. Refine both measurement and concept definitions
4. Specify the new syntheis resulting from the research
5. Define explanatory coherence of the synthesis to make a scientific inference

must be described to define this further. These are concerned with the questions:

(1) What can be considered an appropriate explanation for making scientific inference?

(2) What certainty in scientific inference should be expected?

15.2.1 Types of acceptable explanations

The assumptions are (1) that causal and/or organizational explanations are made and should motivate research, although they may be incomplete as research proceeds, and (2) that as research proceeds explanations may be given that unify apparently different parts of knowledge.

Causal explanations   In ecology when we say A causes B we mean there is some direct transfer of energy or mass, or a direct stimulus producing a response, or a genetic property influencing or determining something. Many causal explanations in ecology depend upon scientific information established in botany, zoology, geology, and meteorological and contributing subjects. Sometimes ecologists have to research causes themselves. For example, we depend upon our basic biology that all respiring organisms require oxygen but may have to make specific analysis and measurement of oxyen demand when studying polluted systems.

Despite the provision of causes from biology, physics and chemistry, finding the correct cause is not always easy. Causal explanations are produced in response to questions that ask why or how something comes about. The causal explanation should be sought that would, if true, provide the deepest understanding. Lipton (1991) refers to this as the Loveliest Potential Explanation. Explanatory power is the guide to scientific inference, and the likeliest explanation (the most warranted relative to the total information) must be distinguished from the loveliest (the explanation which, if correct, gives most understanding).

What should not be chosen is the likeliest explanation – which begs the question of how something can be considered likely. The equations for predicting forest yield class from environmental variables (Blyth and Mac-Leod 1981a,b; Chapter 9) became increasingly more likely, in a statistical sense, as more variables were included and as unique equations were developed for many different slightly varying circumstances. But the variation between those equations, the very thing that made each of them more "likely," also made them less "lovely" because taken together they did not offer a causal explanation for the variation in forest productivity. Also, many of the variables used expressed conditions such as "depth to mottling" in the soil – and there is no direct causal or organizational explanation of how that may influence growth. A unified causal explanation for this problem would contain axioms about how trees of that species grow and it would connect axioms describing how differences may occur on different sites. What the investigation of Blyth and MacLeod did was to establish interesting differences so that why-type questions could be defined and gave important information about the range of variation to be expected.

Ecologists may not be able to define particular causes precisely – yet the assumption that they exist drives research. Ashley Steel made the assumption that small mammals do den, nest, and feed and that requirement for those processes would provide causal explanations for the distributions that she found (Chapter 4). David Janos used physiological information on the symbiotic relationship of vesicular–arbuscular mycorrhizae (VAM) and forest tree species as an axiom in developing a theory to explain high canopy tree species diversity. In both these cases further advance would be helped by a more precise understanding of the underlying causal reasons, e.g., the habitat requirements and preferences of different species of small mammal and energetic costs relative to nutrient gain for VAM-infected trees.

Organizational explanations For many ecological questions these are the important requirement. They begin with the observation of a structure or pattern, and questions are asked about how that may either determine a causal relationship or result from one. Research into an organization–cause relationship (frequently referred to as structure–function in ecology) tends to involve successive improvement in the depth or detail of the explanation. Such explanations are symmetric, involving improvement in understanding of both structure and cause. An organizational explanation is frequently an answer to a question that asks "what" something is.

Many ecological theories contain a connection between causal and organizational explanation. For example, the particular role of VAM in the mineral nutrition of trees on lowland tropical forest soils defines a causal reason that may in turn be a component of a structural explanation of tree species diversity (Chapter 10). The organization of an intertidal rocky shore, when its structure is defined as numbers of different species of sessile animals, has a causal explanation in terms of the number, type, and quality of herbivore preyed on by starfish (Chapter 11).

Unifying explanation In ecology we face two important problems as we seek coherent explanations. First, that we can never be sure that the causes and organizational reasons we may use are correct. This is not that they may be wrong but that we have not defined them effectively. In this sense an explanation may be incomplete. Second, that with integrative concepts our aim may not be discovery of a completely new theory but development and refinement of an existing explanatory theory. The practical solution to both these problems is to build unifying explanations – chains of dependent reasoning that can be represented as theory networks.

As knowledge is built over time then the dependence between component pieces is more clearly defined and measured. Independent, or even apparently conflicting, pieces become integrated. Causal or organizational relationships are the subject of unification – but specifying precisely what elements of explanations are unified is the important part of generalization. The unification of top-down/bottom-up theories through developing a theory about their domains is an example (Chapter 11) but the process also occurs within development of individual pieces of research, e.g., in the development of the theory network of Fig. 4.2 from Fig 4.1.

Unification is a difficult process for ecology, where actual systems may have some similarities and some differences. Integrative concepts attempt to provide this unification, but many have been in existence long before the detailed research that should be part of the explanation they provide. The need felt for the unified explanation pre-dates the research, so developments in definition of the integrative concepts may be essential. For example, explanations for high compared to low diversity of canopy tree species in tropical lowland rainforest is likely to be different to that for herbivores on rocky shores. The concept definitions *and* names for the cases need to be distinguished. Ultimately a theory that unifies all explanations of diversity may be defined but that should not be implied by the use of the same concept names for different cases.

At least in some cases, unification leads not to a reduction in the number of axioms required but a more precise definition both of integrative concepts and of under what conditions they can be used to organize information. In this way unification may be the reduction of *independent* pieces of information. This seems to be the case in the unification of top-down and bottom up theories.

We construct explanations that use causal and organizational reasons and we judge their effectiveness on how well they unify the individual parts of investigations made with existing networks. In some cases a model may be used in attempts to integrate pieces of knowledge but models should not be accepted without questions as unifying explanations because different types of modeling introduce different rules of construction that influence how the theory is expressed or assessed.

15.2.2   Certainty in scientific inference

Most scientists have both a realist and idealist component in their philosophies. As realists they believe:

(1)   There is a real world.
(2)   Scientific methods inform us about it.
(3)   Science aims to give us, in its theories, a literally true story of what the world is like; and acceptance of a scientific theory involves the belief that it is true (Chapter 9).

The extent to which we hold this realism, and particularly the confidence with which we hold (2) and the relative caution we apply to (3), is influenced by our idealism, what we expect science to do. van Fraassen (1980) defines scientific realism by item (3) of this definition. This results in a firm acceptance of particular theories, and scientific questions may be preferred to, or formulated in a way to use, a particular theory. The scientist's ideal is that research results in true theories. However, belief that ecological theories are true, rather than being the best possible explanations so far, does not seem justified either historically or logically. Ecological theories are underdetermined and incompletely specified, so despite any set of observational successes many issues will remain unresolved that may become important in further research using a particular theory. Some scientists, having analyzed difficulties with ecology, suggest that research objectives should be less ambitious with regard to ecological theory (Chapter 16). van Fraassen himself argued (not specifically for ecology) that no more could be required of a theory than that it be empirically correct – that it has met every test. This is constructive empiricism and in this case the scientist's ideal is dependence upon measurement and observation.

CONSTRUCTIVE EMPIRICISM

Science aims to give us theories that are empirically adequate; and acceptance of a theory involves a belief only that it is empirically adequate (van Fraassen 1980). Theories should contain no more than is necessary to explain observations – and no more should be claimed for theories than that is what they do.

The word *constructive* indicates that scientific activity is one of constructing theories that are empirically adequate, not that we have discovered the truth. But this philosophy is not adequate for ecological research. Empirical adequacy does not provide a sufficient standard and, in any case, is not easy to define. The model of temperature and solar radiation influences on shoot extension (Chapter 12) fitted the first 35 days of a data set explaining about 80 percent of the variation and by that standard was empirically adequate. Similar models, with more terms, could fit slightly better, and yet none of the models fitted the second half of the data set. Similar problems occurred in fitting regression models to environmental effects on forest yield (Chapter 10). So, in practice, empirical adequacy, while valuable, is not sufficient.

An explanation has to take into account causal processes and yet there may be no justification for invoking causal processes on the basis of measured data alone. However, there is sufficient trust that some theories are sufficiently reliable that they are used with no checking. For instance, that plants grow using the products of photosynthesis, that herbivores live by eating plants, the laws of physics and chemistry. The use we make of these is such that we do appear to accept that they are true. In ecology we use integrative concepts to describe properties that we think may exist, e.g., causes of diversity or persistence. There can be no direct empirical assessment of integrative concepts, only upward inference in construction of a theory based on at least some propositions about causal processes that we consider to be sufficiently sound that they are unlikely to change.

Neither scientific realism nor constructive empiricism is adequate for the full range of ecological research. A more appropriate philosophy is that given by Ellis (1985).

PRAGMATIC REALISM

> Science aims to provide the best explanatory account of natural phenomena; acceptance of a scientific theory involves the belief that it belongs to such an account.

This emphasizes that scientific theories must explain particular phenomena rather than exist as independent constructions. A theory is not accepted as truth, but is the best explanation so far of a phenomenon. With this philosophy the scientist's ideal is that the process of research is a continuous discourse between development of theory and assessment of that theory against data and alternative theories and that objective knowledge grows out of this discourse. While a subject is actively researched there can be no unequivocal definition of what is, or is not, objective knowledge. Once a theory has had numerous confirming instances, is used as the basis of other studies, and has become so accepted that it is no longer researched, then I think we do adopt a position of scientific realism towards it. As the chances of something being revised decreases, so we may gradually alter our approach towards that of scientific realism – but by then we are studying new problems and having to adopt pragmatic realism towards them.

The focus of pragmatic realism is neither theories nor data but explanatory accounts of natural phenomena. During analysis the standpoint is "What is the question?" rather than "How can I use this theory?" or "How can I explain this data?" Explanations are developed to answer particular questions and the natural constraint produced by a question about contrasting situations facilitates an explanation. The type of question is crucial (Chapter 10). While it may be reasonable to require undergraduates to write an essay on "Why are tropical lowland rainforests diverse?" this question is not posed in a form that makes a suitable research question. Some comparison must be incorporated into the question so that questions are typically of the type "Why this and not that?" It is by having the constraint that a comparison enables an explanation to be assessed. In ecology, preparatory descriptive research may be needed to define a phenomena before such an analysis is possible.

15.3 Principles of Progressive Synthesis

The important step in Progressive Synthesis is not merely aiming for explanations but defining, for ecology, what the components of an explanation should be and, particularly, in defining how explanatory coherence can be assessed and so scientific inference made.

Progressive Synthesis defines how scientific arguments can be constructed. However, research is speculative. Even when faced with the same general problem, different scientists formulate different types of questions and use different investigating techniques. What should be consistent is the quality of the scientific argument and this is not determined by the techniques of investigation, but by the principles used in conducting science.

15.3.1   Principle I: Continuous application of just and effective criticism

Just criticism focuses on the scientific argument, whether it is logically constructed and adequately supported. Criticism in scientific research is not subjective argumentation but reasoned discussion. It is an essential process and scientific research is undermined if, for example, an editor publishes a paper without effective peer review, a referee attempts to prevent a paper from being published for his or her own perceived gain, or condemns a research proposal to stifle competition. These things do sometimes happen, and some of the processes that lead to them are discussed in Section III. The problem lies not only in the direct effect of each infraction but in the disillusionment they can spread.

Effective criticism requires the rigorous compilation and analysis of point and counterpoint for any concept, postulate, axiom, or complete theory. Four stages of criticism are recognized, each associated with stages in the progressive development of objective knowledge. These are:

(1)   *Direct analysis* is applied to individual investigations through laboratory discussions, workshops, conferences, and manuscript review by colleagues involved with, or close to, the work.

(2)   *Testing through direct repetition*, i.e., additional work using the results of an investigation, either directly or indirectly. This is stronger when not conducted by the investigator who made the initial discovery or claim, or someone with a close academic relationship.

(3)   *Refining through extended use* involves use and critical analysis by scientists in related but not identical fields, so that criticism is made from a larger scientific social circle.

(4)   *Standpoint criticism* is made on the basis of work conducted from conspicuously different standpoints where the objectives of investigators, as well as their methods, may be very different.

The technical details of criticism must depend upon particulars of the science. However, in considering a piece of science it is important to judge which parts of it have been subjected to each of these four stages. The difficulty in depending upon criticism as the fundamental principle is that continuous attempts are made to avoid it.

15.3.2   Principle II: Precision is required in defining axioms and concepts, postulates and data statements, and theories

Imprecise definitions of concepts, or use of multiple, and sometimes very different meanings with the same name, are a major problem in ecological research (for a discussion, see Chapter 16). Two functional classifications of concepts are introduced: (1) according to their knowledge status, i.e., concepts by research, imagination, measurement and intuition, (Chapter 3); and (2) according to their role in ecological theory, i.e., natural, functional and integrative concepts (Chapter 10). Identifying the functional status of concepts is necessary for precise definition and allowing us to parse a research problem.

These classifications are important because in ecology we frequently have to transform vague concepts, sometimes using ordinary language, into clearer concepts for which rules of use become rigorous. This is the process of explication.

EXPLICATION

The transformation of an inexact prescientific concept, the explicandum (often a common language term), into a new exact concept, the explicatum (a precise scientific meaning). Although the explicandum can not be given in exact terms, it should be made as clear as possible by informal explanations and examples. (Based on Carnap 1950.)

The intention of the knowledge status-based classification of concepts (Chapter 3) is to provide a framework for the process of explication when one is approaching a new investigation or experimental study. We necessarily name things, either with common language names, or according to results from previous investigations that we think are likely to apply. Analysis proceeds by making concepts by imagination operational so that data statements can be constructed using concepts by measurement. Yet new results may require additional definition according to the newly discovered part and kind relationships, e.g., the examples in Chapters 4 and 5. Development of definitions is essential if theory is to grow. Despite its difficulties, and almost certain incompleteness in any research, it is essential to carry this process of explication through to comprehensive and precise definition.

"There is a temptation to think that, since the explicandum cannot be given in exact terms anyway, it does not matter much how we formulate the problem. But this would be quite wrong. On the contrary, since even in the best case we cannot reach full exactness, we must, in order to prevent the discussion of the problem from becoming entirely futile, do all we can to make at least practically clear what is meant as the explicandum." (Carnap 1950.)

Redefinition of concepts on the basis of research findings is a major way that scientific knowledge progresses, and it is essential that this be explicit.

The intention of the classification of concepts according to their role in ecological theory is to distinguish measurable from non-measurable concepts and identify how reasoning can proceed between them for ecological theories. Many integrative concepts have common language names. The process of explication for them requires thorough development.

Research results can change the definition of a concept or may require a concept to be divided into different names with separate meanings. In this sense, research is a process where both the goals and language of the science develop together (Shapere 1982). Shapere suggests that "for science, there are no conditions governing the application of a term which are immune from revision in the light of further experience". However, as science progresses, scientists gradually come to depend upon certain concepts for which there is at least pragmatic acceptance of a universal definition – but there can be no guarantee that such concepts may not come under scrutiny again. Shapere (1982) presents a discussion against the case that concepts in science have fixed meaning, although this marks science as different from other activities (Weitz 1988).

The integrative concept of *species* illustrates this point. Most field ecologists can be satisfied with a pragmatic definition of *species*, e.g., "All plants in the area I study can be classified into *species* according to the accepted flora for this region". This accepts whatever rules the compilers of the flora have made – and how scientific disputes have been resolved in making those rules. In contrast, for some ecologists their whole research may be concerned with the structure and function of apparent populations of plants or animals and how, or possibly if, they function as a species. For them the definition of species is often the most crucial part of their work. Shrader-Frechette and McCoy (1993) describe four *species* concepts: genetic, evolutionary, phylogenetic, and ecological. Controversy among scientists using the definitions has been bitter.

Use of different definitions of *species* requires defense against the charge of relativism. The field ecologist using a flora is not disconnected from species scientists – but the details of the latter's work may not impinge on that of the former. They use different over-arching axioms specifying different parts of the complete knowledge network to explain different things. But their work could be connected if, for example, the flora is rewritten, or there is a new evolutionary finding with ecological implications. There is a "chain-of-reasoning" connection (Shapere 1982), even if that reasoning is not constantly being examined by field ecologists. Field ecologists complain about taxonomists, most often in a half-joking way, when new editions of a flora appear – particularly if they have to relearn what a plant may be called! Nevertheless, the new systems are accepted. When we define a concept in ecology it is essential that we define, or point to, this chain-of-reasoning connection and make the theory definition and connections we require to explain a particular phenomenon. Disputes will be most intense where chains-of-reasoning grate against each other and need resolution. But that is a part of what scientific research is about.

The active change of concepts is not restricted simply to their definition but also to how they may be classified together – and this is part of active research. Shapere (1982) argues that scientific reasoning is not subservient to necessities of language, commenting that "in science, we not only learn, but also learn how to learn; and part of that learning consists of how to talk and think about the world". The definition of no concept under investigation can be taken for granted but its meaning must be defined for the research in hand. Science is not the logical manipulation of entities with fixed definition – but must involve definition of concepts as well as their possible relationships (a pragmatic realist view). The process of refining definitions must continue until a documented position can be taken against further criticism and a measurement system, or the logic of a definition, can be justified. Definition has then been precise for the purpose of the current research.

15.3.3   Principle III: Explicit standards must be used to examine the relation between theory and data

Selecting a standard for assessment is part of research – there is no external arbiter for scientific inference. Two explicit suggestions are made in this book: (1) constructing and using data statements for postulates, and (2) developing explanatory coherence for theories.

Data statements   Some ecologists, often in the role of reviewers of scientific manuscripts or journal editors, emphasize the use of statistical inference. While this may be understandable as a desire to set an explicit standard, it can lead to inappropriate application of statistical techniques (Chapter 8) or restricting investigations to only those postulates for which a statistical test can be developed and so undervaluing exploratory research. Postulates frequently need to grow in maturity, particularly in relation to measurement accuracy and precision, before they can be effectively exposed to arbitration with data. The data statement:

(1)   defines the scientific procedure to be used in investigating a postulate,
(2)   specifies the measurements to be made for each concept of a postulate, and
(3)   specifies the requirements of the data for any statistical test to be applied.

Its purpose is to facilitate assessment of a postulate by making explicit the technical process of constructing a hypothesis for a postulate. Where an intended hypothesis test is found by exploratory research to be inappropriate, a postulate may be reconstituted as one or more different postulates using further conceptual analysis or it may be revised and reexamined with further exploratory research (Chapter 8). In most cases, exploratory research is required *before* a test can be designed.

The intention must be to specify unambiguous formal predictions for rival postulates that can be assessed against data. It is at this stage that hypotheses may be developed, but the use of statistical inference requires progressive refinement and maturity of a postulate. In some cases, where different measurements may be used of the same concept, a number of hypothesis tests might be made of the same postulate. Statistical inference can not be used as an absolute and sole criterion for assessment.

Developing explanatory coherence   Postulates must be assessed relative to data *and* to their explanatory value. They are not assessed in isolation, but as part of the theory to which they belong. Assessment is a progressive and not a single procedure. We can never prove that a postulate is true and we come to prefer one postulate to another as an explanation. Using contrastive questions is particularly powerful in this regard because it brings the focus onto explaining something specific. But a difficulty in ecology is obtaining precise definition and measurement so there is usually a question about just how secure an inference is, both for experimentally constructed and for natural contrasts.

It is rare that a comprehensive theory can be falsified through a single counter-observation. A comprehensive theory is one with generality, designed to apply to more than one specific ecological community, and with a set of interlocking axioms. The five questions asked in assessing explanatory coherence are:

(1)   How acceptable are the individual propositions when tested against data?
(2)   Are concept definitions consistent throughout the theory network?
(3)   Are part and kind relationships consistent throughout the theory network?
(4)   Are the individual propositions simple?
(5)   Does the complete explanation apply to broad questions?

An important purpose of having explicit standards for assessment is that their use can be more easily criticized. For data statements this may be straightforward e.g., criticism may fall on measurement procedures, or statistical analysis. For explanatory coherence concept definitions require critical scrutiny rather than acceptance – how they have been arrived at, and their use in construction of theory networks. An important assessment may be of how things are found out, a criticism of the heuristic itself not just the apparent results. A criticism of heuristics is likely to compare different approaches towards solving, or giving radically different explanations for, the same problems.

Even when the emphasis in scientific inference falls upon assessment against data there are reasons to expect it to be an incremental process. Three levels occur (after Whewell 1858, cited by Laudan 1981e):

(1) *Explaining the observations to date.* This minimum requirement is weak, although postulates explaining much data are clearly preferred to those explaining little, or to postulates that are contingent on many others.

(2) *Predicting future results.* This is a stronger requirement than (1). But given the fallacy of affirming the consequent it is not sufficient to encourage unequivocal acceptance. Incorrect postulates may give correct predictions. The research of Blyth and MacLeod (1981a,b) (Chapter 9) illustrated the use of prediction. They constructed equations from one set of data and tested them by making predictions about a second set. However, it is clear from Blyth and MacLeod's work that prediction alone is insufficient. Improving predictive equations increases "likeliness" but not "loveliness" of an explanation, i.e., that it has not captured causal or organizational reasons. The generality of such equations remains questionable, i.e., they would not be effective outside the very specific areas from which data was collected. This was the reason why Blyth and MacLeod called for research into growth processes – which would produce "lovelier" explanations.

The role of prediction in the scientific method has received considerable attention from philosophers of science (for a review of some issues, see Lipton 1991). There seems general agreement amongst them that considering the time when a datum or experiment is used in theory development is not important. Making predictions can be important for what they can tell you about the process of theory construction rather than the content of a particular theory (Lipton 1991).

There can be two explanations for a fit between theory and data. That the theory is (approximately) true, or that the theory was designed to fit the available data, i.e., it accommodates the existing data, and the way in which Blyth and MacLeod were able to develop predictive equations for different environments shows how that accommodation may be done. (When scientists try a number of different transformations of data to improve the fit of a relationship and simply take the best one, with no causal justification, then they are designing theory to fit the data.) Similarly, systems simulation models may accommodate to the data as the number of parameters to be fitted increases (Chapter 12). A test based on prediction can be a test of how well this accommodation process has been conducted. Its value lies not in any special virtue of prediction but in the special liability of accommodating data during theory construction (Lipton 1991). Lipton (1991) provides a robust rebuttal to the idea that prediction more important than explanation.

> "Popper valued predictions because they can lead to refutations, but his model rarely applies in the history of science (Darwin's theory was explanatory, it did not make, nor depend upon, predictions). Typically scientists do not react to a failed prediction by abandonment of their theories, but instead try to improve and adapt them. A Popperian scientist (if there were any) would be like a person who threw a car away because it did not start one morning. Scientists typically abandon a theory only when one with greater explanatory coherence comes along."

(3)   *Revealing results of a different kind.* This means that the postulate has a broader generality than the information on which it is based and that it has increased the explanatory power of the whole theory it belongs to. For example, the postulate that late July reduction in foliage and shoot growth rate reduces photosynthesis rate (Chapter 6) is predictive in the sense of (2) and can be assessed by measuring when growth rates are manipulated. But the underlying causal processes also predict change in carbohydrate compounds in the tissues, in the sense of (3). If these changes were observed, then the explanatory value of the postulate would be increased. This can be the value of bold postulates. However, deriving testable hypotheses for bold postulates may take considerable exploratory research, e.g., design of an effective sampling system and method of assay for tissue carbohydrates (Chapter 6).

The Principle of Assessment excludes arguments that are solely metaphysical. This is not intended to be a disparaging distinction – but it is an important one. Metaphysical arguments are about how things might be, and they can have interesting consequences. It is not until an assessment is proposed, and a method for making it has been developed, that the argument changes to being scientific and progressive. Metaphysical arguments can, quite legitimately, be proposed and countered *ad infinitum* as long as they are logically consistent. Scientific arguments must continuously be exposed to assessment through measurement. In ecology we depend upon abstract reasoning to make definitions of, and inference about, integrative concepts. That process can be distinguished from metaphysics by its strong dependence on reasoning from empirical results and acceptance of the need to postpone a decision pending further investigations.

15.4   Components of the method of Progressive Synthesis

The method of Progressive Synthesis comprises five components (Table 15.1). Some describe what scientists regularly do, or aim for, while others prescribe procedures defined in previous chapters. Descriptive and

Table 15.1. *The five components of the method of Progressive Synthesis*

How to reach the Goal	(1)	Analyze the question and seek to use contrastive techniques to focus the research
	(2)	Expect to use different techniques of investigation as theories develop and new types of questions are asked
	(3)	Refine both measurement and concept definitions
What happens as you reach the Goal?	(4)	Specify the new synthesis resulting from the research
The Goal of Scientific Research	(5)	Define explanatory coherence of the synthesis to make a scientific inference

prescriptive aspects of these components are given here, the most likely difficulties associated with each are discussed, and some alternative approaches are identified.

15.4.1   Component 1: Analyze the question and seek to use contrastive techniques to focus the research

Many scientists consider literature review a first task of research. Progressive Synthesis is prescriptive in requiring this to be carried further to the more formal definition of axioms, postulates, and their component concepts, in relation to a particular question. Through such definition the structure of ecological knowledge relevant to a research question can be identified, its interconnections specified, and a judgement made about what is needed to answer a new question and determine the explanatory coherence of relevant theory. Conceptual and propositional analysis is the technique by which investigators can analyze difficulties in:

understanding the relation between general theories and particular investigations,
defining what information a measurement can provide, and
defining and using integrative concepts.

Integrative concepts, unmeasurable directly, are often central to ecological theories and in particular may provide the way that unified explanations can be sought. Yet they must be investigated for particular circumstances and using research that does make measurements. This requires *upward inference*, the process of making inference about unmeasurable (integrative) concepts from measurable (natural and functional) ones. For ecologists this is the aspect of synthesis where the theory network is reorganized and the position and definition of integrative concepts at the top of the network are reconsidered.

A further reason why formal definition is important is that theories may be underdetermined (Dancy and Sosa 1992) and incompletely specified. Underdetermination is where apparently different theories appear to fit the available current evidence, e.g., the two postulates describing diversity in tropical lowland rainforests (Chapter 10). Incomplete specification is typical for ecological theories that depend upon unstated assumptions about either basic or peripheral subjects. In the photosynthesis example (Chapter 6) no definition was made of underlying causal processes of CO_2 diffusion (a basic assumption that is unlikely to influence the explanation being sought) or of translocation (a peripheral assumption about mechanisms and rates of translocation made that may influence whether the Postulate 1 accounts for the observations).

The problems of underdetermination and incomplete specification are closely related and always present in research. Underdetermination is present because in an active research program there are always alternatives to some part of the current best explanation and insufficient data to resolve between them. Incomplete specification is present for two reasons. (1) Ecological theories depend upon biological, physical, and chemical theories that have to be assumed and yet may not actually hold in precisely the way they were expected to, e.g., the photosynthesis example in Chapter 6. (2) We test ecological theories by applying them in circumstances different from from those for which they were first established, e.g., the top-down/bottom-up example in Chapter 11, and we do not know the domain structure of the theory before we start to investigate it. Incomplete specification can give rise to the Duhem–Quine problem of invoking auxiliary information to explain a result rather than rejecting a particular synthesis.

It is frequently the scientist's task to resolve crucial problems of underdetermination by specifying concepts more comprehensively. But at the start of research, scientists must judge what can be considered as basic or peripheral assumptions, depending upon the question posed. In ecology, where many concepts of all types still do not have formal and universally accepted definitions, it is essential to define what you mean (Principle II) and to be critical about those definitions (Principle I).

Failure to specify domains can lead to claiming too much (see Chapter 11), such as when research results from laboratory or controlled environment are extrapolated to field conditions. Favorable conditions may have been established in the laboratory or controlled environment for observing or measuring the dynamics of a particular system. The default position should be: do *not* assume that "propositions from such work are correct until demonstrated otherwise". The requirement is to establish that there are field situations where predictions can be made and tested and laboratory or controlled environment research is not complete in its significance for ecology until the former has been done.

We use analysis of the domains of integrative concepts in the process of making upward inference leading to broader and/or more unified explanations. There are two components to the domain of an integrative concept:

(1)   Specification of the problems answered by constructing the integrative concept, i.e., the extent of what the concept seeks to explain. An important attribute of integrative concepts is their generality – but this must be won through research and not simply asserted.

(2)   The set of functional concepts that an integrative concept uses, or that impinge upon it, must be defined.

In using integrative concepts we are attempting to specify general theories – initially without knowing either whether the theory is reasonable or, if it is, what concepts are needed to specify it completely. If one instance is found, say, of top-down control, or of VAM facilitating canopy tree diversity, that does not constitute evidence that top-down control always exists or that VAM always facilitate diverse canopy tree communities. What is required is to explore a range of contrasting conditions from which the domains of the theories can be worked out, i.e., under what conditions they hold. Contrastive questions give focus to research, which is particularly important when incomplete specification is a recurrent (and expected) problem, as it is in much of ecology. When the contrasts are close then incomplete specification in the theory is likely to apply to both parts of the contrast, i.e., fact and foil, equally.

Contrastive questions consist of a fact to be explained, e.g., diversity of some tropical lowland forests, and a foil that helps to explain it, i.e., other tropical lowland forests that are not so diverse. Different foils may be used to explain, and so allow inferences about, different aspects of the fact. The choice of foil is governed by what aspect of the fact needs to be studied, and foils may be chosen that help to determine which of the competing explanatory hypotheses is correct. Not all contrasts make sensible contrastive questions.

The contrast in photosynthesis rates between late June and late July gives rise to the question "Why does the low photosynthesis rate (P, the fact to be explained) occur in late July rather than a high photosynthesis rate (Q, the foil producing the contrast)?" It assumes that only P or Q are possible and simply seeks a causal explanation of the difference, not to describe a complete theory for photosynthesis rates in either case, although the research may contribute to one. Of course the strategy of using contrastive questions first requires the existence of a contrast. One of the most important features of the photosynthesis research was to confirm and define the difference between rates at the two times (Fig. 6.9) after the first observation (Fig. 6.3). Background knowledge about theory may strongly influence where a contrast may be looked for – but need not dictate it.

There is less focus in non-contrastive questions. For example "What controls photosynthesis rate of large trees?" is a non-contrastive question of the type "What causes P?" The tendency is to respond to such questions by falling back on knowledge about a related subject. In this case we might define what is known about photosynthesis of tree seedlings or other plants, and then attempt to use the hypothetico-deductive method to postulate how this situation (mature trees) is different from the body of theory. But this type of approach tends to dictate the type of explanation in terms of an existing theory about something that may be quite different and so provide no direction about how an explanation might be found.

Some ecologists claim to use the hypothetico-deductive (H-D) method. A weakness of the H-D method is that there need be no empirical constraint on the formulation of postulates. The H-D method emphasizes the origin of postulates from *existing* theory or metaphysical argument. Examples appear in the literature where investigators have sought confirming instances for such postulates – but demonstrating that something can apparently occur lacks power in reasoning unless conditions when it does not occur are also studied. Answering questions of the type "Why P?" (which can be produced from the H-D method) rather than "Why P not Q?" builds incomplete knowledge. Successive use of the H-D method does not automatically correct the imbalance.

Using contrasts is essential for establishing causal and organizational explanations. In causal explanations the difference condition between fact and foil identifies the cause. A difference condition is necessary for causal contrastive explanations but it is not always sufficient. For example, another condition favoring the foil could possibly have offset a difference favoring the fact. Lipton (1991) suggests some additional criteria about the resulting explanation produced to govern selection among causally relevant differences:

(1) A good explanation will tell something new, i.e., it contributes to scientific understanding.
(2) Explanations are preferred where the foil would have occurred if the corresponding cause had occurred.
(3) We sometimes do not know whether a cause is necessary for the effect, but when there are differences that supply a necessary cause, and we know they do, we prefer them.

Contrastive analysis can also be used when organizational explanations are sought. Typically, though, ecological organization needs considerable research in description and definition, whether as population or community structure. Where organization is complex, then it can be difficult to define its influence upon cause in a way that allows experiments. The first investigation into the effects of coarse woody debris on animal populations (Chapter 4) was to seek an organization, i.e., old piles and young piles, piles on cobble bars and piles in the forest, and a real value of that classification was that it defined a series of contrasts.

Ideally there should be only one difference between fact and foil, the difference condition. The central requirement for a contrastive question is that fact and foil have a similar history, against which differences stand out. When the histories are too different and there are many differences then isolation of a cause or causes becomes difficult.

Three types of contrastive investigations have been illustrated:

(1) *Naturally observed contrasts:* Classification of debris piles into young and old, and within the forest and on cobble bars, are examples (Chapter 4). The classifications are (a) based upon postulates of what may be important for animal use, and, (b) required considerable work to develop, particularly the old/new classification. A difficulty in the research, typical of many naturally observed contrasts, was determining whether the differences were attributable to only one cause, e.g., age of pile and how that can be defined. Another difficulty was establishing a logical and progressive sequence of fact and foil relationships. Inference based upon differences between pile types was useful in trying to determine whether debris piles are important for animal use of the riparian zone.

(2) *Unreplicated experiment:* Paine's (1966, 1974) experiment in removing starfish was an unreplicated experiment. The force of the experiment was: (a) the magnitude of its effect; (b) that it could be explained by the theory being developed; and (c) that there was considerable ancillary information collected – some of which was designed to investigate the effect of the treatment itself, e.g., study of edge effects. In practice, because of the variability of rocky shores, it was essential, though difficult, to find a large enough area where removal of predator could be applied (to provide the fact) and then contrasted with a similar area where the only difference was no removal (to provide the foil).

(3) *Replicated experiments:* Contrastive questions are the basis of controlled replicated experiments. The fact is equivalent to the treatment condition and the foil to the control. But it is **not** correct to suppose that contrastive questions apply most effectively in controlled replicated experiments. While a problem with naturally occurring contrasts is that fact and foil may differ in more than only one condition this can also apply to attempts at controlled experimentation. For example, exclosure experiments designed to remove one animal from a system typically can have additional unintended, and sometimes unrecognized, effects due to the exclosure process. For example, experimental exclusion of a major grazing herbivore (say sheep) from a pasture may certainly result in a larger standing plant biomass, but may also result in the exclosure

becoming a refuge for small rodents, which can have specific effects on the plant community. Recall that the purpose of replication is to estimate variance in response, both of the system itself, and that due to small, unintended, differences in the application of the treatment. Replication can be invaluable – but it does not make a poorly designed or applied treatment into a good one. Use of controlled environments frequently permits replication but this does not replace the requirement to understand the effects of the controlled environments themselves.

Supposedly model systems, e.g., microcosms, greenhouse ecosystems, flour beetle cultures, can be beguiling studies because they may well exhibit some apparently interesting dynamics. However, the question of most importance is their relevance, since they may not have many of the ancillary, or as yet undefined, relationships that may set the domains of different ecological systems. For these model systems it can be difficult to establish a close naturally occurring contrast (a foil to the system's fact), so the problem of incomplete specification, which is held at bay by the contrastive technique, then requires an answer through specification of a complete theory for the artificial system. Model systems may seem like experiments but they are not. They are *synthetic constructions* (Chapter 6). True experiment is based upon limited differences between fact and foil.

Progressive Synthesis is clearly descriptive in suggesting the use of contrastive analysis because many scientists use it regularly. But it is prescriptive in emphasizing that (1) analysis based on contrasts provides a more appropriately focused approach than the H-D method and (2) that there is no guaranteed superiority of experimentally induced contrasts over naturally occurring ones, both require rigorous analysis.

15.4.2 Component 2: Expect to use different techniques of investigation as theories develop and new types of question are asked

No single technique – whether survey, descriptions of a field situation, experiment, or developing simulation models – is always best. Each method may provide explanations for particular types of question. And the types of question may change as investigation proceeds. Where a strategy of using contrastive questions is followed, then a progressive use of different techniques can be expected through the different phases of investigation. The examples in Chapter 5 (changes in Alaskan vegetation), Chapter 10 (causes of canopy tree species diversity), and Chapter 11 (top-down and bottom-up control), each contain information that has been obtained using a range of techniques.

Sometimes practicing scientists emphasize the value of particular techniques of investigation, are ambitious for them, and make verbal and occasionally written arguments that such techniques are superior, e.g., "Experimentation is the Key to Success in Ecology!" But such ambitions are really assertions that some types of question are more worthwhile than others, or that there are preferred ways of expressing ecological questions. The continued use of different investigative techniques by ecologists generally suggests a commonsense approach of technique matched to fit the problem.

15.4.3 Component 3: Refine both measurement and concept definitions

Many measurements made in ecology are partial representations of their associated concepts from research or concepts by imagination. This incompleteness is recognized by researchers, and refinement of both measurement and the concept take place over time. In part this reflects the difficulties of making some measurements, but it also reflects the intrinsic variability in many ecological concepts. An essential part of research can be the simultaneous refinement of measurement and change in the definition of a concept to match what is actually measured. This has advantages, provided that the change in definition is explicitly understood.

In some sciences measurements are taken as comprehensive and ideal representations of concepts and theories are written to reflect this. But ecology has important integrative concepts frequently implying organizational properties that can not be measured directly or may have their details defined in different ways in different instances.

15.4.4 Component 4: Specify the new synthesis resulting from the research

Ecological knowledge does increase. Despite legitimate concerns about multiple different definitions of concepts (McIntosh 1985, Shrader-Frechette and McCoy 1993), and the slow rate of achievement of ecological research (Peters 1991), there is growth of ecological knowledge. However, this can involve periods of overextension in what the theory claims to explain, subsequent curtailment, and then reconstruction (Chapter 11). This change can be viewed from two perspectives: changes in the conceptual and propositional structure, and changes in the explanatory properties of the whole theory.

1. Changes in conceptual and propositional structure that makes a new synthesis

A theory can be represented as a network of nodes and links between them defined by classification of concepts and how they are used in propositions

(Chapter 3). Research may change a theory in different ways (Table 15.2), developing increased confidence in an existing network, extending an existing theory network, or creating major change in a theory network. The examples provide a guide for recognizing these. These changes have different effects on what may be explained by the confirmed, extended, or restructured theory and so have different impacts on the state of scientific inference.

Procedures for developing and empirical testing of postulates are discussed in Section I. Some additional comments can be made, specifically about the formation of new concepts, since this controls both extension and reorganization of a theory network that is needed to explain puzzling new evidence or answer a difficult question. A new concept may be composed of hitherto unconsidered parts, or may belong, with other concepts, as a new kind. If a concept has an important property, it is worth asking what might be responsible for that, and developing a postulate (usually for a cause or organization) is a possible approach.

When an explanation is accomplished, the links in existing theory that made it possible are strengthened. Frequent use of a particular theory in making explanations gradually builds up the strength of its conceptual definitions and propositions.

2. Changes in the explanatory properties of a whole theory

Examples of the three types of changes in the explanatory properties of a theory are given in Table 15.2 and represent the types of scientific inference that may be made. Some represent an increase in certainty of existing theory rather than an increase in explanatory capacity. Of course, for many ecological theories we hope to make upward inference. Put at its simplest we can ask: "How many specific cases does it take to make a general theory?"

Sometimes the answer is that particular instances have some similarities and some differences. Consequently a general theory must have not only some core natural, functional, and/or integrative concepts and specified relationships between them, but a set of propositions that determines how these must be used in different circumstances, i.e., ecological theories must *themselves* define the domain structure of their science. This makes ecology a challenging science. There is no certainty about the domain structure for a theory – which must be part of the research. In some cases the operational domain may be a particular spatial or temporal scale but at other times may be defined as application to certain kinds of concept. So domains have to be defined through concept definitions and this is part of the study, not implicit or external to it.

So, sometimes we see top-down control of community structures and sometimes bottom-up control and within both of those instances there is variation in both the way that the control operates and the extent of the influences. The complete theory says what controls these two processes, shows

Table 15.2. *Types of conceptual and propositional change and their effects on the explanatory properties of a theory network. (After Thagard 1992)*

Type of change	Effect on explanatory properties	Example
Developing increased confidence in an existing theory network		
Adding a new instance of a concept with the same part, kind and/or propositional relationships	This increases confidence in the theory by building up the details of its description. This is more valuable than a straightforward confirmatory result, e.g., repeating results in a similar situation	Chapter 10. Finding a new instance of tropical lowland rainforest with high tree diversity that had few species of VAM fungi but where at least some fungi and trees were species different from those found previously would add additional natural concepts and so would be more than just a replication
Extending an existing theory network		
Adding a new proposition that extends or adapts a causal or organization explanation	This is confirmatory – working within an existing framework but not changing the nature of the explanation	Chapter 10. A confirmatory result for Postulate 1.2, which states that continued mycorrhizal infection of trees is more likely where there are active mycorrhizal associations, i.e., infection sources are more plentiful and/or more effective than where VAM do not exist. This would support the need for continued active association (organizational explanation) based on an established causal relationship (the plant nutrient – carbohydrate exchange). It would fill in part of the theory
Adding a new proposition that introduces a new causal or organization explanation	This increases the breadth of the explanation – allowing it to rest more securely on causal features	Chapter 4. Imagine demonstration that one or more species of small mammal were an obligate feeder (or showed major preference) for species of fungi that, within the riparian zone, were only found in piles of CWD. This would increase the causal explanation for the importance of CWD

Table 15.2. *(cont.)*

Type of change	Effect on explanatory properties	Example
Adding a new part relationship to a concept	This dissects the theory and can help to identify causal relationships that may apply to some parts and not other parts. In this way the explanation becomes more coherent because it can be seen more clearly how something works	Chapter 5. Use of *Picea mariana* and *Picea glauca* – as separately responding species (parts of the forest) rather than considering all *Picea* species as indicators of the treeline dissects the theory and (possibly) will lead to one of greater explanatory power
Adding a new kind relationship to a concept	This increases classification within the theory and is valuable if it enables a causal or organizational explanation to be developed	Chapter 4. The classification of CWD piles into different kinds provided an organizational structure that may lead to a causal explanation
Adding a completely new concept	This makes explanations possible where previously they were not	Chapter 10. Development of the natural concept of VAM along with the associated functional concepts of carbon economy, lack of host plant–fungus specificity, method of infection, enabled an explanatory theory for diversity to be postulated

A major change in a theory network

Collapsing part of a kind hierarchy, abandoning a previous distinction	By reorganizing the classification system within a theory then a new type of explanation is offered which may be more powerful, i.e. explains more	Chapter 4. For example, as a conjecture, if further research into the use made of CWD piles by small mammals revealed a preference for new piles over old then the reasons for that difference may become apparent. If this produced a new classification using that reason whether based on shelter, food supply, or some other requirement of the animals, then that would be an advance over a classification of piles based on age

continued

Table 15.2. *(cont.)*

Type of change	Effect on explanatory properties	Example
Reorganizing a new conceptual hierarchy by branch jumping; that is, shifting a concept from one branch to another	This involves a substantial change because the classification system changes. Something that was thought to have one type or set of properties is recognized as having a different set or type	Chapter 10. Recognizing the category of symbiotic-mutualistsic fungi in addition to the longer established parasitic fungi created a new branch in the functionally oriented classification of fungi. Establishing some root invading fungi in that category led to the theory of mycorrhizae
Tree switching, that is, changing the organizing principle of the hierarchical tree of concepts and axioms defining a theory	This major change can lead to a new purpose for a theory	Chapter 5. The theory for climate and species change in Alaska (Fig. 5.4) shows a conceptual hierarchy different from that of the original theory (Fig. 5.2). In particular *Picea* species is shifted from being part of the tree line to being divided into parts, *Picea glauca* and *Picea mariana*, and each have changed relationships with west Alaska and central and east Alaska. The over-arching axiom of the second theory is the reverse of the first theory and the objects to be explained are more complicated

VAM, vesicular-arbuscular mycorrhizae; CWD, coarse woody debris.

how they operate under different conditions, and describes the variation within each of them (Chapter 11).

This view of ecological theories considers their function to be describing how natural things are ordered (Kiester 1980). While a theory may aim for generality, the constraint on its nature and extent must be determined by its capacity to explain actual systems. This pragmatic view of what theory is, and what it is for, is discussed further in Chapter 16.

15.4.5 Component 5: Define explanatory coherence of the synthesis to make a scientific inference

Theories may become more or less coherent as research proceeds and/or new questions or observations are found that the theory should explain. Theories about vegetation and climate change in Alaska during the late Quaternary illustrate how marked changes in coherence can occur. Before the discovery that tree pollen had not been found north of its present distribution the theory was coherent. A few comparatively simple propositions explained much. However, work following the failure to find northward movement has not yet produced a theory with a high degree of coherence; the lack of explicit connections between components is illustrated in Fig. 5.4. The new task is to connect these pieces so that everything is again coherent: the meteorological theory of climate change, the historical sequence of patterns of tree distributions, and the eco-physiology of the tree species.

Scientific explanations are answers that provide scientific understanding. The question to be answered must have (1) a background theory, (2) a set of possible answers that can be envisaged, and (3) conditions that can be established for identifying a satisfactory answer and so scientific understanding. The purpose of conceptual and propositional analysis and the construction of data statements (Chapters 3, 4, and 8) is to establish these three conditions and the way that integrative concepts can be defined in terms of functional and natural concepts is illustrated in Chapters 10 and 11. Questions must be rejected as currently unsuitable for research if these conditions can not be met – not because the question may not be worthwhile, but because in its present form it is not amenable to scientific analysis. Finding the "right" question to answer is essential – and making a propositional and conceptual analysis is the practical way to examine the suitability of a question for research. If you can not find a way to construct a data statement for a postulate, then you may be trying to investigate too much too soon. For example, it may not be possible to connect an integrative concept with appropriate functional concepts, or you may simply lack suitable measurement and/or sampling techniques.

How do we judge which is the best explanation and so make scientific inference?

(1) *Only by having alternatives in the first place.* The importance of having alternative postulates is described in Chapter 7. The case made there is that, since no system of logic and investigation for judging a single postulate is infallible, we are in a better position if we judge between among postulates in relation to a specific question. This also means that contrastive studies are important because they limit what must be explained. We seek to say that a postulate, or theory, explains why

such-and-such occurs in one place or condition but not in another. The importance of considering alternative theories is discussed in Chapters 7 and 11.

(2) *Only by detailed specification of precisely why one explanation can be considered to be best.* It is this detail that defines the inference to be made. An explanation may be best for some questions but not for others. We should not force generalities.

(3) *Only by being prepared to suspend judgement.* It may take some time before one explanation can be preferred over another and an inference made. We should not leap to conclusions.

15.5 Discussion

Progressive Synthesis is a methodology for ecological research problems and a way that different techniques of investigation can be used. It is not advanced as a philosophy in itself but of course, as with any methodology of science, it does have an underlying philosophy. Progressive Synthesis does not concentrate on either development or test of theories; instead, through attempting to provide explanations theory will be developed and tested as a progressive series of syntheses. Nor is the objective to gather data and see what can be made of it – the whole study is focused around explaining a particular question or observation.

The underlying philosophy of pragmatic realism permits different types of explanations, as outlined in this chapter, without assuming that these explanations must represent an absolute truth, or conceding that they are simply instruments that systematize knowledge but have no other value. Research may increase the explanatory power of theories, "truth" being an unrealizable limit to that process. In this way research is progressive.

Progressive Synthesis is recommended as a practical methodology for research in ecology but like all scientific methodologies it does have an underlying philosophy. Recall that a philosophy seeks to be correct and aims to do this by demarcating itself from other philosophies. Here are some of the demarcations:

(1) Progressive Synthesis asserts criticism and seeks demarcation from relativism. This is a considerable challenge (Section III). It is recognized that social groupings are essential to progress but that criticism must be made from progressively wider groups.

(2) A demarcation is defined from the philosophies of empiricism, rationalism, strong inference, experimentalism, and falsificationism though *not* with their techniques of investigation that may be applied to particular research problems and yield valuable results. The standpoint is that

these philosophies of science, each on their own, are not sufficient for ecological science as a whole.

(3) The nature of ecological knowledge to be discovered is a subject for research, not something to be assumed. Reductionism and holism are philosophies which presuppose that particular types of explanation will be found.

(4) There is demarcation from any type of rationalism that asserts a universal domain for concepts. The domain of a theory is as much to be investigated as its core propositions. Defining the part and kind relationships for different concepts, and particularly when they are used in different situations, is an important way the subject advances. This is not to assert that some concepts with universal definition and application may not be found for ecology. But at least some concepts in ecology entail subdivisions, i.e., there is some part of meaning common to the concept in all circumstances, but some parts need further definition for use in particular applications.

With these demarcations in mind it is important to consider two questions. First, "Is Progressive Synthesis only a methodology for established research programs?" The focus on explanations for an evolving series of questions begs the question of whence a first question for a new program comes. It can come from speculation – but speculation that attempts to *explain* some observed feature. For example, Hairston *et al.* (1960) ask "Why is the world green?" and attempt to explain that. They propose a basic truth that held universally under real conditions. In this sense their proposal is metaphysics (Blackmore 1983). Of course new lines of research can also follow an unexpected observation or result, as in the photosynthesis example in Chapter 6. Following that observation it was necessary to speculate about possible causes in order to develop potential explanations. The boundary between metaphysics and science is reached when analysis is made to investigate in particular circumstances and the methods in this chapter can then be applied no matter how new the question is.

The second question is: "Is pragmatic realism a possible philosophy while conducting research?" There is some evidence for division between ecologists into empiricism and rationalism. Lawton (1991) identified barriers between ecologists who do theory and ecologists who make observations and do experiments and considered these barriers as a reason for lack of progress in ecology. Karieva (1990) gave examples of the disconection between what he termed empiricists and theoreticians in development of understanding of population dynamics in spatially complex environments. Lawton (1991) also listed some social and organizational reasons that may contribute to isolation between ecologists and so, I suggest, a tendency to lean towards one of these poles.

An important difficulty seems to be that technical specialization is accompanied by marked differences in philosophy. In the examples given by Karieva (1990), theory is equated with development of a mathematical model and examining its properties. This certainly requires developing personal competence, and that takes time. Similarly, the techniques of estimating animal populations and their successful field application take time to learn and develop. Karieva suggests:

> "The challenge for empiricists is to investigate more rigorously the roles of spatial subdivision and dispersal in natural communities. The challenge for theoreticians is to make the empiricist's job easier; this can best be done by delineating when spatial effects are most likely to be influential, and by offering guidance on how to design experiments."

Whilst this may seem sound advice, which could be translated to apply to a number of problems in ecology, the question arises "Why does that advice have to be given at all?". I suggest the reason is a difference in philosophies between the groups. Whether this is innate, learned as skills are developed, or reinforced by social separation may not matter. Injunctions that the two groups should talk together are likely to be insufficient. To make progress, e.g., to work cooperatively on the design of an experiment as Karieva suggests, they have to see problems in similar terms. That requires at least accommodation to each others' philosophies, or, I suggest, placing the question and its explanation as central and adopting pragmatic realism as at least a shared working philosophy.

15.6 Further reading

What constitutes an explanation is still a topic of research and discussion among philosophers of science. Salmon's (1990) account illustrates difficulties and successive developments in defining this. The approach taken here is that we depend upon causal and organizational reasons in constructing the details of any explanation but use a unifying method of explanation to organize them into more comprehensive theory. Salmon suggests that causal and unifying explanations are compatible. Kitcher (1988) describes unifying explanations in terms of defined patterns of reasoning, e.g., that Darwin provides a pattern of reasoning about the evolution of plants and animals through natural selection.

16 Criticisms and improvements for the scientific method in ecology

Summary

The problems encountered in ecological research, and difficulties in solving them, have led to substantial criticisms of the subject, and particularly its methods. These criticisms, and some proposals for improvement, are discussed. Criticisms of general methods and objectives for a science are most frequently based on the critic's ideals. As science has developed to study new problems, particularly those not associated with the physical and chemical sciences, our understanding of what can be studied, and how best it can be done, have changed. Some criticisms of ecology are based on the ideals, for both objectives and methods, more suitable for the physical and chemical sciences.

Progressive Synthesis has two ideals. (1) Progress in ecology requires the development of theory for integrative concepts. This is an ideal for the subject of ecology, enabling construction of scientific explanations for questions about communities, ecosystems, and populations. (2) Progress in ecology requires dominance of the behavioral norms of science over the counternorms. This is an ideal for research methodology. The use of upward inference as a process of reasoning is made by discussion and debate and may not be resolved definitively by clear-cut measurement or experiment. This requires that social processes in research must proceed effectively and not be restricted or biased.

16.1 Introduction

Criticisms made of ecology have sometimes been acrimonious about its achievements, or perceived lack of them, and about its methods. Improved analysis or new, and hopefully improved, research can meet valid criticism of a particular investigation or research finding. But criticisms about general levels of achievement, or the methods used, are more fundamental and can not be responded to through additional research. Such criticisms often reflect the ideals of the critic and the response by someone being so criticized is usually a justification based on their own ideals.

Understanding a critic's ideals is essential for understanding criticism itself. Althusser (1990) suggests that scientists have two parts to their philosophy. First, a shared realism: belief in a real world, the objectivity of scientific knowledge, and the effectiveness of scientific methodology. Second, an idealism that can differ markedly among scientists. This idealism is often about the future content of scientific knowledge (it will show us such and such), or the purpose of scientific investigation (it will enable us to find out such and such). A scientist's idealism influences the details of his or her realist approach because it provides the foundation assumptions determining the approach to research.

Lewontin (1969) detected contrasting ideals among ecologists. Type 1 ecologists seek properties that are invariant across all ecological systems (ergodic) and involve no special considerations about place or development. Type 2 ecologists study properties that do involve such considerations. McIntosh (1985) suggests that a question for ecologists is "Are there any ergodic properties?" and considers that ecology is torn between these different conceptions of the subject. Historically the methodological criteria of successful physical and biological sciences, those that Type 1 ecologists might hope to use, have been difficult to apply to ecological phenomena. "Current ecology, even in its limited scientific context is a battleground between those urging a 'hard science', reductionist, 'imperial' approach and those arguing a holistic organismic, if not truly arcadian approach" (McIntosh 1985).

These contrasting ideals about ecology run parallel with contrasting views of how research does, and should, operate. One view, more generally held with Type 1 ideals, is that science is a cumulative progression developing in logical sequence from antecedent ideas. Hypotheses are derived from theory, they are tested and either accepted or rejected. There can be crucial experiments and little or no ambiguity exists in the evidence and there are no disputes among scientists in accepting the evidence.

Crane (1972) summarizes an alternative view:

> "The new model of scientific change, . . . allows a much greater degree of ambiguity, controversy, and discontinuity in the development of scientific knowledge. Science is not a continuous line of development but is marked by numerous shifts from one theory to another and breaks in continuity in which subjects are neglected for a period and then reinvestigated from a new point of view."

Certain central concepts, or themata (Holton 1981), utilized in a wide variety of the fields of a subject, provide continuity. The multiplicity of effort, combined with the open structure of science, permits the potential for every idea to have a hearing and ensures endless opportunities for production of new ideas that compete for survival.

The contrast between these approaches provides a framework within which some of the main criticisms of ecology and its research methods can be discussed and suggestions for improvement are reviewed. These criticisms and suggestions illustrate a clash of ideals. Progressive Synthesis is discussed in the light of these criticisms and suggestions and its ideals are defined.

16.2 Criticisms of ecological research

16.2.1 There has been lack of progress in ecology

Weiner (1995) suggests that ecologists are still debating many of the same issues that they were decades ago with no major advances in that understanding of fundamental ecological processes. Peters (1991) considers that ecology is not a science that can meet the challenge of environmental problems and that traditional questions and goals work against forging a new and effective ecological science. In particular he considers ecological questions to be weak.

> "Unlike scientific hypotheses they tell us little about the world around us. Some refer to entities which we cannot identify unambiguously, others call only for definitions, or are phrased in terms that can be answered by circumlocutions and discursive statements of personal opinion." (Peters 1991.)

These are criticisms that see ecology as the type of science of Crane's (1972) alternative view but have the ideal that it should be Type 1 science.

One of McIntosh's (1985) principal concerns was that attempts to approach ecology as a Type 1 science, what he terms "self-conscious ecology", have led to neglect, perhaps even dismissal, of previous research, and so to lack of progress. He notes, "Among the more startling assertions in the contemporary ecological literature are 'Community ecology is in its infancy' (Pianka 1980) and 'Recently ecologists have expanded the scope from studies of single species populations to include analysis of broader assemblages of several co-occurring species loosely defined as communities' (Peterson 1975)." McIntosh describes the study of communities or assemblages as one of the oldest concerns identifiable as ecological. Mathematical ecology has been indicted by a number of critics (e.g., McIntosh 1985, Peters 1991, Weiner 1995) as not only not contributing, or even obfuscating ecological problem solving, but ignoring much previous work. As a consequence, no progress appears to have been made.

Progress has been made in ecology – but it does not always produce simple or clarifying explanations of the type that might be expected following the Type 1 view of scientific achievement. Revision of the simple Northern Hemisphere-wide northward tree movement some 9 ka BP to 6 ka BP resulted in a more complex theory not yet fully defined (Chapter 5). Theories of the cause of high species diversity may differ between habitat types (Chapters 10). The resolution between top-down and bottom-up control of relative numbers in herbivore and predator populations was not the success or failure of one of those theories but understanding of their domains of operation (Chapter 11). Even theories of how photosynthesis rate is controlled seem probably more complex for whole trees in the forest than for seedlings in the laboratory (Chapter 6).

16.2.2   No general theory has emerged

Macfadyen (1975) discusses the ideals driving the need for general theory, and the tensions they meet. He makes two suggestions. As individual scientists, ecologists should "resist the fissiparious tendencies towards exclusive specialization that are now so strong . . . Ecology must remain a united science" (Macfadyen 1975). We should encourage the emergence of ecology as a quantitative study of systems. He quoted from Watt (1971) "if we do not develop a strong theoretical core that will bring all parts of ecology back together we shall all be washed out to sea in an immense tide of unrelated information". These are two ideals: ecology should be a united science and have a theoretical core expressed quantitatively.

But despite these ideals there is substantial opinion that no general theory has emerged. According to Shrader-Frechette and McCoy (1993) certainly there is not one sufficient for making conservation or environmental decisions. For example, they assert that ecologists have not developed an uncontroversial, general theory of community ecology capable of providing the specific predictions for environmental problem solving. They suggest this is primarily because concept definitions are not consistent among researchers and because ecologists are divided on what gives communities their structure or holds them together.

Both Peters (1991) and Weiner (1995) are particularly critical of mathematical ecology and what they consider to be its pretension to provide general ecological theory. Peters (1991) was concerned that use of the term "theoretical" to indicate tautologies is at variance with the definition of theory as a predictive construct. Tautologies are purely logical constructs describing the implications of given premises and never reveal more than these premises contain – they say the same thing in different words. Weiner (1995) suggests that, in those cases where theoretical models have made predictions, they have been rejected. While models have generated some testable hypotheses, mathematical developments in ecology seem to grow more and more remote from empirical predictions.

Both Weiner (1995) and Peters (1991) attack Lotka–Volterra type predator–prey models and their use in representing ecological theory. Weiner suggests the durability of Lotka–Volterra type models is because they are believed, on deductive grounds, to "capture the essence of the system," or have "interesting dynamics" but generation of testable predictions does not seem the primary goal. Peters considers:

> "[no model] applies everywhere and we lack rules which indicate which model to apply to which populations. Therefore we approach the world accepting that the model will either fit or not. In the former case the model applies and in the latter, some other model is required. Our expectations are always satisfied."

There are important differences of opinion about how general theory might be constructed. An important and successful sequence of Type 1 science has been the discovery of empirical relationships, their gradual confirmation as laws, and subsequently by their use as rules for generating explanations. (Chapter 3 gives an example.) There has been debate over whether this sequence does, or could, apply in ecology. McIntosh (1985) notes that, although ecological laws were proposed in the last century, "few, if any have survived careful scrutiny". Loehle (1988) suggests propositions of the type:

> *Energy is dissipated across any trophic transfer.*
> *No population can increase without bound.*

to be laws. (For further discussion, see Shrader-Frechette and McCoy 1990 and Loehle 1990.) Similarly, Pickett *et al.* (1994) propose that complex conditional statements should be considered as laws.

Ecology has few if any laws in the sense of statements generalized from empirical findings that can be applied universally to provide explanations. The expansion of water as it approaches freezing provides what is termed a covering law (Hempel and Oppenheim 1948) that is used to explain broken pipes and leaking automobile radiators whenever freezing weather occurs. While ecology does not have laws of this type it does have extensive theory in which integrative concepts have an organizing function. This theory can be used to construct scientific explanations but they are likely to require statements about the domains of application of particular pieces of information. The propositions suggested by Loehle as laws could be used as over-arching axioms. Their utility would depend on just how effectively axioms derived from them could be developed for use in specific circumstances. If such statements are accepted as part of ecological theory, then there is a considerable body of it.

Much of the concern about ecology is in the expectation of the type of knowledge it produces. An important difficulty is in the use of words *general* and *unified*. The discussion of top-down/bottom-up theory (Chapter 11) illustrates that increase in generality can occur **if** that is defined as increased understanding of where, when, and under what conditions particular components of a theory apply. In that example, understanding the domains of top-down and bottom-up together define a more general theory than either on its own and so with greater explanatory power. The request for a unified theory is more exacting because it requires a continuous, logically related construction that could be used to develop scientific explanations of all ecological problems. This is an ideal about knowledge, as well as a presupposition about the structure of ecological knowledge yet to be discovered.

What characteristics would a unified general theory have? Perhaps it could show how a scientific explanation could be constructed for any ecological question. (A *complete* unified general theory would actually be able to do that.) However, it is not clear how we would use a unified general theory in research. For example, I could suggest that we do have a unified general theory of *species* (see discussion of Principle II of Progressive Synthesis in Chapter 15). We can attempt to answer many different types of questions about *species* but it is unlikely that an explanation for any one would require use of the unified general theory. We know a great deal about *species* but the explanations for the questions we ask need not use all that knowledge.

Similarly for ecology. We might ask, "What would be the ecological effects of global climate change?" After analysis of what changes are considered, producing a substantive answer requires first developing more focused questions. General, textbook-type knowledge might be used in selecting such general questions but not in providing questions that can be used to construct scientific explanations. Another complaint about lack of unity in ecology might be that research in one area of the subject does not make sufficient appropriate use of developments in another, e.g., that explanations for problems about populations are constructed without effective reference to ecosystem knowledge, or vice versa. If such complaints are justified then that is a social problem for ecologists but not necessarily a weakness of the academic subject matter.

Macfadyen's (1975) second ideal was that ecological theory should have a quantitative core. However, criticisms of the use of mathematical models in ecology are particularly strong. There are two answers to this criticism. First, that the use of simple dynamic systems models, e.g., of the Lotka–Volterra type, that do not lead to empirical investigations should be considered to be metaphysics and not theoretical ecology – regardless of the claims made, particularly through the titles of books and papers. Ecologists do engage in metaphysics, i.e., speculation about what might be, given what they know. That is how they formulate and organize ideas and decide what might be investigated. But the practice of science is defining investigations, carrying them out, and producing a synthesis of the results with previous knowledge. If there has been undue respect for metaphysics, when accompanied by, or wrapped in, a mathematical argument, it is probably because some ecologists have an ideal for "general" theory, and that since some other sciences define what is general with mathematical equations so should ecology. But these are ideals.

Second, that quantitative models of various types can be useful in research but each type introduces its own axioms of mathematical technique. Chapter

12 itemized difficulties with different types of mathematical models in providing explanations. Many concepts in ecology may remain partly defined during considerable periods of investigation. Effective mathematical description requires conceptual clarity but will not bring it about of its own accord, and progress with models has sometimes been made at the cost of making inadequate definitions.

16.2.3 Ecological concepts are inadequate

One of the most consistent criticisms is the continuing failure of ecologists to define concepts and hold to those definitions (McIntosh 1985, Loehle 1988, Peters 1991, Shrader-Frechette and McCoy 1993, Grimm and Wissel 1997). The view is that, until consistency can be found, development of the subject will be restricted. Peters (1991) contends that "Because ecologists confuse theory and concept, many conceptual relations in ecology are no longer stages in the development of theory, but alternatives to theory". Shrader-Frechette and McCoy (1993) note Mayr's (1982, 1988) emphasis that recent progress in evolutionary biology is the result of clarifying conceptual differences. And they consider this to be the most important key to progress in community ecology – but judge that effort toward conceptual precision is not going well.

McIntosh (1985) writes that, while it had been fashionable to deplore the construction of concepts in place of falsifiable theories, concepts expressing complex aspects of ecology have been valuable in its development as a science. He notes, "Later critiques commonly called for more rigorous and exact, preferably mathematical, definitions and spoke of mathematical metaphors, ignoring the fact that the beauty of a mathematical symbol (X) is that it has no metaphorical overtones. It means just what it is said to mean – no more, no less".

Shrader-Frechette and McCoy (1993) analyzed ecologists' use of community concepts and showed both definitions and use to be ambiguous and often inconsistent. "Not only have they used different terms to represent the same *community* and *stability* concepts, but ecologists have employed the same terms to stand for different concepts." These authors provide comprehensive tables of specific concepts given different definitions in ecology textbooks and it is a salutary and invaluable experience to read them.

Shrader-Frechette and McCoy (1993) identified three types of problem that could lead to differences in definition of the concepts community and stability: (1) measurement, (2) determining the scale over which the concepts should be judged, and (3) variation in meanings attributed by ecologists. The first two problems could lead to interesting science through comparison and contrasts of measuring the same concept in different ways, or examining it at different scales. But variation in meaning may indicate inconsistency or incoherence. If progress is to be made in clarifying concepts then, at some minimal level, scientists must mean the same thing when they talk about the same concepts. Scientific progress presupposes the ability to isolate semantic from non-semantic problems, and Shrader-Frechette and McCoy suggest that such isolation does not appear to be possible given the degree of variability of meanings.

The procedure of defining both the status of concepts (research, imagination, and measurement; Chapter 3) and their role in ecological theory (natural, functional, and integrative; Chapter 10) is explicitly designed to facilitate rigorous analysis and lead to an effective synthesis. It acknowledges that we have different types of concept, each requiring particular types of definition and having particular roles in the development of theory and providing explanations. Research would be helped by the following. (1) Precise, consistent, and persistent naming, definition, and use of concepts. For example, referring to *tropical lowland rainforest canopy species diversity*, if that is what is meant, rather than, for instance, talk about *diversity*. *Diversity* should refer to a comprehensive theory defining what similarities and differences there are in how diversity is controlled in different types of ecological system. (2) Identify when research has led to a change in meaning and, if required, define new concepts explicitly.

Perhaps some of the attitudes of Lewontin's (1969) Type 1 scientists, who seek universal explanations in ecology, can be explained by their observation of how physics and chemistry have been successful in defining concepts by their measurements. In both ecology and the physical sciences, theories are proposed to explain unobserved entities, and at various times these have caused just as much fuss and resistance in the physical sciences as we see in ecology at present. In the case of the theory of matter, the successive theories {atom}, {electron, proton, neutron}, {pion, meson . . .}, {S particles} have involved continuing reduction into components, which at first were unobserved, and then the remorseless development of measurements of those components. The concepts have become defined by their measurements – even where measurements are uncertain. Similarly in biology the concept of a gene was introduced to explain some observations of inheritance. The actual existence of genes was doubted, but these doubts were overcome before the structure of DNA was elaborated. The detailed measures we now have of DNA and our increasing understanding of how it functions provide a connection between the processes of inheritance and the processes of biochemistry.

In ecology we have two circumstances marking a difference from this progressively more detailed analysis based upon development of measurements. First, many apparently natural units of observation have different properties requiring different measurements to be made of them. An animal may be an item of prey, a predator, a vector of disease, etc. The importance of each measurement is defined by a particular theory so the natural unit has many different properties, each requiring different measurement, and the unity is provided by theories about different aspects of ecology. Second, integrative ecological concepts have complex content (Chapters 10 and 11) and their properties must be described using sets of functional concepts not by single measurements. They are defined by their theories and by their domains. This can make ecology a difficult science because two aspects of its theories must be studied, i.e., "What are the principal axioms of the theory?" and "Under what conditions does the theory hold – what is its domain?" Information about the domain must become part of the theory.

16.2.4 Ecologists fail to test their theories

Just how ecological theories can be assessed has caused extensive debate, some reflecting division between Type 1 and Type 2 ideals. Weiner (1995) suggests that, if falsifiability is rejected as a criterion for ecological theory, then the claim of ecology to be an empirical science must also be rejected. This is a firm Type 1 opinion – yet falsification is not easy to apply (Chapter 7), is certainly no universal panacea to the problems of ecology, and can be used only in tandem with confirmatory studies. While we use data in assessment we must also consider how they were collected (the conditions specified in the data statement) and the implications of confirmation or rejection of the postulate on other parts of the theory, i.e., what is entailed in a scientific inference.

Peters (1991) suggests "Ecologists have repeatedly acted as though an intangible abstraction is more appropriate for theoretical analysis than a clearly defined variable or relationship" and he prefers one that could be assessed empirically. But Loehle (1988) details some difficulties. He suggests that progress towards theory maturation might be more important, and that many of the inconclusive debates in ecology involve immature theories. It is difficult to conduct a conclusive test when a theory is immature because its predictions will be vague or even contradictory.

A further problem is just what might be tested. For example, Shrader-Frechette and McCoy (1990, 1993) made the Quinean interpretation of the Duhem–Quine thesis (Chapter 11), taking it to indicate that theories must be accepted or rejected as units. Their concern was that ecology does not have competing complete whole theories that would facilitate this. On the other hand, Loehle (1988) took the Duhemian approach that a theory usually consists of components such as assumptions, concepts, and definitions that can be examined and tested. In particular he considers that realizing that theories have internal structure can help us to design tests that do more than merely falsify a theory by showing which particular components of the theory need modification.

Difficulties with the hypothetico-deductive (H-D) method and falsification (Chapters 7 and 15) have been discussed by ecologists. Peters (1991) is optimistic and considered that the H-D theories can be distinguished from tautologies by asking whether the premises or conclusions could be proven false by observation. But Shrader-Frechette and McCoy (1993) list some imperfections in the H-D method (see Chapter 7). They rejected the idea of value-free objectivity but consider that a hypothesis can be accepted if, despite repeated attempted falsifications, it survives, or it has survived, intelligent debate and criticism by the scientific community. Empirical confirmability is not the only test of objectivity.

The approach taken throughout this book is that the development and assessment of theories is progressive and that change in concept definition through the development of its defining theory, possibly combined with renaming particular concepts, is an essential process. Scientific inference is incremental, not made solely on the basis of acceptance or rejection of postulates through testing against data. Some specific procedures are used to help this, e.g., holding concepts, the definition and use of domains, maturation of postulates (Chapter 8).

A major difficulty in ecology is in approaches to, and understanding of, scientific inference. There is lack of consistency in meaning, and unfortunately, interchangeable use of the terms *theory* and *hypothesis* (Chapter 3), and mathematically based metaphysics is often labeled as theory – which it is not. A further difficulty is with what constitutes a scientific inference where integrative concepts are used. The definition of scientific explanation and illustrations of how scientific inference can be made throughout Section II show that theories can be tested and continually refined. But this must be done in the context of constructing scientific explanations for particular questions rather than attempting to develop a general theory in a universal domain. In practice we can anticipate that detailed, precise theories will arise through empirical investigation of particular systems and that defining their domain structure is required in order to define what is general between them.

16.3 Suggestions made for improving ecological research

Underlying these four criticisms is the consistent theme of an inadequate relationship between theory and data – a concern that seems to be intense for ecology but really is as old as science itself. Suggestions for ecology seem to be motivated both to improve on subjective descriptions and depart from metaphysical speculation. But they take different approaches in how this should be achieved. In this section the suggestions of Loehle (1988, 1990), Peters (1991), Shrader-Frechette and McCoy (1993) and Pickett *et al.* (1994) are discussed. Similarities and differences between these suggestions for suitable research objectives of ecology illustrate some important difficulties.

16.3.1 Suitable research objectives for ecology

Peters' (1991) is the most radical suggestion. His response to the inadequacy in ecological theory is to concentrate on seeking statistical relationships. For him why-type questions lead to infinite regress, i.e., more "why-type?" questions and no answers. They should be replaced with questions that ask "How much?" "How many?" or "Where?" So, rather than seeking causal explanations, which might be answers for why-type questions, he prefers to seek simple regularity. Peter's (1991) view is that we must learn to pose testable hypotheses and put bounds on the knowledge we seek by attempting to predict something.

For Peters the objective is simple, empirically based theories. His ideology is that such theories exist and can be found. He gives examples, many of them based upon regression equations, and suggests that an aim of ecology should be to simplify the variables used in such equations. So, his view is the opposite of that of Blyth and MacLeod (1981a,b), who, after just such an attempt, suggested that progress should be made by attempting a more causal understanding (Chapter 9). Peters specifically declares himself to be an instrumentalist.

INSTRUMENTALISM

Theories are merely instruments, tools, or calculating devices for deriving some observation statements (predictions) from other observation statements (data). Consequently, there is no question as to the truth or falsity of these theories – they can be neither true nor false. *Instrumentalism* is thus opposed to most realist theories of science. Propositions are used in investigations, are neither true nor false but are characterized only as effective or ineffective, and judged relative to whether or not their assertion (prediction) is warranted. Ideas and practice work together as instruments; ideas relate experiences, making predictions possible, and in turn are tested by experience. (After Flew 1984.)

For Peters, realism is inappropriate for ecology because it demands too much too soon. He suggests that we should seek reduction in residual error, not plausibility of a causal explanation. The end-point is defined by the pure error associated with measuring the response variable, similar to statistical empiricism (Chapter 9). However, Peters does not entirely turn his back on explanations, suggesting that explanation differs from prediction only by the order in which theory and observation are invoked. He also notes that, for some purposes, additional criteria are needed beyond prediction.

Shrader-Frechette and McCoy (1993) present many of the same concerns about ecology as Peters (1991) but their ideas about an appropriate objective differ. Certainly they consider that "The random nature of many ecological events, the problem of replication, and the importance of disturbance all suggest that perhaps the best candidates for ecological laws are statistical." They also suggest that such relationships could have characteristics essential to scientific laws, e.g., universality, and that such laws would be evidence that ecology provides genuine scientific explanations. However, for them, the state of ecological theory is inadequate for problems of conservation, their particular concern. They state: "Insofar as ecology is applied to cases of practical problem solving, we believe it is primarily a science of case studies and rough generalizations, rather than a science of general theory and exceptionless empirical laws."

Shrader-Frechette and McCoy (1993), although perhaps despairing of ecological theory for practical purposes, do not adopt an extreme instrumentalist position. For them empirical confirmation is not always required to establish scientific objectivity. It requires being free from intentional bias, that there be good reasons for a postulate, e.g., new evidence or better interpretations, and that the postulate is better able to provide better explanatory power than its competitors and can survive sustained criticism, debate, and amendments. So for them, objectivity is tied more to practice than to rules.

While suggesting that general theory should be deemphasized, and natural history emphasized, Shrader-Frechette and McCoy (1993) acknowledge that there have to be guides to what items of natural history have to be accounted for, or should be looked at.

NATURAL HISTORY

"The sciences, as botany, mineralogy or zoology, dealing with the study of all objects in nature: used especially in reference to the beginnings of these sciences in former times." (*Random House Unabridged Dictionary*, 2nd edition.)

Shrader-Frechette and McCoy (1993) suggest an empirical approach, which, while not theory-free, puts a focus on case studies. Most research designs for case studies have five distinct components, the questions, the hypotheses, the units of analysis, the logic linking data to hypothesis, the criteria for interpretation. They consider that the principal shortcomings of the case study method are potential bias of the investigators and little basis for scientific generalization. Their particular concern is practical investigation for conservation problems. But they do argue for

"a science of ecology based on practical knowledge of taxa rather than on general theory, for a science of ecology based on case studies, practices, and natural-history information, and for a rationality that is both ethical and scientific . . ."

They suggest that this would represent a mean between the non-testable, "soft" corroboration of early ecological studies and the optimistic null models of the most recent work. The early studies often expected too little of ecological method; the later emphasis on testability sometimes expects too much of it.

They give an example of control of vampire bats, which used a series of detailed observations of the animals' natural history. But really it depended on ecological theory. That there were at least sufficient previous examples of animal control based on critical stages in life cycle or behavior to encourage the idea (i.e., give a theory for) that a similar type of control system may be found for vampire bats (i.e., there was enough generality for a specific application). So the problem is, "How much theory is needed?" And, given that generalizations contained in theories are all incomplete, what, and how much, needs to be "added back," as they put it, from natural history observations. They consider that explanatory and predictive power ought not to be rejected as ecological ideals, even though, in their view, there are controversies over what these are.

They suggest that, while general ecological theory is weak, "there are numerous lower-level theories in ecology that provide reliable predictions". And these are the types of theory that can be used where predictions are required for fewer rather than more taxa. However, they also say, in discussing one of their examples, that "the mechanisms discussed in the vampire-bat study operate in accord with general laws of nature". This is a weakness in their argument, since the laws are called "general" – but of nature, rather than ecology – yet that begs the question. Any "laws" in nature are developed by humans as part of science, and the science of nature is ecology. In practice there are few laws in the strict sense of being based directly on empirical generalizations, but there are certainly axioms that can be used.

Shrader-Frechette and McCoy (1993) state that every scientific judgement at some point requires an appeal to cases and consider "if ecology turns out to be a science of case studies and practical applications, it is not obvious that this is a defect. Ecology may not be flawed because it must sacrifice universality for utility and practicality, or because it must sacrifice generality for the precision gained through case studies." Many ecologists do pay attention to the particular nature of the organism they study and site they investigate – and still build up an ecological theory containing generalizations. The successive investigations in the top-down research program, and David Janos' investigations into mycorrhizal influence on lowland tropical rainforests, are intensive investigations, sustained over years, with understanding of how the systems studied are both similar to, and different from, others.

Revision of fire management policy in the forests of the western USA is the result of sustained scientific research that has constructed detailed understanding of why, and how, fire affects forests of different types (Agee 1993). The theory on which fire management is based is a synthesis of biology, the chemistry and physics of fires, and ecosystem analysis. Many experiments and detailed investigations requiring extensive and intensive measurements have been made. The general theory of forest fires, how and under what circumstances they start, spread, and result in forest regeneration and particular species compositions, comprises many axioms, some technical and mathematical, some biological and ecological. In particular Agee (1993) defines the domain structure of both functional and integrative concepts, i.e., what modifies axioms of the theory through the region. It is an excellent example of ecological theory used as the basis for resource management science, but it is certainly not natural history – even though the author refers to it as such!

Despair with ecology led Peters (1991) to promote instrumentalism, and Shrader-Frechette and McCoy (1993) to natural history, but neither approach is likely to prove adequate, since both restrict the scope of questions and explanations that can be sought, particularly by ignoring or excluding integrative concepts.

Pickett *et al.* (1994) make perhaps the most ambitious claims for a unified ecology and propose an approach to reach it. They urge that the objective for ecology should be integration of its component paradigms.

> "Specifically, integration is the explicit joining of two or more areas of understanding into a single conceptual-empirical structure."

They identify four "disciplinary paradigms":

(1)   Focusing on pools and fluxes of materials and energy.
(2)   Enumeration of ecological entities (plant and animal demography) and study their behavior and genetics.
(3)   Studies of historical processes and effects as in paleoecology and evolutionary processes.
(4)   Studies of systems that adjust quickly to environmental ecology (instantaneous ecology), such as niche partitioning.

They suggest that (1) and (2) represent opposite poles as do (3) and (4), forming two axes that are orthogonal, and that a target for integration is to work at the intersection of the axes. They also list questions that may stimulate progress towards integration:

> How does processing of material and energy constrain the structure of entities?
>
> How does the structure of entities constrain processing and fluxes?

Integration of current fields of knowledge is clearly an ideal, and one that makes at least some suggestion of what the content of future ecology should be, i.e., it should embrace present theories and how they are linked. The problem is that giving ideals about the future content of ecology a dominant role in scientific approach may generate questions that are unreal. This can result in metaphysical speculation and in turn use of the H–D method.

The restricted objective advanced by Peters (1991) is insufficient for ecology. We need integrative concepts despite our difficulties in defining and using them. The difficulty he notes with why-type questions must, and can, be kept in check by the contrastive method. The restriction proposed by Shrader-Frechette and McCoy (1993) of depending upon natural history, rather than theory, underrates much ecology and discounts important developments, e.g., science of the type reported in Chapters 10 and 11. The ambitions suggested by Pickett *et al.* (1994) are unnecessary for the development of the subject, although individual scientists may well wish to pursue them. Progressive Synthesis assumes no particular objectives other than that theory can be developed through empirical investigations and upward inference can be made about integrative concepts.

16.3.2   Forms of reasoning that should be used

Peters' (1991) instrumentalism requires both deduction, to decide what to do next, and induction, seeking further examples in the predictive scheme. He did, however, place emphasis on the H–D method and specifically excluded attempts to seek explanations. He identifies them with "cause", "mechanism", and "understanding" and says that these are "elusive and ethereal goals that have made ecology a discipline of insoluble questions rather than a source of information and hypotheses about our environment". For Peters, with the exception of prediction, all criteria for judging scientific theories are relative. Shrader-Frechette and McCoy (1993) place emphasis on the need to minimize Type II errors. They considered this a new emphasis in ecology.

One of the principal difficulties is that making scientific inference is incremental, often requiring synthesis of results obtained by different techniques of investigation. This can be confusing and some scientists pursue the idea that there is a golden method, sometimes statistical inference, sometimes the H–D method. But these have limitations and the approach in this book is that there are different forms of reasoning, each appropriate for particular types of question, and that which method of reasoning should be used is constrained by explanations being sought and improving the scientific inference that can be made.

16.3.3   The relation between concepts and theories

Peters (1991) describes the need for concepts to be operational. This idea is related to his instrumentalism. His specific concern is that many ecological concepts, particularly those classified in this book as integrative, can not be made operational.

OPERATIONALISM

The theory that entities, processes, and properties are definable in terms of the set of measurements, operations and experiments by which they are detected. The principal is to establish the meaning of scientific concepts in strict accordance with the practice of scientific research and experimentation. (After Flew 1984.)

Loehle (1988) claims that concepts in ecology can be made operational, and so theories can be investigated. If this can be done then explanations can be obtained through research. This provoked a counter from Shrader-Frechette and McCoy (1990), and then a reply (Loehle 1990). It is an important argument. Loehle's (1988) idea that higher-level theories (those using integrative concepts) can be investigated empirically depends upon this type of reduction. Making concepts operational is the first requirement that can enable other aspects of maturing theories such as elaborating the mathematical structure, checking robustness of both measurements and analysis, and developing new experimental and statistical techniques. Shrader-Frechette and McCoy (1990) oppose Loehle's (1988) proposal on four grounds, which I summarize, along with his response.

First, they argue that the imprecision of ecological theories prevents them from being reduced. Those higher-level phenomena can not be interpreted in terms of lower-level phenomena. For them there is at present no precise theory at the higher level and ecologists views on such concepts as *diversity* and *succession* are vague and inconclusive. Without such precision then they consider there could be no formulation of lower-level component theories. Loehle (1990) considers that lack of precise higher theory is only a statement about the state of the art, not about whether theory reduction is possible in principle, that theory reduction is a useful tool and one that might be employed. Reductions might not be complete, but they could nevertheless be valuable if they are sufficient to work on.

Second, Shrader-Frechette and McCoy (1990) argue that by suggesting that some classes of question need particular types of explanation Loehle misunderstood what a scientific explanation is, i.e., that it must be general and not just apply to some cases. Shrader-Frechette and McCoy specify that scientific explanations must follow the form of argument-plus-law, the covering law model (Fig. 16.1a), and since laws are general, i.e., they hold universally (see Chapter 3), then scientific explanations must also hold generally (Lambert and Brittan 1987). Loehle counters that ecology may need both ideographic/historical explanations (of the kind "What happened here was . . .") as well as nomothetic (proposing or prescribing a law of nature) ones.

Third, they consider that Loehle had misidentified certain fundamental ecological principles as "ecological laws," and he acknowledges this. Shrader-

Fig. 16.1. The developmental relationship between measurable concepts, laws, and theories. (a) As traditionally viewed in the philosophy of science, particularly for subjects such as physics and chemistry, where laws are usually considered to exist, as empirical generalizations, and establishment of laws is considered to precede development of theory to explain them. (b) In ecology, where propositions can be developed into axioms through repeated investigation under different circumstances, the domain of application must be specified, and both measurements and concept definitions may have core similarities and particular differences across the domain. The development of axioms as components of theories, rather than laws, progresses as measurements and concept definitions are refined.

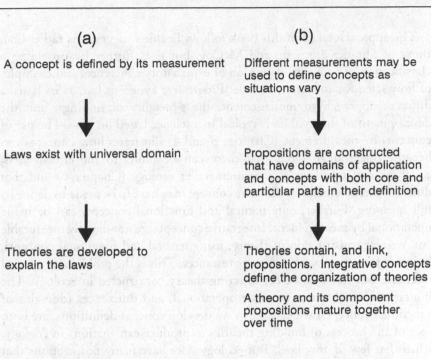

(a)

A concept is defined by its measurement

↓

Laws exist with universal domain

↓

Theories are developed to explain the laws

(b)

Different measurements may be used to define concepts as situations vary

↓

Propositions are constructed that have domains of application and concepts with both core and particular parts in their definition

↓

Theories contain, and link, propositions. Integrative concepts define the organization of theories

A theory and its component propositions mature together over time

Frechette and McCoy (1990) state that this was important because in their view laws, rather than scientific principles, at the higher level are necessary for theory reduction to occur. Loehle (1990) claims that the principles he suggested (such as the two quoted previously in 16.2.2) could be investigated and would be adequate to organize partitioning reductions. Pickett *et al.* (1994) make a claim similar to Loehle's, proposing that statements of multiple conditionals such as the following should be considered as laws:

> *If an open site becomes available, and if species become available differentially at that site or species perform differentially at that site, then vegetation structure or composition will change through time.*

In my view neither this proposition, nor the two quoted from Loehle, are laws as defined in Chapter 3 – but they are the type of proposition that can be an over-arching axiom (rather than Loehle's use of principle). The important point is that Pickett *et al.*'s proposition contains concepts of *open site, available differentially, perform differentially, vegetation structure, vegetation composition, change.* Each needs precise definition and these definitions will contain some similarity and some relevant differences between the general meaning in the theory of succession and its application in particular situations.

Fourth, Shrader-Frechette and McCoy (1990) state that Loehle was wrong to deny that theories are accepted or rejected as units. Loehle does not accept this and the argument comes down to the contrasting views taken about the significance of the Duhem–Quine thesis discussed in Chapter 11.

The approach taken in this book follows Loehle's suggestions rather than those of Shrader-Frechette and McCoy, but goes further by providing a classification of concepts, definition of explanatory coherence, and examples of how scientific inference is made. Progressive Synthesis has, as its basis, a different approach to measurement, the generality of findings, and the development of theory than is typical in a science based on laws. The use of concepts by measurement (Chapters 3 and 4) illustrates how concepts can become operational – but that there can be limits to the effectiveness, accuracy, and precision of measurement in ecology (Chapter 6) and that different measurements of the same concept may have to be made to define its full meaning. Further, only natural and functional concepts can be made operational by measurement. Integrative concepts are not directly measurable and require construction of theory, using natural and functional concepts, building from studies of particular instances. This is the problem of upward inference but not a cause for rejecting theory constructed in ecology. The limitations to what can be made operational, and differences (domains of functional concepts) we may see as we develop concept definitions, are both part of the process of building theories to produce explanations in ecology, which has few, if any, laws. But ecology does have many propositions that increase in value as over-arching axioms for use in future research as both concept definitions are made more precise and domains of application are understood.

The idea that scientific explanations must be based on a covering law dates from Hempel and Oppenheim (1948). However, philosophers of science have subsequently developed ideas on what can constitute a scientific explanation and there is certainly no consensus that the covering law model is exclusively correct (Salmon 1990). The view given in the previous paragraph is accommodated by more recent developments in what constitutes a scientific explanation (Thagard 1992).

16.4 Ideals and strategy of Progressive Synthesis

Kantorovich (1988) suggests that a fundamental assumption of philosophers of science was that science is a system for seeking truth; it is an "inference machine". But their failure to define an inferential system for science and the observation that apparently non-rational processes are an integral part of scientific activity, has made philosophers question this assumption. Their tradition has been to seek for a definition of the scientific method in logic or strategy of investigation. But some now feel that their traditional goal is no longer realizable as the development of a single method, whether it is the inductive approach of logical positivism, falsificationism, statistical empiricism, etc. Kantorovich (1988) writes: "it is not believed as widely as before that the source of scientific rationality can only be found in some system of formal logic or methodology. The philosophy of science, however, has not yet settled on a new, widely accepted path".

Shapere (1980, 1984) offers an explanation of the failure of the traditional approaches. They have all held one type of assumption in common: "namely that there is something which is presupposed by the knowledge-acquiring enterprise, but which is itself immune to revision or rejection in the light of any new knowledge or beliefs acquired". Assumptions that are so strongly held are ideals. Shapere (1980) lists four categories of inviolable assumptions. For the purposes of discussion the order of Shapere's (1980) list has been revised.

1. An ideal about the way the world is

Some approaches to scientific method make claims about the way the world is which must be accepted before inquiry can begin or further belief can be acquired. Shapere (1984) gives a number of principles that have been used, e.g., the principle of uniformity of nature – that nature always acts in the same way, the principle of simplicity of nature. One aspect of the idea of simplicity, that there are only a finite number of different kinds of thing, underlies the principle of induction, which can be stated as "The future will be like (obey the same laws as) the past". All methodologies that accept induction make at least some assumptions of simplicity and/or regularity. In ecology, the principle that "everything is connected to everything else" has been advanced (for a review, see Peters 1991) as well as the counter-assertion by Peters that simple theories exist and finding them is the purpose of research. Shrader-Frechette and McCoy's (1993) case study method for solving conservation problems uses the assumption of simplicity (rough theory exists sufficient to be used) with potential complexity in the natural history.

The assumption of Progressive Synthesis is that we do not know *a priori* the way the world is. Any theory, theoretical entity, or observation can be considered, provided it can be defined and one can work toward some observation and assessment. It is important to reject both uniformity and simplicity as principles of nature that can be depended on in designing research, since there are already counter-examples to both. Establishing just what connections are important to explain what phenomena is a matter for research rather than assuming that everything is connected to everything else or that simple relationships exist. However, Progressive Synthesis does assume that there is sufficient knowledge to start with an axiom–postulate–data statement approach, although of course exploratory research rather than formal testing may be necessary for some problems.

2. An ideal about a rule of learning

There are infallible rules – Shapere gives the example of deductive or inductive logic. The assertion of the importance of falsification (Weiner 1995) is an example of such an approach suggested for ecology. In Progressive Synthesis no single form of logic or rule of learning is advocated as superior – there is a methodology but no inviolably correct technique of investigation. Problems may require different forms of logic: propositional, probabilistic, deductive reasoning, inductive reasoning.

3. An ideal of unchangeable conceptual definitions

Certain concepts can not be abandoned, modified, or replaced. In ecology this issue surfaces in a way different from that of the physical sciences. The present problem with integrative concepts in ecology, as both Grimm and Wissel (1997) and Shrader-Frechette and McCoy (1993) demonstrate so graphically, is lack of consistency between researchers. If each scientist has his or her own definition – then there is complete relativism and a concerted research program will not be possible.

Progressive Synthesis explicitly recognizes that concepts can, do, and must, change their meaning and this is a major way that scientific knowledge grows, but it is essential that meanings be closely defined. Conceptual development can include using "holding concepts" or partially specified domains as instrumental variables in the development of theories.

4. An ideal of method

This claims that there is a method (i.e., technique of investigation) that can be applied that will not be altered as the result of anything discovered through investigation. Shapere (1984) gives the example of Kant who held that knowledge, and the seeking of it, could not exist without "forms of intuition" and "categories". The latter included the idea that every event must have a cause. Peter's instrumentalism is a strenuous claim for a preferential technique in ecology that should be considered as the scientific method for ecology.

Progressive Synthesis explicitly does not elevate any technique of investigation, or assessment process, either that they are best, or in how they are conducted. For example, experiment is not the pinnacle of method, nor disconfirmation the acme of perfection in logical approach. Each process has essential values for science – and each has difficulties in application and limitations in what they can tell. Methods of experimentation in ecology, and their quantitative analysis, have evolved and seem likely to continue to do so. Even if a dominant strategy for research develops within a research program, as predator removal or exclusion experiments became in the top-down research program, then sooner or later some other approach may be necessary to bring in a different type of knowledge.

However, for Progressive Synthesis Shapere's requirement that there should be no assumption of method immune to revision is more difficult to refute than the other three types of assumption he describes. Progressive Synthesis is defined as a methodology based on two ideals.

Ideal 1:   Progress in ecology requires the development of theory for integrative concepts.

This is an ideal for the subject of ecology. Concepts such as *diversity, stability, top-down control* motivate research but were not clearly defined when first proposed. Much important description can, and does, take place without apparent use or concern about integrative concepts. However, it is an essential ideal for the development of ecological theory. Paine's (1966, 1974) analysis of rocky shore communities focused on empirical analysis of a particular intertidal zone (How can I explain the vertical distribution of mussels here?). But the result was the introduction of the concept of keystone species and extending use of that concept required development of substantial theory. Now, when we use the concept of keystone species, we do not think solely of *Pisaster ochraceus* but about a number of predators in different environments, each with some similarities and some differences in how they control the numbers of other species in their environments. The proposed strategy for pursuing this ideal is the interconnected use of natural, functional, and integrative concepts (Fig. 10.1) and the definition of domains. Theory development is progressive and may grow to account for more observations.

A component of this ideal, and the way that theory can develop, is use of causal explanations. However, it is a matter for research just what may constitute a causal explanation, depending upon the question being asked. For example, whether herbivore response to plant material is best calculated as a function of total energy consumed or nitrogen concentration in the material, depends upon the question being examined, rather than there being some ultimate cause. Furthermore, causes are not the only focus of explanations – organizational and unifying explanations can also be constructed. Instrumental variables, such as *use* (Chapter 4) can play an essential role in investigations and do not necessarily have to be resolved as far as a cause in order to be valuable. Nevertheless, despite these relaxed requirements in Progressive Synthesis, there is still the assumption that research is working towards finding causes and this rests on the philosophy of pragmatic realism.

Lambert and Brittan (1987) raise two problems about seeking explanations that apply to Progressive Synthesis: (1) that why-questions must be used, and (2) that there is no definitive way of determining what Lambert and Brittan term a "telling" answer to a why-type question. The explicit use and development of theory answer these two criticisms – so a theory is required as a starting point and this may come from previous scientific investigation, or from metaphysical speculation. Questions that use theories may be how- what- or when-questions – but they must aim at some direct or implied causal or organization structure in the theory.

Ideal 2:   Progress in ecology requires dominance of the behavioral norms of science over the counternorms.

This is an ideal for research methodology. Cooperative social interaction is essential for the development and spread of technologies and theories. But it is clear from the work reviewed in Section III that social systems can slow, or even distort, progress unless the essential institutions of science, particularly reviewing systems, maintain the norms of behavior. This is particularly important for a subject like ecology that depends upon *upward inference* as a system of reasoning rather than measurement as an apparently acid test. Reasoning is subject to dispute, and dispute can handily be resolved by mustering social support. Social groupings can define, for their own purposes, not only what is considered correct but also what will be researched next. The proposed strategy for pursuing this ideal is explicit recognition and use of four levels of criticism, standpoint criticism being particularly important for ecology.

Progressive Synthesis has fixed principles. If, for example, criticism was replaced by a principle of consensus, then that would be a sufficient shift that the new activity no longer had the purpose of science. When scientists themselves adopt consensus as being more important than criticism then science may take on a restricted perspective. Only certain things are worked on and/or there are "right" and "wrong" ways to do things defined by social fashions. Of course governmental institutions rate consensus as important, and they fund much of science. Seeking consensus among scientists (the aggregation approach, Chapter 14) may be an effective way that science can be used by society for defining what are the important unsolved problems for society to which it can contribute. But adopting this approach, and accepting the decisions it may reach is a political decision anyway. However, if consensus seeking is extended to identification of strategies for solution of a scientific problem then there is a conflict, and possibly less likelihood of a solution. Of course, not all government funding of science is for applied problems and it seems particularly important that basic research should not be governed by a consensus approach (Lawton 1994). The principles of criticism, precise definition, and explicit assessment are to minimize relativism that can be difficult to overcome in a subject such as ecology.

16.5 Further reading

Some of the issues discussed in this chapter have been explored extensively in the philosophy of science literature. For further discussion on instrumentalism and rationalism see Suppe (1974). Ernst Mach (1838–1916), a physicist and philosopher developed an extensive critique of aspects of the theoretical physics of his time (Mach 1976) based on concern about what could be known and to remove untestable propositions. His approach continues to provoke discussion and a case for instrumentalism in modern biology is made by Rosenberg (1994), who suggests that the special complexities of biology are too great for the construction of comprehensive theory. Feyerabend (1975) gives a rattling good discussion on why science can not, and should not, be run according to fixed rules associated with falsificationism, empiricism, or rationalism, and discusses assumptions associated with each. However, the dispute over whether or not there is a fixed method for scientific research continues to be hotly debated, though the nature of the debate has changed, as is described in the exchange between Worrall (1988, 1989) and Laudan (1989).

Appendix: Suggestions for instructors

The material in this book has been used in graduate teaching in three ways: as a required introductory course of scientific method to Master's students, as a discussion course with mid-program doctoral students, and as an intensive week or 10-day residential course for mixed-level graduate students.

Introductory course in scientific method

For Master's students the course has been taught toward the end of the student's first year when they are preparing for their first thesis-oriented research. Course readings start at Chapter 1 and the objective is for students to develop a written research plan that they further discuss and refine with their faculty supervisory committees. The sequence of homeworks in Box A1 require a progression in thinking about individual research problems. Effective reading in a 10-week teaching period focuses on Chapters 1–8 and 15, with lectures on scientific inference.

Box A1. Series of homeworks used in an introductory course in scientific method

Six-minute talk on your potential research question

Each person will make a 6-minute description on what their research will be about. This description should contain three things.

(1) *A description of the background to the research (2 minutes).* This might include something on the motivation for the work, who else is studying this problem, or is interested in related problems at the university or research institute.

(2) *A statement of what you think the scientific question is (2 minutes).* This is likely to vary between individuals. There are instances where theories are incomplete, where they have failed, or where new types of measurement are needed.

(3) *A statement of what you think you do not know (2 minutes).* This is where you describe what your own first work in the research planning process is likely to be. It should focus on what you, personally, think you must do. Of course it will relate to how you have stated the scientific question, but it may include statements of what you may need to research in order to develop a research plan. This may be investigation of part of the scientific literature, possibly how a technique of measurement works, or whether you can find a site with particular characteristics.

You can expect that this homework will take considerable time. You can prepare by first writing out statements under each of these three headings. If you find other headings are more appropriate for your particular work then change the structure to suite them. There are two challenges:

(1) Deciding what you will and will not say, and placing what you will say in an order. The advantage of this exercise is that you are allowed, in fact required, to say what you do not know about a problem. There is no expectation to be correct about the future results of a research problem.

(2) Making a verbal presentation lasting just 6 minutes. You must practice this before coming to class. Notes are permitted but you must not read a script. To give a verbal presentation rather than a reading requires that you have mastered the subject sufficiently to be fluent in your choice of language as you describe the science.

Homework 1. Background and introduction to the research question

This is a written description of the question you will be researching. For some people this may not yet be clear and the purpose of this homework is that, by writing, you should start the process of thinking on paper about just what is a suitable research question. Writing something down leads you to question what you have written and so develop ideas further. This means that it can take a considerable time to write just a few pages. Include key references but I do not expect a completely documented review of scientific papers. Describe where your ideas come from, and what are their strengths, weakness and importance? Ashley Steel's version (Chapter 4, page 75) is along the lines of what you should prepare, although in most cases it should be 4 or 5 pages.

Homework 2. Propositional and conceptual analysis and theory network

This is the analysis of your research problem using the process defined in Chapter 3 and illustrations in Chapters 4 and 5. You must list your axioms, classify and define concepts, list postulates, and present a network diagram. The important part of the text of this homework is the justification of axioms and definition of concepts. This homework becomes a formal description of the logic of your research question. This may take up to 10 pages to describe, including the network diagram. The sooner you start on this homework the better! It is best if you can discuss this with your faculty supervisor and others.

Homework 3. Specification of data statements

Having laid out what you want to research (postulates), and the reason for researching them, in Homework 2, the purpose of this homework is to define at least one data statement. You must list and describe the three different parts of the data statement. There should be a full justification, including that for any statistical analyses that you intend to use. Examples of data statements are given in Chapters 2 and 4. Pay particular attention to definition and possible research required into measurement effectiveness, accuracy and precision (Chapter 6) and exploratory research (Chapter 8), if it is needed.

Homework 4. How scientific inference will be made

This is a description of the synthesis that you think most likely (Introduction to Section II). You should specify the types of answer you may obtain under different conditions, and this should include possible alternatives. While you may be using statistical inference in some data statements, in this homework you must go beyond that. List the types of change you expect to see (Table 15.2). Are you using any integrative concepts and what might be the significance of your work for their definition and use?

The initial talk is valuable in introducing each person's work to the whole class. Usually there is a diversity of types of problem in a class that helps to illustrate that different techniques of investigation are suitable for different problems.

Two class-work exercises in critical analysis have been particularly useful. The first is analysis of a scientific paper for its *methods*. We tend to read for content and not method and so may miss methodological difficulties. The second is discussion with a practicing consulting statistician. This discussion takes place after the students have read Chapter 8. It is most effective to do this after the students have taken a course in statistical inference – during such a course the students concentration is on mastering the technical aspects of statistics.

Two exercises are valuable in constructive learning. First, the exercise in confirmation and falsification (Chapter 7) is extremely valuable, particular for students to be the person setting the rule and watching others struggle with the reasoning exercise.

Second, there are two extensive discussion periods where students, in groups of three or four, discuss each others' proposals. This is best done in association with home-works 3 and 4. Each student is required to read the proposals of others in the group.

Mid-program course for doctoral students

Reading and discussion courses in scientific method can be stimulating. However, a course with a title such as "What makes a good thesis?" in which individual student research plans are read, developed, and discussed by the whole group sustains interest. Readings from this book can be used when relevant points arise and usually start with the Introduction to Section II and Chapter 15.

Residential course

It is valuable for students to come to the course prepared to give an introductory talk about their own research. It is also important to illustrate a complete cycle of research through discussion with a research scientist working at the course location, particular-ly someone who is interested in the process of science and prepared to discuss the pros and cons of alternative approaches.

References

Abrams, P. A. (1991). The predictive ability of peer review of grant proposals: the case of ecology and the US National Science Foundation. *Social Studies of Science*, **21**, 111–132.

Adam, P. (1990). *Saltmarsh Ecology*. Cambridge: Cambridge University Press.

Agee, J. K. (1993). *Fire Ecology of Pacific Northwest Forests*. Washington, DC: Island Press.

Alexander, I. J. (1989). Systematics and ecology of ectomycorrhizal legumes. In *Advances in Legume Biology*, ed. D. H. Stirton and J. L. Zarucchi, pp. 607–624. Monograph in Systematics, **29**, Missouri Botanical Garden.

Allen, M. F. (1991). *The Ecology of Mycorrhizae*. Cambridge: Cambridge University Press.

Althusser, L. (1990). *Philosophy and the Spontaneous Philosophy of the Scientists*. London: Verso.

Altman, D. G. (1982). How large a sample? In *Statistics in Practice*, ed. S. M. Gore and D. G. Altman, pp. 6–8. London: British Medical Journal.

Amlaner, C. J. and Macdonald, D. W. (eds.) (1980). *A Handbook on Radio Tracking*. Oxford: Pergamon Press.

Anderson, P. M. and Brubaker, L. B. (1994). Vegetation history of northcentral Alaska: a mapped summary of late-Quaternary pollen data. *Quaternary Science Reviews*, **13**, 71–92.

Anon. (1990). *Research Student and Supervisor: An Approach to Good Supervisory Practice*. Washington, DC: Council of Graduate Schools.

Anon. (1993). *The Chicago Manual of Style*. Chicago: University of Chicago Press.

Anon. (1994). *World Translation Index*, 8. Delft: International Translations Centre (Inc.).

Aphalo, P. J. and Jarvis, P. G. (1991). Do stomata respond to relative humidity? *Plant, Cell and Environment*, **14**, 127–132.

Aphalo, P. J. and Jarvis, P. G. (1993). Separation of direct and indirect responses of stomata to light: results from a leaf inversion experiment at constant intercellular CO_2 molar fraction. *Journal of Experimental Botany*, **44**, 791–800.

Arkebauer, T. J., Weiss, A., Sinclair, T. R. and Blum, A. (1994). In defense of radiation use efficiency: a response to Demetriades-Shah *et al.* (1992). *Agricultural and Forest Metereology*, **68**, 221–227.

ASTM (1984). *Standard Practice for Evaluating Fate Models of Chemicals*. Standard E 978–84. Philadelphia: American Society for Testing and Materials.

ASTM (1992). *Standard Practice for Evaluating Mathematical Models for the Environmental Fate of Chemicals*. Standard E 978-92. Philadelphia: American Society for Testing and Materials.

Ayala, F. J. (1974). Introduction to studies in the philosophy of biology. In *Studies in the Philosophy of Biology, Reduction and Related Problems*, ed. F. J. Ayala and T. Dobzhansky, pp. vii–xvi. Berkley: University of California Press.

Babbage, C. (1830). *Reflections on the Decline of Science in England and Some of its Causes*. London: Fellowes and Boothe.

Baldocchi, D. (1994). An analytic solution for coupled leaf photosynthesis and stomatal conductance models. *Tree Physiology*, **14**, 1069–1079.

Barber, B. (1952). *Science and the Social Order*. New York: Collier.

Barber, B. (1961). Resistance by scientists to scientific discovery. *Science*, **134**, 596–602.

Barnard, C. I. (1968). *The Functions of the Executive*. Cambridge, MA: Harvard University Press.

Bayliss, G. T. S. (1975). The magnolioid mycorrhiza and mycotropy in root systems derived from it. In *Endomycorrhizas*, ed. F. E. Sanders, B. Mosse and P. B. Tinker, pp. 373–389. London: Academic Press.

Bazerman, C. (1988). *Shaping Written Knowledge*. Madison, WI: University of Wisconsin Press.

Becker, R. A., Chambers, J. M. and Wilks, A. R. (1988). *The New S Language*. Pacific Grove, CA: Wadsworth and Brooks.

Begon, M., Firbank, L. and Wall, R. (1986). Is there a self-thinning rule for animal populations? *Oikos*, **46**, 122–124.

Bella, D. A. (1996). The pressure of organizations and the responsibilities of university professors. *BioScience*, **46**, 772–778.

Berkson, W. (1976). Lakatos One and Lakatos Two. In *Essays in Memory of Imre Lakatos*, ed. R. S. Cohen, P. K. Feyerabend and M. W. Wartofsky, pp. 39–54. Boston Studies in the Philosophy of Science **39**. Dordrecht: D Reidel.

Black, C. C. (1971). Ecological implications of dividing plants into groups with distinct photosynthetic production capacities. *Advances in Ecological Research*, **7**, 87–114.

Blackmore, J. T. (1983). Should we abolish the distinction between science and metaphysics? *Philosophia*, **12**, 393–400.

Blake, R. M., DuCasse, C. J. and Madden, E. H. (1960). *Theories of Scientific Method: The Renaissance Through the Nineteenth Century*. Seattle, WA: University of Washington Press.

Bloor, D. (1976). *Knowledge and Social Imagery*. London: Routledge and Kegan Paul.

Blyth, J. F. and MacLeod, D. A. (1981a). Sitka spruce (*Picea sitchensis*) in north-east Scotland. I. Relationship between site factors and growth. *Forestry*, **54**, 41–62.

Blyth, J. F. and MacLeod, D. A. (1981b). Sitka spruce (*Picea sitchensis*) in north-east Scotland. II. Yield prediction by regression analysis. *Forestry*, **54**, 63–73.

Boguslaw, R. (1968). Values in the research society. In *Challenge to Reason*, ed. C. W. Churchman, pp. 51–66. New York: McGraw-Hill.

Bolhàr-Nordenkampf, H. R. and Lechner, E. G. (1988). Winter stress and chlorophyll fluorescence in Norway spruce (*Picea abies* Karst.). In *Applications of Chlorophyll Fluorescence*, ed. H. K. Lichtenthaler, pp 173–180. Dordrecht: Kluwer.

Bormann, F. H. and Likens, G. E. (1979). *Pattern and Process in a Forested Ecosystem*. New York: Springer-Verlag.

Bowen, C. D. (1963). *Francis Bacon. The Temper of a Man*. Boston, MA: Little, Brown & Co.

Box, G. E. P. and Jenkins, G. M. (1970). *Time Series Analysis, Forcasting and Control*. San Francisco: Holden-Day.

Boylan, M. (1986). Monadic and systemic teleology. In *Current Issues in Teleology*, ed. N. Rescher, pp. 15–25. Lanham, MD: University Press of America.

Brannigan, A. (1981). *The Social Basis of Scientific Discoveries*. Cambridge: Cambridge University Press.

Braunwald, E. (1987). On analysing scientific fraud. *Nature*, **325**, 215–216.

Brennan, P. and Croft, P. (1994). Interpreting the results of observational research: chance is not such a fine thing. *British Medical Journal*, **309**, 727–730.

Broad, C. D. (1925). *The Mind and Its Place in Nature*. London: Kegan Paul, Trench and Tribner.

Brooks, T. A. (1986). Evidence of complex citer motivations. *Journal of American Society for Information Science*, **37**, 34–36.

Brown, J. H. (1981). Two decades of homage to Santa Rosalis: toward a general theory of diversity. *American Zoologist*, **21**, 877–888.

Brubaker, L. B., Anderson, P. M. and Hu, F. H. (1999). Vegetation ecotone dynamics in southwest Alaska during the Late Quaternary. *Quarterly Science Review*, in press.

Buzzelli, D. E. (1993a). The definition of misconduct: a view from NSF. *Science*, **259**, 584–585, 647–648.

Buzzelli, D. E. (1993b). NSF's approach to misconduct in science. *Accountability in Research*, **3**, 215–221.

Buzzelli, D. E. (1993c). Plaigiarism in science: the experience of NSF. *Ethics and Policy*, **13**, 6–7.

Byerly, R. Jr and Pielke, R. A. (1995). The changing ecology of United States science. *Science*, **269**, 1531–1532.

Carey, A. B. and Johnson, M. L. (1995). Small mammals in managed, naturally young and old growth forests. *Ecological Applications*, **5**, 336–355.

Carmean, W. H. (1975). Forest site quality evaluation in the United States. *Advances in Agronomy*, **27**, 209–269.

Carnap, R. (1950). *Logical Foundations of Probability*. Chicago: University of Chicago Press.

Carpenter, S. R., Cottingham, K. L. and Stow, C. A. (1994). Fitting predator–prey models to time series with observation errors. *Ecology*, **75**, 1254–1264.

Carpenter, S. R. and Kitchell, J. F. (1987). The temporal scale of variance in limnetic primary production. *American Naturalist*, **129**, 417–433.

Carpenter, S. R. and Kitchell, J. F. (1988). Consumer control of lake productivity. *BioScience*, **38**, 764–769.

Carpenter, S. R. and Kitchell, J. F. (1993a). Simulation models of the trophic cascade: predictions and evaluations. In *The Trophic Cascade in Lakes*, ed. S. R. Carpenter and J. F. Kitchell, pp. 310–331. Cambridge: Cambridge University Press.

Carpenter, S. R. and Kitchell, J. F. (1993b). Synthesis and new directions. In *The Trophic Cascade in Lakes*, ed. S. R. Carpenter and J. F. Kitchell, pp. 331–350. Cambridge: Cambridge University Press.

Carpenter, S. R. and Kitchell, J. F. (eds.) (1993c). *The Trophic Cascade in Lakes*. Cambridge: Cambridge University Press.

Carpenter, S. R., Kitchell, J. F. and Hodgson, J. R. (1985). Cascading trophic interactions and lake productivity. *BioScience*, **35**, 634–639.

Carr, B. (1987). *Metaphysics: An Introduction*. London: Macmillan.

Cartwright, N. (1983). *How the Laws of Physics Lie*. Oxford: Oxford University.

Caswell, H. (1988). Theory and models in ecology: a different perspective. *Ecological Modelling*, **43**, 33–44.

Chalmers, A. F. (1982). *What is this thing called Science?* Second edition. Milton Keynes: Open University Press.

Chamberlin, T. C. (1965). The method of multiple working hypotheses. *Science*, **148**, 754–759.

Chapman, H. H. (1926). Factors determining natural reproduction of longleaf pine on cutover lands in LaSalle Parish, La. *Bulletin Yale School of Forestry*, **16**, 1–44.

Chapman, H. H. (1932). Is the longleaf type a climax? *Ecology*, **13**, 328–334.

Cherrett, J. M. (1989). *Ecological Concepts*. 29th Symposium of the British Ecological Society. Oxford: Blackwell.

Chubin, D. E. (1996). Reculturing science: politics, policy, and promises to keep. *Science and Public Policy*, **23**, 2–12.

Churchman, C. W. (1948). *Theory of Experimental Inference*. New York: Macmillan.

Cicchetti, D. V. (1991). The reliability of peer review for manuscript and grant submissions: a cross-disciplinary investigation. *Behavioral and Brain Sciences*, **14**, 119–186.

Clark, W. C., and Majone, G. (1985). The critical appraisal of scientific inquiries with policy implications. *Science, Technology and Human Values*, **10**, 6–19.

Clements, F. E. (1905). *Research Methods in Ecology*. Lincoln, NE: University Publishing Company.

Cleveland, W. S. (1985). Graphical perception and graphical methods for analyzing scientific data. *Science*, **229**, 828–833.

Cleveland, W. S. (1993). *Visualizing Data*. Summit, NJ: Hobart Press.

COHMAP Members (1988). Climate changes of the last 18,000 years: observations and model simulations. *Science*, **241**, 1043–1052.

Cole, J., Lovett, G. and Findlay, S. (eds.) (1991). *Comparative Analysis of Ecosystems*. New York: Springer-Verlag.

Cole, J. R. and Cole, S. (1972). The Orgtega hypothesis. *Science*, **178**, 368–375.

Cole, J. R. and Singer, B. (1991). A theory of limited differences: explaining the productivity puzzle in science. In *The Outer Circle*, ed. H. Zuckerman, J. R. Cole and J. T. Bruer, pp. 277–310. New York: W. W. Norton and Company.

Cole, S., Rubin, L. and Cole, J. R. (1978). *Peer Review in the National Science Foundation*. Washington, DC: National Academy of Sciences.

Connell, J. H. and Lowman, M. D. (1989). Low-diversity tropical rain forests: some possible mechanisms for their existence. *American Naturalist*, **134**, 88–119.

Connor, E. F. and Simberloff, D. (1979). You can't falsify ecological hypotheses without data. *Bulletin of the Ecological Society of America*, **60**, 154–155.

Costantino, R. F., Cushing, J. M. Dennis, B. and Desharnis, R. A. (1995). Experimentally induced transitions in the dynamic behaviour of insect populations. *Nature*, **375**, 227–230.

Cowling, E. B. (1992). The performance and legacy of NAPAP. *Ecological Applications*, **2**, 111–116.

Crane, D. (1972). *Invisible Colleges: Diffusion of Knowledge in Scientific Communities*. Chicago: Chicago University Press.

Crocker, R. L. and Major, J. (1955). Soil development in relation to vegetation and surface age at Glacier Bay, Alaska. *Journal of Ecology*, **43**, 427–448.

Cronin, B. (1984). *The Citation Process*. London: Taylor Graham.

Cronin, B. and Overfelt, K. (1994). Citation-based auditing of academic performance. *Journal of the American Society for Information Science*, **45**, 61–72.

Culliton, B. J. (1987). Integrity of research papers questioned. *Science*, **235**, 422–423.

Currie, D. J. and Paquin, V. (1987). Large-scale biogeographic patterns of species richness of trees. *Nature*, **329**, 326–327.

Dalziel, T. R. K., Wilson, E. J. and Proctor, M. V. (1994). The effectiveness of catchment liming in restoring acid waters at Loch Fleet, Galloway, Scotland. *Forest Ecology and Management*, **68**, 107–117.

Dancy, J. and Sosa, E. (1992). *A Companion to Epistemology*. Oxford: Blackwell.

Darden, L. and Maull, N. (1977). Interfield theories. *Philosophy of Science*, **44**, 43–64.

Darwin, F. (1925). *The Life and Letters of Charles Darwin*. Volume I. New York: Appleton.

Dayton, P. K. (1971). Competition, disturbance, and community organization: the provision and subsequent utilization of space. *Ecological Monographs*, **41**, 351–389.

Dayton, P. K. (1975). Experimental evaluation of ecological dominance in a rocky intertidal algal community. *Ecological Monographs*, **45**, 137–159.

Deans, J. D. (1979). Fluctuations of the soil environment and fine root growth in a young Sitka spruce plantation. *Plant and Soil*, **52**, 195–208.

Deleuze, G. and Guattari, F. (1994). *What is Philosophy?* New York: Columbia University Press.

Demetriades-Shah, T. H., Fuchs, M., Kanemasu, E. T. and Flitcroft, I. D. (1992). A note of caution concerning the relationship between cumulated intercepted solar radiation and crop growth. *Agricultural and Forest Meteorology*, **58**, 193–207.

Demetriades-Shah, T. H., Fuchs, M., Kanemasu, E. T. and Flitcroft, I. D. (1994). Further discussions on the relationship between cumulated intercepted solar radiation and crop growth. *Agricultural and Forest Meteorology*, **68**, 231–242.

Diggle, P. J. (1990). *Time-series: A Biostatistical Introduction*. Oxford, Clarendon.

Dilworth, C. (1996). *The Metaphysics of Science: an Account of Modern Science in Terms of Principles, Laws and Theories*. Dordrecht, Kluwer Academic.

Dixon, W. J. and Massey, F. J. (1957). *Introduction to Statistical Analysis*. New York: McGraw-Hill.

Drake, J. A., Flum, T. E., Witteman, G. J., Voskuil, T., Hoylman, A. M., Creason, C., Kenny, D. A., Huxel, G. R., Larue, C. S. and Duncan, J. R. (1993). The construction and assembly of an ecological landscape. *Journal of Animal Ecology*, **62**,

117–130.

Duhem, P. (1962). *The Aim and Structure of Physical Theory*. English translation by Philip P. Wiener of the second French edition of 1914, *La Théory Physique: Son Objet, Sa Structure*. Princeton, NJ: Princeton University Press.

Dunbar, M. J. (1980). The blunting of Occam's Razor, or to hell with parsimony. *Canadian Journal of Zoology*, **58**, 123–128.

Dupré, J. (1993). *The Disorder of Things*. Cambridge, MA: Harvard University Press.

Dyson, F. J. (1988). *Infinite in all directions: Gifford lectures given at Aberdeen, Scotland, April–November 1985*. New York: Harper Row.

Eberhardt, L. L. and Thomas, J. M. (1991). Designing environmental field studies. *Ecology*, **61**, 53–73.

Ehrlich, P. R. and Daily, G. C. (1993). Science and the management of natural resources. *Ecological Applications*, **3**, 558–560.

Elliot, K. J. and White, A. S. (1994). Effects of light, nitrogen, and phosphorus on red pine seedling growth and nutrient use efficiency. *Forest Science*, **40**, 47–58.

Ellis, B, (1985). What science aims to do. In *Images of Science*, ed. P. M. Churchland and C. A. Hooker, pp. 48–74. Chicago: University of Chicago Press.

Elton, C. S. (1958). *The Ecology of Invasion by Animals and Plants*. London: Methuen.

Erwin, E., Gendin, S. and Kleiman, L. (1994). *Ethical Issues in Scientific Research*. New York: Garland Publishing.

Feyerabend, P. K. (1975). *Against Method*. London: NLB.

Field, C. B. (1988). On the role of photosynthetic responses in constraining the habitat distribution of rainforest plants. *Australian Journal of Plant Physiology*, **15**, 343–358.

Finnerty, J. D. (1979). Cycles in Canadian lynx. *American Naturalist*, **114**, 453–455.

Fisher, R. A. (1966). *The Design of Experiments*. Eighth edition. Edinburgh: Oliver and Boyd.

Flew, A. (1984). *A Dictionary of Philosophy*. Second edition. New York: St Martins Press.

Ford, E. D., Milne, R. and Deans, J. D. (1987a). Shoot extension in *Picea sitchensis*. I. Seasonal variation within a forest canopy. *Annals of Botany*, **60**, 531–542.

Ford, E. D., Milne, R. and Deans, J. D. (1987b). Shoot extension in *Picea sitchensis*. II. Analysis of weather influences on daily growth rate. *Annals of Botany*, **60**, 543–552.

Ford, E. D. and Sorrensen, K. A. (1992). Theory and models of inter-plant competition: a spatial process. In *Individual-based Models and Approaches in Ecology: Populations, Communities, and Ecosystems*, ed. D. L. DeAngelis and L. J. Gross, pp. 363–407. London: Routledge, Chapman & Hall.

Foster, R. F. (1937). *Francis Bacon. Essays, Advancement of Learning, New Atlantis and Other Pieces*. New York: Odyssey Press.

Fowler, H. W. (1996). *The New Fowler's Modern English Usage*. Oxford: Clarendon Press.

Francis, R. C. and Hare, S. R. (1994). Decadal-scale regime shifts in the large marine ecosystems of the North-east Pacific: a case for historical science. *Fisheries Oceanography*, **3**, 279–291.

Frank, B. (1985). On the root-symbiosis-depending nutrition through hypogenous fungi of certain trees. (Translated by J. M. Trappe from *Berichte der deutschen botanischen Gesellschaft*, **3**, 128–145, 1845). In *Proceedings of the Sixth North American Conference on Mycorrhizae*, ed. R. Molina, pp. 18–25. Corvallis, OR: Forest Research Laboratory, Oregon State University.

Fraser, L. H. (1988). Top-down vs bottom-up control influenced by productivity in a North Derbyshire, UK, dale. *Oikos*, **81**, 99–108.

Fretwell, S. D. (1977). The regulation of plant communities by the food chains exploiting them. *Perspectives in Biology and Medicine*, **20**, 169–185.

Frey, D. (1975). Biological integrity of water: an historical perspective. In *The Integrity of Water*, ed. R. K. Ballentine and L. J. Guarraia, pp. 127–139. Washington, DC: Environmental Protection Agency.

Fritts, H. C. (1976). *Tree Rings and Climate*. London: Academic Press.

Galilei, G. (1962). *Dialogue Concerning the Two Chief World Systems*. Translated by S. Drake. Berkley: University of California Press.

Giere, R. N. (1988). *Explaining Science*. Chicago: University of Chicago Press.

Gilbert, G. N. (1977). Referencing as persuasion. *Social Studies of Science*, **7**, 113–122.

Gillies, D. (1993). *Philosophy of Science in the Twentieth Century*. Oxford: Blackwell.

Gilpin, M. E. (1973). Do hares eat lynx? *American Naturalist*, **107**, 727–730.

Givinish, T. J. (1987). Comparative studies of leaf form: assessing the relative roles of selective pressures and phylogenetic constraints. *New Phytologist*, **106** (Suppl.), 131–160.

Glass, A. D. M. (1989). *Plant Nutrition: An Introduction to Current Concepts*. Boston, MA: Jones and Bartlett.

Golley, F. B. (1993). *A History of the Ecosystem Concept in Ecology*. New Haven, CT: Yale University Press.

Gonzalez, A., Lawton, J. H., Gibert, F. S., Blackburn, T. H. and Evans-Freke, I. (1998). Metapopulation dynamics, abundance, and distribution in a microecosystem. *Science*, **281**, 2045–2047.

Goodstein, D. (1995). Peer review after the big crunch. *American Scientist*, **83**, 401–402.

Greenland, S. (1990). Randomization, statistics and causal inference. *Epidemiology*, **1**, 421–429.

Grime, J. P. (1979). *Plant Strategies and Vegetation Processes*. New York: Wiley.

Grimm, V. and Wissel, C. (1997).

Babel, or the ecological stability discussions: an inventory and analysis of terminology and a guide for avoiding confusion. *Oecologia*, **109**, 323–334.

Grubb, P. J. (1992). A positive distrust in simplicity – lessons from plant defences and from competition among plants and among animals. *Journal of Ecology*, **80**, 585–610.

Grünbaum, A. (1976). Is the method of bold conjectures and attempted refutations justifiably the method of science? *British Journal for the Philosophy of Science*, **27**, 105–136.

Guston, D. H. (1996). Principal-agent theory and the structure of science policy. *Science and Public Policy*, **23**, 229–240.

Hacskaylo, E. (1985). Advances and quiescence - patterns of modern research on ectomycorrhizae. In *Proceedings of the Sixth North American Conference on Mycorrhizae*, ed. R. Molina, pp. 40–47. Corvallis, OR: Forest Research Laboratory, Oregon State University.

Hadley, D. B. and Mower, R. G. (1990). Evaluating tree canopy color using computerized microdensitometry. *Journal of the American Society for Horticultural Science*, **115**, 189–194.

Haefner, J. W. (1996). *Modeling Biological Systems*. New York: Chapman & Hall.

Hafner, E. M. and Presswood, S. (1965). Strong inference and weak interactions. *Science*, **149**, 503–510.

Hagen, J. B. (1992). *An Entangled Bank. The Origins of Ecosystem Ecology*. New Brunswick, NJ: Rutgers University Press.

Haines-Young, R., Green, D. R. and Cousins, S. (1993). *Landscape Ecology and Geographic Infromation Systems*. New York: Taylor and Francis.

Hairston, N. G., Smith, F. E. and Slobodkin, L. B. (1960). Community structure, population control, and competition. *American Naturalist*, **94**, 421–425.

Hairston, N. G. Sr (1989). *Ecological Experiments*. Cambridge: Cambridge University Press.

Hall, C. A. S. (1988). An assessment of several of the historically most influential theoretical models used in ecology and of the data provided in their support. *Ecological Modelling*, **43**, 5–31.

Hall, D. J., Cooper, W. E. and Werner, E. E. (1970). An experimental approach to the production dynamics and structure of freshwater animal communities. *Limnology and Oceanography*, **15**, 839–928.

Hall, M. S. (1988). *Getting Funded: A Complete Guide to Proposal Writing*. Third edition. Portland, OR: Continuing Education Publications, Portland State University.

Hamilton, D. P. (1991). Research papers: who's uncited now? *Science*, **251**, 25.

Hamilton, L. S. (1960). Color standardization for foliage description in forestry. *Journal of Forestry*, **58**, 187–194.

Hansen, J. and Beck, E. (1994). Seasonal changes in the utilization and turnover of assimilation products in 8-year-old Scots pine (*Pinus sylvestris*) trees. *Trees*, **8**, 172–182.

Harding, S, (1991). *Whose Science? Whose Knowledge?* Ithaca, NY: Cornell University Press.

Hardwick, R. C. (1986). Physiological consequences of modular growth in plants. Philosophical Transactions of the Royal Society, London, B, Biological Science, **313**, 161–173.

Harley, J. L. (1985). Mycorrhiza: the first 65 years; from the time of Frank till 1950. In *Proceedings of the Sixth North American Conference on Mycorrhizae*, ed. R. Molina, pp. 26–33. Corvallis, OR: Forest Research Laboratory, Oregon State University.

Harvey, P. H., Colwell, R. K., Silvertown, J. W. and May, R. M. (1983). Null models in ecology. *Annual Review of Ecology and Systematics*, **14**, 189–211.

Hastings, A., Hom, C. L., Ellner, S., Turchinet, P. and Godfray, H. C. J. (1993). Chaos in ecology: is mother nature a strange attractor? *Annual Review of Ecology and Systematics*, **24**,

1–33.

Hawkins, D. K. (1997). Hybridization between coastal cutthroat trout (*Oncorhynchus clarki clarki*) and steelhead (*O. mykiss*). Unpublished Ph.D. thesis, University of Washington.

Hawkins, D. K. and Foote, C. J. (1998). Early survival and development of coastal cutthroat trout (*Oncorhynchus clarki clarki*), steelhead (*Oncorhynchus mykiss*), and their hybrids. *Canadian Journal of Fish and Aquatic Science*, **55**, 2097–2104.

Hawkins, D. K. and Quinn, T. P. (1996). Critical swimming velocity and associated morphology of juvenile coastal cutthroat trout (*Oncorhynchus clarki clarki*), steelhead trout (*Oncorhynchus mykiss*), and their hybrids. *Canadian Journal of Fish and Aquatic Science*, **53**, 1487–1496.

Hawksworth, D. L. (ed.) (1995). *Biodiversity Measurement and Estimation*. London: Chapman & Hall.

Heal, W. O. and Grime, J. P. (1991). Comparative analysis of ecosystems: past lessons and future directions. In *Comparative Analyses of Ecosystems*, ed. J. Cole, G. Lovett and S. Findlay, pp. 7–23. New York: Springer-Verlag.

Hempel, C. G. and Oppenheim, P. (1948). Studies in the logic of explanation. *Philosophy of Science*, **15**, 135–175.

Henning, D. H. (1970). Natural resources administration. *Public Administration Review*, **30**, 134–140.

Herold. A. (1980). Regulation of photosynthesis by sink activity – the missing link. *New Phytologist*, **86**, 131–144.

Herold, A. and McNeil, P. H. (1979). Restoration of photosynthesis in pot-bound tobacco. *Journal of Experimental Botany*, **30**, 1187–1194.

Hicks, D. M. and Katz, J. S. (1996). Where is science going? *Science, Technology and Human Values*, **21**, 379–406.

Hilborn, R. and Mangel, M. (1997). *The Ecological Detective*. Monograph in Population Biology, **28**. Princeton, NJ: Princeton University Press.

Ho, L. C. (1988). Metabolism and compartmentation of imported sugars in sink organs in relation to sink strength. *Annual Review of Plant Physiology and Molecular Biology*, **39**, 355–378.

Höberg, P. (1986). Soil nutrient availability and tree species composition in tropical Africa. *Journal of Tropical Ecology*, **2**, 359–372.

Holland, P. W. (1986). Statistics and causal inference. (With discussion.) *Journal of the American Statistical Association*, **81**, 945–970.

Holling, C. S. (1973). Resilience and stability of ecological systems. *Annual Review of Ecology and Systematics*, **4**, 1–23.

Holton, G. (1973). *Thematic Origins of Scientific Thought: Kepler to Einstein*. Cambridge, MA: Harvard University Press.

Holton, G. (1981). On themata in scientific thought. In *On Scientific Thinking*, ed. R. D. Tweney, M. E. Doherty and C. R. Mynatt, pp. 313–315. New York: Columbia University Press.

Holyoak, K. J. and Thagard, P. (1995). *Mental Leaps: Analogy in Creative Thought*. Cambridge, MA: MIT Press.

Horrobin, D. F. (1982). Peer review: a philosophically faulty concept which is proving disastrous for science. *Behavioral and Brain Sciences*, **5**, 217–218.

Hu, F. S., Brubaker, L. B. and Anderson, P. M. (1993). A 12000 year record of vegetation change and soil development from Wien Lake, central Alaska. *Canadian Journal of Botany*, **71**, 1133–1142.

Hubbell, S. F. and Foster, R. B. (1986). Biology, chance, and history and the structure of tropical rain forest tree communities. In *Community Ecology*, ed. J. Diamond and T. Case, pp. 314–329. New York: Harper Row.

Huey, R. B. (1987). Reply to Stearns: some acynical advice for graduate students. *Bulletin of the American Ecological Society*, **68**(2), 150–153.

Hull, D. L. (1988). *Science as a Process*. Chicago: University of Chicago Press.

Hunter, M. D. and Price, P. W. (1992). Playing chutes and ladders: heterogeneity and the relative roles of bottom-up and top-down forces in natural communities. *Ecology*, **73**, 724–732.

Hunter, M. D., Varley, G. C. and Gradwell, G. R. (1997). Estimating the relative roles of top-down and bottom-up forces on insect herbivore populations: a classic study revisited. *Proceedings of the National Academy of Sciences, USA*, **94**, 9176–9181.

Hurlbert, S. H. (1984). Pseudoreplication and the design of ecological field experiments. *Ecological Monographs*, **54**, 187–211.

Huston, M. A. (1994). *Biological Diversity*. Cambridge: Cambridge University Press.

Hutchinson, G. E. (1959). Homage to Santa Rosalia; or why there are so many kinds of animals. *American Naturalist*, **93**, 145–159.

Inhaber, H. and Alvo, M. (1978). World science as an input output system. *Scientometrics*, **1**, 43–64.

Jacobsen, L., Perrow, M. R., Landkildehus, F., Hjírne, M., Lauridsen, T. L. and Berg, S. (1997). Interactions between piscivores, zooplanktivores and zooplankton in submerged nacrophytes: preliminary observations from enclosure and pond experiments. *Hydrobiologia*, **342/343**, 197–205.

Jaech, J. L. (1985). *Statistical Analysis of Measurement Errors*. New York: John Wiley and Sons.

Janos, D. P. (1980a). Versicular-arbuscular mycorrhizae affect lowland tropical rain forest plant growth. *Ecology*, **61**, 151–162.

Janos, D. P. (1980b). Mycorrhizae influence tropical succession. *Biotropica*, **12** (Suppl.), 56–64.

Janos, D. P. (1983). Tropical mycorrhizas, nutrient cycles and plant growth. In *Tropical Rain Forest: Ecology and Management*, ed. S. L. Sutton, T. C. Whitmore and A. C. Chadwick, pp. 327–345. Oxford: Blackwell.

Janos, D. P. (1985). Mycorrhizal fungi: agents or symptoms of tropical community composition? In *Proceedings of the Sixth North American Conference on Mycorrhizae*, ed. R. Molina, pp. 98–103. Corvallis, OR: Forest Research Laboratory, Oregon State University.

Janos, D. P. (1987). VA mycorrhizas in humid tropical ecosystems. In *Ecophysiology of VA Mycorrhizal Plants*, ed. G. R. Safir, pp. 107–134. Boca Raton, FL: CRC Press.

Janos, D. P. (1988). Mycorrhizal applications in tropical forestry: are temperate-zone approaches appropriate? In *Trees and Mycorrhiza*, ed. F. S. P. Ng, pp. 133–188. Kuala Lumpur: Forest Research Institute.

Janos, D. P. (1992). Heterogeneity and scale in tropical vesicular-arbuscular mycorrhiza formation. In *Mycorrhizas in Ecosystems*, ed. D. J. Read, D. H. Lewis, A. H. Fitter and I. J. Alexander, pp. 276–282. Wallingford, Oxon: C.A.B. International.

Janos, D. P. (1996). Mycorrhizas, succession, and the rehabilitation of deforested lands in the humid tropics. In *Fungi and Environmental Change*, ed. J. C. Frankland, N. Magan and G. M. Gadd, pp. 129–162. British Mycological Society Symposium, **20**. Cambridge: Cambridge University Press.

Janse, J. M. (1896). Les endophytes radicaux de quelques plantes javanaise. *Annales du Jardin Botanique de Buitenzorg*, **14**, 53–212. (Read in a translation by K. D. Doak.)

Janzen, D. H. (1970). Herbivores and the number of tree species in tropical forests. *American Naturalist*, **104**, 501–528.

Jarvis, N. J., Mullins, C. E. and MacLeod, D. A. (1983). The prediction of evapotranspiration and growth of Sitka spruce from meteorological records. *Annales Geophysicae*, **1**, 335–344.

Jarvis, P. G. and Leverenz, J. W. (1982). Productivity of temperate deciduous and evergreen forests. In *Physiological Plant Ecology IV*, ed. O. L. Lange, P. S. Nobel, C. B. Osmond and H. Ziegler, pp. 233–280. Berlin: Springer-Verlag.

Jarvis, P. G. and Sandford, A. P. (1985). The measurement of carbon dioxide in air. In *Instrumentation for Environmental Physiology*, ed. B. Marshall and F. I. Woodward, pp. 29–57. Society for Experimental Biology, Seminar Series, **22**. Cambridge: Cambridge University Press.

Johnstone, D. J. (1987). Tests of significance following R. A. Fisher. *British Journal of Philosophy of Science*, **38**, 481–499.

Kantorovich, A. (1988). Philosophy of science: from justification to explanation. *British Journal for the Philosophy of Science*, **39**, 469–494.

Kantrowitz, A. (1975). Controlling technology democratically. *American Scientist*, **63**, 505–509.

Karieva, P. (1990). Population dynamics in spatially complex environments: theory and data. *Philosophical Transactions of the Royal Society, London B*, **330**, 175–190.

Karieva, P. (1994). Higher order interactions as a foil to reductionist ecology. *Ecology*, **75**, 1527–1528.

Karieva, P. (1995). Predicting and producing chaos. *Nature*, **375**, 189–190.

Keddy, P. A. (1987). Beyond reductionism and scholasticism in plant community ecology. *Vegetatio*, **69**, 209–211.

Keener, M. E., DeMichele, D. W. and Sharpe, P. J. H. (1979). Sink metabolism: a conceptual framework for analysis. *Annals of Botany*, **44**, 659–669.

Keller, E. F. (1984). *Reflections on Gender and Science*. New Haven, CT: Yale University Press.

Keller, E. F. (1991). The wo/man scientists: issues of sex and gender in the pursuit of science. In *The Outer Circle*, ed. H. Zuckerman, J. R. Cole and J. T. Bruer, pp. 227–236. New York: W. W. Norton and Company.

Kerans, B. L. and Karr, J. R. (1994). A benthic index of biotic integrity (B-BI) for rivers of the Tennessee Valley. *Ecological Applications*, **4**, 768–785.

Kersting, K. (1997). On the way to eternity, the success of an aquatic laboratory microecosystem. *Aquatic Ecology*, **31**, 29–35.

Kiester, A. R. (1980). Natural kinds, natural history and ecology. In *Conceptual Issues in Ecology*, ed. E. Saarinen, pp. 345–356. Dordrecht: D. Reidel.

Kinry, J. R. (1994). A note of caution concerning the paper by Demetriades-Shah *et al.* (1992). *Agricultural and Forest Meteorology*, **68**, 229–230.

Kirschbaum, M. V. F. and Farquahar, G. D. (1984). Temperature dependence of whole-leaf photosynthesis in *Eucalyptus pauciflora* Sieb. Ex Spreng. *Australian Journal of Plant Physiology*, **11**, 519–538.

Kitcher, P. (1988). Explanatory unification. In *Theories of Explanation*, ed. J. Pitt, pp. 167–187. Oxford: Oxford University Press.

Klayman, J. and Ha, Y.-W. (1987). Confirmation, disconfirmation and information in hypothesis testing. *Psychological Review*, **94**, 211–228.

Klayman, J. and Ha, Y-W. (1989). Hypothesis testing in rule discovery: strategy, structure and content. *Journal of Experimental Psychology*, **15**, 596–604.

Klebanoff, A. and Hastings, A. (1994). Chaos in three species food chains. *Journal of Mathematical Biology*, **32**, 427–451.

Knorr-Cetina, K. D. (1981). *The Manufacture of Knowledge*. Oxford: Pergamon Press.

Kohler, R. E. (1991). *Partners in Science*. Chicago: University of Chicago Press.

Kolber, Z. S., Barber, R. T., Coale, K. H., Fitzwater, S. E., Greene, R. M., Johnson, K. S., Lindley, S. and Falkowski, P. G. (1994). Iron limitation of phytoplankton photosynthesis in the equatorial Pacific Ocean. *Nature*, **371**, 145–149.

Konikow, L. F. and Bredehoeft, J. D. (1992). Ground-water models cannot be validated. *Advances in Water Resources*, **15**, 75–83.

Kostoff, R. N. (1994). Federal research impact assessment: state-of-the art. *Journal of American Society for Information Science*, **45**, 428–440.

Kot, M., Schafler, W. M., Truty, G. L., Graser, D. J. and Olsen, L. F. (1988). Changing criteria for imposing order. *Ecological Modelling*, **43**, 75–110.

Kraemer, H. C. and Thiemann, S. (1987). *How Many Subjects?* Newbury Park, CA: Sage Publications.

Kramer, P. S. and Boyer, J. S. (1995). *Water Relations of Plants and Soils*. San Diego: Academic Press.

Kubie, L. S. (1953). Some unsolved problems of the scientific career. *American Scientist*, **41**, 596–613; **42**, 104–112.

Kuhn, T. S. (1970). *The Structure of Scientific Revolutions*. Second edition. Chicago: University of Chicago Press.

Kuhn, T. S. (1977). *The Essential Tension*. Chicago: University of Chicago Press.

Lakatos, I. (1970). Falsification and the methodology of scientific research programs. In *The Methodology of Scientific Research Programs*, ed. J. Worrall and G. Currie, pp. 8–101. Philosophical Papers, Volume I, *Imre Lakatos*. Cambridge: Cambridge University Press.

Lambert, K. and Brittan, G. G. Jr (1987). *An Introduction to the Philosophy of Science*. Third edition. Atascadero, CA: Ridgeview.

Lanner, R. M. (1964). Temperature and the diurnal rhythm of height growth in pines. *Journal of Forestry*, **62**, 493–495.

Laski, H. J. (1931). The limitation of the expert. *Fabian Tracts*, **235**, 1–14.

Latto, J. (1994). Evidence for a

self-thinning rule in animals. *Oikos*, **69**, 531–534.

Laudan, L. (1981a). The sources of modern methodology: two models of change. In *Science and Hypothesis*, pp. 6–19. Dordrecht: D. Reidel.

Laudan, L. (1981b). Hume (and Hacking) on induction. In *Science and Hypothesis*, pp. 72–85. Dordrecht: D. Reidel.

Laudan, L. (1981c). The clock metaphor and hypotheses: the impact of Descartes on English methodological thought, 1650–1670. In *Science and Hypothesis*, pp. 27–58. Dordrecht, Holland: D. Reidel.

Laudan, L. (1981d). Thomas Reid and the Newtonian turn of British methodological thought. In *Science and Hypothesis*, pp. 86–110. Dordrecht: D. Reidel.

Laudan, L. (1981e). William Whewell on the consilience of inductions. In *Science and Hypothesis*, pp. 163–180. Dordrecht: D. Reidel.

Laudan, L. (1989). If it ain't broke, don't fix it. *British Journal for the Philosophy of Science*, **40**, 369–375.

Lawton, J. H. (1991). Ecology as she is done, and could be done. *Oikos*, **61**, 289–290.

Lawton, J. H. (1994). Peer reviw, co-evolution and tortoises. *Oikos*, **69**, 361–363.

LeBlanc, H. and Wisdom, W. A. (1976). *Deductive Logic*. Boston, MA: Allyn and Bacon.

Leigh, E. G. Jr (1968). The ecological role of Volterra's equations. In *Some Mathematical Problems in Biology*, ed. M. Gerstenhaber, pp. 1–61. Providence, RI: American Mathematical Society.

Leverenz, J. W. (1988). The effects of illumination sequence, CO2 concentration, temperature and acclimation on the convexity of the photosynthetic light response curve. *Physiologia Plantarum*, **74**, 332–341.

Leverenz, J. W. and Hällgren, J.-E. (1991). Measuring photosynthesis and respiration of foliage. In *Techniques and Approaches in Forest Tree Ecophysiology*, ed. J. P. Lassoie

and T. M. Hinckley, pp. 303–328. Boca Raton, FL: CRC Press.

Levins, R., and Lewontin, R. (1980). Dialectics and reductionism in ecology. *Synthese*, **43**, 47–78.

Lewis, D. (1986). *Philosophical Papers, Volume II*. Oxford: Oxford University Press.

Lewontin, R. C. (1969). The meaning of stability. *Brookhaven Symposium in Biology*, **22**, Report of a Symposium held 26–28 May 1969 on Diversity and Stability in Ecological Systems, pp. 13–24. Upton, NY: Biology Department, Brookhaven National Laboratory.

Lindeman, R. L. (1942). The trophic-dynamic aspects of ecology. *Ecology*, **23**, 399–418.

Linder, S. and Lohammer, T. (1981). Amount and quality of information on CO2-exchange required for estimating annual carbon balance of coniferous trees. In *Understanding and Predicting Tree Growth*, ed. S. Linder, pp. 73–87. Studia Forestalia Suecica, **160**, 73–87.

Lipton, P. (1991). *Inference to the Best Explanation*. London: Routledge.

Liu, M. (1993). Progress in documentation. The complexity of citation practice: a review of citation studies. *Journal of Documentation*, **49**, 370–408.

Livingston, R. J. (1991). Historical relationships between research and resource management in the Apalachicola River estuary. *Ecological Applications*, **1**, 361–382.

Locke, L., Spirduso, W. W. and Silverman, S. J. (1993). *Proposals that Work: A Guide for Planning Dissertation and Grant Proposals*. Third edition. Newbury Park, CA: Sage Publications.

Loehle, C. (1987). Hypothesis testing in ecology: psychological aspects and the importance of theory maturation. *Quarterly Review of Biology*, **62**, 397–409.

Loehle, C. (1988). Philosophical tools: potential contributions to ecology. *Oikos*, **51**, 97–104.

Loehle, C. (1989a). Catastrophe theory in ecology: a critical review and an

example of the butterfly catastrophe. *Ecological Modelling*, **49**, 125–152.

Loehle, C. (1989b). Why women scientists publish less than men. *Bulletin of the Ecological Society of America*, **68**, 495–496.

Loehle, C. (1990). Philosophical tools: reply to Shrader-Frechette and McCoy. *Oikos*, **58**, 115–119.

Long, S. P. and Hällgren, J.-E. (1985). Measurements of CO2 assimilation by plants in the field and the laboratory. In *Techniques in Bioproductivity and Photosynthesis*, ed. J. Coombs, D. O. Hall, S. P. Long and J. M. O. Scurlock, pp. 62–94. Oxford: Pergamon Press.

Lonsdale, W. M. (1990). The self-thinning rule: dead or alive? *Ecology*, **71**, 1373–1388.

Losee, J. (1993). *A Historical Introduction to the Philosophy of Science*. Third edition. Oxford: Oxford University Press.

Lotka, A. J. (1925). *Elements of Physical Biology*. Baltimore, MD: Wilkins and Wilkins.

Loucks, O. L. (1992). Forest response research in NAPAP: potentially successful linkage of policy and science. *Ecological Applications*, **2**, 117–123.

Ludlow, M. M. and Jarvis, P. G. (1971). Photosynthesis in Sitka spruce (*Picea sitchensis* (Bong.) Carr.). I. General characteristics. *Journal of Applied Ecology*, **8**, 925–953.

Ludwig, D., Jones, D. D. and Holling, C. S. (1978). Qualitative analysis of insect outbreak systems: the spruce budworm and forest. *Journal of Animal Ecology*, **47**, 315–332.

Lundmark, T., Hällgren, J.-E. and Heden, J. (1988). Recovery from winter depression of photosynthesis in pine and spruce. *Trees*, **2**, 110–114.

Macfadyen, A. (1963). The contribution of the microfauna to total soil metabolism. In *Soil Organisms*, ed. J. Doeksen and J. Van der Drift, pp. 3–17. Amsterdam: North-Holland.

Macfadyen, A. (1964). Energy flow in

ecosystems and its exploitation by grazing. In *Grazing in Terrestrial and Marine Environments*, ed. D. J. Crisp, pp. 3–20. Oxford: Blackwell.

Macfadyen, A. (1975). Some thoughts on the behaviour of ecologists. *Journal of Animal Ecology*, **44**, 351–363.

Mach, E. (1976). *Knowledge and Error*. Translation by T. J. McCormick and P. Foulkes. Dordrecht: D. Reidel.

MacRoberts, M. H. and MacRoberts, B. R. (1986a). Quantitative measures of communication in science: a study at the formal level. *Social Studies of Science*, **16**, 151–172.

MacRoberts, M. H. and MacRoberts, B. R. (1986b). Another test of the normative theory of citing. *Journal of the American Society for Information Science*, **38**, 305–306.

MacRoberts, M. H. and MacRoberts, B. R. (1987). Testing the Ortega hypothesis: facts and artifacts. *Scientometrics*, **12**, 293–295.

MacRoberts, M. H. and MacRoberts, B. R. (1989). Problems of citation analysis: a critical review. *Journal of the American Society for Information Science*, **40**, 342–349.

MacRoberts, M. H. and MacRoberts, B. R. (1996). Problems of citation analysis. Scientometrics, **36**, 435–444.

Magnusson, W. E. (1997). Teaching experimental design in ecology, or how to do statistics without a bikini. *Bulletin of the Ecological Society of America*, **78**, 205–209.

Magurran, A. E. (1988). *Ecological Diversity and its Measurement*. Princeton, NJ: Princeton University Press.

Mahoney, M. J. (1976). *The Scientist as Subject: The Psychological Imperative*. Cambridge, MA: Ballinger Publishing Company.

Mahoney, M. J. (1977). Publication prejudices: an experimental study of confirmatory bias in the peer review system. *Cognitive Therapy and Research*, **1**, 161–175.

Maier, C. A. and Teskey, R. O. (1992). Internal and external control of net photosynthesis and stomatal conductance of mature eastern white pine (*Pinus strobus*). *Canadian Journal of Forest Research*, **22**, 1387–1394.

Malloch, D. W., Pirozynski, K. A. and Raven, P. H. (1980). Ecological and evolutionary significance of mycorrhizal symbioses in vascular plants. *Proceedings of the National Academy of Sciences, USA*, **77**, 2113–2118.

Margolis, H. (1993). *Paradigms and Barriers*. Chicago: Chicago University Press.

Marshall, E. (1986). San Diego's tough stand on research fraud. *Science*, **234**, 534–535.

Marshall, J. K. (1977). Conclusion: belowground ecosystem research – present and future. In *The Belowground Ecosystem: A Synthesis of Plant-Associated Processes*, Range Science Department Science Series No. 26, ed. J. K. Marshall, pp. 349–351. Fort Collins, CO: Colorado State University.

Martin, J. H., Coale, K. H., Johnson, K. S. and 41 additional authors. (1994). Testing the iron hypothesis in ecosystems of the equatorial Pacific Ocean. *Nature*, **371**, 123–129.

Matson, P. A. and Carpenter, S. R. (1990). Special feature. Statistical analysis of ecological response to large scale perturbations. *Ecology*, **71**, 2037.

Matson, P. A and Hunter, M. D. (1992). Special feature. The relative contributions of top-down and bottom-up forces in population and community ecology. *Ecology*, **73**, 723.

Matson, P. A., Potvin, C. and Travis, J. (1993). Special feature. Statistical methods: an upgrade for ecologists. *Ecology*, **74**, 1615–1616.

Maull, N. L. (1977). Unifying science without reduction. *Studies in the History and Philosophy of Science*, **8**, 143–162.

May, R. M. (1981). Models for two interacting populations. In *Theoretical Ecology. Principles and Applications*. Second edition, ed. R. M. May, pp. 78–104. Oxford: Blackwell Scientific Publications.

Mayr, E. (1982). *The Growth of Biological Thought: Diversity, Evolution, and Inheritance*. Cambridge, MA: Harvard University Press.

Mayr, E. (1988). *Toward a New Philosophy of Biology: Observations of an Evolutionist*. Cambridge, MA: Harvard University Press.

McGrath, J. E. and Altman, I. (1966). *Small Group Research: A Synthesis and Critique of the Field*. New York: Holt, Rinehart and Winston.

McIntosh, R. P. (1980). The background and some current problems of theoretical ecology. *Synthese*, **43**, 195–255.

McIntosh, R. P. (1985). *The Background of Ecology*. Cambridge: Cambridge University Press.

McNutt, R. A., Evans, A. T., Fletcher, R. H. and Fletcher, S. W. (1990). The effects of blinding on the quality of peer review. *Journal of the American Medical Association*, **263**, 137–176.

Mead, R. (1988). *The Design of Experiments*. Cambridge: Cambridge University Press.

Medawar, P. B. (1963). Is the scientific paper a fraud? *The Listener*, **70**, 377–378.

Mellor, D. H. and Oliver, A. (eds.) (1997). *Properties*. Oxford: Oxford University Press.

Menge, B. A. (1992). Community regulation: under what conditions are bottom-up factors important on rocky shores? *Ecology*, **73**, 755–765.

Menge, B. A., Daley, B. A., Wheeler, P. A., Dahloff, E., Sandford, E. and Straub, P. T. (1997). Benthic-pelagic links in rocky intertidal communities: bottom-up effects on top-down control? *Proceedings of the National Academy of Sciences, USA*, **94**, 14530–14535.

Menge, B. A. and Sutherland, J. P. (1976). Species diversity gradients: synthesis of the roles of predation, competition, and temporal heterogeneity. *American Naturalist*, **110**, 351–369.

Mentis, M. T. (1988). Hypothetico-deductive and inductive approaches in ecology. *Functional Ecology*, **2**, 5–14.

Merton, R. K. (1957a). Priorities in scientific discovery: a chapter in the sociology of science. *American Sociological Review*, **22**, 635–659.

Merton, R. K. (1957b). *Social Theory and Social Structure*. Revised and enlarged edition. Glencoe, IL: The Free Press.

Merton, R. K. (1961). Singletons and multiples in scientific discovery: a chapter in the sociology of science. *Proceedings of the American Philosophical Society*, **105**, 470–486.

Merton, R. K. (1963). The ambivalence of scientists. *Bulletin of the Johns Hopkins Hospital*, **112**, 77–97.

Merton, R. K. (1968). The Matthew effect in science. *Science*, **159**, 56–63

Miller, D. J. and Hersen, M. (1992). *Research Fraud in the Behavioral and Biomedical Sciences*. New York: John Wiley and Sons.

Miller, M. L. and Gale, R.P. (1986). Professional styles of federal forest and marine fisheries resource managers. *North American Journal of Fisheries Management*, **6**, 141–148.

Milne, R., Ford, E. D. and Deans, J. D. (1983). Time lags in the water relations of Sitka spruce. *Forest Ecology and Management*, **5**, 1–25.

Milsum, J. H. (1966). *Biological Control Systems Aanalysis*. New York: McGraw-Hill.

Mitroff, I. I. (1974a). Norms and counter-norms in a select group of the Apollo moon scientists: a case study of the ambivalence of scientists. *American Sociological Review*, **39**, 579–595.

Mitroff, I. I. (1974b). *The Subjective Side of Science: A Philosophical Inquiry into the Psychology of the Apollo Moon Scientists*. Amsterdam: Elsevier.

Montalenti, G. (1974). From Aristotle to Democritus via Darwin: a short survey of a long historical and logical journey. In *Studies in the Philosophy of Biology: Reduction and Related Problems*, ed. F. J. Ayala and T.

Dobhansky, pp. 3–19. Berkley: University of California Press.

Monteith, J. L. (1994). Validity of the correlation between intercepted radiation and biomass. *Agricultural and Forest Meteorology*, **68**, 213–220.

Morris, L. A., Moss, S. A. and Garbett, W. S. (1993). Competitive interference between selected herbaceous and woody plants and *Pinus taeda* L. during two growing seasons following planting. *Forest Science*, **39**, 166–187.

Moss, B. (1986). Originality in ecology: one editor's reply. *Bulletin of the British Ecological Society*, **17**, 171–173.

Moss, D. N. (1962). Photosynthesis and barrenness. *Crop Science*, **2**, 366–367.

Mosse, B. (1985). Endotrophic mycorrhizae (1885–1950): the dawn and the middle ages. *Proceedings of the Sixth North American Conference on Mycorrhizae*, ed. R. Molina, pp. 48–55. Corvallis, OR: Forest Research Laboratory, Oregon State University.

Muir, F. (1976). *The Frank Muir Book. An Irreverent Companion to Social History*. London: Heinman.

Munsell (1976). *Munsell Book of Color*. Baltimore, MD: Munsell Color Company.

Murdoch, W. M. (1966). "Community structure, population control, and competition" – a critique. *American Naturalist*, **100**, 219–226.

Murray, B. G. (1992). Research methods in physics and biology. *Oikos*, **64**, 594–596.

Murray, J. D. (1989). *Mathematical Biology*. Springer-Verlag, Berlin.

Musgrave, A. (1976). Method or madness? In *Essays in Memory of Imre Lakatos*, ed. R. S. Cohen, P. K. Feyerabend and M. W. Wartofsky, pp. 457–491. Boston Studies in the Philosophy of Science, **39**. Dordrecht: D. Reidel.

Myers, G. (1985). Texts as knowledge claims: the social construction of two biology articles. *Social Studies of Science*, **15**, 593–630.

Nagel, E. (1961). *The Structure of*

Science. New York: Harcourt, Brace and World.

Nakamura, F. and Swanson, F. J. (1994). Distribution of coarse woody debris in a mountain stream, western Cascade Range, Oregon. *Canadian Journal of Forest Research*, **24**, 2395–2403.

NAPAP (1991). *NAPAP 1990 Integrated Assessment*. Washington, DC: United States Government Prinring Office.

National Science Board (1997). *National Science Board and National Science Foundation Staff Task Force on Merit Review*. Final Recommendations. NSB/MR–97–05.

National Science Foundation (1989). *Important Notice to Presidents of Colleges and Universities and Heads of other National Science Foundation Grantee Organizations*. Notice No. 107.

National Science Foundation (1990). *Survey Data on the Extent of Misconduct in Science and Engineering*. Office of the Inspector General Oversight Study, OIG 90–3214.

National Science Foundation (1994). Office of Inspector General. *Semiannual Report to Congress*, **11**, 30–31.

National Science Foundation (1995a). *Grant Proposal Guide*. (NSF 95–27). Arlington, VA: National Science Foundation.

National Science Foundation (1995b). *NSF in a Changing World*. (NSF 95–24). Arlington, VA: National Science Foundation.

National Science Foundation (1995c). *Proposal Forms Kit*. (NSF 95–28). Arlington, VA: National Science Foundation.

National Science Foundation (1997a). Office of Inspector General. *Semiannual Report to Congress*, **15**, 34–49.

National Science Foundation (1997b). *New criteria for NSF proposals*. Notice No. 121. Arlington, VA: National Science Foundation.

Nevitt, G. A., Dittman, A. H., Quinn,

T. P. and Moody, W. J. Jr (1994). Evidence for a peripheral olfactory memory in imprinted salmon. *Proceedings of the National Academy of Science, USA*, **91**, 4288–4292.

Newbery, D. M., Alexander, I. J., Thomas, D. W. and Gartlan, J. S. (1988). Ectomycorrhizal rain forest legumes and soil phosphorus in Korup National Park, Cameroon. *New Phytologist*, **109**, 433–450.

Newman, E. I. (1986). Originality in ecology, the editor's choice. *Bulletin of the British Ecological Society*, **17**, 127–129.

Nickles, T. (1977). Introduction. In *Discovery, Logic and Rationality*, ed. T. Nickles, Boston Studies in the Philosophy of Science, **56**, pp. 1–59. Dordrecht: D. Reidel.

Nola, R. (1987). The status of Popper's theory of scientific method. *British Journal for the Philosophy of Science*, **38**, 441–480.

Norberg, R. A. (1988). Theory of growth geometry of plants and self-thinning of plant populations: geometric similarity, elastic similarity, and different growth rates of plant parts. *American Naturalist*, **131**, 220–256.

Northrop, F. S. C. (1983). *The Logic of the Sciences and the Humanities*. Woodbridge, CT: Ox Bow Press.

Odeh, R. E. and Fox, M. (1975). *Sample Size Choice*. New York: Marcel Dekker.

Odum, E. P. (1953). *Fundementals of Ecology*. Philadelphia: W.B. Saunders.

Odum, H. T. (1957). Structure and productivity of Silver Springs, Florida. *Ecological Monographs*, **27**, 55–112.

Odum, H. T. and Odum, E. P. (1955). Trophic structure and productivity of a windward coral reef community on Eniwetok Atoll. *Ecological Monographs*, **25**, 291–320.

Oksanen, L., Fretwell, S. D., Arrunda, J. and Niemela, P. (1981). Exploitation ecosystems in gradients of primary productivity. *American Naturalist*, **118**, 240–261.

Olson, J. S. (1964). Gross and net production of terrestrial vegetation. *Journal of Ecology*, **52** (Suppl.), 99–118.

O'Neil, R. V., DeAngelis, D. L., Waide, J. B. and Allen, T. F. H. (1986). *A Hierarchical Concept of Ecosystems*. Monograph in Population Biology, **23**. Princeton, NJ: Princeton University Press.

ORB (NAPAP Oversight Review Board). (1991). *The Experience and Legacy of NAPAP*. Washington, DC: National Acid Precipitation Assessment Program.

Ord, J. K. (1979). Time-series and spatial patterns in ecology. In *Spatial and Temporal Analysis in Ecology*, ed. R. M. Cormack and J. K. Ord, pp. 1–94. Fairland, MD: International Cooperative Publishing House.

Oreskes, N., Shrader-Frechette, K. and Belitz, K. (1994). Verification, validation, and confirmation of numerical models in the earth sciences. *Science*, **263**, 641–646.

Orians, G. H., Dirzo, R. and Cushman, J. H. (1996). *Biodiversity and Ecosystem Processes in Tropical Forests*. Berlin: Springer-Verlag.

Ovington, J. D. (1961). Some aspects of energy flow in plantations of *Pinus sylvestris* L. *Annals of Botany*, **25**, 12–20.

Paine, R. T. (1966). Food web complexity and species diversity. *American Naturalist*, **100**, 65–75.

Paine, R. T. (1969). The *Pisaster–Tegula* interaction: prey patches, predator food preference and intertidal community structure. *Ecology*, **50**, 950–961.

Paine, R. T. (1971). A short-term experimental investigation of resource partitioning in a New Zealand rocky intertidal habitat. *Ecology*, **52**, 1096–1106.

Paine, R. T. (1974). Intertidal community structure. Experimental studies on the relationship between a dominant competitor and its principal predator. *Oecologia* (Berlin), **15**, 93–120.

Pao, M.L. (1985). Lotka Law – A testing procedure. *Information Processing and Management*, **21**, 305–320.

Pao, M. L. (1986). An empirical examination of Lotka Law. *Journal of the American Society for Information Science*, **37**, 26–33.

Pascual, M. A. and Kareiva, P. (1996). Predicting the outcome of competition using experimental data: maximum likelihood and Bayesian approaches. *Ecology*, **77**, 337–349.

Pearsall, W. H. (1964). The development of ecology in Britain. *Journal of Ecology*, **52** (Suppl.), 1–12.

Pearson, E. S. and Hartley, H. O. (1976). *Biometrika Tables for Statisticians*. Volume I. London: Biometrika Trust, University College London.

Peitgen, H.-O. and Richter, P. H. (1986). *The Beauty of Fractals: Images of Complex Dynamic Systems*. Berlin: Springer-Verlag.

Perry, C. R. (1991). *The Fine Art of Technical Writing*. Hillsboro, OR: Blue Heron.

Peters, D. P. and Ceci, S. J. (1982). Peer-review practices of psychology journals: the fate of published articles, submitted again. *Behavioral and Brain Sciences*, **5**, 187–255.

Peters, R. H. (1991). *A Critique for Ecology*. Cambridge: Cambridge University Press.

Peterson, B. J., Deegan, L., Helfrich, J., Hobbie, J. E., Hullar, M., Moller, B., Ford, T. E., Hershey, A., Hiltner, A., Kipphut, G., Lock, M. A., Fiebig, D. M., McKinley, V., Miller, M. C., Vestal, J. R., Ventullo, R. and Volk, G. (1993). Biological responses of a tundra river to fertilization. *Ecology*, **74**, 653–672.

Peterson, C. H. (1975). Stability of species and of community for the benthos of two lagoons. *Ecology*, **56**, 958–965.

Phillips, J. (1934). Succession, development, the climax, and the complex organism: an analysis of concepts. Pt 1. *Journal of Ecology*, **22**, 554–571.

Phillips, J. (1935). Succession, development, the climax, and the complex organism: an analysis of concepts. Pt 3. The complex organism. *Journal of Ecology*, **23**, 488–508.

Pianka, E. (1980). Guild structure in desert lizards. *Oikos*, **35**, 194–201.

Pickett, S. T. A., Kolasa, J. and Jones, C. G. (1994). *Ecological Understanding*. San Diego: Academic Press.

Pimm, S. L. (1984). The complexity and stability of ecosystems. *Nature*, **307**, 321–326.

Pinch, T. (1990). The sociology of the scientific community. In *Companion to the History of Modern Science*, ed. R. C. Olby, G. N. Cantor, J. R. R. Christie and M. J. S. Hodge, pp. 87–99. New York: Routledge.

Pisek, A. and Winkler, E. (1958). Assimilationsvermögen und Respiration der Fichte (*Picea excelsa* Link) in verscheidener Höenlage und der Zirbe (*Pinus cembra* L.) an der alpinen Waldgrenz. *Planta*, **51**, 518–543.

Platt, J. R. (1964). Strong inference. *Science*, **146**, 347–353.

Polanyi, M. (1958). *Personal Knowledge: Towards a Post-Critical Philosophy*. New York: Harper and Row.

Poole, C. (1987). Beyond the confidence interval. *American Journal of Public Health*, **77**, 195–199.

Popper, K. R. (1968). *The Logic of Scientific Discovery*. Second edition. New York: Harper and Row.

Popper, K. R. (1972). *Objective Knowledge. An Evolutionary Approach*. Oxford: Oxford University Press.

Popper, K. R. (1982). *The Open Universe: An Argument for Indeterminism*. London: Routledge.

Power, M. E. (1992). Top-down and bottom-up forces in food webs: do plants have primacy? *Ecology*, **73**, 733–746.

Primack, R. B. and O'Leary, V. (1989). Research productivity of men and women ecologists: a longitudinal study of former graduate students. *Bulletin of the Ecological Society of America*, **70**, 7–12.

Putnam, H. (1981). *Reason, Truth and History*. Cambridge: Cambridge

University Press.

Quenette, P. Y. and Gerard, J. F. (1993). Why biologists do not think like Newtonian physicists. *Oikos*, **68**, 361–363.

Quine, W. V. (1951). Two dogmas of empiricism. *Philosophical Review*, **60**, 20–43.

Quine, W. V. (1972). *Methods of Logic*. New York: Holt, Rinehart and Winston.

Quinn, G. P. and Keough, M. J. (1993). Potential effect of enclosure size on field experiments with herbivorous intertidal gastropods. *Marine Ecology Progress Series*, **98**, 199–201.

Quinn, J. R. and Dunham, A. E. (1983). On hypothesis testing in ecology and evolution. *American Naturalist*, **122**, 602–617.

Quinn, T. P. (1993). A review of homing and straying of wild and hatchery-produced salmon. *Fisheries Research*, **18**, 29–44.

Reader, R. J. (1992). Herbivory as a confounding factor in an experiment measuring competition among plants. *Ecology*, **73**, 373–376.

Redfield, G. W. and Crowder, L. B. (1989). Suggestions for grantsmanship in ecology. *Bulletin of the Ecological Society of American*, **70**, 185–189.

Reynolds, J. H. and Ford, E. D. (1999). Multi-criteria assessment of ecological process models. *Ecology*, **80**, 538–553.

Richards, P. W. (1952). *The Tropical Rain Forest*. Cambridge: Cambridge University Press.

Rip, A. and van der Meulen, B. J. R. (1996). The post-modern research system. *Science and Public Policy*, **23**, 343–352.

Ritchie, G. A. and Hinckley, T. M. (1975). The pressure chamber as an instrument for ecological research. *Advances in Ecological Research*, **9**, 165–254.

Romesburg, H. C. (1979). Simulating scientific inquiry with the card game Eleusis. *Science Education*, **63**, 599–608.

Root-Bernstein, R. S. (1984). Creative process as a unifying theme of

human cultures. *Daedalus*, **113**, 197–219.

Root-Bernstein, R. S. (1989a). *Discovering*. Cambridge, MA: Harvard University Press.

Root-Bernstein, R. S. (1989b). Who discovers and invents? *Research Technology Management*, **32**, 43–50.

Rosenberg, A. (1994). *Instrumental Biology or the Disunity of Science*. Chicago: University of Chicago Press.

Roy, R. (1979). Proposals, peer review and research results. *Science*, **204**, 1155–1157.

Russell, M. (1992). Lessons from NAPAP. *Ecological Applications*, **2**, 107–110.

Rykiel, E. J. Jr (1996). Testing ecological models: the meaning of validation. *Ecological Modelling*, **90**, 229–244.

Salmon, W. S. (1973). Confirmation. *Scientific American*, **228** (5), 75–83.

Salmon, W. S. (1990). *Four Decades of Scientific Explanation*. Minneapolis: University of Minnesota Press.

Sanford, L. P. (1997). Turbulent mixing in experimental ecosystem studies. *Marine Ecology Progress Series*, **161**, (0), 265–293.

Sarnelle, O. (1997). *Daphnia* effects on microzooplankton: comparisons of enclosures and whole-lake responses. *Ecology*, **78**, 913–928.

Sattler, R. (1986). *Bio-philosophy*. Berlin: Springer-Verlag.

Schaffer, W. M. and Kot, M. (1986). Chaos in ecological systems: the coals that Newcastle forgot. *Trends in Ecology and Evolution*, **1**, 58–63.

Scheer, B. T. (1939). Homing instinct in salmon. *Quarterly Review of Biology*, **14**, 408–420.

Scheiner, S. M. (1993). MANOVA: multiple response variables and multispecies interactions. In *Design and Analysis of Ecological Experiments*, ed. S.M. Scheiner and J. Gurevitch, pp. 94–112. New York: Chapman & Hall.

Scheiner, S. M. and Gurevitch, J. (eds.) (1993). *Design and Analysis of Ecological Experiments*. New York: Chapman & Hall.

Schenck, N. C. (1985). VA mycorrhizal fungi 1950 to the present: the era of enlightenment. In *Proceedings of the Sixth North American Conference on Mycorrhizae*, ed. R. Molina, pp. 56–60. Corvallis, OR: Forest Research Laboratory, Oregon State University.

Schiff, A. L. (1962). *Fire and Water: Scientific Heresy in the Forest Service*. Cambridge, MA: Harvard University Press.

Schmidt, F. L. (1996). Statistical significance testing and cummulative knowledge in psychology: implications for training researchers. *Psychological Methods*, **1**, 115–129.

Schneider, A. and Schmitz, K. (1989). Seasonal course of translocation and distribution of 14C-labelled photoassimilate in young trees of *Larix decidua* Mill. *Trees*, **4**, 185–191.

Schoener, T. W. (1989). The ecological niche. In *Ecological Concepts*, ed. J. M. Cherrett, pp. 79–113. British Ecological Society, Symposium, **29**. Oxford: Blackwell.

Schuster, J. A. and Yeo, R. R. (1986). Introduction. In *The Politics and Rhetoric of Scientific Method*, ed. J. A. Schuster and R. R. Yeo, pp. ix–xxxvii. Dordrecht: D. Reidel.

Schutz, V. and Ward Whicker, F. (1982). *Radioecological Techniques*. New York: Plenum Press.

Seglen, P. O. (1992). The skewness of science. *Journal of the American Society for Information Science*, **43**, 628–638.

Semlitsch, R. D. (1993). Effects of different predators on the survival and development of tadpoles from the hybridogenetic *Rana esculenta* complex. *Oikos*, **67**, 40–46.

Shapere, D. (1971). The paradigm concept. *Science*, **172**, 706–709.

Shapere, D. (1974). *Galileo*. Chicago: University of Chicago Press.

Shapere, D. (1977). Scientific theories and their domains. In *The Structure of Scientific Theories*, ed. F. Suppe, pp. 518–565. Urbane, IL: University of Illinois Press.

Shapere, D. (1980). The character of

scientific change. In *Scientific Discovery, Logic, and Rationality*, ed. T. Nickles, pp. 61–116. Dordrecht: D. Reidel.

Shapere, D. (1982). Reason, reference, and the quest for knowledge. *Philosophy of Science*, **49**, 1–23.

Shapere, D. (1984). Modern science and the philosophical tradition. In *Reason and the Search for Knowledge*, ed. R. S. Cohen and M. W. Wartofsky, pp. 408–417. Boston Studies in the Philosophy of Science, **78**. Dordrecht: D. Reidel.

Shapin, S. (1995). Here and everywhere: sociology of scientific knowledge. *Annual Review of Sociology*, **21**, 289–321.

Shipley, B. and Keddy, P. A. (1987). The individualistic and community-unit concepts as falsifiable hypotheses. *Vegetatio*, **69**, 47–55.

Shrader-Frechette, K. S. (1986). Organismic biology and ecosystems ecology: description or explanation? In *Current Issues in Teleology*, ed. N. Rescher, pp. 77–92. Lanham, MD: University Press of America.

Shrader-Frechette, K. S. (1994). *Ethics of Scientific Research*. Lanham, MD: Rowman and Littlefield.

Shrader-Frechette, K. S. and McCoy, E. D. (1990). Theory reduction and explanation in ecology. *Oikos*, **58**, 109–114.

Shrader-Frechette, K. S. and McCoy, E. D. (1993). *Method in Ecology*. Cambridge: Cambridge University Press.

Sinderman, C. J. (1982). *Winning the Games Scientists Play*. New York: Plenum Press.

Singer, C. (1959). *A Short History of Scientific Ideas to 1900*. Oxford: Oxford University Press.

Skalski, J. R. and Robson, D. S. (1992). *Techniques for Wildlife Investigations*. San Diego: Academic Press.

Sloep, P. B. (1993). Methodology revitalized? *British Journal for the Philosophy of Science*, **44**, 231–249.

Slowiaczeek, L. M., Klayman, J., Sherman, S. J. and Skov, R. B. (1992). Information selection and

use in hypothesis testing: what is a good question, and what is a good answer? *Memory and Cognition*, **20**, 392–405.

Smith, R.V. (1990). *Graduate Research: A Guide for Students in the Sciences*. New York: Plenum Press.

Smits, W. T. M. (1992). Mycorrhizal studies in dipterocarp forests in Indonesia. In *Mycorrhizas in Ecosystems*, ed. D. Read, D. H. Lewis, A. H. Fitter and I. J. Alexander, pp. 283–292. Wallingford, Oxon: C.A.B International.

Sokal, R. R. and Rohlf, F. J. (1981). *Biometry: The Principles and Practice of Statistics in Biological Research*. Second edition. New York: W.H. Freeman and Company.

Sorrensen-Cothern, K. A., Ford, E. D. and Sprugel, D. G. (1993). A model of competition incorporating plasticity through modular foliage and crown development. *Ecological Monographs*, **63**, 277–304.

Southern, H. N. (1970). Ecology at the cross-roads. *Journal of Ecology*, **58**, 1–11.

Star, S. L. (1983). Simplification in scientific work: an example from neuroscience research. *Social Studies of Science*, **13**, 205–228.

Statistical Sciences, Inc. (1993). *S-PLUS for Windows User's Manual*. Seattle, WA: Statistical Sciences, Inc.

Stearns, S. C. (1986). Some modest advice for graduate students. *Bulletin of the British Ecological Society*, **17**(2), 82–29.

Stearns, S. C. (1987). Some modest advice for graduate students. *Bulletin of the Ecological Society of America*, **68**(2), 145–150.

Steel, E. A. (1993). Woody debris piles: habitat for birds and small mammals in the riparian zone. Unpublished MS thesis, University of Washington.

Steel, E. A., Naiman, R. J. and West, S. D. (1999). Use of woody debris piles by birds and small mammals in a riparian corridor. *Northwest Science*, in press.

Stern, J. E. and Elliott, D. (1997). *The Ethics of Scientific Research*. Hanover, NH: University Press of New

England.

Stewart W. W. and N. Feder. (1987). The integrity of the scientific literature. *Nature*, **325**, 207–214.

Stock, M. W. (1985). *A Practical Guide to Graduate Research*. New York: McGraw-Hill.

Stock, M. W. (1989). Graduate instruction in the research process. In *Discovering New Knowledge About Trees and Forests*, ed. R. A. Leary, pp. 1–4. General Technical Report, USDA. Forest Service, NC–135.

Storer, N. W. (1966). *The Social System of Science*. New York: Holt, Rinehart and Winston.

Strand, M. and Öquist, G. (1988). Effects of frost hardening, dehardening and freezing stress on *in vivo* chlorophyll fluorescence of seedlings of Scots pine (*Pinus sylvestris* L.). *Plant, Cell and Environment*, **11**, 231–238.

Strong, D. R. (1980). Null hypotheses in ecology. *Synthese*, **43**, 271–285.

Strong, D. R. (1992). Are trophic cascades all wet? Differentiation and donor control in speciose ecosystems. *Ecology*, **73**, 747–754.

Strunk, W. and White, E. B. (1979). *The Elements of Style*. New York: Macmillan.

Suppe, F. (1974). Development of the received view. In *Structure of Scientific Theories*, ed. F. Suppe, pp. 16–61. Urbane, IL: University of Illinois Press.

Sykes, R. M. (1973). The trophic-dynamic aspects of ecosystem models. In *Proceedings of the 16th Conference of Great Lakes Research*, pp. 977–988. Ann Arbor, MI: International Association for Great Lakes Reasearch, University of Michigan.

Tagliacozzo, R. (1967). Citations and citation indexes: a review. *Methods of Information in Medicine*, **6**, 136–142.

Tansley, A. G. (ed.) (1911). *Types of British Vegetation*. London: Cambridge University Press.

Tansley, A. G. (1935). The use and abuse of vegetational concepts and terms. *Ecology*, **16**, 284–307.

Tansley, A. G. (1939). *The British Isles and their Vegetation*. London: Cambridge University Press.

Tart, C. (1972). States of consciousness and state-specific sciences. *Science*, **176**, 1203–1210.

Taubes, G. (1986). The game of the name is fame. But is it science? *Discover*, **7**(12), 28–52.

Taylor, L. R. (1961). Aggregation, variance and the mean. *Nature*, **189**, 732–735.

Teal, J. M. (1957). Community metabolism in a temperate cold spring. *Ecological Monographs*, **27**, 283–302.

Thagard, P. (1992). *Conceptual Revolutions*. Princeton, NJ: Princeton University Press.

Thorpe, W. H. (1974). Reductionism in biology. In *Studies in the Philosophy of Biology, Reduction and Related Problems*, ed. F. J. Ayala and T. Dobzhansky, pp. 109–138. Berkley: University of California Press.

Ting, I. P. (1985). Crassulacean acid metabolism. *Annual Reviews in Plant Physiology*, **36**, 595–622.

Tornay, S. C. (1938). *Ockham*. La Salle, IL: Open Court.

Trappe, J. M. and Berch, S. M. (1985). The prehistory of mycorrhizae: A.B. Frank's predecessors. In *Proceedings of the Sixth North American Conference on Mycorrhizae*, ed. R. Molina, pp. 2–17. Corvallis, OR: Forest Research Laboratory, Oregon State University.

Trudgill, S. T. (1988). *Soil and Vegetation Systems*. Second edition. Oxford: Oxford University Press.

Tufte, E. R. (1983). *The Visual Display of Quantitative Information*. Cheshire, CT: Graphics Press.

Tukey, J. W. (1977). *Exploratory Data Analysis*. Reading, MA: Addison-Wesley.

Tukey, J. W. (1980). We need both exploratory and confirmatory. *American Statistician*, **34**, 23–25.

Ueno, O. and Takedo, T. (1992). Photosynthetic pathways, ecological characteristics, and the geographical distribution of Cyperacea in Japan. *Oecologia*, **89**, 195–203.

Ugolini, F. C. and Edmonds, R. L. (1983). Soil biology. In *Pedogenesis and Soil Taxonomy*. I. *Concepts and Interactions*, ed. L. P. Wilding, N. E. Smeck and G. F. Hall, pp. 193–231. Amsterdam: Elsevier.

Underwood, A. J. (1997). *Experiments in Ecology*. Cambridge: Cambridge University Press.

United States Office of Scientific Research and Development (1945). *Science: The Endless Frontier. A Report to the President by Vannevar Bush*. Washington, DC: United States Government Printing Office.

van Fraassen, B. C. (1980). *The Scientific Image*. Oxford: Clarendon Press.

Van Vallen, L. (1976). Dishonesty and grants. *Nature*, **261**, 2.

Van Vallen, L. and Pitelka, F. A. (1974). Commentary – intellectual censorship in ecology. *Ecology*, **55**, 925–926.

Vitousek, P. M. (1984). A general theory of forest nutrient dynamics. *Rapporter, Institutionen för Ekologi och Miljovard, Sveriges Lantbruksuniversitet*, **13**, 121–135.

Volterra, V. (1926). Variazionie fluttuazioni del numero d'individui in specie animali conviventi. *Memorie della Academia nazionale dei Lincei*, **2**, 31–113. (Translation, "Variations and fluctuations of a number of individuals in animal species living together".) In *Animal Ecology*, ed. R.N. Chapman, pp. 409–408, 1931. New York: McGraw Hill.

Walsh, W. H. (1975). *Kant's Criticism of Meta-physics*. Chicago: University of Chicago Press.

Ward, D. (1993). Foraging theory, like all other fields of science, needs multiple working hypotheses. *Oikos*, **67**, 376–378.

Watson, D. J. (1968). *The Double Helix*. New York: Atheneum.

Watt, K. E. F. (1966). The nature of systems analysis. In *Systems Analysis in Ecology*, ed. K. E. F. Watt, pp. 1–14. New York: Academic Press.

Watt, K. E. F. (1971). Dynamics of populations: a synthesis. In *Dynamics of Populations: A Synthesis*, ed. P. J.

den Boer and G. R. Gradwell, pp. 568–580. Wageningen: Centre for Agricultural Publishing and Documentation.

Weiner, J. (1995). On the practice of ecology. *Journal of Ecology*, **83**, 153–158.

Weinstein, M. S. (1977). Hares, lynx and trappers. *American Naturalist*, **111**, 806–808.

Weitz, M. (1988). *Theories of Concepts*. London: Routledge.

Weller, D. E. (1987). A reevaluation of the − 3/2 power rule of plant self-thinning. *Ecological Monographs*, **57**, 23–43.

Weller, D. E. (1991). The self-thinning rule: dead or unsupported? – A reply to Lonsdale. *Ecology*, **72**, 747–750.

Westoby, M. (1981). The place of the self-thinning rule in population dynamics. *American Naturalist*, **118**, 581–587.

White, J. (1981). The allometric interpretation of the self-thinning rule. *Journal of Theoretical Biology*, **89**, 475–500.

White, T. R. C. (1978). The importance of relative shortage of food in animal ecology. *Oecologia* (Berlin), **33**, 71–86.

Whitmore, T. C. (1984). *Tropical Rain Forests of the Far East*. Second edition. Oxford: Clarendon Press.

Widnall, S. E. (1988). American Association for the Advancement of Science presidential lecture: voices from the pipeline. *Association Affairs*, 1740–1745.

Wiegert, R. G. and Owen, D. F. (1971). Trophic structure, available resources and population density in terrestrial vs. aquatic ecosystems. *Journal of Theoretical Biology*, **30**, 69–81.

Wiegolaski, F. E. (1966). The influence of air temperature on plant growth and development during the period of maximal stem elongation. *Oikos*, **17**, 121–141.

Williams, P. J. Le B. and Egge, J. E. (1998). The management and behaviour the mesocosms. *Estuarine Coastal and Shelf Science*, **46**, 3–14.

Worrall, J. (1988). The value of a fixed methodology. *British Journal for the Philosophy of Science*, **39**, 263–275.

Worrall, J. (1989). Fix it and be damned: a reply to Laudan. *British Journal for the Philosophy of Science*, **40**, 376–388.

Wright, H. E. J. Jr (1967). A square rod piston sampler for lake sediments. *Journal of Sedimentary Petrology*, **37**, 975–976.

Yamamura, K. (1990). Sampling scale dependence of Taylor's power law. *Oikos*, **59**, 121–125.

Yoda, K., Kira, T., Ogawa, H. and Hozumi, K. (1963). Self thinning in overcrowded pure stands under cultivated and natural conditions. *Journal of the Institute Polytechnique of Osaka City University Series D*, **14**, 107–129.

Zar, J. H. (1996). *Biostatistical Analysis*. Third edition. Upper Saddle River, NJ: Prentice-Hall.

Zeide, B. (1987). Analysis of the 3/2 power law of self-thinning. *Forest Science*, **33**, 517–537.

Zuckerman, H. (1991). The careers of men and women scientists: a review of current research. In *The Outer Circle*, ed. H. Zuckerman, J. R. Cole and J. T. Bruer, pp. 27–56. New York: W. W. Norton and Company.

Glossary

Analysis: see *Scientific analysis.*

Axiom: An *axiom* is a proposition assumed to be true on the basis of previous research, observations, or information, and is used in defining the working part of the theory that is the foundation for the research. Typically, an axiom specifies that something does or does not occur, or that one thing does or does not influence another; or it defines a mathematical relationship.

> Page: p. 48. See also: *Over-arching axiom; Concept from research;* examples of construction and use in Chapters 4, 5 and 6.

Bias: Faults in study design, or in the misapplication of a statistical test, lead to incorrect estimation of a quantity or a misrepresentation of the relationship between A and B.

> Page: p. 205. See also: this definition applies to statistical tests; biased and unbiased are also qualities of measurements, see 6.2.

Bold postulate: A *bold postulate* has a high information content, and its confirmation, or rejection, advances the theory substantially.

> Page: p. 155. See also: *Postulate;* Popper's falsificationist philosophy, see 7.4.

Causal reasons in ecology: When we say *b* causes *c* we mean there is some direct transfer of energy or mass, or a direct stimulus that produces a response, or a genetic property that influences or determines something. Causal reasons are produced in response to questions that ask why something comes about.

> Page: p. 198. See also: *Causal scientific explanation.*

Causal scientific explanation: *Causal scientific explanations* have the structure and purpose of *scientific explanations* and:

(1) Must be based on causal and/or associated organizational reasons. (Defined in Chapter 7.)

(2) They are consistent. Under the same conditions the causal process will produce the same effect.

(3) They are general. To explain a kind of event causally is to provide some general explanatory information about events of that kind.

(4) When experiments are possible a designed manipulation or intervention of the causal process produces a predictable response.

> Page: p. 275. See also: *Scientific explanation; Causation;* discussion in 15.2.1.

Causation: If *c* and *d* are two actual events or occurrences such that *d* would not have occurred without *c*, then *c* is the cause of *d*. If *c, d, e* are a finite sequence of particular events such that *d* depends causally on *c*, and *e* on *d*, then this sequence is a causal chain. One event is a *cause* of another if, and only if, there exists a causal chain leading from the first to the second. (Adapted from Lewis 1986.)

> Page: p. 196. See also: *Causal explanation; Causal reasons in ecology;* why a correlation may not indicate causation, see 7.8.

Closed system: In a *closed system* all the components that influence the system's function can be defined.

> Page: p. 137. See also: *Open system, System.*

Codified knowledge: The scientific literature contains *codified knowledge;* that is, knowledge that has a recorded structure and is used to define axioms and concepts that determine the logic of the theory.

> Page: p. 52. See also: *Uncodified knowledge;* example of defining codified and uncodified knowledge, in Chapter 4.

Complete system reduction: A *complete system reduction* asserts that the functioning of a system can be expressed as a simple relation or function between measurements of a few inputs to and outputs from the system.

> Page: p. 247. See also: *Partitioning reduction; Reductionism.*

Concept: A *concept* is any object or idea to which we can give a name and define, and so enable things to be understood in a particular way.

Page: p. 46. See also: for concepts classified according to knowledge status, see *Concept by imagination, by intuition, by measurement, from research;* for concepts classified according to their role in ecological theory, see *Functional concepts, Holding concept, Integrative concepts, Natural concepts.* Difficulties in naming and defining concepts in ecology are discussed in 10.2, 15.3.2, 16.2.3.

Concept by imagination: Concepts by imagination are ideas used in developing postulates. They may arise through logical reasoning from within the theory under examination, through comparative reasoning with other theories, or by considering something that has not been measured before.

Page: p. 61. See also: *Postulate,* examples of construction and use of concept by imagination in Chapter 4.

Concept by intuition: Concepts by intuition are non-specialist ideas used in a theory. These concepts are taken from a non-scientific context, usually have no rigorous definition, and may be important at the start of an investigation. They are often made more precise as research programs develop.

Page: p. 64.

Concept by measurement: Concepts by measurement are data used to examine the logical outcome of a postulate. It is most important not to assume that measurements represent exactly the concepts for which they are designed. Measurements may be limited in effectiveness, accuracy, and precision and more than one concept by measurement may give information about the same concept from research or concept by imagination.

Page: p. 62. See also: *Data statement, Measurement unit, Principles of measurement for new concepts;* examples of development and use of concept by measurement in Chapters 4 and 6.

Concept from research: Concepts from research are the ideas we use in describing the more established parts of a theory, i.e., the axioms.

Page: p. 59. See also: *Axiom;* examples of construction and use of concept from research in Chapter 4.

Confidence in the test of a statistical hypothesis: The *confidence in the test of a statistical hypothesis* is $1 - \alpha$. By convention α, the specified significance level is usually set at 0.05 for ecological investigations.

Page: p. 214. See also: *Power of a test, Hypothesis.*

Confounding: Confounding is when alternative explanations can be given for a study result. While A may be interpreted as influencing B there is an alternative explanation that the design or measurements of the study did not eliminate.

Page: p. 204.

Conjunction: The conjunction of propositions is expressed as "and"; a conjunction is true if, and only if, all of its constituent propositions are true.

Page: p. 172. See also: *Disjunction;* rules of propositional logic, see 7.2.

Constructive empiricism: Science aims to give us theories that are empirically adequate; acceptance of a theory involves a belief only that it is empirically adequate (van Fraassen 1980). Theories should contain no more than is necessary to explain observations – and no more should be claimed for theories than that that is what they do.

Page: p. 473. See also: *Empiricism, Pragmatic realism, Scientific realism.*

Contradiction: The conjunction or implication of any proposition with its own negation is false.

Page: p. 178. See also: *Negation;* rules of propositional logic, 7.2.

Contrastive question: A *contrastive question* asks "Why this and not that?" It seeks to compare what needs to be understood, the fact, with something, the foil, that is similar in most respects but different in an important component. Construction of good contrastive questions requires an understanding of background knowledge (axioms) and a postulate of the difference between fact and foil.

Page: p. 195. See also: *Scientific explanation, Why-type question.*

Control procedure: A *control procedure* allows effects other than those being studied to be removed from influencing the system.

Page: p. 163. See also: *Controlled analytical experiment; Experimental analysis of ecological systems.*

Controlled analytical experiment: In a *controlled analytical experiment,* the performance of the treated system is compared with that of a system not treated, i.e. the treatment control.

Page: p. 160. See also: *Control procedure; Experimental analysis of ecological systems; Response-level experiment; Treatment; Treatment control.*

Criticism: Criticism asserts:

(1)   For a correct beginning in research, both *axioms* and the

data on which they depend must be known.

(2)   The decision about what to research and how to conduct the research is acknowledged to be subjective.

(3)   The results of research must be exposed to criticism that is broad (not limited to similar thinkers), unrestricted (not constrained by considerations that have nothing to do with the research), and fair (not motivated by desire for personal attack).

Page: p. 265. See also: *Relativism; Scientific criticism;* for a discussion of the central role of criticism in science, see 15.3.1; difficulties in applying criticism, 13.2.2, 13.3.2, 13.4, 14.2.

Crucial experiment: A *crucial experiment* is established between two theories, T_1 and T_2, if T_1 predicts one thing will occur and T_2 predicts it will not, and this difference can be attributed to a specified difference between T_1 and T_2.

Page: p. 339. See also: *Duhem–Quine thesis; Experimental analysis of ecological systems; Problemshift.*

Data statement: A *data statement*

(1)   defines the scientific procedure to be used in investigating a postulate,

(2)   specifies the measurements to be made for each concept of a postulate, and

(3)   specifies the requirements of the data for any statistical test to be applied.

Page: p. 53. See also: *Concept by measurement; Hypothesis; Postulate;* for description of constructing a data statement, see Chapters 2, 4, and 8.

Deduction: Deduction is the application of the rules of propositional logic in an argument. Propositional logic assigns statements as true or false and the conclusions made do not claim more information than is already contained in the statement of the theory.

Page: p. 171. See also: *Induction;* rules of propositional logic, see 7.2.

Degrees of freedom: The larger a sample is, the more general confidence we can have that statistics calculated from it represent values for the whole universe of observations. But if calculating one statistic uses another statistic that has already been calculated, then there is repeated use of the same information. To account for this "double use," we impose a restriction on the second statistic. If n is the sample size, and k the number of independent restrictions, then the *degrees of freedom* are defined as $n - k$. The sample mean has no restrictions placed on it, because you simply add all the values together and divide by the number of values (i.e., no other statistic is used in calculat-

ing the mean). But the sample variance has $n - 1$ degrees of freedom, because the mean is used in calculating it.

Page: p. 225. See also: *Confidence in the test of a statistical hypothesis.*

Deterministic mathematical model: A model is *deterministic* when it is assumed that all possible behaviors are determined by the set of equations comprising the model.

Page: p. 355. See also: *Model explanations; Systems simulation model; Statistical model* of a stochastic process.

Difference condition for contrasts: To explain why P occurred rather than Q we must cite a causal difference between P occurring and Q not occurring consisting of a cause of P and the absence of a corresponding event in the case of Q not occurring. (Adapted from Lipton 1991.)

Page: p. 198. See also: *Contrastive question.*

Disjunction: The *disjunction* of propositions is expressed as "or"; a disjunction is true if any of its propositions are true.

Page: p. 173. See also: *Conjunction;* rules of propositional logic, see 7.2.

Domain of a functional concept or proposition: The *domain of a functional concept or proposition* is defined as:

(1)   the set of conditions that cause variation in a functional concept or proposition, or

(2)   the conditions under which a functional process may influence an ecological system.

Page: p. 288. See also: *Domain of an integrative concept; Functional concept; Proposition.*

Domain of an integrative concept: The *domain of an integrative concept* has two components:

(1)   Specification of the extent of what the concept seeks to explain.

(2)   Specification of the set of functional concepts that an integrative concept uses.

This is the requirement to make a formal specification of the theory that defines the concept.

Page: p. 289. See also: *Domain of a functional concept or proposition; Integrative concepts.*

Duhem–Quine thesis: The complex structure of theories, including their auxiliary information as well as specified axioms and postulates, leads to difficulties in testing theories because rejection of a single postulate may be declined in favor of appealing to alterations in linked propositions, or in previously unspecified auxiliary information.

Page: p. 340. See also: *Crucial experiment.*

Dynamic systems models: A *dynamic systems model* has a defined mathematical form, most usually differential equations, but not including logical switches or stochastic forcing.

Page: p. 354. See also: *System.*

Elementary probability law: The *elementary probability law* can be expressed by an integral of the form

$$P = \int_a^b p(x)\mathrm{d}x$$

where P is the probability that an observation occurs between a and b. $p(x)$ is the distribution function of the observations. This function may take various forms, but the one most commonly considered is the normal distribution function, for which $p(x)$ has the form

$$p(x) = \frac{1}{\sqrt{2\pi\sigma^2}} e^{-\frac{(x-\mu)^2}{2\sigma^2}}$$

which when plotted has a bell-shaped curve that approaches the axis asymptotically in both directions. (*Asymptotically* means that the curve gets closer and closer to the axis but never actually reaches it so that values further from the mean are less and less likely but never impossible.) The values μ, the "true" mean, and σ^2, the "true" variance are defined in terms of $p(x)$:

$$\mu = \int_{-\infty}^{\infty} xp(x)\mathrm{d}x$$

$$\sigma^2 = \int_{-\infty}^{\infty} x^2p(x)\mathrm{d}x - \mu^2.$$

Page: p. 219. See also: *Statistical inference.*

Emergence: Emergence is the theory that the characteristic behavior of the whole could not, even in theory, be deduced from the most complete knowledge of the behavior of its components, taken separately or in other combinations, and of their proportions and arrangements in this whole.

Page: p. 243. See also: *Reductionism*; discussion in 10.5.

Empiricism: The *empiricist* considers that experience and measurements are fundamental to understanding and determine the axioms we adopt. Nothing can be known unless its existence can be inferred from experience or measurement.

Page: p. 251. See also: *Constructive empiricism*; difference from *Rationalism.*

Equivalence relations and symmetry: An element of a proposition is said to be equivalent to itself (the identity relationship): equivalence is represented by the symbol $\equiv$. A proposition is said to be symmetrical when its truth-value is unchanged by reversing the order of its elements, i.e., $\{p$ is related to $q\} \equiv \{q$ is related to $p\}$, such as in $p + q \equiv q + p$. Equivalent terms may be substituted in a proposition without changing its truth-value, i.e., if $r \equiv p$, then $\{r$ is related to $q\} \equiv \{p$ is related to $q\}$.

Page: p. 173. See also: *Proposition*; rules of propositional logic, see 7.2.

Exemplar: "The concrete problem-solutions that students encounter from the start of their scientific education whether in laboratories, on examinations, or at the ends of chapters in science texts" (Kuhn 1970). To this can be added the technical problems, and their solutions, that students examine in the scientific literature and laboratory during their research training.

Page: p. 316. See also: difference from *Positive heuristic* and *Hard core knowledge*, see 11.3.2.

Experiment: An *experiment* involves deliberate action(s) by an investigator, treatment(s), imposed upon an ecological system in such a way that the system response to the treatment can be observed and/or measured by contrasting it with a condition of no treatment, the control(s). Controls must be designed so that changes not due to the intended effects of the treatment may be accounted for. Treatment(s) and control(s) may be observed at the same time on different samples (a synchronic contrast) or the treatment may be applied at a particular time to create a before (control) and an after (treatment) (a diachronic contrast).

Page: p. 136. See also: *Controlled analytical experiment*; *Experimental analysis of ecological systems*; *Response-level experiment*; *Crucial experiment.*

Experimental analysis of ecological systems: In the *experimental analysis of ecological systems* treatments are designed to discover the functional relationships between organisms and/or between organisms and their environment. Because the nature of the response is under investigation, attention must be paid to precisely how the treatment causes any effects. Generally measurement of a number of response variables may be required to detect and interpret this.

Page: p. 138. See also: *Controlled analytical experiment*; *Response-level experiment.*

Experimental unit: The *experimental unit* is the system to which the treatment is applied, and for which there is a corresponding control.

Page: p. 159. See also: *Experiment*, and avoiding confusion with *Measurement unit.*

Explanation: See: *Causal scientific explanation; Scientific explanation.*

Explanatory coherence: If a theory gives an effective scientific explanation of a why-question it is *coherent*. This requires that:

(1) Individual propositions are acceptable, including that they have been tested with data statements.

(2) Concept definitions are consistent throughout the theory network.

(3) Part and kind relationships are consistent throughout the theory network.

(4) There are not *ad hoc* propositions that include or exclude special circumstances.

(5) Generally theories with fewer rather than more propositions are favored as more coherent explanations.

(6) The explanation applies to broad questions and circumstances.

Page: p. 275. See also: *Scientific explanation; Scientific inference.*

Explication: The transformation of an inexact prescientific concept, the explicandum (often a common language term), into a new exact concept, the explicatum (a precise scientific meaning). Although the explicandum can not be given in exact terms, it should be made as clear as possible by informal explanations and examples. (Based on Carnap 1950.)

Page: p. 476. See also: discussions of difficulties in naming and defining ecological concepts in Chapters 10 and 16.

Exploratory analysis: Exploratory analysis is used to refine a postulate so that a testable hypothesis can be constructed for it or it is reconstituted into postulates for which hypothesis tests can be developed. There are two processes (Fig. 8.3):

(1) Developing a scientific procedure and set of measurements. This involves research to define parts (i) and (ii) of the data statement:

  (i) Define the scientific procedure to be used in investigating a postulate,

  (ii) Specify the measurements to be made for each concept of a postulate.

The result is that you define the alternatives to be examined in a hypothesis, and how it might illuminate the importance of a postulate in a complete theory. You establish the formal conditions of the test and this requires construction of a null hypothesis.

(2) Satisfying the logic required for statistical inference. This involves research to specify part (iii) of the data statement:

  (iii) specify the requirements of the data for any statistical test to be applied. The result of this process

is that you ensure that the mathematical assumptions on which the test is based are met. For example, for a *t*-test this requires:

- defining the universe of observations,
- ensuring independence between samples,
- ensuring the measured variable has, or can be transformed to a normal distribution,
- estimating the variance of the measure and calculate the sample size necessary to
- detect a specified difference in means.

Page: p. 208. See also: *Statical inference; Data statement; Hypothesis.*

Falsification: A postulate is *falsified* when its predictions are shown by empirical evidence not to be true.

Page: p. 155. See also: discussion that *falsification* can occur when a test expected to turn out true is found to be false *and* one that is expected to turn out false turns out to be true, see 7.6.

Functional concepts: Functional concepts define properties of natural concepts or express relationships between two or more natural concepts. Direct measurements of natural concepts can be made to define functional concepts.

Page: p. 281. See also: *Integrative concepts; Natural concepts.*

Hard core knowledge: Hard core knowledge is a collection of theoretical knowledge to be used and explored over time but not generally challenged.

Page: p. 328. See also: discussion of the methodology of scientific research programs in Chapter 11.

Holding concept: A *holding concept* expresses the existence of a phenomenon or relationship without defining it sufficiently to be the subject of practical research. An important objective of conceptual and propositional analysis is to refine the description of holding concepts to the point where they can be studied. Typically, holding concepts assert:

(1) an undefined collective description of a set of natural concepts,

(2) the existence of a functional concept, or,

(3) a functional concept that defines or qualifies an integrative concept.

Page: p. 76. See also: *Concept.*

Hypothesis: The term *hypothesis* is reserved for use where a specific test is designed for a postulate. The test may be examination of a logical outcome, or it may be statistical. In a statistical argument, the construction of hypotheses takes particular forms (Chapter 8). A hypothesis test must be specified in a data statement, and it may be possible to construct more than one hypothesis test for a postulate.

Page: p. 54. See also: *Postulate, Sample statistic, Statistical inference.*

Hypothetico-deductive method: The method of creating scientific theory from which the results already obtained could have been deduced and which also entail new predictions that can be verified or refuted by observation or experiment. (After Flew 1984.)

Page: p. 183. See also: discussion of differences from contrastive questions, see 7.8.

Implication: A compound proposition which asserts that an antecedent condition entails (requires or necessarily implies) a consequent condition is called an *implication* (symbolized as $p \Rightarrow q$, or "if p, then q," where p is the antecedent and q the consequent).

Page: p. 174. See also: *Induction*; see rules of propositional logic, 7.2.

Independence: Two events, x_1 and x_2, are *independent* if, when p_1 is the probability that x_1 lies within an interval I_1 and p_2 is the probability that x_2 lies within an interval I_2, then the probability of the *joint* event (x_1 in I_1 and x_2 in I_2) is p_1 times p_2.

Page: p. 221. See also: *Statistical inference.*

Induction: An inference is *inductive* if it passes from particular statements such as accounts of the results of observations or experiments, to universal statements such as theories. (Adapted from Popper 1968.) In practice if a series of instances of a phenomenon are found then induction is the process of stating some aspect of generality from that. This may be based on empirical evidence, i.e., inference from repeated occurrences of an observation, or based on the hypothetico-deductive method, i.e., investigation of a conjecture based on an existing theory.

Page: p. 176. See also: *Deduction, Hypothetico-deductive method*; rules of propositional logic, see 7.2.

Inference: see *Scientific inference, Statistical inference.*

Instrumentalism: Theories are merely instruments, tools, or calculating devices for deriving some observation statements (predictions) from other observation statements (data). Consequently, there is no question as to the truth or falsity of these theories – they can be neither true nor false. *Instrumentalism* is thus opposed to most realist theories of science. Propositions are used in investigations, are not true or false but are characterized only as effective or ineffective and judged relative to whether or not their assertion (prediction) is warranted. Ideas

and practice work together as instruments; ideas relate experiences, making predictions possible, and in turn are tested by experience. (Adapted after Flew 1984.)

Page: p. 507. See also: contrast with *Scientific realism, Pragmatic realism.*

Integrative concepts: Integrative concepts are theoretical constructions about the organization or properties of ecological systems. They can not be measured directly, and both their definition and detailed description must be synthesized from studies of a number of systems.

Page: p. 283. See also: *Natural concepts, Functional concepts.*

Large-scale perturbation experiment: Large-scale perturbation experiments attempt analytical experimentation to large ecological systems, e.g., a whole river or forest. These are diachronic contrasts but the large scale limits, or may completely remove, the possibility of replication and or the construction of controls. This complicates assessment.

Page: p. 164. See also: *Experimental analysis of ecological systems.*

Law: A *law* is an empirical relationship between two or more concepts, established by measurement, and asserted to be universally true. A *law* can be used as a rule of inference.

Page: p. 50. See also: discussion of laws and the construction of theories to explain them and why this may not be an effective descrition of the progress of scientific understanding in ecology, see Chapter 16.

Law of the excluded middle: All propositions are true or false. No proposition is both true and false, and no proposition is neither true nor false.

Page: p. 172. See also: rules of propositional logic, see 7.2.

Measurement unit: The *measurement unit* defines the response to the treatment. The measurement unit and experimental unit are not always the same.

Page: p. 159. See also: *Experimental unit.*

Metaphysics: The process of analyzing the nature of reality and the fundamental structure of our thought about reality.

Page: p. 238.

Methodology: A system of techniques of investigation, methods for applying the techniques, and general principles for how the methods should be used in scientific inquiry.

Page: p. 4. See also: *Progressive synthesis.*

Model: See *Deterministic mathematical model, Systems simulation model, Statistical model of a stochastic process.*

Model calibration: A test of a model with known input and output information that is used to adjust or estimate factors for which data are not available (ASTM 1984).

Page: p. 383. See also: *Systems simulation model.*

Model explanations: Model explanations are based on the principle of analogy. A model describes important features in a simplified representation of a system and can be used to illustrate how interactions may take place to produce particular outcomes. By studying a model ecologists attempt to understand how actual systems behave under different conditions.

Page: p. 352. See also: discussion of how models can be used in constructing *Scientific explanations* in Chapter 12.

Model sensitivity: The degree to which the model result is affected by changes in a selected parameters.

Page: p. 383. See also: *Systems simulation model.*

Natural concepts: Natural concepts define and/or classify measurable or observable entities or events in the ecological world.

Page: p. 281. See also: *Functional concepts; Integrative concepts.*

Natural history: The sciences, as botany, mineralogy or zoology, dealing with the study of all objects in nature: used especially in reference to the beginnings of these sciences in former times. (*Random House Unabridged Dictionary*, second edition.)

Page: p. 508.

Negation: The negation or denial of a proposition always has a truth value opposite to that of the proposition: in cases of double negation, the original truth-value is maintained.

Page: p. 173. See also: *Falsification;* rules of propositional logic, see 7.2.

Objective knowledge: Objective knowledge is knowledge that has been researched and then scrutinized by, and debated among, scientists. Such knowledge is independent of a single person, and therefore not subjective. In contrast to some common uses, the term "objective knowledge" does not imply absolute or permanent knowledge but simply the most reliable current knowledge. The content of objective knowledge changes as scientific investigation continues, new information is discovered, and the network of knowledge both grows and changes.

Page: p. 43. See also: *Scientific inference; Scientific methodology; Scientific criticism.*

Observations: Observations are numbers from single measurements.

Page: p. 220.

Open question: At the level of a theory, an *open question* defines an objective that cannot yet be defined in terms of specific axioms and postulates.

Page: p. 118.

Open system: An *open system* exchanges matter and/or energy across its boundaries and/or is influenced by external stimuli. The components of an open system, and/or their functioning, may change in response to an external event or stimulus.

Page: p. 137. See also: *Closed system; System.*

Operationalism: The theory that entities, processes, and properties are definable in terms of the set of measurements, operations and experiments by which they are detected. The principle is to establish the meaning of scientific concepts in strict accordance with the practice of scientific research and experimentation. (After Flew 1984.)

Pages: p. 512. See also: contrast with *Upward inference, Integrative concepts.*

Organizational reasons in ecology: These begin with the observation of a structure or pattern, and questions are asked about how that may either determine a causal relationship or result from one. Research into an organization–cause relationship (frequently referred to as structure–function in ecology) tends to involve successive improvement in the depth or detail of the explanation. This improvement is usually symmetric, i.e., involving improvement in scientific understanding of both organization and cause.

Page: p. 198. See also: *Causal reasons in ecology; Scientific explanations.*

Over-arching axiom: An *over-arching axiom* is a fundamental proposition, used as an axiom, which states broad assumptions of the theory and cannot be challenged directly by single investigations.

Page: p. 49. See also: *Axiom.*

Over-arching postulate: An *over-arching postulate* is a general question, stated in propositional form, which can not be answered by a single investigation. It is a speculation motivating a program of research and the development of a theory.

Page: p. 50. See also: *Postulate.*

Parsimony: No more causes should be proposed than are necessary to explain a phenomenon. Occam's Razor asserts the principle of *parsimony* in making propositions (Tornay 1938): *frustra fit plura quod potest fieri per pauciora* [it is vain to do through more what can be done with less].

Page: p. 242. See also: *Statistical model of a stochastic process.*

Partitioning reduction: A *partitioning reduction* is made when a scientist considers that a system is complex and that explanation must be developed through studying partitions of the system. This can be motivated by conceptual difficulties, lack of resources, or development of techniques and skills that encourage a scientist to focus on some component.

Page: p. 250. See also: *Complete system reduction.*

Positive heuristic: A heuristic is an art of discovery in logic, so a *positive heuristic* is an indicator of what to look for in a program of scientific research and how to find it. It may include a way of doing particular types of experiment, or constructing models of a particular type requiring particular types of data.

Page: p. 329. See also: *Hard core knowledge.*

Postulate: A *postulate* is a conjecture written in the form of a proposition. It is untested, or considered sufficiently uncertain to be the subject of further direct investigation. Even though there may be inconsistency in the literature about a theory, you can identify that and construct a postulate(s) to investigate it.

Page: p. 49. See also: *Over-arching postulate; Concept by imagination.*

Power of a test: The *power of a test* is $1 - \beta$; the probability a hypothesis will be concluded as false if it is false.

Page: p. 214. See also: *Statistical inference.*

Pragmatic realism: Science aims to provide the best explanatory account of natural phenomena; and acceptance of a scientific theory involves the belief that it belongs to such an account.

Page: p. 474. See also: *Scientific realism; Empiricism; Constructive empiricism.*

Principles of measurement for new concepts:

(1) *There must be a postulate under investigation.* Measurements do not exist in isolation from the scientific research they serve. Extensive research can be devoted to developing and testing a new measurement – but the efficiency, accuracy and precision required are determined by the scientific objectives defined by the postulate.

(2) *Each new measurement must be specified in a data statement.* It is essential to establish the relationships among the data that are required to test a postulate and how

measurements are to be used to collect the data, and what that use can achieve. This may require exploratory analysis.

(3) *More than one measurement of a new concept should be investigated.* The philosophy of multiple measurement recognizes that a concept by measurement is not equivalent to a concept by imagination or from research. Instances where a single measurement is sufficiently well understood that multiple measurements need not be considered have usually been established by consistent use. In research using new concepts it can be important to use different measures.

(4) *The accuracy and precision of a measurement should be investigated explicitly.* A measurement should not be assumed to have both the required accuracy and precision. Although some measurements are very precise, some may be more exploratory and generally informative.

Page: p. 135. See also: *Measurement unit; Concept by measurement; Postulate; Data statement.*

Problemshift: There can be no such things as crucial experiments or investigations that overthrow whole research programs. What an experiment or investigation may do is to cause a *problemshift*. A new theory incorporates the content of an old one and adds something new. Use of problemshift rather than revolution and paradigm change is based upon both criticism and continuity.

Page: p. 320. See also: *Hard core knowledge; Positive heuristic.*

Progressive Synthesis: Progressive Synthesis has three principles:

I. Criticism. Standards must be applied to ensure just and effective criticism.

II. Definition. Precision is required in defining concepts, axioms and postulates, and data statements.

III. Assessment. Explicit standards must be used to examine the relation between theory and data.

Progressive Synthesis has five component methods:

(1) Analyze the question and seek to use contrastive techniques to focus research.

(2) Expect to use different techniques of investigation as theories develop and new types of question are asked.

(3) Refine both measurement and concept definitions.

(4) Specify the new synthesis resulting from the research.

(5) Define explanatory coherence of the synthesis to make scientific inference.

Page: p. 277. See also: *Methodology; Pragmatic realism.*

Progressive theory development: The development of a theory is *progressive* if its generality increases. This may manifest as:

(1) increases in the number of circumstances in which the theory holds,

(2) increased evidence that what is measured represents concepts from research; and

(3) evidence that the theoretical constructions used are sufficiently detailed that they do explain what is true in common and what may be different in particular examples.

Page: p. 127. See also: *Regressive theory development.*

Proposition: A *proposition* asserts a relationship between a set of concepts. A proposition can be classified as true or false using propositional logic, or frequently, in ecological research, as probably true or probably false using statistical inference. There are two types of proposition: those we consider to be sufficiently true to use as the basis of future research (axioms), and conjectures (postulates) which we investigate.

Page: p. 48 and 172. See also: *Axiom; Postulate.*

Pseudoreplication: Pseudoreplication occurs where more than one measurement is made within an experimental unit and yet the measurements are considered as separate (independent) indicators of the experimental treatment(s) or the control(s).

Page: p. 161. See also: *Replicate.*

Random sample: A sample is drawn at random to ensure that there is no bias influencing what is to be measured. A formal definition of *random sample* has the following necessary and sufficient conditions:

(1) The observations of the sample are from the same universe in the sense that they obey the same probability law.

(2) The observations are independent draws. (Recall definition of "independence".)

Page: p. 222. See also: *Statistical inference.*

Rationalism: There are four characteristics of *rationalism*:

(1) The belief that it is possible to obtain by reason alone a knowledge of the nature of what exists,

(2) the view that knowledge forms a single system which

(3) is deductive in character, and,

(4) the belief that everything is explicable, that is, that everything can in principle be brought under a single system. (After Flew 1984.)

Page: p. 254. See also: the alternative philosophy of *Empiricism.*

Reductionism: Any doctrine that claims to reduce the apparently more sophisticated and complex to the less so. (After Flew 1984.)

Page: p. 244. See also: the alternative of *Emergence.*

Regressive theory development: Development of a theory may be *regressive* if the theory becomes unable to incorporate the conclusions from some investigations without making *ad hoc* adjustments that invoke special exceptions for particular circumstances.

Page: p. 128. See also: *Progressive theory development.*

Relativism: The antithesis of criticism is *relativism*. The relativist argues that all truth is relative: axioms are relative to the data used to support them, and data are relative to the axioms and postulates that sparked the investigation. There is no correct starting point for a scientific argument; the correctness of procedure changes from individual to individual or from social group to social group. Because no absolute criteria of "right and wrong" can be found either in science or morals for the relativist, the choice of means to one's ends changes widely: no point of reference can stabilize this constant change in plans of action. If relativism is carried far enough, it leads to an anarchism that denies any criteria in choosing means or solving problems.

Page: p. 265. See also: *Criticism; Scientific criticism.*

Replicate: In analytical experiments, *replicates* are systems assumed to have identical properties and receiving identical treatments, either as treatments or controls. The purpose of replication is to estimate variation in response to a treatment due to unknown differences in the system itself and/or the application of the treatment.

Page: p. 160. See also: *Experiment; Treatment control.*

Response-level experiment: In *response-level experimentation*, treatments are designed to discover the magnitude of an already established or likely effect. Examples are experiments to find out how much the growth of a crop or an individual plant may respond to a fertilizer, how much certain animals may grow in response to a new diet, or what concentration of a pollutant kills 50 percent of test organisms. The general methodological principles of response-level experimentation are:

(1) Few (usually one) measures of response, e.g., yield of a crop receiving a fertilizer, growth rate of an animal given a new diet, mortality of an organism exposed to a pollutant. Additional measures usually inform about the nature of the response, e.g., quality of the crop, fat content of the animal.

(2) Multiple treatments in the same experiment, e.g., different quantities of the same type of fertilizer, diets that differ in a few components, or different concentrations of the same pollutant.

(3) Extensive replication designed to ensure that quantitative differences between treatments could be estimated, e.g., difference in yield or growth rate or percentage killed.

Page: p. 139. See also: *Experimental analysis of ecological systems; Controlled analytical experiment.*

Revolution, see *Scientific revolution.*

Robustness: A statistical test is *robust* if its validity is not seriously threatened by moderate deviations from the underlying mathematical assumptions. (After Zar 1996.)

Page: p. 226. See also: *Statistical inference.*

Sample: A *sample* is a number of observations from a universe. A sample can be defined by various statistics, e.g., mean, variance, median. The sample can then be represented as a point in n-dimensional space, where n is the number of statistics used, e.g., successive samples of small mammals from a repeatedly randomized trapping line would have different means and variances. Each time that a sample is taken it occupies a different point in n-dimensional space, i.e., the sample can be represented by the different statistics that you calculate from it.

Page: p. 220. See also: *Universe.*

Sample statistic: A *sample statistic* is any mathematical function calculated from the observations: For example, for observations, $x_1, x_2, x_3, \ldots x_n$, the sample mean, $\bar{x}$, is defined as,

$$\bar{x} = \frac{\sum_{i=1}^{n} x_i}{n}$$

The *sample variance*, s^2, an estimate of the unknown universe variance, is defined,

$$s^2 = \frac{\sum_{i=1}^{n} (x_i - \bar{x})}{n - 1}$$

Page: p. 224. See also: *Statistical inference; Observations.*

Scientific analysis: *Scientific analysis* results in:
(1) Definition of the theory to be used to answer a specific question.
(2) Definition of the postulates to be examined.
(3) Definition of the techniques of investigation to be used.
(4) Definition of measurements to be made.
(5) Definition of how inference will be made.

Page: p. 16. See also: *Synthesis.*

Scientific criticism: The degree of objectivity that we can accord knowledge changes as scrutiny and debate pass through

four stages:
(1) Direct analysis.
(2) Testing through direct repetition.
(3) Refining through extended use.
(4) Standpoint criticism.

The first three stages are made in the context of justification and focus on how things are researched. The fourth stage, standpoint criticism, analyzes the context of discovery, what things are researched, and what approaches are taken.

Page: p. 425. See also: *Criticism; Standpoint criticism.*

Scientific explanation: A scientific explanation answers a why-type question constructed so that it:
- specifies the topic of concern,
- defines the contrasting set of alternatives in the question, and,
- defines the explanatory relevance required.

A scientific answer to a why-type question requires a synthesis that decides between contrasting alternatives and may include:
- developing increased confidence in an existing theory network,
- extending an existing theory network, or
- a major change in a theory network. (After Lambert and Brittan 1987, Thagard 1992.)

Page: p. 272 See also: *Causal scientific explanation; Explanatory coherence; Scientific inference; Why-type question;* discussion of types of scientific explanation, see 15.2.1.

Scientific inference: A *scientific inference* is made for a specified scientific question using the following procedures and standards:
(1) A *synthesis* must be made of new results with existing theory.
(2) This synthesis must provide a *scientific explanation* of why something exists or occurs.
(3) The scientific explanation provided must be *coherent*, explaining both new and previously obtained information. The explanation may increase in coherence as new *scientific understanding* is obtained.

Page: p. 269. See also: the component terms used in this definition are defined in the Introduction to Section II and illustrated throughout that section; the relationship between statistical inference and scientific inference is discussed in Chapter 8.

Scientific method: The purpose of *scientific method* is to place the subjective process of developing new ideas into a logical framework of challenge and questioning to develop objective knowledge.

Page: p. 43. See also: *Methodology.*

Scientific paradigm: A *scientific paradigm* has theoretical, social and technical components. When first introduced, the theoretical component provides a new view of some problem. This is adhered to by a group of scientists with a common attitude, ideal, and educational experience using shared scientific techniques and exemplars to solve problems defined as important by the group.

Page: p. 311. See also: *Exampler; Scientific revolution.*

Scientific realism: Scientific realism has three tenets:
(1)   There is a real world.
(2)   Scientific methods find out about the real world.
(3)   Science aims to give us, in its theories, a literally true story of what the world is like; and acceptance of a scientific theory involves the belief that it is true.

Page: p. 236. See also: *Pragmatic realism;* discussion in 15.2.2.

Scientific revolution: Scientific revolutions have three characteristics:
(1)   Previous awareness of anomalies within a paradigm.
(2)   Gradual, and simultaneous, recognition of observational and conceptual problems.
(3)   Development of a new paradigm, often with resistance from scientists who support the prior paradigm.

Page: p. 313. See also: *Scientific paradigm.*

Scientific understanding: To say a scientific explanation yields *scientific understanding* is to say that it shows or exhibits some new piece of information relevant to what the explanation requires. (After Lambert and Brittan 1987.)

Page: p. 276. See also: *Explanatory coherence; Synthesis.*

Standpoint: A *standpoint* is a subjectively held viewpoint about an issue justified by an individual person or group's value system.

Page: p. 432. See also: *Scientific criticism.*

Statistical experimentalism: Statistical experimentalism attempts to resolve the differences between empiricism, rationalism, criticism, and relativism by requiring the existence of some "ideal" against which scientific progress can be measured. There is an objective truth, we approach this through continued use of experiments but we can not expect to reach it because experiments operate on samples or subsets of a complete system and produce answers within certain probabilities (Churchman 1948).

Page: p. 266. See also: *Empiricism; Rationalism; Criticism; Relativism.*

Statistical inference: To make a *statistical inference:*

(1)   A sampling or measurement protocol must be designed.
(2)   A measurement must be selected.
(3)   A statistical hypothesis must be constructed which may be focussed on investigation or experiment.
(4)   The requirement of the statistical theory used in testing the statistical hypothesis must be satisfied.

Page: p. 205. See also: *Scientific inference; Exploratory analysis.*

Statistical model of a stochastic process: A *statistical model of a stochastic process* represents dependence of successive and/or neighboring events in response to variation in an external influence on the process. These models are parsimonious, using the fewest number of parameters capable of explaining quantitative variation in some observed data.

Page: p. 369. See also: *Systems simulation model; Deterministic mathematical model; Model explanations.*

Stochastic process: A *stochastic process* is a continuous causal process in time, space, or both, responding to variation in an external influence, and producing a varying series of measured states or events.

Page: p. 369. See also: *Statistical model of a stochastic process.*

Stratification: Experimental units may be divisible into groups for which you expect to see treatment response and variation to behave similarly within groups but dissimilarly between groups. Identifying the potential sources of variation and sampling from within these groups is the process of *stratification.*

Page: p. 161. See also: *Experimental unit; Replicate; Treatment.*

Synthesis: Combining constituent elements into a single or unified entity. Research introduces new concepts, changes the status of existing concepts, or defines new relationships between concepts.

Page: p. 271. See also: *Scientific inference; Scientific analysis.*

Synthetic construction: Synthetic constructions seek to design and construct all or part of an ecological system. The synthesized system is usually closed. It sometimes comprises organisms that represent primary producers, herbivores, decomposers, or predator–prey combinations, may be exposed to treatments or constructed under different conditions designed to investigate the ecological relationships and overall properties of the ecological system.

Page: p. 165. See also: *System; Closed system.*

System: A *system* has component parts that interact to achieve or maintain a particular property in a variable environment.

Sometimes this property is called homeostasis, which maintains the constancy or directed purpose of a function, status, or process when there is an external disturbance, e.g., thermoregulation in animals. Sometimes the property is called dynamic, which emphasizes the change within the system, e.g., the way that animal behavior may change in different environments and maintain important life functions.

Page: p. 137. See also: *Closed system; Open system.*

Systems simulation model: Systems simulation models are hybrids using the flexibility of computer programing languages to represent changes in states and conditions, as well as changes in rates, by different types of mathematical function, empirically based relationships, and incorporating stochastic forcing.

Page: p. 378. See also: *System; Deterministic mathematical model; Statistical model of a stochastic process; Model calibration; Model sensitivity; Validation.*

Teleology: Teleology (Gr. *telos* end, *logos* a discourse) in biology is the interpretation of biological structure or function in terms of purpose.

Page: p. 240.

Theory: A *theory* has two parts:

(1) The working part of a theory is represented as a logical construction comprising propositions, some of which contain established information (axioms) while others define questions (postulates), as diagrammed in Fig. 3.2 and summarized in Table 3.1. The working part of a theory provides the information and logical basis for making generalizations.

(2) Theories also contain a motivational and/or speculative component that defines a general direction for investigation or type of question that can be answered. An overarching axiom or postulate frequently specifies this component.

Page: p. 43. See also: *Axiom; Postulate; Law,* discussion of the distinction between theory, postulate and law in Chapter 3.

Transitivity: If the consequent of a true implication is itself the antecedent of another true implication, then it is true that the antecedent of the first necessarily entails the consequent of the second (i.e., if $p \Rightarrow q$ and $q \Rightarrow r$, then $p \Rightarrow r$).

Page: p. 178. See also: rules of propositional logic, Chapter 7.2.

Treatment: In analytical experiments, a *treatment* is designed to manipulate a system and produce an observable or measurable effect.

Page: p. 160. See also: *Experiment; Treatment control.*

Treatment control: Inclusion of a control for non-intended treatment effects allows the designed effects of the treatment to be estimated more precisely.

Page: p. 164. See also: *Experiment; Experimental analysis of ecological systems.*

Truth status of implications: An implication, e.g., $p \Rightarrow q$, is false only if the antecedent is true and the consequent false (i.e., observe p and *not-q*); otherwise it is true. From this, it follows that:

(1) a true antecedent occurs only with a true consequent, if p true then q must be true;

(2) a false consequent always indicates a false antecedent, *not-q* $\Rightarrow$ *not-p*;

(3) a false antecedent has no necessary bearing on the truth value of the consequent, if *not-p* we can deduce nothing about q; and

(4) a true consequent has no necessary bearing on the truth-value of the antecedent, if q we can deduce nothing about p.

Page: p. 174. See also: rules of propositional logic, see 7.2.

Uncodified knowledge: For practical investigations there is also *uncodified knowledge* – that is, knowledge about the particular field site, population, species, or environmental conditions that informs how a theory may be applied in a particular circumstance and that you must discover for yourself through preliminary research.

Page: p. 52. See also: *Codified knowledge.*

Universe: A *universe* is the set of all observations that could possibly be obtained that follow the probability law.

Page: p. 221. See also: *Sample; Observation.*

Upward inference: The process of developing inference for an over-arching theory from a set of specific investigations and where the theory contains concepts that do not have a direct equivalent concept by measurement.

Page: p. 280. See also: *Integrative concept.*

Validation: A test of the model with known input and output information that is used to assess that the calibration parameters are accurate without further change. Preferably, it should represent a condition different from that used for model calibration. (After ASTM 1992.)

Page: p. 384. See also: *Systems simulation model.*

Why-type question: Why-type questions ask about functioning of ecological systems on the basis of some observed phenomenon. Knowledge-that something happened is descriptive. Knowledge-why is explanatory.

Page: p. 273. See also: *Scientific explanation, Contrastive explanation.*

Author index

Subject index